Actual Problems of Ecology and Environmental Management: Cooperation for Sustainable Development and Environmental Safety (APEEM 2020)

AA004462

E3S Web of Conferences Volume 169 (2020)

Moscow, Russia
23 - 25 April 2020

Editors:

Margarita Redina
Tatiana Ledashcheva
Alexandr Nikol'skii
Natalia Chernykh
Evgenii Nesterov
Alexandr Khaustov
Sofia Lamarton
Alexandr Myaskov
Ioan Bica
Devard Stom
Saulius Mickevicius
Konstantin Lebedev
Roman Jashchenko
Susanna Cherchesova
Yurong Jang
Xiangkai Li

ISBN: 978-1-7138-1291-3

Printed from e-media with permission by:

Curran Associates, Inc.
57 Morehouse Lane
Red Hook, NY 12571

Some format issues inherent in the e-media version may also appear in this print version.

No changes have been made to the content of these proceedings. There may be changes to pagination and minor adjustments for aesthetics.

Printed with permission by Curran Associates, Inc. (2021)

For additional information, please contact EDP Sciences – Web of Conferences
at the address below.

EDP Sciences – Web of Conferences
17, Avenue du Hoggar
Parc d'Activité de Courtabœuf
BP 112
F-91944 Les Ulis Cedex A
France

Phone: +33 (0) 1 69 18 75 75
Fax: +33 (0) 1 69 28 84 91

contact-edps@webofconferences.org

Additional copies of this publication are available from:

Curran Associates, Inc.
57 Morehouse Lane
Red Hook, NY 12571 USA
Phone: 845-758-0400
Fax: 845-758-2633
Email: curran@proceedings.com
Web: www.proceedings.com

TABLE OF CONTENTS

CONCENTRATION OF HEAVY METALS AND METALLOIDS IN THE BOTTOM SEDIMENTS OF THE VILKITSKY ISLAND'S LAKES (THE KARA SEA) 1
Agbalyan Elena, Krasnenko Aleksander, Shinkaruk Elena

MIGRATION OF URANIUM IN THE SYSTEM SOIL-PLANT (MODEL EXPERIMENTS) 7
Bondareva Lydia

ZINC AND CADMIUM ACCUMULATION IN DIFFERENT PARTS OF WILD PLANTS OF THE ASTERACEAE FAMILY AND TRITICUM AESTIVUM 13
Chaplygin Victor, Mandzhieva Saglara, Litvinov Yuri, Kravtsova Natalia, Sherstnev Aleksey, Chernikova Natalia, Deryabkina Irina

LEAD COMPOUNDS IN BOTTOM SEDIMENTS OF THE SEVERSKY DONETS FLOODPLAIN 19
Burachevskaya Marina, Nevidomskaya Dina, Tsitsuashvili Victoria, Rajput Vishnu, Bren Dmitry

GEOCHEMICAL STUDIES OF SURFACE BOTTOM SEDIMENTS AND SOILS OF SMALL WATERCOURSES 24
Nesterov Evgeni, Belyakov Timofei, Dubrova Stanislav, Egorov Pavel

DIFFUSE BIOGENIC LOAD FROM THE CATCHMENT AREA OF SMALL WATERCOURSES 29
Nesterov Evgeni, Frumin Grigori, Egorov Pavel, Lyubimov Alexander

INTEGRAL ASSESSMENT OF HEAVY METAL POLLUTION IN DON RIVER ESTUARY SOILS 36
Nevidomskaya Dina, Minkina Tatiana, Fedorov Yuri, Nazarenko Olga, Kravtsova Natalya, Litvinov Yuri

MIGRATION OF CYANTRANILIPROLE IN FRACTURED SOILS: CALIBRATION OF PESTICIDE LEACHING MODEL BY USING EXPERIMENTAL DATA 41
Belik Aleksandra, Kokoreva Anna, Kolupaeva Victoria

RISK ASSESSMENT OF ICE-MELTER REAGENTS FOR URBAN PLANTS 47
Baeva Yulia, Chernykh Natalia, Shmerko Sofia, Stepnova Varvara

ASSESSMENT OF HEAVY METALS IN MODERN BOTTOM SEDIMENTS ON THE KARADAG SEASHORE 52
Drygval Anna, Drygval Polina, Stanis Elena, Maltsev Vladimir

HEAVY METALS IN THE ORGANIC SOIL HORIZON OF THE NORTHERN TAIGA FORESTS RESULTING FROM INDUSTRIAL AIR POLLUTION 57
Evdokimov Alexander

ASSESSMENT OF THE GENOTOXICITY OF WATER AND SOIL IN THE PLACES OF STORAGE OF RESERVES OF OLD PESTICIDES BY ALLIUM-TEST 62
Cherednichenko Oksana, Nuraliev Serikbay, Bekmanov Bakytzhan, Pilugina Anastassiya, Dzhansugurova Leila, Baigushikova Gulnara

SPECIFICITY OF ACCUMULATION OF HYDROCARBONS IN VARIOUS COMPONENTS OF GEOSYSTEMS 69
Khaustov Aleksandr, Redina Margarita

GEOCHEMICAL FEATURES OF SOILS FROM THE MID-URAL ORE DEPOSIT 77
Dziuba Ekaterina, Andreev Dmitriy, Buzmakov Sergey, Khotyanovskaya Yuliya

IDENTIFICATION OF THE STATE OF THE SOIL-PLANT SYSTEMS ON THE RUDN-UNIVERSITY CAMPUS (BASED ON PAH CONCENTRATIONS) 90
Khaustov Aleksandr, Redina Margarita, Kenzhin Zhandos, Gabov Dmitry, Yakovleva Evgeniya

DANGEROUS GEODYNAMIC PROCESSES OF THE EASTERN ORENBURG 98
Nesterenko Maksim, Tsviak Aleksey, Kapustina Olga, Nesterenko Aleksey, Nikiforov Sergey

ANALYSIS OF DYNAMIC OF TROPOSPHERIC OZONE AND NITROGEN DIOXIDE IN THE HIGHEST REGION OF THE QUITO CITY 103
Salazar Cristian, Mullo Freddy

ACCUMULATION OF HEAVY METALS IN AQUATIC VEGETATION OF THE BARGUZIN RIVER 110
Shiretorova Valentina G., Zhigzhitzhapova Svetlava V., Dylenova Elena P., Radnaeva Larisa D.

GEOECOLOGICAL SITUATION IN COASTAL AND SEA AREA OF THE WESTERN YAMAL 114
Shirokov Roy

ECOTOXICOLOGICAL HAZARD ASSESSMENT OF DEICING CHEMICALS FOR HIGHER PLANTS 124
Fedorchenko M.D., Mikhaylichenko K. Yu., Ushakova O.V., Adarchenko I.A.

USING REMOTE SENSING FOR MONITORING THE DYNAMIC OF SOIL ORGANIC CARBON CONCENTRATION IN LAKE VALENCIA BASIN, VENEZUELA, BASED ON LANDSAT 8 DATA 128
Arias Govín Israel Jose, Stanis Elena V., Latushkina Elena N., Ospanova Aigul

STUDY OF CYANTRANILIPROLE SORPTION IN SOILS OF RUSSIA 135
Kolupaeva Victoria, Nyukhina Inna, Belik Aleksandra
DISTRIBUTION OF HEAVY METALS IN ANTHROPOGENIC SOILS: ON THE EXAMPLE OF THE INDUSTRIAL ZONE OF MOSCOW 141
Latushkina Elena, Oleynik Anna
CONTENT OF HEAVY METALS IN HAPLIC CHERNOZEM UNDER CONDITIONS OF AGROGENESIS 145
Medvedeva Anna, Buryukova Olga, Kucherenko Alexey, Ilchenko Yaroslav, Chaplygin Victor, Rajput Vishnu D., Kizilkaya Ridvan
METODOLOGICAL ASPECTS IN THE STUDYING OF SOIL PARTICLE SIZE DISTRIBUTION UNDER CONTAMINATION AND AFTER RECLAMATION 154
Zamulina Inna, Burachevskaya Marina, Mandzhieva Saglara, Bauer Tatiana, Barakhov Anatolii, Mazarji Mahmoud
INTERDISCIPLINARY INTEGRATION ON SUSTAINABLE DEVELOPMENT ISSUES 160
Ermakov Dmitry Sergeyevich, Ermakov Alexander Sergeyevich
CONTAMINATED SITES INVESTIGATION. OBJECTIVES AND METHODS 165
Bica Ioan
METHODOLOGY OF THE GEOECOLOGICAL FUNCTIONAL ZONING OF TERRITORIES SUBJECTED TO ANTHROPOGENIC DISTURBANCE DUE TO WASTE DISPOSAL 176
Dubrova Stanislav, Chelinho Sónia, Zelenkovskiy Pavel, Podlipskiy Ivan
IMPROVING RADIO-ECOLOGICAL MONITORING OF POTENTIALLY RADON-HAZARDOUS TERRITORIES 182
Lashchenova Tatiana, Karl Lidiya, Marenny Albert, Kiselev Sergey
POSSIBILITIES OF USING SENTINEL-2 SATELLITE DATA TO ESTIMATE CROP COVER AND LAI OF WINTER WHEAT AT DIFFERENT STAGES OF DEVELOPMENT 187
Mullo Feddy, Prudnikova Elena
FAST SCREENING OF SOME FLAVONOIDS CONTENT IN RAW PLANT MATERIALS: OPPORTUNITIES OF 1H NMR SPECTROSCOPY 192
Kalabin Gennady, Vasil'ev Vasilii, Ivlev Vasilii, Babkin Vasilii
IMPACT OF BLACK YEASTS ON THE DURABILITY OF POLYURETHANE FOAM IN TROPICAL CONDITIONS IN VIETNAM 198
Nguyen Thu Hoai, Do Thi Thu Hong, Ngo Cao Cuong, Do Tat Thinh, Sereda V.N.
DEWATERING AS A PRIMARY TREATMENT OF FAECAL SLUDGE IN INDIVIDUAL RESIDENTIAL SECTOR (A TECHNOLOGIES REVIEW) 204
Basamykina Alena, Kharlamova Marianna, Mada Sharon Yeukai
PRODUCTION OF HIGH PROTEIN SUPPLEMENTARY FEED FROM LANDFILL GAS 213
Klevanova Elena, Lukanin Alexander
MUNICIPAL WASTE MANAGEMENT IN TOWNS OF SWITZERLAND: A FOREIGN ECOLOGIST'S VIEW 219
Mamadzhanov Roman, Zakirova Yulia, Umarov Mykhadi
DEVELOPMENT OF THE TECHNOLOGY FOR PROCESSING PLANT BREEDING BY-PRODUCTS TO OBTAIN BIOSORBENT 225
Bauer Tatiana, Sushkova Svetlana, Fedorenko Aleksei, Lobzenko Ilya, Barakhov Anatoly, Butova Vera, Tusat Ekrem
TECHNOLOGY OF LANDSCAPING IN ARID ZONES BY USING BIOHUMUS FROM SHEEP WOOL 232
Ydyrys Alibek, Yeszhanov Birlikbay, Baymurzaev Nurlan, Sharakhmetov Sayat, Mautenbaev Askar, Tynybekov Bekzat, Baidaulet Tilek
EVALUATION OF THE BIOLOGICAL ACTIVITY OF MEADOW CHERNOZEM SOILS AFTER THE APPLICATION OF BIOCHARS WITH DIFFERENT PYROLYSIS TEMPERATURES IN A MODEL EXPERIMENT 236
Zinchenko Vladislav, Gorovtsov Andrey, Sushkova Svetlana, Antonenko Svetlana, Fedorenko Elena, Pogonyshev Pavel, Barakhov Anatoly
THE EFFECT OF NO-TILL TECHNOLOGY ON THE MINERAL NITROGEN CONTENT IN THE LOWER DON CHERNOZEM 240
Medvedeva Anna, Buryukova Olga, Ilchenko Yaroslav, Minkina Tatyana, Kamenev Roman, Gülser Coºkun
IDENTIFYING PRIORITY MANAGEMENT OF ECUADORIAN FORESTS BASED ON THE ENVIRONMENTAL INTEGRATED ASSESSMENT 247
Llerena Silvia, Arias Priscila, Cueva Jhonn, Almeida Georgina, Salazar Cristian
ECONOMIC EFFICIENCY OF THE WATER MONITORING SYSTEM IN NIGERIA 255
Nwankwo Nnabueze Kalu, Zakirova Yulia

DETERMINING THE ENVIRONMENTAL SITUATION AS A FACTOR IN THE DIFFERENTIATION OF HOUSING PRICES IN NUR-SULTAN BY USING THE ANALYTIC HIERARCHY PROCESS 262
Akhmedinova Kamila

PALAEOGEOGRAPHICAL HERITAGE OF THE PERMIAN PERIOD IN THE NETWORK OF PROTECTED AREAS AND MUSEUMS OF THE PERM REGION 268
Glazyrina Yuliya, Zhuzhgova Larisa, Buzmakov Sergey, Sannikov Pavel

THE ROLE OF SPECIALLY PROTECTED NATURAL AREAS IN SOLVING ENVIRONMENTAL PROBLEMS OF THE RUSSIAN ARCTIC (USING THE EXAMPLE OF VAYGACH ISLAND) 274
Aleinikova Anna, Byashkin Maksim, Gaivoron Tatiana, Mainasheva Galina, Marsheva Natalia

COGNITIVE ANALYSIS OF THE SOCIO-ECOLOGICAL AND ECONOMIC SYSTEM OF THE REGION FOR CHOOSING A SUSTAINABLE DEVELOPMENT STRATEGY (ON THE EXAMPLE OF OIL-PRODUCING REGIONS OF RUSSIA) 279
Ledashcheva Tatiana, Pinaev Vladimir

SUSTAINABILITY AS A BASIC PROPERTY OF THE PHENOMENON OF LIFE ORGANIZED IN SUPRAORGANISMAL SYSTEMS 286
Nikol'skii Alexander, Belovezhets Konstantin, Vanisova Elena

ECOSYSTEMS TRANSFORMATION AND ASSESSMENT OF BIRD RESOURCES IN THE REPUBLIC OF TATARSTAN (RUSSIA) 293
Rakhimov Ilgizar, Ibragimova Kadrya

ABOUT THE REASONS FOR THE RARE PLANT SPECIES DISAPPEARANCE IN THE UPPER VOLGA REGION 298
Borisova Elena, Konotop Nikita, Vinogradova Julia, Markov Dmitriy, Sivukhin Alexey

SEGETAL COMPLEX – PIONEER STAGES OF SYNTROPHIC COMMUNITIES IN THE MESHCHOVSKIY OPOLE OF THE NON-CHERNOZEM ZONE OF RUSSIA 303
Burlutskiy Valeriy A., Mazurov Vladimir N., Peliy Alexander F., Semeshkina Polina S., Borodina Ekaterina S., Diop Ameth

DO DIVERSITY AND BIO-MORPHOLOGICAL SPECTRUM OF PSAMMOPHYTES MATTER IN RESTORING VEGETATION OF SAND PITS? 308
Dabbagh Alaa

THE STUDY OF THE ZOOBENTHOS OF THE TSRAUDON RIVER BASIN (THE TEREK RIVER BASIN) 313
Dzhioeva Ia E., Cherchesova Susanna K., Navatorov Oleg A., Lamarton Sofia F.

PRODUCTIVITY DYNAMICS OF OAK FORESTS OF THE CRIMEAN PENINSULA 317
Gorbunov Roman

ASSESSMENT OF ECOLOGICAL STATUS OF STREET PLANTINGS IN LIPETSK CITY 323
Parakhina Elena, Silaeva Zhanna, Kiseleva Ludmila, Zhidkova Elena

ANTHROPOGENIC DYNAMICS AND RECOVERY OF DISTURBED LANDSCAPES AT THE FAR EAST RUSSIA PERMAFROST AREA 328
Tikhmenev Pavel E., Smirnov Andrey A., Tikhmenev Evgeniy A., Stanchenko Galina V.

MANGANESE AND COBALT LEVELS IN HAIR AND URINE OF RUDN UNIVERSITY FIRST-YEAR STUDENTS 333
Kirichuk Anatoly A.

BIOLOGICAL CONTAMINATION OF SOILS IN URBANIZED ECOSYSTEMS BY TOXOCARA SP. EGGS. 337
Maslennikova Olga, Erofeeva Viktoriya

ABOUT THE ROLE OF ENVIRONMENTAL FACTORS IN CARCINOGENESIS 343
Pinaev Sergey K., Pinaeva Olga G., Chizhov Alexey Ya.

PHYSIOLOGICAL EVALUATION OF THE CARDIOLOGY SYSTEM OF NONRESIDENT STUDENTS AT DIFFERENT TRAINING PERIODS 348
Mayorova Yana, Glebov Viktor, Erofeeva Viktoriya, Yablochnikov Sergey, Laver Bogdan

EFFECTS OF ALMATY CITY ECOLOGICAL FACTORS ON STUDENTS BLOOD INDICES 353
Baidaulet Tilek, Ydyrys Alibek, Imanaliyeva Moldir, Askarova Zifa, Srailova Gulziya, Murzakhmetova Maira, Tuleukhanov Sultan

ENVIRONMENTALLY-INDUCED ALTERNATIVE ONCOGENESIS: EROS ARROWS 357
Pinaev Sergey K., Pinaeva Olga G., Chizhov Alexey Ya.

IMPACT OF URBAN ENVIRONMENT ON STUDENTS' DAILY RHYTHMS AND SLEEP PATTERNS 361
Piven Elena A., Piven Nikolay P., Bushuev Nikolay N.

COMPARATIVE ANALYSIS OF THE APPLICATION OF EDUCATIONAL TECHNOLOGIES FOR SUSTAINABLE DEVELOPMENT 366
Sokolova Lyudmila I., Ermakov Dmitry S.

IMPLEMENTATION TRAJECTORIES OF ENVIRONMENTAL EDUCATION FOR SUSTAINABLE DEVELOPMENT IN FORMAL, NON-FORMAL AND INFORMAL EDUCATION BASED ON EDUSCRUM PROJECT MANAGEMENT METHODOLOGY 372
Ryazanova Natalya, Naumov Vladimir, Kamennykh Natalia

TOXIC COMMUNICATION AS A RISK FACTOR WHEN PROMOTING HEALTH, SAFETY & ENVIRONMENTAL CULTURE IN AN ORGANIZATION 378
Pavlova Elena, Terentyeva Elena, Matevosyan Armine

"GREEN NEOLOGY" AND ITS ROLE IN ENVIRONMENTAL EDUCATION (A MEDIA DISCOURSE APPROACH) 383
Guslyakova Alla, Valeeva Nailya, Vatkova Olga

REGIONAL CASE STUDY OF SUSTAINABLE DEVELOPMENT GOALS IMPLEMENTATION: INFORMAL EDUCATION AND THIRD MISSION OF UNIVERSITIES 391
Vologzhina Sayana, Ryazanova Natalya, Eroshenko Vasily

LEGAL ASPECT OF THE IMPLEMENTATION OF THE UN SUSTAINABLE DEVELOPMENT GOALS IN THE FIELD OF ENVIRONMENTAL MANAGEMENT IN RUSSIA 397
Mukhlynina Maria, Vedysheva Natalia

FOREST CONSERVATION IN RUSSIA: PROBLEMS OF IMPLEMENTATION 401
Mukhlynin Dmitry

ON THE IMPLEMENTATION OF THE STATE ENVIRONMENTAL POLICY IN LAW-MAKING AND LAW-ENFORCEMENT ACTIVITIES IN RUSSIA 406
Mukhlynina Maria

RENEWABLE ENERGY AS A FACTOR OF SUSTAINABLE DEVELOPMENT AND REGULATION OF ECOLOGICAL PROBLEMS IN AFRICA 411
Nefedova Liudmila, Solovyev Alexander, Rafikova Yulie, Solovyev Dmitriy

Author Index

edp sciences

Web of Conferences

Statement of Peer review

In submitting conference proceedings to *Web of Conferences*, the editors of the proceedings certify to the Publisher that

1. They adhere to its **Policy on Publishing Integrity** in order to safeguard good scientific practice in publishing.
2. All articles have been subjected to peer review administered by the proceedings editors.
3. Reviews have been conducted by expert referees, who have been requested to provide unbiased and constructive comments aimed, whenever possible, at improving the work.
4. Proceedings editors have taken all reasonable steps to ensure the quality of the materials they publish and their decision to accept or reject a paper for publication has been based only on the merits of the work and the relevance to the journal.

Title, date and place of the conference

Actual Problems of Ecology 2020 - Actual Problems of Ecology and Environmental Management: cooperation for sustainable development and environmental safety
April 23-25, 2020
Russia, Moscow, Faculty of Ecology of Peoples Friendship University of Russia (RUDN University)

Proceedings editor(s):

Margarita Redina, Chairman of the organizing committee
Tatiana Ledashcheva, Deputy Chairman of the organizing committee
Alexandr Nikol'skii, Chairman of the scientific committee
Natalia Chernykh
Alexandr Khaustov
Alexandr Myaskov
Devard Stom
Konstantin Lebedev
Susanna Cherchesova
Evgenii Nesterov
Sofia Lamarton
Ioan Bica
Saulius Mickevicius
Roman Jashchenko
Yurong Jang
Xiangkai Li

Date and editor's signature

May 10, 2020

/M. Redina /T. Ledashcheva

Concentration of heavy metals and metalloids in the bottom sediments of the Vilkitsky Island's lakes (the Kara Sea)

Elena Agbalyan, *Aleksander* Krasnenko, and *Elena* Shinkaruk*

Arctic Research Center of the Yamal-Nenets autonomous district, 629000 Salekhard, Russia

Abstract. We performed this study in order to obtain relevant data about the actual ecological status of the lakes in remote areas of the Russian Arctic. The objective was to determine the level of technogenic pollution of bottom sediments in the lakes on the Vilkitsky island in the Kara Sea. To this end, we examined 8 of the island's small lakes. The content of heavy metals and metalloids (Ti, V, Cr, Mn, Fe, Co, Ni, Cu, Zn, Ga, As, Y, Nb) in the samples of bottom sediments was determined using the method of Energy Dispersive X-ray Fluorescence (EDXRF). The total pollution value in the examined bottom sediments varied from Zc = 9.5 to Zc = 15.3. The maximum level of pollution was found in Lake 4 located in an anthropogenically disturbed area due to the levels of Cr, Co, Ni, As, Y, Nb.

1 Introduction

Vilkitsky Island lies in the eastern part of the Kara Sea in the Arctic zone of the Russian Federation. The components of its natural environment have a substantial record of being used in human activity. Today the island's northeast coast features a functioning polar station as well as a number of abandoned military objects. The anthropogenic impact is associated with mechanical disturbance and chemical pollution of the area's surface soils. The sources of soil pollution are contained in solid natural and industrial waste. Works are currently in progress to clear and restore the island's anthropogenically disturbed areas [1, [2].

The hydrographic pattern of the island is represented by thermokarst lakes, pits, swamps and streams flowing from them, which form estuary coastland closer to the sea. The island has unique wetlands of the West Siberian Arctic island tundras and is a breeding ground for water fowl [2].

Our team assessed the ecological condition of the Vilkitsky Island's water bodies. Bottom sediments are an integral indicator of anthropogenic load on water bodies. Accumulating pollutants from the aquatic environment, they act as a potential source of secondary pollution.

We studied eight lakes located in the northeastern and southeastern parts of Vilkitsky Island, performing a thorough assessment of the granulometric and chemical composition

* Corresponding author: elena1608197@yandex.ru

of bottom sediments employing the sieve method. The pH value of the aqueous extract of bottom sediments was determined with the use of the potentiometric method, and electrical conductivity (SEC), with the use of the conductometric method.

The aim of the work was to study the level of technogenic pollution of bottom sediments of the lakes of Vilkitsky Island.

2 Materials and methods

We studied eight lakes and one stream located in the northeastern and southeastern parts of Vilkitsky Island, performing a thorough assessment of the granulometric and chemical composition of bottom sediments employing the sieve method. The pH value of the aqueous extract of bottom sediments was determined with the use of the potentiometric method, and electrical conductivity (SEC), with the use of the conductometric method.

The content of heavy metals and metalloids (Ti, V, Cr, Mn, Fe, Co, Ni, Cu, Zn, Ga, As, Y, Nb) was determined by X-ray fluorescence energy dispersive analysis (XPAED) using a BRA 135-f spectrometer (NPP NPO JSC Petrel, Saint Petersburg). We conducted these studies using the chemical analytical laboratory of the Nadym State Scientific Institute of Yamalo-Nenets Autonomous District's Scientific Center for Arctic Studies.

To assess the pollution levels of bottom sediments, we used the background values obtained for the tundra region of the Yamalo-Nenets Autonomous District that are presented in a study by T. Kremlyova (2015).

We assessed pollution levels in bottom sediments using the Kc concentration factors and the total pollution index, Zc. The concentration coefficient characterizes the amount of bottom sediment elements relative to their background levels. Our calculations include elements with Kc values of at least 1.5. The total pollution value is the sum of the concentration coefficients of the corresponding elements; it reflects the additional excess of the background level and characterizes the level of technogenic pollution.

3 Results

The granulometric composition is an important characteristics of bottom sediments. The ability of bottom sediments to concentrate and retain trace elements depends on the particle size. Due to the large surface area, fine-grained bottom sediments are the main places where inorganic and organic components accumulate [4].

The granulometric composition of the bottom sediments of the examined lakes varied within a small range of values, from <0.1 mm to 1 mm, with the sand fraction prevailing in all of the bottom sediment samples.

The sample collection point on Lake No. 1 shows, primarily, the accumulation of medium (70.0%) and fine (26.4%) sands; the share of fractions smaller than 0.1 mm is 2.8%. The maximum content of thin sands and silty sands was found in the bottom sediments of Lake No. 3 and unnamed lake No. 4 (6.6% and 4.9%, respectively). Lake No 2 shows the biggest amount of small sand fractions (0.1 - 0.25 mm) amounting to 11.6% in the bottom sediments, whereas Lake No. 3 shows the amount of 18.4%, and Lake No. 4 – the amount of 20.4%.

The gross composition of lake sediments corresponds to the genesis of the rocks of the catchment basin (deposits of laidas, watten, sea shores). Bottom sediments B-01, B-02, B-03 and B-04 show a more pronounced ability to accumulate trace elements.

The pH values of water extracts of bottom sediments ranged from 6.33 pH units to 6.74 pH units. The reaction of the environment of bottom sediments was slightly acidic.

A high variability of UEP values was found in the bottom sediment water extracts (from 10.5 μS / cm to 280.2 μS / cm). In two of the examined lakes, a significant content of basic ions was found in the water extracts of the bottom sediments.

We noted a significantly wide range of variation in the concentrations of heavy metals in the bottom sediments (Table 1). Zn and Cu were detected in one of the B-04 samples (20.8 mg/kg and 1.4 mg/kg, respectively), which had been taken in an anthropogenically disturbed area. The content of heavy metals in the remaining samples of the bottom sediments was within the detection limits. Ni was detected in samples B-03 and B-04 in quantities of 1.5 mg/kg and 25.5 mg/kg, respectively. The concentration of Ni in the bottom sediments of Lake No. 4 exceeds the background value by 1.8 times.

Table 1. Content of heavy metals and As in the bottom sediments of the lakes of Vilkitsky Island (mg/kg)

Water body	Code	Cr	Ni	Cu	Zn	As
Unnamed lake No. 1	B-01	49.9	BLD	BLD	BLD	23.5
Unnamed lake No. 2	B-02	48.4	BLD	BLD	BLD	25.85
Unnamed lake No. 3	B-03	61.8	1.5	BLD	BLD	24.3
Unnamed lake No. 4	B-04	97.0	25.5	1.4	20.8	20.4
Unnamed lake No. 5	B-05	34.6	BLD	BLD	BLD	24.1
Unnamed lake No. 6	B-06	48.7	BLD	BLD	BLD	28.6
Unnamed lake No. 7	B-07	64.5	BLD	BLD	BLD	26.7
Unnamed lake No. 8	B-08	47.9	BLD	BLD	BLD	28.4
Stream 1	B-p1	31.6	BLD	BLD	BLD	24.8
Stream 2	B-p2	45.6	BLD	BLD	BLD	27.2
M±SD		53.0±16.7	13.5±9.8	-	-	25.4±2.3
Tundra zone background*		37.8	14.2	9.4	29.8	2.9
Clark**		83	58	47	83	17

Note: BLD – below detection limit, * - background values (as per [3]), ** clarks of the elements in the earth's crust (as per [5])

The content of As in the bottom sediments varies from 20.4 mg/kg in the sample B-04 to 28.6 mg/kg in the sample B-06. The concentrations of As in all samples are higher than the background value.

The average content of Cr in the bottom sediments is 53.0 ± 16.7 mg/kg. The concentration of Cr varies from 31.6 mg/kg in the sample B-p1 to 97 mg/kg in the sample B-04. The content of Cr exceeds the background Cr level by 2 times in the sample and B-04 and by 1.3 times in the samples B-03 and B-07.

The content of Ti in the samples of the bottom sediments B-01, B-02, B-03, B-04 exceeds the background value (Table 2). The maximum concentration of Ti was found in the sample B-04, whereas the minimum content of this element was registered for the sample B-p1.

The V concentrations in the bottom sediments vary widely from 0.2 mg/kg to 53.8 mg/kg. The maximum V content was found in the sample B-04. The V contents in the bottom sediments of Lakes No. 1, 3, 4 and 6 are above the background level.

The average concentrations of Mn and Fe in the bottom sediments are 275.3 ± 98 mg/kg and 8854.7 ± 8144.6 mg/kg, respectively. The maximum concentration of Mn was found in the samples B-07 (453.3 mg/kg) and B-04 (390.8 mg/kg). The maximum content of Fe was registered for the in sample B-04. The content of Mn and Fe in the bottom sediments of Lakes No. 3, No. 4 and No. 6 is higher than the background values.

Table 2. Content of heavy metals in the bottom sediments of the lakes of Vilkitsky Island (mg/kg)

Water body	Code	Ti	V	Mn	Fe	Co	Ga	Y	Nb
Unnamed lake No. 1	B-01	3,417.1	11.6	332.6	8,742.3	BLD	5.02	12.1	10.2
Unnamed lake No. 2	B-02	5,861.1	2.6	223.7	BLD	BLD	3.85	9.9	14.3
Unnamed lake No. 3	B-03	4,315.7	16.5	331.0	11,472.1	2.23	5.24	12.7	14.6
Unnamed lake No. 4	B-04	6,558.9	53.8	390.8	27,519.9	9.90	7.36	19.7	16.8
Unnamed lake No. 5	B-05	858.3	BLD	107.5	1,016.0	BLD	1.85	7.5	14.7
Unnamed lake No. 6	B-06	2,397.6	23.8	330.4	20,870.3	BLD	3.53	9.1	13.1
Unnamed lake No. 7	B-07	2,726.3	5.6	453.3	7301.5	0.97	5.62	13.4	19.9
Unnamed lake No. 8	B-08	1,752.0	BLD	168.8	6158.2	0.57	1.63	8.7	14.5
Stream 1	B-p1	184.5	BLD	181.4	1,064.4	BLD	2.23	7.6	14.3
Stream 2	B-p2	2,448.9	0.2	233.0	4,402.8	BLD	3.01	10.0	17.5
M±SD		3,052.1 ± 1,847.4	11.4± 15.4	275.3 ± 98.7	8,854.7 ± 8,144.6	25.4± 2.3	3.9± 1.7	11.1 ±3,3	15.0± 2,4
Tundra zone background*		6,000	51.9	460	20,600	2.9	7.2	11,4	7,5
Clark**		4,500	90	1,000	46,500	17	19	29	20

Note: BLD – below detection limit, * - background values (as per [3]), ** clarks of the elements in the earth's crust (as per [5])

The content of Co in the bottom sediments of Lakes No. 1, No. 2, No. 5 and No. 6 and the stream are below the detection limit. In other water bodies, the Co concentrations differed by 17 times (samples B-08 and B-04) and 4.4 times (samples B-03 and B-04). The maximum Co concentration was registered for the bottom sediments of Lake No. 4 (9.9 mg/kg).

The concentration of Ga in the bottom sediments of the examined lakes varies from 1.63 mg/kg to 7.36 mg/kg. The maximum concentration of Ga was registered for the bottom sediments of Lake No 4, and the minimum concentration was detected in Lake No 8. The bottom sediments of Lakes No. 1, 2, 3, 4 and 7 showed increased concentrations of Ga in relation to the background value.

The average content of Y in the bottom sediments is 11.1 ± 3.3 mg/kg. The maximum concentration of Y was registered for the bottom sediments of Lake No. 4 (19.7 mg/kg), and the minimum concentration of 7.5 mg/kg is observed in Lake No. 5. The bottom sediments of Lakes No. 1, 3 , 4 and 7 showed a content of Y exceeding the background value.

The content of Nb in the bottom sediments of the lakes varies insignificantly from 10.2 mg/kg to 19.9 mg/kg. The maximum concentrations of Nb were registered for the bottom sediments of Lakes No. 7 and No. 4. The average Nb content in the bottom sediments is

15.0 ± 2.4 mg/kg. The background Nb content in the bottom sediments of the lakes of the tundra zone of the Yamalo-Nenets Autonomous District is two times lower [3].

The calculated concentration coefficients relative to the baseline values showed an increased content of As (KAs from 7.0 to 9.8), Cr (KCroN№№3 = 1.6; KCrozN№4 = 2.6; KCrozN№7 = 1.7), Ni (KNiozN№ 4 = 1.8) and Y (KYozN№4 = 1.7) in the bottom sediments (Table 3). The total pollution value in the bottom sediments was found to vary from Zc = 9.5 to Zc = 15.3. The average pollution level was established for all of the examined water bodies, with the exception of Lake 1. The highest level of pollution was found in Lake 4, located in the anthropogenically disturbed area, due to the accumulations of Cr, Co, Ni, As, Y, and Nb there.

Table 3. Concentration coefficients and total pollution value of the bottom sediments on Vilkitsky Island.

Code	Ti	V	Cr	Mn	Fe	Co	Ni	Cu	Zn	Ga	As	Y	Nb	**Zc**
B-01	0.6	0.2	1.3	0.7	0.4	0.0	0.0	0.0	0.0	0.7	8.1	1.1	1.4	**9.5**
B-02	1.0	0.0	1.3	0.5	0.0	0.0	0.0	0.0	0.0	0.5	8.9	0.9	1.9	**10.8**
B-03	0.7	0.3	1.6	0.7	0.6	0.3	0.1	0.0	0.0	0.7	8.4	1.1	1.9	**11.9**
B-04	1.1	1.0	2.6	0.8	1.3	1.5	1.8	0.2	0.7	1.0	7.0	1.7	2.2	**15.3**
B-05	0.1	0.0	0.9	0.2	0.0	0.0	0.0	0.0	0.0	0.3	8.3	0.7	2.0	**10.3**
B-06	0.4	0.5	1.3	0.7	1.0	0.0	0.0	0.0	0.0	0.5	9.8	0.8	1.7	**11.5**
B-07	0.5	0.1	1.7	1.0	0.4	0.1	0.0	0.0	0.0	0.8	9.2	1.2	2.7	**12.6**
B-08	0.3	0.0	1.3	0.4	0.3	0.1	0.0	0.0	0.0	0.2	9.8	0.8	1.9	**11.7**
B-p1	0.0	0.0	0.8	0.4	0.1	0.0	0.0	0.0	0.0	0.3	8.6	0.7	1.9	**10.5**
B-p2	0.4	0.0	1.2	0.5	0.2	0.0	0.0	0.0	0.0	0.4	9.4	0.9	2.3	**11.7**

4 Discussion

There are no regulatory documents establishing the levels of maximum allowable concentrations of heavy metals in bottom sediments. There is information about regional standards approved by the Order of the Department of Natural Resources Regulation, Forestry Affairs and Development of the Oil and Gas Industry of the Yamalo-Nenets Autonomous District dated 27 March 2017 No. 348 "On Establishing Environmental Quality Standards" Background Contaminants in Snow Cover, Surface Sediments of Water Bodies and Vegetation within the territory of the Yamalo-Nenets Autonomous District." Comparison of the measured concentrations of the elements in the bottom sediments of the water bodies of Vilkitsky Island with the values established by the environmental quality standard reveals an increase in the content of Mn in Lake No. 7.

The 2017 studies of Vilkitsky Island showed that most of the water bodies situated within the anthropogenic zone are moderately polluted. The water body located on the border of a military unit and an industrial waste landfill can be classified as polluted [2]. Furthermore, our data indicate the average degree of pollution of most of the examined lakes.

The bottom sediments of the lakes of Vilkitsky Island are As-rich. According to some authors, the soils of the Yamal-Nenets Autonomous District are characterized by an increased content of As, which indicates an increased regional background value [6].
Increased concentrations of Mn in the surface waters of the Autonomous District are typical of that area and have to do with the geochemical features of the territory [7]. Mn enters the

environment in the process of chemical weathering of rocks and minerals, and has a high mobility in wetlands [8].

5 Conclusion

The island of Vilkitsky in the Kara Sea is currently uninhabited, with its natural landscapes previously disturbed by human activity.

The analysis of the ecological condition of the Vilkitsky Island water bodies was carried out on the basis of geochemical criteria for the pollution of bottom sediments. The bottom sediments of the examined lakes are represented by sand fractions. The environment of bottom sediments has demonstrated a slightly acidic reaction; we have also observed and confirmed UEP variability in the water extracts of bottom sediments.

In accordance with the total contamination value, the average pollution value for each of the examined lakes (except one) has been determined on base of the extreme As concentrations that are naturally conditioned. The most extreme pollution levels have been found in the lake located in the area that's home to the abandoned military objects.

As of today, works to clean the island of Vilkitsky are underway. In the years ahead, reclamation of the disturbed and contaminated land will be mandatory.

The research was supported by the Non-Commercial Partnership "Russian Center of Arctic Exploration".

References

1. R.A. Kolesnikov, V.M. Makeev, E.N. Romanova, E.V. Rozhkovsky, R.D. Vostrikov, Science Bulletin of the Yamalo-Nenets Autonomous Okrug, **3(96)**, 11-19 (2017).
2. R.A. Kolesnikov, V.M. Makeev, E.N. Romanova, V.I. Sturman, K.A. Zhurkina, Engineering survey. **12(5-6)**, 32-41 (2018)
3. T.A. Kremleva, *Geochemical factors of resistance of water systems to anthropogenic loads: Dissertation*, 260 (Moscow, 2015)
4. V.A. Dauvalter, *Geoecology of bottom sediments of lakes*, 224 (MSTU Publishing House, Moscow, 2012)
5. A.P. Vinogradov, Geochemistry, **7**, 555-571 (1962)
6. V.M. Tomashunas, E.V. Abakumov, Hygiene and Sanitation, **93(6)**, 26-31 (2014).
7. E.V. Agbalyan, E.V. Shinkaruk, Successes of the Modern Natural Science, **7**, 45-51 (2019)
8. T.I. Moiseenko, G.V. Kalabin, V.Y. Khoroshavin, Bulletin of the RAS. Geographical series, **2**, 49-58 (2012)

E3S Web of Conferences **169**, 01002 (2020) https://doi.org/10.1051/e3sconf/202016901002
APEEM 2020

Migration of uranium in the system soil-plant (model experiments)

Lydia Bondareva*
Federal scientific centre named after F.F. Erisman, Mytischi, Moscow region, Semashko street, 2 141014, Russia

Abstract. Migration of uranium-238 in the system soil-plant at an example of a flower crop, marigold (Tagetes erecta) was studied. The above-mentioned components were used to treat the soil after planting. Using the method of sequential fractionation, soil extracts were obtained, with the content of uranium amounting to ~2 %. Upon the introduction of oil and uranium into the experimental system, the portion of uranium increased in the root parts of the plants (μg/sample): from 0.81 to 9.43, and in the above-ground parts (μg/sample): from 0.78 to 17.1; with the introduction of the oilfield water there was the increased extraction of uranium which had earlier been present in the soil while the uranium content in the root and above-ground parts of the plants in the systems containing oil, oilfield water was considerably higher than in the control system (from 1.17 to 6.73 μg as compared with 0.78 in the above-ground parts of the plants and from 9.92 to 12.87 as compared with 0.81 μg in the root parts). Upon the introduction of uranium combined with oil the radionuclide content was significantly different from the content in the root parts (17.1 in the above-ground parts and 9.43 μg in the root parts)..

1 Introduction

Technogenic radionuclides due to tests of nuclear weapons, operation of enterprises of the nuclear fuel cycle, radioisotope sources used in industry, etc. have long been considered to be the most dangerous. By the present time, it has been established that it is natural sources of radiation rather than technogenic ones which play the main role in public exposure.

Oil, gas and oilfield water, being in contact with the ground, dissolve and contain many chemical substances including natural radionuclides such as uranium-238, radium-226, thorium-232 and potassium-40. During oil extraction, these radioactive substances are carried to the soil surface where their amount can vary from the one insignificantly exceeding the background level up to the values which are dangerous for employees of the industry or the environment.

In soils uranium-238 exists in the form of UO_2^{2+}, which forms highly soluble complexes with the organic matter in soils, increasing its mobility in soil and, consequently, its bioaccessibility to the root system of plants [1, 2].

* Corresponding author: lydiabondareva@gmail.ru

Development of lands, including agricultural ones is known to be an important link in a biological chain transporting natural radionuclides into the organisms of agricultural animals and, further to humans. In this connection, there appears the necessity to study uranium migration along trophic chains. One of the links of this trophic chain is the intake of uranium from soil to plants [3].

The study of the migration ways of natural radionuclides in environmental objects allows one to preliminary estimate the impact of radiation on human health.

The main aim of the study is to study the migration of uranium-238 in the system soil-plant at an example of a flower crop, African marigold (*Tagetes erecta).*

2 Objects and methods of research

2.1 Objects of research

Soil (grey forest, paddy-field soil with the average acidity level (pH 5,1 - pH 5,5)) was collected near the forest and residential buildings in the campus of the Siberian Federal University by the envelope method.

African marigolds (*Tagetes erecta* L.) are eye-catching flower plants which, besides being rather decorative, can be used as medical plants decontaminating the soil in which they grow. They belong to one of the most widely spread flowers planted in summer. The plant, depending on the species and variety can have a height ranging from 15–20 to 70–80 cm, and its root system is fibrous. Marigold leaves, when rubbed, have a specific odour which is characteristic of the whole plant. Marigolds are unpretentious and grow fast. All their varieties are grown from seeds to obtain seedlings to be planted into the ground. Marigolds bloom for the whole summer until the first frost, and afterwards they quickly die [3, 4]. In [4-6] it was also shown that marigolds (*Tagetes erecta* L.) are a phytoremedy under the conditions of pollution by oil and oil sludge [6].

Crude oil from the deposit in Western Siberia (Tomsk Region) was used in the experiment. The uranium content in this oil was ~ 12.5 mg per 100 g of oil. The salts contained were as follows (mg/l): NaCl - 1523, $NaHCO_3$ - 2820, Na_2CO_3 - 168.7, Na_2SO_4 - 10.5, $CaCl_2$ - 56.9, $MgCl_2{\cdot}6H_2O$ - 35.5. The degree of mineralization of the mineralized water was 4614.6 mg/l. According to the data from [2], the uranium content in different types of oil varies within two orders of magnitude: $n\times10^{-4}$ - $n\times10^{-2}$ %, i.e. up to 10-15 mg per 100 g of oil. Formation water (oilfield water) extracted with the same oil was also used in the experiment; the formation water after extraction was separated from the oil using a separating funnel. This water is known to have high mineralization, 15-17 g/l, and to contain dissolved uranium. As it follows from [1], the average uranium content in lime scale (mineral residue) of the ground water in the Tomsk Region is 1.9 mg/kg.

2.2 Methods of experiment

The soil was cleaned to remove coarse fragments: plant residues, stones and other foreign matter. Soil samples with the weight of 300,0 g were placed into containers (plastic cups) with a drainage system. A total of 8 samples was prepared.

At the first stage of the experiment the systems were regularly irrigated with settled tap water to bring them to the state of natural humidity and prevent the subsequent complete soil drying. After four weeks of irrigation, seedlings of *Tagetes erecta* L were planted into the containers (three seedlings per each container). Then, after two weeks the following solutions were introduced into the containers (three simultaneous treatments for each system): 1) water (control), 2) 15 ml of oil, 3) 15 ml of oilfield water, 4) 15 of oil and 480 μg of uranium in the form of salt solution of $UO_2(NO_3)_2$.

The containers were covered with polythene bags to create a microclimate in each system and prevent the evaporation of the introduced solutions. After a week, the systems were irrigated with usual water and after additional five days, the containers were opened.

Soil from the roots was collected separately, dried at 80 °C in a chamber dryer to the air dry state. The remaining soil was mixed and divided into two portions. Before preparing the soil samples for the analysis one portion was dried at a temperature of 80 °C in the chamber dryer, the other one was frozen at -5 °C in a freezer.

The plants were divided into the above-ground and root parts and dried in the chamber dryer at 80 °C.

Each plant part was subjected to extraction using ammonium-acetate solution. For this, a portion of the dry biomass (2 g) was mixed with 20 ml of 1M CH_3COONH_4 and left for 24 h with repeated stirring. Then, the solid fraction was filtered using a blue ribbon filter; the obtained volume of the extract was measured. The samples dried on the filter were subsequently dissolved by the wet ashing.

Porous water of the samples under study was separated by centrifugation, to be followed by the study of the decanted oil. Water soluble and exchangeable forms of the fixed elements were successively extracted from the soil samples.

Water soluble forms (fraction 1) were obtained by mixing and shaking a portion of the soil under study (10 g) with 100 ml of distilled water for 24hours, with the ratio soil:solution being 1:10.

Exchangeable forms (fraction 2) were obtained by mixing and shaking a portion of the soil under study (10 g) with 100 ml of the ammonium acetate buffer solution (1M CH_3COONH_4) for 24 hours, with the ratio soil:solution being 1:10.

All the liquid fractions were analyzed by the method of mass spectrometry with inductively coupled plasma using an Agilent 7500 mass spectrometer by Agilent Technology (USA).

3 Results and discussion

3.1 Study of the soils

Using the method of sequential fractionation, the relative mobility of uranium and its binding to soil components were estimated (table 1).

Table 1. Comparison of the experimental systems with the control.

	Variant*	Average	Reliability of differences with the control (p)
Water-soluble fraction	Control	0.024 ± 0.011	-
	System 2	0.030 ± 0.019	no
	System 3	0.034 ± 0.023	no
	System 4	0.189 ± 0.053	<0.05
Exchangeable fraction	Control	0.014 ± 0.006	-
	System 2	0.022 ± 0.008	no
	System 3	0.066 ± 0.029	<0.05
	System 4	0.202 ± 0.089	<0.05

Water-soluble and exchangeable forms of compounds are considered to be "mobile", i.e. accessible to plants and thus, actively involved in the cycle of substances, being the most dangerous to the ecosystem.

The uranium content in the water-soluble soil fraction in the root rhizosphere is lower than in the exchangeable fraction. The maximum value of uranium in the exchangeable fraction is observed in the sample with the introduction of uranium (system 4).In the untreated soil samples, the maximum value of uranium accounts for the exchangeable fraction in the sample with oil and uranium, amounting to 4.53 μg, which is 0.81% of the total. Thus, in all the samples, the maximum yield of uranium is associated with the exchangeable fraction. The highest concentrations of uranium are characteristic of the samples with the introduction of oil.

The yield of uranium in soil extracts from the soil with the addition of oil and uranium was insignificant, amounting to 0.97% of all uranium in the soil. This indicates that most of uranium remained in the soil in the form of non-exchangeable and poorly soluble compounds, including those bound to clay minerals both as a result of chemical processes and physical sorption processes occurring both on the surface of minerals and in the interlayer space.

Since the model systems were periodically irrigated, a part of the uranium was removed from the system with drainage water. In columns with the soil with the introduced solution of uranium salt, a high yield of uranium was observed after irrigation, up to ~ 0.20 μg. The total uranium content in the drainage water of the system with the natural uranium content did not exceed 0.05 μg.

A statistical analysis of the data on the uranium content in the soil extracts from the experimental systems was performed. For water-soluble fractions: in the control system (without the introduction of uranium) the average value of the extracted uranium (μg) was 0.024 ± 0.011 (within the range of 0.006-0.037); in the system with the introduction of oilfield water (without initial concentration of uranium in oilfield water), the average value was 0.030 ± 0.019 (within the range of 0.011-0.056); in the system with the addition of crude oil the average value was 0.034 ± 0.023 (within the range of 0.010-0.059); and in the system with the introduction of uranium and crude oil it amounted to 0.189 ± 0.054 (within the range of 0.127-0.259). For exchangeable fractions: in the control system (without the introduction of uranium) the average value of the extracted uranium (μg) was 0.014 ± 0.006 (within the range of 0.008-0.024); in the system with the addition of oilfield water, this value amounted to 0.022 ± 0.008 (within the range of 0.008-0.029); in the system with the introduced crude oil it was 0.066 ± 0.029 (within the range of 0.041-0.111); and in the system with the introduction of crude oil and uranium the average value was equal to 0.202 ± 0.047 (within the range of 0.095-0.308).

When extracting a water-soluble fraction, no statistically significant differences were found in the uranium yield between the control and systems 2, 3, but the differences between the control and system 4 were statistically significant at $p < 0.05$.

When extracting the exchangeable fraction, the differences between the control and system 2 turned out to be statistically unreliable. Statistically significant ($p < 0.05$) differences in the uranium yield were found between the control and systems 3 and 4.

Thus, in the course of the experiment, the uranium content in the drainage water, as well as in the water-soluble and exchangeable fractions of the soils was determined. Upon the addition of uranium salts into the soil systems, up to 1% of the uranium introduced flew out of the column immediately after the introduction. The interaction of uranium with the soil particles caused its binding by the water-soluble and exchangeable soil fractions, involving ~ 2% of the introduced uranium.

The maximum value of the total uranium (VI) content for the soil fractions of the root rhizosphere was obtained in the sample with the addition of oil and uranium, which amounted to 36% of all the uranium extracted from the soil, while the minimum was obtained in the sample with the addition of oil, amounting to 12% of the uranium extracted from the soil.

The soil ability to retain uranium was estimated by the value of the interphase distribution coefficient K_D - the ratio between the specific activities of the solid phase and pore solution of the soil in the state of interphase equilibrium. The calculation results are shown in Table 2.

Table 2. Coefficients of the uranium distribution between its content in the pore water and solid residue (soil).

№	System	K_D
1	Soil - Control	17000
2	Soil + Oil	10100
3	Soil + Oilfield water	11000
4	Soil+U+Oil	9850

The results presented in Table 2 show that the introduction of oil into the system contributes to the redistribution of uranium, firmly fixed in the soil fragments, into more soluble forms

Study of the plants

We had next results: uranium content in the extract (acetate-ammonium extract) of the root and above-ground parts of the experimental plants (mobile forms), as well as the uranium content in the root and above-ground parts of the plant biomass (per 1 g); portion of the accumulated uranium in the above-ground and root systems in the control is almost the same, amounting to ~ 50%.

With the external load, i.e. when oil, oilfield water and a mixture of uranium-oil are added, a change in the ratios between the content of U-238 in the root and above-ground parts of the plants is observed. For example, with the oilfield water and oil introduced, the uranium content in the root part is much higher than in the above-ground part. And at the same time, the introduced oilfield water increases the uranium content in the root part up to 89%, which is 8 times higher than the content in the above-ground part of the experimental system under consideration. When oil was added, the observed difference is lower than 2 times (34 and 66%, for the root and above-ground parts, respectively). However, in this system the plants concentrate the uranium present in the initial soil sample to a higher extent - about 20 μg/dry soil, which is slightly different from the uranium content accumulated by the plants in the system with uranium and oil added. Thus, the organic substances present in oil can be assumed to contribute to the formation of uranium forms which are bio-accessible for the root system, which consequently leads to greater accumulation of U-238 by the root system and further, by the entire plant biomass.

When both uranium and oil were introduced into the experimental system, an increase in the portion of U-238 was observed in the above-ground part of the plant (~ 65%), which is noticeably higher as compared to its content in the root system (~ 35%).

The root system after the experiment with the addition of oilfield water was badly rotten and partially deformed.

4 Conclusion

Thus, having analyzed all the results obtained, we made the following conclusions.

- the maximum part of uranium, 98%, binds to soil components (organic matter, clay minerals) after introduction.

- only 2% of uranium is extracted from the soil together with mobile fractions and this amount is accumulated in the plant. The mobility of uranium depends on the moisture content in the soil. In moist soils, its mobility is higher.

- when oil and oilfield water are added, there is an increased extraction of uranium, which was present in the soil earlier, and the uranium content in the root and above-ground parts of plants in the systems with the addition of oil and oilfield water is much higher than in the control system (from 1.17 to 6.73 μg as compared to 0.78 in the above-ground part of the plants and from 9.92 to 12.87 as compared to 0.81 μg in the root part of the plants), while when uranium is added with oil, the radionuclide content noticeably differs from the content in the root part of the plants (17.1 μg in the above-ground part and 9.43 μg in the root part of the plants).

References

1. S.I. Arbuzov, L.P. Rikvanov, *Geochemistry of radionuclides* (Politecnic Institute, Tomsk, 2011).
2. A.S. Abdulaeva, *Radioactivity of the maintain ores - South of the Russia: Ecology and Development* (Makachkala, 2012).
3. S.N. Gashev, T.I. Moiseenko, Vestnik TumGU, **12**, 26-36 (2012).
4. E.K. Druusma, J.L. Carroll, *Environmental compartments* (Springer, Berlin Heidelberg, New York, 1996).
5. *Genetic effects of radiation* (New York: United Nations, 1986)
6. K. B. Gongalsky, Environ, Monit. Assess., **89**, 197-219 (2003)

Zinc and cadmium accumulation in different parts of wild plants of the *Asteraceae* family and *Triticum aestivum*

Victor Chaplygin, *Saglara* Mandzhieva*, *Yuri* Litvinov, *Natalia* Kravtsova, *Aleksey* Sherstnev, *Natalia* Chernikova, and *Irina* Deryabkina

Southern federal university, 344090 Rostov-on-Don, Russia

Abstract. The regularities of the accumulation and distribution of metals (Zn and Cd) in various parts of wild herbaceous plants and the agricultural crop of the impact zone in Rostov region (Russia) was considered. A large accumulation of elements in wild herbaceous plants of the *Asteraceae* family was noted compared with *Triticum aestivum*. The pollution of *Triticum aestivum* by Cd was observed. Wild herbaceous plants of the *Asteraceae* family were contaminated mostly with Zn. It was found that *Artemisia austriaca* and *Ambrosia artemisiifolia* accumulated Zn and Cd in the aboveground part. *Achillea nobilis* accumulated Cd mainly in the root system. Closer location monitoring plots to the source of anthropogenic load and the location of the northwest winds prevailing in the region were main influencing factors leading to higher pollution.

1 Introduction

Excessive accumulation of heavy metals (HM) in crops, which is used for feed and food purposes poses a serious threat to human and animal health [1, 2]. The investigation of the HM accumulation by plants is one of the most important and, at the same time, the most difficult aspects of environmental monitoring. Many HMs are essential microelements for all plants. For example, Cu is a part of a number of proteins and enzymes, mainly redox, Zn is necessary for photosynthesis and participates in nitrogen metabolism, and Cd activates a number of zinc-dependent enzymes: tryptophan oxygenase, carboxypeptidase. However, their high concentrations are dangerous for the vital functions of plants [3, 4]. Different plant species are characterized by selective absorption of trace elements from the soil and localization of them in various organs. It must be noted that each plant has a specific type of resistance mechanism to technogenic pollution due to the action of physiological barriers. The selectivity of HM accumulation by plants depends on the level of technogenic load and the type of plant species [5]. Wild plant species are much better adapted to anthropogenic stress than crops. To understand how adaptation mechanisms are formed it is necessary to study and compare the sustainability of wild and cultivated plants to HM

* Corresponding author: msaglara@mail.ru

accumulation in soil. In this regard, territories subject to long-term industrial pollution are important objects for studying the distribution of trace elements in plant communities [6].

Accumulation of HM in plants is especially relevant in agricultural regions of Rostov region (Russia) where industrial enterprises are often closely adjacent to agricultural land [7]. Under such conditions the studying of the HM translocation into plants is of the greatest importance, making it possible to predict the effects of technogenic pollution on the environment.

The most common HM in industrial emissions related to the first hazard class are Zn and Cd. These metals are the most toxic elements that have multiple toxic effects on plants [8-10]. So, Zn excess in the soil leads to a significant decrease in the ratio of root biomass to plant shoot biomass, which is caused by inhibition of root growth. The toxic concentration of Cd in plants leads to impaired enzyme activity and transpiration, inhibition of photosynthesis, and recovery of NO_2 to NO. With an excess of metal in plants, growth retardation occurs, the root system is damaged and leaf chlorosis develops. In addition, these elements demonstrate similar geochemical behavior and have additive properties, replacing each other in the physiological processes in plants [5, 11].

The aim of this work was to analyze the accumulation of zinc (Zn) and cadmium (Cd) by various organs of soft wheat (*Triticum aestivum*) and wild plants of the *Asteraceae* family which grown in the Rostov region.

2 Objectives and methods

Monitoring sites No. 1 - 4 were located at a distance of 1.6 to 2.2 km in various directions from the Novocherkasskaya State District Power Station (NSDPS), the largest fuel and energy complex in the Rostov Region (Russia) (Fig. 1). The most common species of wild herbs in the studied area represents by *Asteráceae* family: ragweed (*Ambrosia artemisiifolia* L.), Austrian wormwood (*Artemisia austriaca* Pall. Ex. Wild) and noble yarrow (*Achillea nobilis* L.). Despite its proximity to NSDPS there are fields sown with oilseed and cereal crops and grazing. Additional sites no. 1a, 2a, 3a, 4a were laid to study *Triticum aestivum*. These sites located on the farmland adjacent to monitoring sites of No. 1, 2, 3, and 4 which further situated on virgin lands or fallow lands. The soil of the monitoring plots in fallow and arable areas is represented by Haplic Chernozem.

Plants species were sampled annually in the first half of July 2015-2019 in the phase of full ripeness of *Triticum aestivum*, as well as mass flowering of numerous representatives of the steppe forbs. The sampling period was chosen due to the fact that the maximum entry of elements into the plants takes place at this stage [12]. The raw weight of each plant species was at least 2 kg. The roots were sampled together with soil to avoid loss of the root system mass. The plants were dried to an air-dry state. Plants aboveground part (stems and leafs) and roots were separated from each other and ground up. Prior to grinding, the roots were cleared of soil particles. In the samples of plants Zn and Cd, which are present in the emissions of NSDPS, were determined [13]. The mineralization of plant samples was performed by the dry ashing method [14, 15]. For analysis, a 1-gram oven-dried at 65 °C plant sample was used. Samples were ashed for 6 hours in a muffle furnace at a temperature of 500 °C. Ash was dissolved in 5 ml of a 20% HCl solution and after 2-3 minutes filtered through a paper filter. Zn and Cd concentration was determined on an atomic absorption spectrophotometer (Quant 2-AT, Kortek, Russia). All studies were performed in triplicate.

The HM pollution in plants was assessed by comparing the concentration of elements in plants with the maximum permissible level (MPL) for forage grasses [16].

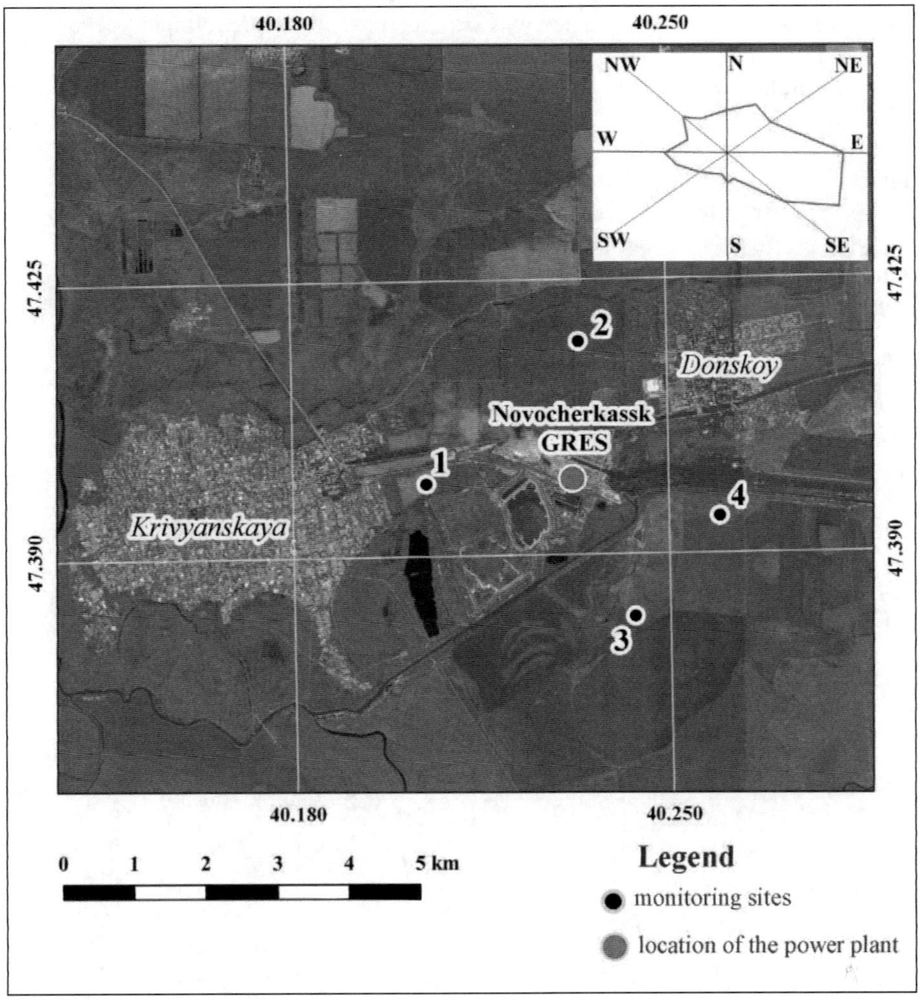

Fig. 1. Monitoring sites location.

3 Results and discussion

The monitoring of territories around NSDPS showed steady Cd pollution of plants. The excess of the MPL in *Triticum aestivum* straw for Cd was found to be at sites No. 1a and No. 2a (2.2-5.3 times, respectively) (Table 1). The Cd is known to be relatively easily transmitted from the roots into the aboveground parts, leading to relatively large amounts to penetrate the assimilate in storage organs [9, 17]. The Zn and Cd mainly accumulated in the root system of *Triticum aestivum*.

Artemisia austriaca is characterized by the accumulation of Zn and Cd in the aboveground part (Table 1). For this type of plant, an excess of MPL by Zn was found to be 1.8 times at site No. 4. Moreover, an excess of MPL by Cd at sites No. 1 and No. 4 were 2.4 and 1.1 times, respectively.

For *Achillea nobilis*, Cd pollution was observed at monitoring sites No. 1 and No. 2, accounting for1.5-3.3 MPL, respectively. It was also found that the MPL for Zn exceeded by 1.9 times at site No. 4. *Achillea nobilis* accumulated Cd mainly in the root system (Table 1).

Plants of *Ambrosia artemisiifolia* were observed to be contaminated with Zn at sites No. 1 and No. 4 with a factor of 2.4 and 1.4 MPL, respectively. For this plant, Cd at sites No. 1

was found to be 2.4 MPL. This species was characterized by a predominant accumulation of Zn and Cd in the aboveground part (Table 1).

By the accumulation of metals in the aboveground part, plants can be represented in the following orders:

Zn: *Ambrosia artemisiifolia* > *Achillea nobilis* > *Artemisia austriaca* > *Triticum aestivum*

Cd: *Triticum aestivum* > *Achillea nobilis* > *Artemisia austriaca* > *Ambrosia artemisiifolia*

For the content of HM in plant roots, the sequences were found to be as follows:

Zn: *Achillea nobilis* > *Ambrosia artemisiifolia* > *Triticum aestivum* ≥ *Artemisia austriaca*

Cd: *Triticum aestivum* > *Achillea nobilis* > *Artemisia austriaca* > *Ambrosia artemisiifolia.*

Table 1. The content of heavy metals in the aboveground part and roots of plants at monitoring sites, mg/kg.

Sites No.	**Plant**	**Zn**		**Cd**	
		The aboveground part	**Roots**	**The aboveground part**	**Roots**
1a (1.6 NW)	*Triticum aestivum*	24±2	46±4	1.60±0.12	1.80±0.05
1 (1.6 NW)	*Artemisia austriaca*	51±5	118±9	0.73±0.07	2.00±0.06
	Achillea nobilis	19±3	109±10	1.00±0.08	2.12±0.07
	Ambrosia artemisiifolia	122±13	98±9	0.73±0.08	1.26±0.12
2a (1.5 N)	*Triticum aestivum*	14±1	30±1	0.53±0.04	0.70±0.09
2 (1.5 N)	*Artemisia austriaca*	28±3	20±1	0.18±0.01	0.13±0.02
	Achillea nobilis	48±6	42±3	0.44±0.04	0.13±0.02
	Ambrosia artemisiifolia	40±4	35±3	0.24±0.02	0.08±0.01
3a (2.2 SE)	*Triticum aestivum*	8±1	7±1	0.31±0.02	0.44±0.03
3 (2.2 SE)	*Artemisia austriaca*	11±1	8±1	0.21±0.02	0.15±0.01
	Achillea nobilis	37±3	24±2	0.11±0.01	0.18±0.02
	Ambrosia artemisiifolia	20±2	18±2	0.19±0.02	0.10±0.01
4a (2.1 E)	*Triticum aestivum*	7±1	20±1	0.16±0.02	0.09±0.01
4 (2.1 E)	*Artemisia austriaca*	90±11	7±1	0.34±0.04	0.15±0.02
	Achillea nobilis	93±10	34±3	0.12±0.01	0.16±0.02
	Ambrosia artemisiifolia	72±6	18±1	0.17±0.02	0.05±0.01
MPL [13]		50	-	0.30	-

The data obtained on the Cd content on monitoring sites No. 1, No. 1a, No. 2, and No. 2a were similar to the results obtained by [18], where high concentrations of Cd - up to 1.7 mg/kg - were noted in *H. perforatum* samples.

A comparative analysis of the studied species of herbaceous plants showed that for *Triticum aestivum* and *Achillea nobilis* there was HM accumulation at the root system. In wild plants, Zn accumulated mainly in the aboveground part. *Ambrosia artemisiifolia* characterized by the highest Zn content in the aboveground part, in comparison with other studied plants. *Achillea nobilis* accumulated more Cd among other wild plants. Due to the selectivity of the element accumulation, different types of plants accumulated various amounts of HM in their organs [6]. It is indicated that a similar high variability of the elemental chemical composition of plants even on the same sites can be associated with different ages of phytomass, rooting depth, mobilization/immobilization of metal by roots in the soil, foliar absorption, etc [5]. Due to the different physiological significance of trace elements, the selectivity of their absorption from soils, their content in plants and various organs varies significantly [19].

4 Conclusion

It is established that monitoring sites No. 1 and No. 2, as well as adjacent to them (No. 1a and No. 2a) were the most contaminated. This is originally connected with two factors as follows: firstly, the closer location of these sites to the source of anthropogenic load and secondly, the location of the northwest winds prevailing in the region. It was observed Zn content in plants was higher than that of Cd content. Higher content of Zn in the aboveground part was established for the considered wild-growing herbaceous plants in comparison with the studied agricultural crop. Higher content of Cd was found in the aboveground part and roots of *Triticum aestivum* plants. *Triticum aestivum* accumulates the studied elements mainly in the root system, while wild herbs were characterized by their accumulation in the aboveground part. These differences are associated with the physiological features of the studied plant species and their differentiation in terms of families belonging.

The research was financially supported by the project of the Ministry of Science and Higher Education of the Russian Federation (State assignment in the field of scientific activity, Southern Federal University, 2020) and the grant from the President of Russian Federation, no. MK-2818.2019.5.

References

1. M. Roy, L.M. McDonald, Land Degradation & Development **26(8)**, 785-792 (2015)
2. S.A. Anjum, M. Tanveer, S. Hussain, M. Bao, L.C. Wang, I. Khan, E. Ullah, S.A. Tung, R.A. Samad, B. Shahzad, Environmental science and pollution research **22(21)**, 17022–17030 (2015)
3. A.P.G.C. Marques, A.O.S.S. Rangel, P.M.L. Castro, Critical reviews in environmental science and technology **39**, 622–654 (2009)
4. P.J. White, P.H. Brown, Annals of Botany **105**, 1073–1080 (2010)
5. A. Kabata-Pendias *Trace elements in soils and plants* (NW: CRC Press, Taylor and Francis Group, 2011)
6. T.M. Minkina, S.S. Mandzhieva, V.A. Chaplygin, O.G. Nazarenko, A.Y. Maksimov, I.V. Zamulina, M.V. Burachevskaya, S.N. Sushkova, Arid Ecosystems **8(3)**, 190-202 (2018)

7. V.A. Chaplygin, S.S. Mandzhieva, T.M. Minkina, S.N. Sushkova, R. Kizilkaya, C. Gulser, I.V. Zamulina, N.E. Kravtsova, I.P. Lobzenko, N.P. Chernikova. Environmental Geochemistry and Health (2019). DOI: 10.1007/s10653-019-00411-6.
8. E. Andresen, H. Küpper, *Cadmium: from toxicity to essentiality* (Springer, Dordrecht, 2013)
9. Z.J. Zhao, Z.R. Nan, Z.W. Wang, Y.M. Yang, M. Shimizu, Journal of Arid Land **6(1)**, 59-68 (2014)
10. I. Shtangeeva, A. Viksna, V. Grebnevs, Environmental geochemistry and health **42(1)**, 209-219 (2020)
11. I. Yermakov, N. Maksimov, A. Evmenyeva, M. Breygina, Plant Reproduction, **31(4)**, 357-365 (2018).
12. V.B. Ilyin, A.I. Syso, *Heavy metals and nonmetals in the soil-plant system* (Siberian branch RAS publishing House, Novosibirsk, 2012) (in Russian).
13. T.M. Minkina, S.S. Mandzhieva, G.V. Motuzova, V.A. Chapligin, S.N. Suchkova, Yu.A. Fedorov, S.I. Kolesnikov, T.V. Bauer, Environmental Engineering and Management Journal, **13(5)**, 1307-1315 (2014)
14. *Methodological guidelines on the determination of heavy metals in agricultural soils and crops* (Moscow, TsINAO, 1992) (in Russian).
15. G. Du Laing, F.M.G. Tack, M.G. Verloo, Analytica Chimica Acta, **497**, 191–198 (2003)
16. *Provisional maximum permissible levels (MPL) for some chemical elements and gossypol in forage for farm animals and feed additives* (Moscow, 1987) (in Russian)
17. V.B. Ilyin, G.A. Garmash, N.Yu. Garmash, Agrochemistry, **6**, 90-100 (1985) (in Russian)
18. D.N. Badea, Revista de Chimie (Bucharest), **66(8)**, 1132-1136 (2015).
19. V.K. Kashin, L.L. Ubugunov, Agrochemistry **4**, 68-76 (2012).

https://doi.org/10.1051/e3sconf/202016901004

Lead compounds in bottom sediments of the Seversky Donets floodplain

Marina Burachevskaya*, *Dina* Nevidomskaya, *Victoria* Tsitsuashvili, *Vishnu* Rajput, and *Dmitry* Bren

Southern federal university, Academy of biology and biotechnology, 344090 Rostov-on-Don, Russia

Abstract. Investigations on heavy metals in water bodies are extremely important in a comprehensive assessment when the anthropogenic activities disturb the natural environment. The present paper studies the composition of Pb compounds in highly contaminated bottom sediments of a natural settling basin of a chemical plant at the Atamanskoye lake in the floodplain of the Seversky Donets River by sequential fractionation. The predominance of the residual Pb fraction, which is represented by layered silicates (hydromuscovite, montmorillonite, kaolinite) is shown. Organic matter and Fe-Mn (hydr)oxides also play an important role in metal interaction. High mobility of Pb is noted, which associated with imposing high environmental risks in the studied area.

1 Introduction

Large enterprises of the chemical industry exert great pressure on the environment through their emissions including gaseous, liquid, and solid wastes. The most hazardous pollutants which are entering into terrestrial and aquatic ecosystems daily are heavy metals. Despite organic pollutants undergoing degradation processes, metals are only capable of redistribution between the individual components of water systems and the environment; thus remaining in the environment. In this regard, particular attention must be paid to floodplain landscapes, due to their position placed at the lower part of the cascade landscape-geochemical system to collect heavy metals [1, 2, 3]. Bottom sediments are considered to be an integral indicator of the man-caused load on the water body. In this case, bottom sediments are a kind of depot for heavy metals, keeping the long-term anthropogenic impact on the ecosystem. A change in the physicochemical equilibrium in aqueous solutions can provoke the secondary contamination of adjacent media, taking into account that the buffer capacity of bottom sediments (especially that of the upper layers) is limited.

The toxicity of an element is determined by the form of its location, as well as the composition and properties of the deposit medium [4, 5]. Chemical methods of sequential extraction are widely used to evaluate the forms of metal compounds in soils and bottom sediments [3, 6, 7, 8].

* Corresponding author: marina.0911@mail.ru

The purpose of the paper was to study the fractional composition of Pb in highly polluted bottom sediments. Pb belongs to the class I environmental hazard, hence it becomes very important to study its mobility and concentration in the highly contaminated bottom sediments.

2 Materials and methods

Sampling was done in the bottom sediments located at the Atamanskoye lake in the floodplain of the Seversky Donets River, the main tributary of the Don River. The study area experiences many years of anthropogenic stress due to more than 40 years serving as a natural settling basin for enterprises in the industrial area of Kamensk-Shakhtinsky, Rostov Region. According to the literature report [9], there are 180 thousand tons of flowing sludge, 343 thousand tons of stiff sludge, and more than 444 thousand tons of soft plastic sludge in the lake. All of the above deposits are contaminated with several metals and metalloids, such as As, Li, Hg, Cd, Zn, Cu, Pb, Sn, Mn, Cr, as well as organometallic compounds. The physic-chemical characteristics and the total metal content in the sample of bottom sediments are presented in Tables 1 and 2. The study of the properties of bottom sediments was carried out by generally accepted methods [5]. The total content of Pb in the samples was determined by x-ray fluorescence method.

Table 1. Physico-chemical properties of bottom sediments (aleuropelitic silt, surface horizon)

C_{org}, %	pH_{H2O}	Silt and clay (<0.01 mm), %	Silt (<0.001 mm), %	CEC, cmol/kg	$CaCO_3$, %	$MgCO_3$, %
4.1	8.3±0.8*	69.6	17.6	21.3±2.0	1.8	0.9

* ±Standard deviation

Table 2. Total chemical composition of bottom sediments (aleuropelitic silt, surface horizon), mg/kg

SiO_2	Al_2O_3	Fe_2O_3	CaO	MgO	K_2O	P_2O_5	Pb
51.8 ±5.4*	8.9 ±0.8	3.2 ±0.2	12.1 ±1.1	2.2 ±0.1	1.5 ±0.1	0.3 ±0.001	1580 ±120

* ±Standard deviation

Sequential extraction can assess the degree of contamination and, hence, the contamination hazard level. Furthermore, these methods meet all the necessary requirements for implementation over a wide area: simple, applicable to all kinds of sediments, low cost, and easily understandable and comparable results. To study the composition of Pb compounds in bottom sediments the Tessier sequential fractionation method was used [8]. This widely-known method is often used to assess the fractional composition of heavy metals. It is initially focused on the analysis of bottom sediments and provides the isolation of five fractions of metal compounds: exchangeable (1M $MgCl_2$), bound to carbonates (1M CH_3COONa), bound to Fe-Mn (hydr)oxides (0.04M $NH_2OH•HCl$ in 25% CH_3COOH), bound to organic matter (0.02 M HNO_3 + 30% H_2O_2, pH 2, then 3.2 M CH_3COONH_4 in 20% HNO_3), and residual (HF + $HClO_4$, then $HNO_{3conc.}$). The method provides detailed information on the nature, occurrence forms, biological and physicochemical activity, mobility, and transportation of metal. The analysis of the Pb content in the extracts was carried out by atomic absorption spectrophotometry (AAS).

The tests were made in triplicate. Statistical data were done using the STATISTICA 2010 program.

3 Results and discussion

Content of in Pb the study samples located on the territory of the natural settling basin of the industrial wastewater of the chemical plant, exceeds in 158-fold the clarke by the metal (10 mg/kg) in the lithosphere was detected [10] (Table 2).

Fractionation of Pb compounds in the study sample of bottom sediments was shown in Table 3. The dominance of the residual fraction (49% of the total content of all fractions) could be inferred from the data. This fraction contains mainly primary and secondary minerals, which may hold trace metals with in their crystal structure [8]. Mineralogical analysis [11] showed the predominance of layered silicates (hydromuscovite, montmorillonite and kaolinite) (Table 4). Layer silicates are characterized by the chemical and energetic heterogeneity of their surface, which is characterized by the presence of structural defects and functional groups of different composition, which could serve as active centers during metal absorption. Such centers are frequently exchangeable cations, surface hydroxyl groups, and oxygen of the tetrahedral sheet. The application of XANES and EXAFS methods makes it possible to identify the heavy metals presence in the octahedral structure of layer silicates [12, 13, 14].

Table 3. Fractional composition of Pb compounds in contaminated bottom sediments, mg/kg

Fractions				
Exchangeable	**Bound to carbonates**	**Bound to Fe-Mn (hydr)oxides**	**Bound to organic matter**	**Residual**
287±30*	100±11	187±19	196±22	769±72

* ±Standard deviation

Table 4. The mineralogical composition of the residual fraction in contaminated bottom sediments, %

Minerals				
Quartz	**Feldspar**	**Hydromuscovite**	**Montmorillonite**	**Kaolinite**
5.0	5.0	58.3	19.0	12.7

Metals may be bound to various forms of organic matter: living organisms, detritus, coatings on mineral particles, etc. The complexation and peptization properties of natural organic matter (notably humic and fulvic acids) are well recognized, as is the phenomenon of bioaccumulation in certain living organisms [8]. The fraction associated with organic matter accounts for 13% (Fig. 1). Pb has a high ability to form strong chelate complexes with organic matter [14, 15]. This is evidenced by the stability constants of metal complexes with humic acids, which determine the high organophilicity towards the metal [16].

The fraction bound to carbonate is the lowest and does not exceed 7% (Table 3). Metals can be co-precipitated by carbonates, enter in their structure, or be sorbed on Fe–Mn oxides, which are deposited on the surface of carbonates [14].

It is noted that Pb bound to Fe and Mn (hydr)oxides - 12% (Table 3). Oxides and hydroxides play an extremely important role in the fixation of heavy metals [16, 17]. Fe oxides and hydroxides have a large active surface due to the ability to form polymolecular films on the surface of secondary clay minerals. Heavy metals ions interact by displacing H^+ ions that enter OH^- or OH^{2-} groups on the oxide surface, or by substitution Fe^{3+} or Fe^{2+} ions, which probably determines the state of many metals [18].

Chemical fractionation showed high values of Pb content in the exchangeable fraction (287 mg/kg or 19% of the total fractions), which is 48 times higher than the maximum permissible concentration of the mobile metal (6.0 mg/kg) [19] (Table 3, Fig. 1). Exchangeable metals are fixed mainly by electrostatic forces on the surface of clay and other minerals, amorphous compounds and organic substances with a low pH [15]. These are the most mobile metal forms. This indicates a large environmental risk to adjacent landscapes.

4 Conclusions

The sequence of distribution of Pb among the fractions were found to be in the order as follows: residual > bound to Fe and Mn (hydr)oxides = bound to organic matter > exchangeable > bound to carbonates.

The mobility of Pb is quite high (19%) in highly polluted bottom sediments, which is the most dangerous consequence of pollutions entered the area by enterprise emissions. The main stabilizers of Pb mobility are layered silicates (hydromuscovite, montmorillonite, kaolinite), organic matter, and Fe and Mn (hydr)oxides.

The study was supported by the Russian Science Foundation grant, project No. 19-74-00085.

References

1. L. Zhang, X. Ye, H. Feng, Y. Jing, T. Ouyang, X. Yu, R. Liang, C. Gao, W. Chen, Marine pollution Bulletin, **54(7)**, 974-982 (2007)
2. A.D. Dukes, R.T. Eklund, Z.D. Morgan, R.C. Layland, Water, Air, & Soil Pollution, **231**, 11 (2020)
3. A. Malsiu, I. Shehu, T. Stafilov, F.Faiku, J. of Environ. and Public Health, ID 3098594 (2020)
4. A. Manceau, M.A. Marcus, N. Tamura, Rev. Mineral Geochem., **49**, 341-428 (2002)
5. P.I. Kurilov, R.P. Kruglyakova, N.I. Savitskaya, P.S. Fedotov, J. of Anal. Chem., **64(7)**, 738-745 (2009)
6. T.M. Minkina, G.V. Motuzova, O.G. Nazarenko, V.S. Kryshchenko, S.S. Mandzhieva, Eur. Soil Sci., **41(7)**, 708-716 (2008)
7. T.M. Minkina, G.V. Motusova, O.G. Nazarenko, S.S. Mandzhieva, *Heavy metal* compounds *in soil: transformation upon soil pollution and ecological significance* (Nova Science Publishers Inc., 2010)
8. A. Tessier, P.G.C. Campbell, M. Bisson, Anal Chem., **51**, 844-850 (1979)
9. V. V. Privalenko, *Study of the regime of pollution of underground water and the environment at the experimental production site in the area of Kamensk-Shakhtinsky in 1993-1997 (report of the engineering and geological party of the Rostov GRE)* (Rostov-on-Don, 1998) (in Russian)
10. A.P. Vinogradov, *Geochemistry of rare and dispersed chemical elements in soils* (Moscow, RAN, 1957) (in Russian)
11. *Methodological Recommendations of the Scientific Council on the Methods of Mineralogical Studies* (no. 158. Moscow, 2008) (in Russian).
12. J. Myers, K. Thorbjornsen, Soil Sediment Cont. **13**, 1–16 (2004).
13. M. J. Kwon, M. I. Boyanov, J. S. Yang, S. Lee, Y. H. Hwang, J. Y. Lee, B. Mishra, and K. M. Kemner, Environ. Pollut. **226**, 346–355 (2017).
14. D.G. Nevidomskaya, T.M. Minkina, A.V. Soldatov, V.A. Shuvaeva, Y.V. Zubavichus, Yu.S. Podkovyrina, J. Soils and Sedim., **16(4)**, 1183-1192 (2016)
15. D.L. Pinskii, T.M. Minkina, Eur. J. Soil Sci., **2**, 59-68 (2013)

16. A. Kabata-Pendias, *Trace elements in soil and plants* (Boca Raton, London, N.Y., CRC Press, 2011)
17. E.V. Mellis, J.C. Casagrande, M.C.P. Cruz, *Conf. proc. 7th international Conference on the Biogeochemistry of Trace Elements*, **V.III**, 20–21 (Uppsala, Sweden, 2003)
18. I.O. Plekhanova, ,O.V. Klenova, Y.D. Kutukova, Eur. Soil Sci., **34**, 496–503 (2001)
19. GN *2.1.7.2041-06. Maximum permissible concentrations of chemicals in the soil: Hygiene standards* (Rospotrebnadzor, Moscow, 2006) (in Russian)

E3S Web of Conferences **169**, 01005 (2020) https://doi.org/10.1051/e3sconf/202016901005
APEEM 2020

Geochemical studies of surface bottom sediments and soils of small watercourses

Evgeni Nesterov*, *Timofei* Belyakov, *Stanislav* Dubrova, and *Pavel* Egorov

Herzen State Pedagogical University of Russia, Geology and Geoecology Department, 198328 Saint Petersburg, Russian Federation

Abstract. Analytical data on samples of bottom sediments of the Novaya river watercourse made it possible to identify areas of pollution, sources of pollution and determine the nature of the influence of the anthropogenic factor on the state of the river. It was found that bottom sediments are divided into slightly polluted, moderately polluted, and dangerously polluted. Soils on the total indicator of pollution belong to the permissible category. The list of defined indicators includes: hydrogen index (pH), total metal content (cadmium, copper, lead, mercury, nickel, zinc), arsenic, petroleum products, benz(a)pyrene, ammonium nitrogen, phosphate ion, organic matter. Based on the results of the interpretation of the laboratory data, it was found that in accordance with the Regional standard, the precipitation the majority of samples are slightly polluted; at some samples, are moderately polluted; at a quarter of samples are dangerously polluted.

1 Introduction

Geochemical studies of surface bottom sediments were carried out on various sections of the small watercourse (the Novaya river, Saint Petersburg) and included determination of the content of organic matter (according to the LOI), phosphorus, nitrogen, macro-and microelements [2].

In the course of the work, an assessment of the general chemical contamination and sanitary condition of bottom sediments was made. The geoecological testing of bottom sediments was carried out in four stages during the calendar year. The quantitative chemical analysis of bottom sediments was carried out according to the certified measurement methods in the Test center of TACIS LLC.

The list of defined indicators includes: hydrogen index (pH), total metal content (cadmium, copper, lead, mercury, nickel, zinc), arsenic, petroleum products, benz(a)pyrene, ammonium nitrogen, phosphate ion, organic matter. Macroelements are those containing more than 0.01%, and microelements - from 10-6 to 0.01%. Macroelements include copper, lead, zinc, nickel, and microelements - cadmium, arsenic.

The heavy metals selected for the analysis are identified as priority pollutants for the Novaya river (Fig. 1).

* Corresponding author: nestem26@mail.ru

 https://doi.org/10.1051/e3sconf/202016901005

Fig. 1. The scheme of sampling

The degree of contamination of bottom sediments was assessed using standardized values of the contents of pollutants in accordance with the recommendations of the normative document: «The norms and criteria for assessing the pollution of bottom sediments in water objects of St. Petersburg. The regional standard». The composition of standard sediments: 10% of the organic matter content and 25% of the clay fraction. The concentrations of pollutants in non-standard bottom sediments were recalculated by their concentrations in standard bottom sediments [3-4, 6-7].

2 Research materials and methods

In accordance with the established Regional standard classifier, the bottom sediments of the water objects of St. Petersburg are divided into 4 levels and V classes. The target level: if the concentration of pollutants is below the target level, the bottom sediments are considered clean. These sediments belong to class 0. The limit level: the concentrations of pollutants up to this level represent the maximum acceptable risk for both human health and nature. Bottom sediments with concentrations of pollutants between the target and limit levels belong to class I. These sediments are considered to be slightly polluted. The control level: under certain conditions, pollutants in sediments with concentrations below this level can have a negative impact on the clean water environment. Bottom sediments with a concentration of pollutants between the limit and control levels belong to class II. They are considered to be moderately polluted. The level requiring intervention: if the concentration of pollutants exceeds the class II control level, it is considered an indicator of hard pollution of bottom sediments. Bottom sediments with a concentration of pollutants between the control level and the level requiring intervention belong to class III. The pollution of these sediments is considered to be moderate to hard. Bottom sediments where the concentration of pollutants exceeds the level that requires intervention are considered to be dangerously polluted. They belong to class IV.

2.1 Ecochemical characteristics of bottom sediments

The results of the interpretation of the laboratory data showed that in accordance with the Regional standard, the precipitation samples collected at stations 9, 10, 11, 13, 14, 15, 16, 17 are slightly polluted; 4, 5, 18 – moderately polluted; 1, 12, 19, 20, 21 – dangerously polluted (Table 1).

Table 1. Nitrogen and phosphorus in the Novaya river basin

Nr.	May 2019	May 2019		July 2019			October 2019		
	P total	N total	N amm.	P total	N total	N amm.	P total	N total	N amm.
1	0.042	11.0 (s)** 11.0 (b)	8.4 (s) 8.2 (b)	0.11	25.0	22.0	-	-	-
2	0.015	5.1	3.8	0.14	18.0	15.0	-	-	-
3	0.047	9.2	7.3	0.09	12.0	7.8	-	-	-
4	0.060	12.0	8.1	0.14	12.0	9.3	5.1	26	18.1
5	0.033	10.2	7.0	0.14	11.0	8.6	-	-	-
6	0.070	5.4	3.7	0.18	17.0	8.5	-	-	-
7	-	7.1	6.6	-	-	-	0.016	4.0	3.2
8	0.130	9.8	7.5	0.040	16.0	7.6	0.25	6.6	5.3
9	0.030	11.0	6.7	0.28	23.0	12.0	0.66	13	10.4
10	1.200	9.2	6.9	5.50	41.0	31.0	3.0	31	24.0
11	0.070	12.0	7.8	0.35	25.0	13.0	0.55	18	15.2
12	0.100	7.0 (s) 8.3(b)	4.6 (s) 6.1 (b)	0.15 (s) 0.16 (b)	19.0 (s) 12.0 (b)	12.0 (s) 9.7 (b)	0.31	21	17.2
13	0.070	7.2 (s) 8.4 (b)	5.6 (s) 6.3 (b)	0.23 (s) 0.09 (b)	15.0 (s) 13.7 (b)	11.0 (s) 7.5 (b)	0.26	21	17.7
14	0.080 (s) 0.240 (b)	8.1 (s) 8.5 (b)	5.9 (s) 6.9 (b)	0.08 (s) 0.10 (b)	13.0 (s) 13.0 (b)	7.4 (s) 6.8 (b)	0.27 (s) 0.26 (b)	22 (s) 22 (b)	19.0 (s) 17.9 (b)
15	0.110	7.7 (s) 9.4 (b)	6,9 (s) 7.6 (b)	0.25 (s) 0.33 (b)	61.0 (s) 18.0 (b)	20.0 (s) 7.2 (b)	<0.10	24	20.0
16	0.160 (s) 0.060 (b)	-	-	0.31 (s) 0.28 (b)	22.0 (s) 24.0 (b)	8.6 (s) 11.0 (b)	0.08 (s) <0.10 (b)	24 (s) 25 (b)	19.7 21.0
17	0.050 (s) 0.080 (b)	9.7 (s) 9.2 (b)	7.7 (s) 7.9 (b)	0.22 (s) 0.10 (b)	16.0 (s) 24.0 (b)	7.2 (s) 7.2 (b)	<0.10 (s) <0.10 (b)	27 (s) 27 (b)	22.0 (s) 22.9 (b)
18	0.080 (s) 0.170 (b)	8.2 (s) 8.0 (b)	6.0 (s) 6.3 (b)	0.09 (s) 0.12 (b)	18.0 (s) 18.0 (b)	10.0 (s) 11.0 (b)	<0.10 (s) <0.10 (b)	24 (s) 24 (b)	19.4 (s) 19.7 (b)
19	0.060 (s) 0.090 (b)	8.8 (s) 10.2 (b)	7.1 (s) 8.7 (b)	0.09 (s) 0.051 (b)	14.0 (s) 26.0 (b)	13.7 (s) 20.0 (b)	0.18 (s) 0.18 (b)	22 (s) 22 (b)	18.3 (s) 18.0 (b)
20	0.460	<1.0	0.39	0.31	3.0	0.18	0.54	1,4	0.93
21	0.220 (s) 1.500 (b)	<1.0	0.31 (s) 0.35 (b)	0.30 (s) 0.33 (b)	13.0 (s) 12.0 (b)	0.18 (s) 0.15 (b)	0.56 (s) 0.51 (b)	<1 (s) <1 (b)	0.57 0.56

* in October, releases 1 and 2 - tamponaded, channel 6 – dry.
** (s) – surface ; (b) – bottom.

In order to obtain more objective information, the selection was made at different agreed stations. So, the samples were taken at station 7 in May and October, and at station 16 – in July and October.

Ecological condition of soils in the water catchment area

To study the sources of pollution of the Novaya river and to calculate the diffuse biogenic load, we conducted an assessment of soil pollution in the catchment area. Soil sampling is performed in accordance with the State standarts GOST 17.4.4.02-84.

2.2 Methods of soil research

The soil pollution was estimated by the gross content of metals (vanadium, chromium, cobalt, nickel, copper, zinc, lead) and arsenic [6]. The determination of the network of testing stations and sampling was carried out in accordance with the recommendations set out in the articles 4.19 and 6.17 of SR (Sanitary Regulations) 11-102-97. The data provided in SR 11-102-97 were used to assess the background state of soils and grounds (Table 2). The assessment of the degree of soil pollution was carried out by comparing the measured values of the content of pollutants with the standards established by HS (hygienic standard) 2.1.7.2041-06 (MPC of chemicals in the soil) and HS 2.1.7.2511-09 (TAC of chemicals in the soil), as well as by the indicator [5].

Table 2. Assessment of the degree of chemical pollution of soils (according to Sanitary Regulations and Norms 2.1.7.1287-03)

Pollution categories	**Σ pollution Indic. (Zc)**	**Soil content (mg/kg)**					
		I danger class		**II danger class**		**III danger class**	
		Organic compounds	**Inorganic compounds**	**Organic compounds**	**Inorganic compounds**	**Organic compounds**	**Inorganic compounds**
Clean	-	from background to MPC	from background to MPC	from background to MPC	from background to MPC	from background to MPC	from background to MPC
Permissible	< 16	from 1 to 2 MPC	from 2 backgr. values to MPC	from 1 to 2 MPC	from 2 backgr. values to MPC	from 1 to 2 MPC	from 2 backgr. values to MPC
Moderately dangerous	16 – 32					from 2 to 5 MPC	from MPC to Cmax
Dangerous	32 – 128	from 2 to 5 MPC	from MPC to Cmax	from 2 to 5 MPC	from MPC to Cmax	> 5 MPC	> Cmax
Extremely dangerous	> 128	> 5 MPC	> Cmax	>5 MPC	> Cmax		

2.3 The results of soil research

In the course of the research, 120 soil samples were collected in the catchment area. The analysis of the test results shows that in 13 samples there is an excess of the MPC value of arsenic, in one sample - of lead, in 18 samples - of zinc, in six samples - of copper, and in one sample - of nickel. In all samples, the integral pollution index does not exceed the standard value (16), the Zc varies from 1 to 11.7, which corresponds to the permissible level of soil pollution.

3 Conclusion

Based on the results of the interpretation of the laboratory data, it was found that in accordance with the Regional standard, the precipitation samples collected at stations 9, 10, 11, 13, 14, 15, 16, 17 are slightly polluted; at 4, 5, 18 - moderately polluted; at 1, 12, 19, 20, 21 – dangerously polluted. Bottom sediments with a concentration of pollutants that exceed the level requiring intervention are considered dangerously polluted. They belong to class IV.

As a result of soil testing, it was shown that the main pollutant of soils in the studied area is zinc, there is an excess of the MPC in 18 selected samples. The integral indicator of pollution does not exceed the standard value. The samples by the degree of chemical contamination, by the total indicator of pollution can be referred to the permissible category of pollution.

Further research may be related to the study of bottom sediments and soils at the full depth of the section, which will allow restoring the temporal dynamics of pollution and formulating scenarios for the development of environmental conditions.

References

1. B. Schütt, Geomorphology, **23(2-4)**, 273-283 (1998)
2. E. Sklyarov, *Interpretation of Geochemical Data: Tutorial. Intermet Engineering*, 288 (2001)
3. E. Nesterov, A. Timirgaleev, E. Maslova, News of Higher Educational Institutions. North Caucasus region. Natural sciences, **2**, 96-99 (2008)
4. M. Postolova, E. Nesterov, N. Popkov, *Issues of Regional Ecology,* **5**, 19 (2017)
5. M. Postolova, E. Nesterov, E. Mingareeva, *Heavy Metals Content in Soils of Various Sampling Times. Geography: Science and Education Development. Collective monograph*, 167-171 (2018)
6. E. Nesterov, P. Egorov, M. Markova, D. Morozov, *International Multidisciplinary Scientific GeoConference Surveying Geology and Mining Ecology Management. Geochemical Indication of Bottom Sediments in the North-West of the East European Plain. SGEM*, **19**, 699-707 (2019)
7. S. Lebedev, M. Kulkova, L. Zarina, E. Nesterov, *Assessment of Long-lived Radionuclides (3h, 14c) and Heavy Metals in Snow Cover of Urban Areas as Indicators of Environmental Pollution (St. Petersburg, Russia). Processes and Phenomena on the Boundary Between Biogenic and Abiogenic Nature. Collection of papers presented at VI International Symposium,* 297-313 (2020)

Diffuse biogenic load from the catchment area of small watercourses

Evgeni Nesterov[1,*], *Grigori* Frumin[2], *Pavel* Egorov[1], and *Alexander* Lyubimov[1]

[1]Herzen State Pedagogical University of Russia, Geology and Geoecology Department, 198328 Saint Petersburg, Russian Federation
[2] Russian State Hydrometeorological University, Ecology and Bioresources Department, 195196 Saint Petersburg, Russian Federation

Abstract. The article systematizes information on quantitative parameters for assessing the modules of the background runoff of nutrients (total phosphorus and total nitrogen) with the catchment of rivers and lakes. A new approach proposed for estimating the modules of the background runoff of nutrients based on mathematical models. A methodology for calculating the modules of the background runoff of nutrients from the catchments of the rivers of the Baltic Sea basin developed. The study of the dependence of the trophic level of the water reservoir on the amount of phosphorus and nitrogen entering it led to the development of the so-called load concept, which is based on the idea of the existence of a quantitative relationship between the amount of phosphorus intake and the reaction of the reservoir. The maximum allowable export to the Gulf of Finland is 4860 t/year of total phosphorus and 106680 t/year of total nitrogen. These values significantly exceed the total load on the Novaya river from its catchment area: 0.072 t/year for total phosphorus and 0.066 t/year for total nitrogen.

1 Introduction

Irresponsible attitude to the ecology of small rivers exacerbates the already serious problem of water on the Earth, the problem that can be generally formulated as a contradiction between the increasing public demand for high quality water, on the one hand, and its continuing deterioration, on the other. This deterioration is most significantly connected with chemical water pollution. The result of this is usually a change in the position of the reservoir on the trophic scale. The importance of the problem of water pollution is so great that the justification of its relevance seems to be a conscious necessity. The study of diffuse pollution of any water body inevitably covers a set of processes of interaction of surface and underground waters in the catchment area with the regime of anthropogenic influence. The object of consideration is usually a catchment area, as a territorial unit that combines local landscapes. In scientific research and design practice, mathematical modeling is the most effective tool for analyzing the processes of diffuse pollution of water objects and water ecosystems [9].

* Corresponding author: nestem26@mail.ru

In the course of the research, an assessment of the diffuse biogenic load from the catchment area of the river Novaya on the river Novaya was made. The upper reaches and part of the middle course of the river are in an area of intensive agriculture. Agricultural land covers 2900 hectares, 90% of the total amount of which are arable land of 2,600 hectares. This category of land represents a certain danger for the hydraulic system due to possible flushing of the treated soil into the system of hydromelioration and therefore in the Novaya riverbed. Fields are a source of systematic ingress of chemical elements into the hydraulic system, along with the rain-washing out of fertilizers and pesticides. Objects in urbanized areas and allotments have a significant impact on the state of the river. Urbanized objects absolutely dominate the catchment area and occupy 2/3 of its area. Almost 54 km^2 of the administrative district is occupied by urban-type buildings and less than 2 km^2 is occupied by rural settlements, while 0.5 km^2 is covered by individual buildings and structures, which is 65.2 and 0.6%, respectively.

The Novaya river catchment area (Fig. 1) is located in the South-Western part of the Kirovsky district of Saint Petersburg and the adjacent part of the Krasnoselsky district.

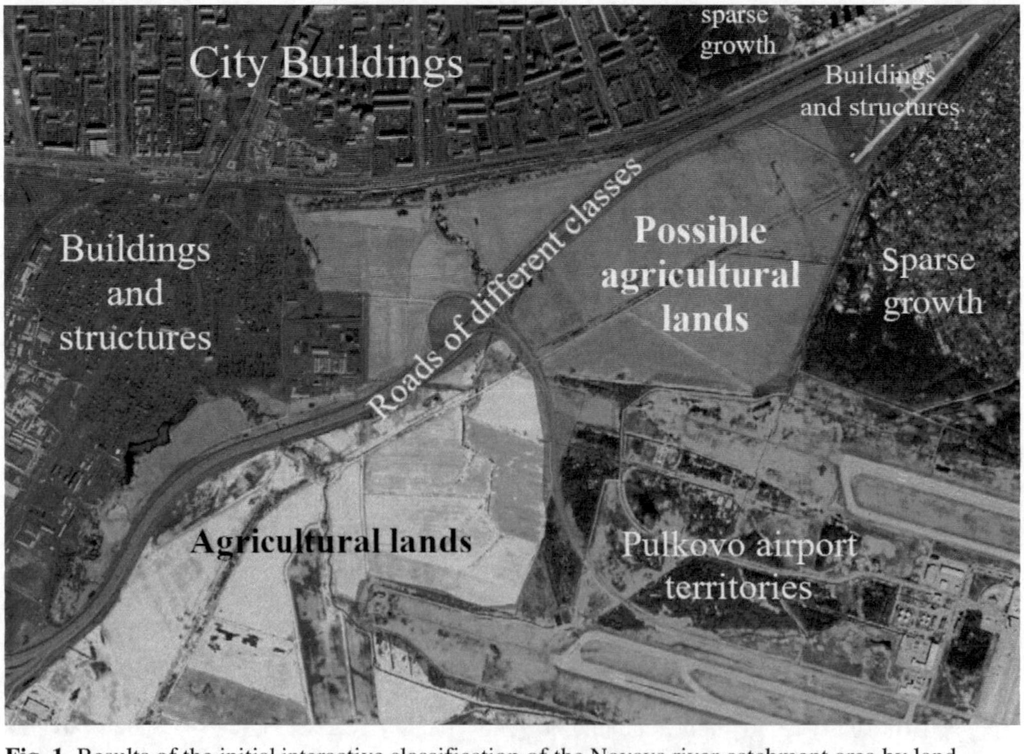

Fig. 1. Results of the initial interactive classification of the Novaya river catchment area by land category

The indicators that characterize the catchment area: the catchment area is 9.7 km^2; the water area of ponds-overflows within the Kirovsky district of St. Petersburg is 0.1%; the catchment area includes megalopolis land - 65.1%; other settlements – 0.03%; industrial and other special purpose land - 0.27 %; agricultural land – 34.9%; forest land – 2.1%; railways - 15 km; ring motorway - 11.4 km; other highways - 25 km; field roads, forest and other special purpose roads - 28 km. To solve the problem of water flow in the Novaya river, the following equation was applied [8]:

$$R = 0.0089 \cdot F, \qquad (1)$$

where R is the water flow for the catchment area = 500 $m^3/(s\ km^2)$; F is the catchment area, km^2. For the Novaya river, F = 9.7 km^2. Then the average annual flow rate R = 0.091 m^3/s. There are fairly close links between the basin area and the length of the main river:

$$F = aLn, \tag{2}$$

where a and n are geographical parameters. For example, for the European part of Russia, a = 0.58, n = 1.75. The length of the Novaya river is L = 5 km, then F = 9.7 km^2. This value was used in subsequent calculations.

2 Research materials and methods

Several indicators were used for calculations. The first indicator is the data on the content of phosphorus (P_2O_5, %) in the soils of the Novaya river catchment area. 120 samples were selected and analyzed for the phosphorus content (the area of one site is 100 m^2). Statistical processing of the analysis results showed the following: the average P_2O_5 content is 0.206655%, the maximum is 0.482022%, the minimum is 0.097119%, and the standard error is 0.0739%. As a result of recalculation, it was found that the surveyed soils of the Novaya river catchment area (0.012 km^2) contain 0.108 kg of phosphorus (0.09% of phosphorus). The phosphorus content in the soils of the entire catchment area of the Novaya river (9.7 km^2) is approximately 0.087 tons.

According to [3], sod-podzolic, predominant-light loamy and medium loamy soils contain 0.14-0.16% of phosphorus. Sod-podzolic light loamy soils developing on moraine loam - 0.09-0.12% of phosphorus. Sod-podzolic sandy loams, underlain by moraine loams, - 0.07-0.12% of phosphorus. Sod-podzolic sandy soils – 0.06-0.08%. The following approach was used to estimate the nitrogen content in the soils of the Novaya river catchment area. Concentrations of 15 chemical elements containing in catchment soils were compared with their abundance in soils (clarkes) (Table 1).

Table 1. Content of chemical elements in soils of the Novaya river catchment area (C) and clarkes, mg/kg.

Element	Clarke	Content	Element	Clarke	Content
S	850	442.6	Ga	30	14.8
Sc	7	9.27	As	5	3.2
V	100	85.6	Rb	100	125.7
Cr	200	52.9	Sr	300	154.5
Co	8	6.92	Ba	500	654.9
Ni	40	23.3	La	40	49.1
Cu	20	23.1	Pb	10	28.5
Zn	50	86.4			

According to the data shown in Table 1.1, the following dependence was established (equation 3, Fig. 2):

$$\ln C = 0.401 + 0.865 \ln\,[\text{clarke}], \tag{3}$$

n = 15; r = 0.926; r^2 = 0.857; $\sigma y(x)$ = 0,584; F_p/F_T = 17.0. Here n is the quantity of chemical elements, r – the correlation coefficient, r^2 – the coefficient of determination, $\sigma y(x)$ – the standard error, F_p/F_T - the ratio of the calculated value of the Fisher criterion to the table value at the 95% of the significance level. The mathematical model (1) is adequate ($F_p > F_T$) and can be used to predict the content of other chemical elements in the Novaya river catchment soils, since $Fp/F_T > 4$.

 https://doi.org/10.1051/e3sconf/202016901006

The prevalence of nitrogen in soils is 1000 mg/kg. Substituting this value in the formula (3), we get the content of this element in the soils of one sample of the Novaya river catchment area 588 mg/kg (0.061%), and for the entire catchment – about 0.057 tons. This result should be regarded as purely indicative and requiring an experimental verification. The total amount of nitrogen in soils varies from 0.02-0.05% in sod-podzolic sandy soils to 0.2-0.5% in chernozems (black earth).

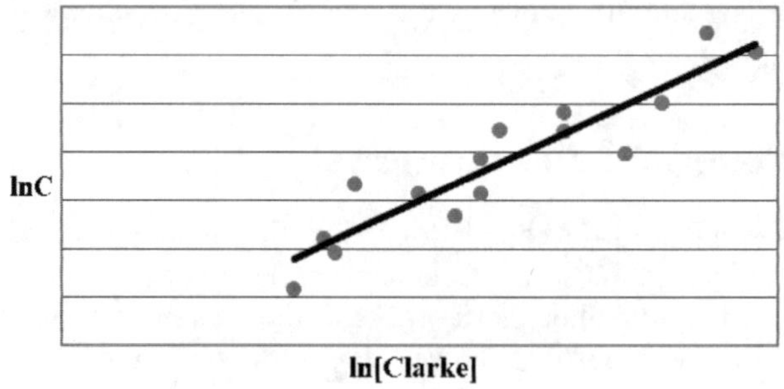

Fig. 2. Relationship between the chemical elements content in the soils of the Novaya river catchment area (C) and the clarkes

Another way to estimate the nitrogen content in the Novaya river catchment area is based on the abundance of nitrogen in soils. The average nitrogen content in one soil sample is 1000 mg/kg, and in the entire catchment area (97,000 samples) is 0.097 t. The second indicator used for calculations is the amount of precipitation (Fig. 3). The total amount of precipitation in St. Petersburg in 2018 was 530 mm. As a rule, most of the biogenic substances received in the catchment from various sources do not reach the closing sections of large rivers, as they are held by the catchment and various links of the hydrographic network. As a result, the total load on the water object is only a part of the biogenic supply to the catchment areas.

The calculations were performed using the ILLM - Institute of Limnology Load Model (Certificate of state registration No. 2014612519 dated 27.02.2014), developed at the Institute of lake science of the Russian Academy of Sciences on the basis of domestic and foreign experience in modeling the removal of biogenic substances from catchment areas [1, 7], as well as HELCOM recommendations for assessing the load on water objects in the Baltic sea basin [4]. The model is designed to solve problems related to the quantitative assessment of the biogenic load generated by point and dispersed sources of pollution, and the forecast of its change under the influence of possible anthropogenic and climatic changes. The result of the simulation is a quantitative assessment of the load on the water reservoir from the catchment area and its individual components. The dispersed (diffuse) load is formed by the removal of substances from the entire catchment area. This is the emission of chemicals from the soil, washing away surface contaminants. In [2], an empirical model developed at the Institute of water ecology and inland fish farming in Germany was used to calculate the retention of chemicals by catchments and their hydrographic network.

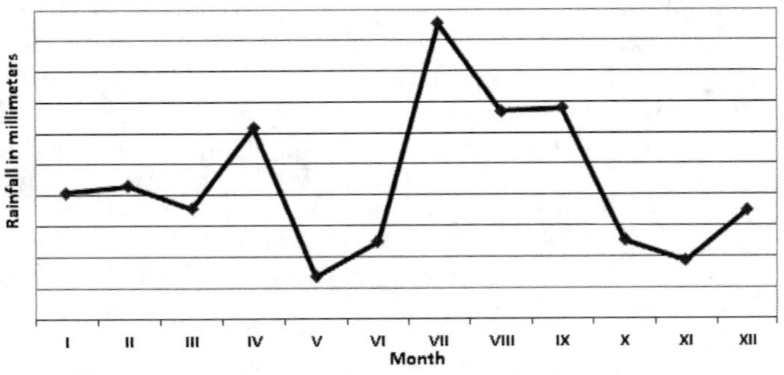

Fig. 3. Dynamics of the annual precipitation in Saint Petersburg in 2018

An increase in the area of water objects, expressed as a % of the total catchment area, leads to an increase in the retention of biogenic elements and, accordingly, to a decrease in the values of removal of substances from the catchment. At the same time, an increase in runoff leads to both an increase in the emission of substances from the soil and a decrease in the retention of biogenic elements by water bodies in the catchment area. It is possible to estimate the retention of chemicals by the catchment and the riverbed network dependence on the characteristics of the flow and the size of the catchment. If the catchment area under study is characterized by high value of water surface share in the total catchment area, then it is appropriate to apply the ratio for calculations.

The values of the empirical parameters a_1, b_1, a_2 and b_2 are set in accordance with the data in Table 2 [1, 6].

Table 2. The values of the empirical parameters

Element	Catchment area	a_1	b_1	a_2	b_2
Phosphorus total	For all catchments	26.6	-1.71	13.3	-0.93
	< 1000 km^2	41.4	-1.93	57.6	-1.26
	1000-10000 km^2	21.7	-1.55	9.3	-0.81
	> 10000 km^2	28.9	-1.80	26.9	-1.25
Nitrogen total	For all catchments	-	-	1.9	-0.49
Nitrogen total	For lakes	-	-	7.28	-0.49

For the Novaya river catchment area W=0.1%, and atmospheric phosphorus fallouts are insignificant. These indicators were excluded in the calculations. Cnat values are given in Table 3 [2]. Forest lands in the Novaya river catchment area make up 2.1% or 0.2 km^2.

Table 3. Phosphorus concentrations (mg/l) in soil waters and primary links of the hydrographic network for various types of underlying surface

Underlying surface	**Mixed territories**	**Natural territories**	**Agricultural lands**	**Urbanized territories**
Phosphorus	0.12	0.05	0.08	0.20

The modulus of phosphorus removal (kg/(km^2/year) from the catchment is L/(1000 A), and its background (natural) component is Lnat/(1000 A).

3 Research results

Calculation of the flow modulus: $q = 0.03171y = 0.03171 \cdot 530 = 16.8$.

Calculation of total phosphorus removal coefficient (Table 2): $R_t = 1 - R_r = 1/(1 + a_1qb_1) = 1/(1 + 26.6 \cdot 16.8\text{-}1.71) = 0.82$. Calculation of the total phosphorus load on the

Novaya river from the catchment area: $L_P = R_t \cdot L_{tot} = 0.0873 \cdot 0.82 = 0.072$ t/year. Calculation of the background component of the total phosphorus load on the Novaya river (Table 3): $L_{nat} = (1- R)[L_a + Cnat \cdot y \cdot (A(1-W/100)/1000] = (1- 0.18)[0.12 \cdot 530 \cdot (0.2/1000)] = 0.010$ t/year. Calculation of the total nitrogen removal coefficient (table 2): $R_t = 1 - R_r = 1/(1 + a_1 q b_1) = 1/(1 + 1.9 \cdot 16.8\text{-}0.49) = 0.68$. Calculation of the total nitrogen load on the Novaya river from the catchment area: $L_N = R_t \cdot L_{tot} = 0.097 \cdot 0.68 = 0.066$ t/year. The results of the study are summarized in Table 4.

Table 4. Diffuse biogenic load from the catchment of the Novaya river

Biogenic element	Total load, kg/year	Background load kg/year	Anthropogenic load, kg/year
Phosphorus total	72	10	62
Nitrogen total	66*	-	-
Zn	50	86,4	

Note. * The result should be considered as indicative.

In work [8], the maximum allowable input (export) and maximum allowable runoff modules of biogenic elements in the Baltic Sea subequatorial are given. The maximum allowable export to the Gulf of Finland is 4860 t/year of total phosphorus and 106680 t/year of total nitrogen. These values significantly exceed the total load on the Novaya river from its catchment area: 0.072 t/year for total phosphorus and 0.066 t/year for total nitrogen.

4 Conclusion

The main sources of pollution of reservoirs with biogenic substances are the flushing of nitrogen and phosphorus fertilizers from fields, the construction of water reservoirs without proper cleaning of the bed, and the discharge of waste waters, including those that have undergone biological treatment. The assessment of the diffuse biogenic load from the Novaya river catchment to the Novaya river showed that the upper reaches and part of the middle course of the river are in the zone of intensive agriculture. This category of land represents a certain danger to the hydraulic system due to the possible flushing of the treated soil into the hydro-reclamation system and, consequently, into the bed of the Novaya river. Fields are a source of systematic ingress of chemical elements into the hydraulic system along with the washing out of fertilizers and pesticides by rain. Objects in urbanized areas and allotments have a significant impact on the state of the river. Urbanized objects absolutely dominate the catchment area and occupy 2/3 of its area. Almost 54 km^2 of the administrative area is covered by urban-type buildings and less than 2 km^2 is occupied by rural settlements, while 0.5 km^2 is covered by individual buildings and structures, which is 65.2 and 0.6%, respectively. The maximum allowable export to the Gulf of Finland is 4860 t/year of total phosphorus and 106680 t/year of total nitrogen. These values significantly exceed the total load on the Novaya river from its catchment area: 0.072 t/year for total phosphorus and 0.066 t/year for total nitrogen.

References

1. H. Behrendt, D. Opitz. Hydrobiology, *Retention of nutrients in river systems: dependence on specific runoff and hydraulic load,* **410**, 111-122 (1999).
2. S. Kondratiev, S. Golosov, I. Zverev, V. Ryabchenko, A. Dvornikov, *Modeling abiotic processes in the catchment system - reservoir (on the example of Lake Peipsi - Pskov),*

104 (Nestor Story Publishing House, Saint Petersburg, 2004).
3. V. Mineeva, *Agrochemistry: Textbook*, 720 (MSU Publishing House, Moscow, 2004).
4. *Guidelines for the compilation of waterborne pollution to the Baltic Sea (PLC-water)*, 80 (HELCOM, Helsinki 2005).
5. S. Kondratiev, *The formation of the external load on water bodies: modeling problems*, 255 (Science Publishing House, Saint Petersburg, 2007).
6. H. Behrendt, R. Dannowski. *Nutrients and heavy metals in the Odra River system. Weissensee Verlag Publ., Germany,* 337 (2007).
7. S. Kondratiev, M. Kazmina, M. Shmakova, E. Markova, Regional ecology, **3-4,** 50–59 (2011).
8. G. Frumin, *Calculation of permissible nutrient load on transboundary water bodies*. 27 (Russian State Hydrometeorological University, Saint-Petersburg, 2013).
9. L. Ratcovitch, V. Markin, I. Glazunova, S. Sokolova, Construction and architecture, **3**, 64-75 (2016).
10. E. Nesterov, P. Egorov, M. Markova, D. Morozov. *SGEM*, **19,** 699-707 (2019).

E3S Web of Conferences **169**, 01007 (2020) https://doi.org/10.1051/e3sconf/202016901007
APEEM 2020

Integral assessment of heavy metal pollution in Don River estuary soils

Dina Nevidomskaya[1],[*], *Tatiana* Minkina[1], *Yuri* Fedorov[1], *Olga* Nazarenko[2], *Natalya* Kravtsova[1], and *Yuri* Litvinov[1]

[1]Southern Federal University, 344006 Rostov-on- Don, Russia
[2]State Center for Agrochemical Service "Rostov", 346735 Rostov Region, Russia

Abstract. Based on the cumulative pollution index, spatial distribution maps were obtained to establish categories of soil pollution taking into account the total content of heavy metals and the mobile forms of metals in the Don River estuarine region. The objects of the study included samples of zonal soils (Chernozem) and intrazonal soils (Fluvisols) from monitoring plots. The total concentrations of Mn, Cr, Ni, Cu, Zn, Pb, and Cd in the soils were determined by X-ray fluorescent scanning spectrometer. Mobile heavy metals compounds were transferred to a solution by extraction of 1 N NH_4Ac, pH 4.8. When calculating the total metal content, it was shown the studied soils had generally an acceptable pollution category but taking into account mobile forms the categories of soils contamination variate up acceptable to extremely dangerous. The most polluted sites are associated with the estuary of small rivers and the branches flowing into the Taganrog Bay, the territory of the Taganrog port and its terminals, and road bridges. In line with hight categories of pollution, the use of the soils for cropping should be limited, and the cultivation of concentrator plants is excluded.

1 Introduction

The territory of the Lower Don covers one of the most densely populated regions of the South of Russia, where associated with high-intensity economic activity in the catchments and water area, contributing to the flow of pollutants with a significant representation by heavy metal (HM) [1, 2]. Of particular significance are the studies of the estuarine areas of large rivers, which are one of the most important links between land and sea. This area is considered a part of a marginal filter or a cascade of barrier zones with avalanche sedimentation, interpenetration, mixing and transformation of river and sea waters. This area is categorized by its differences from each other in the terms of salinity, ionic composition, HM and other substances of natural and anthropogenic origin [3, 4, 5, 6]. Due to this fact, the development and application of indicators reflecting the actual ecological state of the soil cover are relevant [7].

[*] Corresponding author: dnevidomskaya@mail.ru

The purpose of this work was to evaluate HM pollution of soil cover in the Don River estuary and obtain spatial distribution maps by the cumulative pollution index (*Zc*) based on the results of long-term monitoring.

2 Materials and methods

The studies were conducted on the territory of the Don River estuary, which included a part of the river from the top of the delta – the Stanitsa of Razdorskaya to the confluence of the Taganrog Bay in the Sea of Azov [2]. Route and field studies were carried out in 2011-2019. At the monitoring sites, the soil was sampled from a depth of 0-20 cm according to the guidelines [8].

The major background of soil cover consists of low- and medium-humus Haplic Chernozem [9] of different depth on heavy loess-like loams or clays. Fluvisols [9] with different degrees of salinity occur in floodplains of small rivers and especially in the Don River delta [2].

According to the guidelines [10] the main physico-chemical attributes of soils were estimated: the total organic matter, particle size distribution, carbonates, exchangeable Ca^{2+} and Mg^{2+}, pH.

The total content of Mn, Cr, Ni, Cu, Zn, Pb, and Cd in soils was determined by X-ray fluorescence method using a scanning spectrometer "SPECTROSCAN MAKC-GV". Mobile HM compounds were transferred to a solution by extraction of 1 N ammonium acetate buffer (CH_3COONH_4) – AAB at the following conditions: pH 4.8, soil to solution ratio 1 : 10, extraction time 18 hours. Metal content in soil extracts was determined by atomic absorption spectrophotometry (AAS) using a BUCK Scientific 200 spectrophotometer.

The level of soil contamination with HMs was estimated by comparing with maximum permissible concentrations (MPCs) for total and mobile contents [11], and background contents [2].

The cumulative pollution index (*Zc*) was calculated based on the equation represented in Eq. (1) [12]. This equation was used for the integral assessment of the level of HM pollution of soils (1) [12]:

$$Zc = \Sigma Kc - (n - 1) \quad (1)$$

where *Kc* is the concentration factor equal to the ratio of the actual total content of the element in the soil to its background value ($Kc = Cj/Cb$), and n is the number of chemical elements with $Kc > 1$. The HM pollution level of the study soils was compared with their background [2].

To assess the potential HM bioavailability for plants, it is proposed to modify the cumulative pollution index taking into account mobile forms of metals (2):

$$Kc = Mj/Mb \quad (2)$$

where *Mj* is the actual content of mobile forms of HM in soil, and *Mb* is the content of mobile forms of HM in the soil which is taken as the background analogue of unpolluted soil [2].

3 Results and discussion

Comparison of the results of phisical and chemical analyses of Fluvisols from the examined monitoring plots demonstrates that the particle size distribution of soils is rather mixed and is re presented mainly by sandy, sandy loam, and clay-loam species. Fractions of fine and medium sand are predominant. The studied soils have a weakly alkaline or strongly alkaline reaction (7.3–8.9) and a low humus content of 0.2–3.7, which was mainly determined by the qualitative content of the infused and redeposited material composing the inwashed and buried horizons. The high carbonate content from soil covers in some monitoring sites is associated with the presence of biogenic calcite in deposited horizons.

The excesses of total HMs content in soils over regulated MPCs have been established for Zn (up to 3 times), Cr (up to 2 times), Pb (up to 2 times), Cu (up to 1 times). It has been shown that the sources of Zn, Pb, Cu supply are of anthropogenic character due to excessive discharge of industrial and domestic wastewater, aerotechnogenic input. The high Cr level in soils is due to the natural lithogenic factor. Polyelement contamination is observed at the monitoring sites located on the sea edge of the Don River delta, in the estuarine zones of small rivers within the Azov Sea basin.

The content of HMs mobile forms in the soils has significantly exceeded MPCs, that of Cd (up to 10 times), Zn (more than 7 times), Pb (up to 2 times) and Ni (up to 4 times), which is a risk their high degree of bioavailability. The highest content of mobile forms is observed in sandy varieties of Fluvisols. This is related to the low contents of organic matter and fine fractions, which are the main carriers of metals.

With consideration of the total content of HM, the lowest *Zc* is typical of light-textured low-humic monitoring sites located on the coast of the Taganrog Bay and the coastal part of the delta. The highest *Zc* values are typical for Fluvisols with ups content of humus and clay particles facilitate HMs accumulation. Soils of almost all monitoring sites have allowable pollution categories (*Zc* < 16) (Fig. 1) [13].

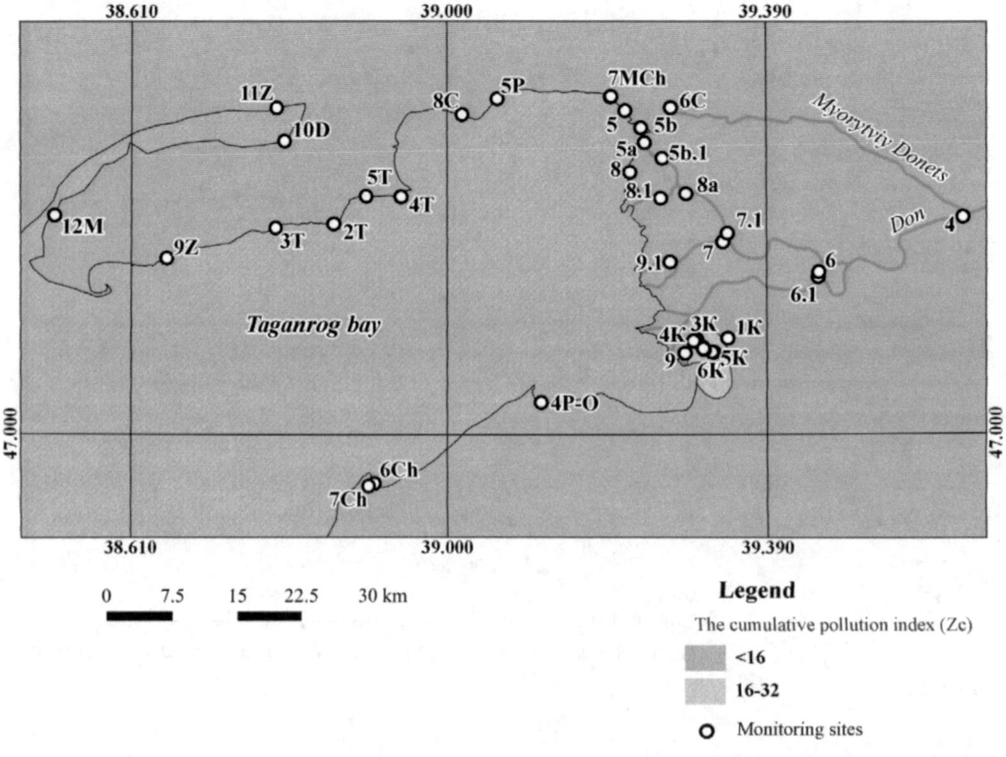

Fig. 1. Distribution map of the integral assessment of soil pollution (*Zc*) taking into account the total content of HM.

Taking into consideration the mobile forms of HM compounds, more significant differences were found in the values of *Zc*, which differed from the cumulative index calculated from the total contents (Fig. 2). Based on the calculation of *Zc* for mobile forms, allowable soil pollution categories ($Zc < 16$) changed to dangerous ($Zc = 32\text{-}128$) and extremely dangerous ($Zc > 128$) ones. The pollution categories established by taking into account the mobile forms of HM compounds reasonably reflected the level of anthropogenic load on the soil cover. According to *Zc*, the most polluted monitoring sites are associated with the estuary of small rivers (Kagalnik, Sambek) and the branches flowing into the Taganrog Bay, the territory of the Taganrog port and its terminals, and road bridges. In line with these categories of pollution, the use of the soils for cropping should be limited, and the cultivation of concentrator plants is excluded.

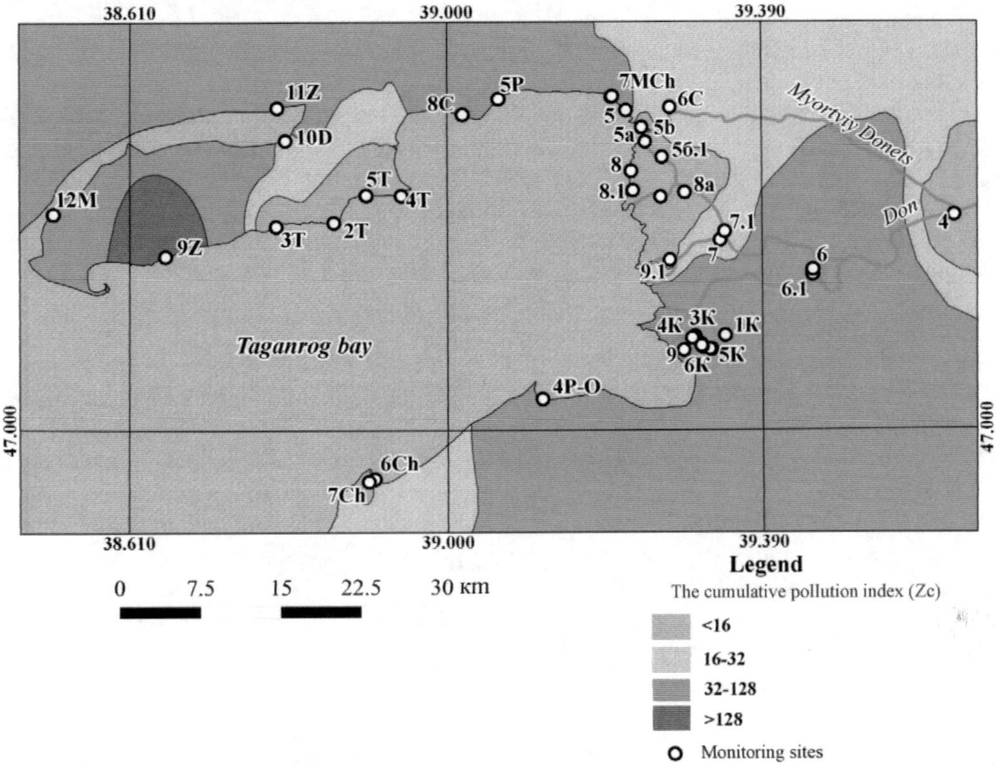

Fig. 2. Distribution map of the integral assessment of soil pollution (*Zc*) taking into account the mobile forms of HM.

4 Conclusions

The use of the proposed cumulative soil pollution index (*Zc*), by taking into account the mobile forms of HM made it possible to identify areas with different categories of pollution in the study area subject to anthropogenic impact. The most polluted sites are associated with the estuary of small rivers and the branches flowing into the Taganrog Bay, the territory of the Taganrog port and its terminals, and road bridges. In line with hight categories of pollution, the use of the soils for cropping should be limited, and the cultivation of concentrator plants is excluded. The risk of soil pollution is higher when buffer properties of soils are low.

The research was financially supported by the project of the Ministry of Science and Higher Education of the Russian Federation within the framework of the state task (Southern Federal University), and the Russian Foundation for Basic Research, project no. 19-05-50097.

References

1. T. Minkina, Yu. Fedorov, D. Nevidomskaya, S. Mandzhieva, M. Kozlova, Arid. Ecosyst. **5**, 4 (2015)
2. T. Minkina, D. Nevidomskaya, T. Pol'shina, Yu. Fedorov, S. Mandzhieva, V. Chaplygin, T. Bauer, M. Burachevskaya, J. Soils Sediments **17**, 5 (2017)
3. G. C. Ray, J. McCormick-Ray, *Estuarine ecosystems* (Encyclopedea of Biodiversity Second Edition, 2013)
4. D. Romic, M. Romic, M. Zovko, H. Bakic, G. Ondrasek, Environ. Geochem. Health., **34** (2012).
5. A. Rovira, R. Ballinger, C. Ibáñez, P. Parker, M. D. Dominguez, X. Simon, A. Lewandowski, B. Hochfeld, M. Tudor, L. Vernaeve, J Soils Sediments, **14** (2014)
6. Yu. Fedorov, I. Dotsenko, A. Mikhailenko, *SGEM*, **1** (2015)
7. T. Minkina, G. Fedorenko, D. Nevidomskaya, T. Pol'shina, A. Fedorenko, V. Chaplygin, S. Mandzhieva, S. Sushkova, T. Hassan, Environ. Geochem. Health, (to be published) https://doi.org/10.1007/s10653-019-00379-3
8. GOST (State Standard) 58595-2019 *Soils. General requirements for sampling.* (Moscow, Rospotrebnadzor, 2019) (in Russian)
9. *IUSS Working Group WRB. World reference base for soil resources 2014, update 2015. International soil classification system for naming soils and creating legends for soil maps. World soil resources reports,* **106** (FAO, Rome, 2015)
10. L. Vorob'eva, *Theory and practice chemical analysis of soils* (Moscow, GEOS, 2006) (in Russian)
11. HS 2.1.7.2041-06. *Maximum permissible concentrations of chemicals in soil: Hygienic standards. Introduction* (Rospotrebnadzor, Moscow, 2006) (in Russian)
12. L. Håkanson, Water Resour., **14** (1980)
13. Guidelines MU 2.1.7.730-99 *Hygienic evaluation of soil in residential areas* (Rospotrebnadzor, Moscow, 1999) (in Russian)

https://doi.org/10.1051/e3sconf/202016901008

Migration of cyantraniliprole in fractured soils: calibration of pesticide leaching model by using experimental data

Aleksandra Belik[1*], *Anna* Kokoreva[1,2], and *Victoria* Kolupaeva[1]

[1]All-Russia Research Institute of Phytopathology, 143050, Bolshie Vyazemy, Russia
[2]Moscow State University, 119991, Moscow, Russia

Abstract. The phenomenon of preferential migration of substances can increase the risk of pesticides. In the first year of the experiment, in 7 days after application cyantraniliprole penetrated to a depth of 25 cm in agrosoddy-podzolic soil. In the next year on the 7th day after application the pesticide was detected at a depth of 15 cm. The pesticide migrated deeper than the unconfigured PERL model took into account. The calibration of the PERL model by using experimental data (soil experimental support) allowed to reduce the error of prediction. The obtained data can be used to create new standard soil and climate scenarios for pesticide leaching models.

1 Introduction

The study of pesticide migration in soils requires an understanding of the mechanisms of substance movement with water flows. In well-structured soils, the presence of macropores and cracks leads to a rapid flow of soil solution [1, 2], resulting in uneven moistening. The solution penetrates to a considerable depth, practically without interacting with the surrounding soil matrix and bypassing most of the soil pore space (stagnant zone) [3]. The phenomenon of preferential migration of substances can increase the risk of pesticides due to the possible pesticide migration in groundwater. Such situation can be observed in soddy-podzolic soils, which are typical for Moscow oblast.

Experimental study of pesticide migration is a difficult process with significant financial costs, therefore the method of mathematical modeling is often used to obtain predicted concentrations of pesticide in environmental objects. However, it is important to know how well the models predict the migration of pesticides in soils with different pore space. This requires checking the adequacy of the models by comparing the predicted values with the experimental data. The models predicting the pesticide behavior in soils have different approaches to describing the migration process. The models with a complex mathematical apparatus have a significant drawback - the abundance of input parameters of experimental support [3]. Therefore, calibration of simpler pesticide leaching models by using the experimentally determining parameters, which can physically reasonably take into account preferential migration, is actual.

* Corresponding author: belikalexandra@gmail.com

In the process of state registration of pesticides in Russia for ecological risk assessment of pesticides the PEARL model is used. Thus, the main goal of this work was to calibrate the PEARL model by using the field studies of cyantraniliprole migration with experimentally determined pore space parameters of the studied soils.

2 Objects and Methods

2.1 Pesticide

Cyantraniliprole is an active substance with insecticidal activity for a wide range of crops. It is non-volatile, moderately persistent ($DT_{50geomean\ lab.}$ = 34.4 days) and medium-mobility ($K_{oc\ average}$ = 241 ml/g) substance [4].

2.2 Soils

An agrosoddy-podzolic soil (Eutric Albic Glossic Retisol (Abruptic, Loamic, Aric, Cutanic) [5] was chosen for the experiments, it's typical soil for Moscow oblast and has preferential flows. Some properties of this soils are shown in Table 1.

Table 1. Some physical and chemical properties of investigated soil.

Hori-zon	Depth, cm	$C_{org.}$, %	pH_w	ρ_b, g/cm^3	ρ_s, g/cm^3	ε, cm^3/cm^3	Particle size, %			K_f, m/day
							Sand	Silt	Clay	
P	0-36	1.07	6.52	1.20	2.6	0.54	10.6	87.9	1.5	0.14
EL	36-42	0.26	6.29	1.35	2.78	0.51	5.2	93.5	1.3	0.14
BEL	42-51	0.23	6.07	1.28	2.72	0.53	10.4	89	0.6	0.43
BT_1	51-80	0.22	5.66	1.43	2.72	0.47	12.7	87.3	0	0.38
BT_2	80-129	0.16	6.18	1.34	-	-	15.4	84.6	0	0.33
BC	>129	0.17	7.15	1.25	-	-	16.8	83.1	0.1	0.16

2.3 Field experiment

The pesticide was applied to a 25 m^2 plot in the spring at doses of 0.74 kg/ha in the first year and 0.20 kg/ha in the second year. Before each application, the soil was treated with glyphosate, all weeds were removed, and soil was digging. Soil samples for the determination of residual pesticide were collected by the drilling method 0, 7, 53, 101 and 351 days after the first treatment, and 0, 7, 14, 29, 64, 102, 343 days after second treatment of soil with the pesticide at vertical intervals of 5 cm; two mixed samples were then composed from 10 individual samples each, and the content of pesticide in them was determined by liquid chromatography. Temperature loggers were installed on the experimental plot at the depths of 5, 15, and 30 cm, which recorded temperature every hour; the water content was determined in soil layers by thermogravimetry at each sampling time. Daily weather data (rainfall, solar radiation, maximum and minimum temperature, wind speed and relative humidity) were obtained directly in the field by using a portable weather station.

2.4 Model PEARL 4.4.4

In this model the solute transport in soils is based on the convective diffusion equation. In the model, the soil is divided into 3 domains: matrix and two macropore domains, each of which includes static and dynamic pores: 1 - interconnected pores passing throughout the profile (main bypass domain) and 2 - separate pores terminating at different depths, (internal catchment domain) [6, 7].

3 Results and discussion

In the first year of the experiment, in 7 days after application cyantraniliprole penetrated to a depth of 25 cm, during the entire first year of the study the pesticide wasn't detected below this layer. According to previous studies [8], cyantraniprole didn't migrate below 20 cm in the studied soils of the Russian Federation. The deeper penetration of the pesticide in our experiment in the first week after application may be due to both the structural features of the soil pore space, and the dry period before the experiment and 10 mm on 5th day. Indeed, one of the mechanisms of activation of preferential migration paths is a combination of factors: lack of precipitation for some time, unsaturated conditions and then heavy rainfall [1, 3, 9].

In the next year on the 7th day after application the pesticide was detected at a depth of 15 cm. Then, on the 29th day, cyantraniprole reached a depth of 20 cm and didn't migrate below. In the second year the weather conditions after application of pesticide were very different from the first year: before the application the soil had enough time for saturation, it rained for several days (Fig. 1).

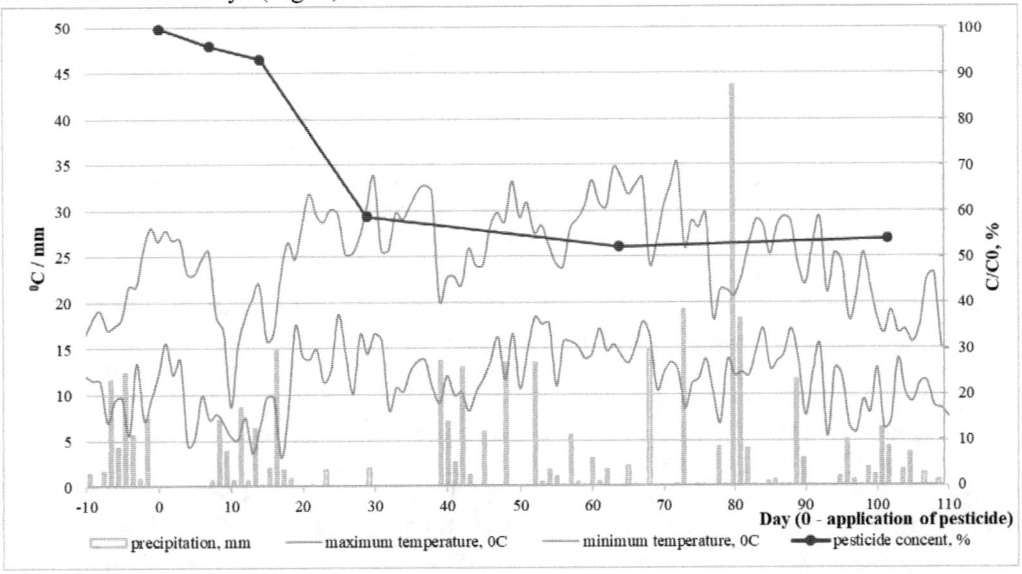

Fig. 1. Dynamics of the total content of cyantraniliprole in investigated soil in the second year.

To work pesticide leaching model, it is necessary to have initial conditions, conditions at the lower and upper boundaries and experimental support (physical and chemical properties of soils, toxicants). The parametrization of the soil experimental support was made based on the soil properties shown in Table 1. Data on the physicochemical properties of cyantraniliprole were borrowed from the conclusions of the European Food Safety Authority [4]. DT_{50} = 49.9 days was used; it was found by the All-Russia Research Institute of Phytopathology for this soil type [10].

Before predicting cyanthraniliprol migration, it is important to evaluate the work of the soil block of the PEARL model. It is necessary to compare the obtained experiment and predicted values. SRMSE (Scaled root mean square error) is often used to assess the adequacy of models describing the behavior of pesticides [11, 12]. Since pesticide degradation takes place together with the migration process, it was necessary to check how accurately PEARL predicts soil temperature from weather data. The comparison of the calculated data with the readings of temperature loggers showed good forecast accuracy (Fig. 2).

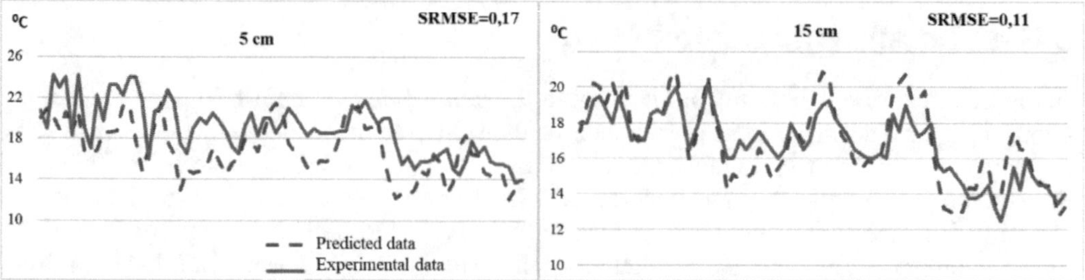

Fig. 2. Experimental and predicted temperature data.

The migration of substances in the soil occurred mainly with water flows, so the next step was to check the quality of the moisture forecast. The significant errors were found that are expected to increase with depth. SRMSEs were 0.07-0.14 in the upper layers and reached 0.34 in the deep layers (30-35 cm). Based on the received data, it becomes necessary to further calibration the model. According to previous studies evaluating the sensitivity of the water block to the input parameters of the PEARL model [13], the greatest influence on the forecast of soil moisture is exerted by approximation parameters of soil water retention curve and filtration coefficient. By sequentially selecting the numerical values of the sensitive parameters, only the filtration coefficient values were calibrated within the limits corresponding to the studied type of soils, it led to reduce the forecast error (for example, for a layer of 30-35 cm to 0.24). The comparison of adjusted and unadjusted models using the Williams–Klute test [14] was shown that the setting of water block of the PEARL model led to an improvement in the forecast: at p-level = 0.029675, the t-value equal to -2.267001.

The next step was to analyze the prediction of the behavior of the pesticide in the soil. In this paper, simulation of the pesticide migration was conducted for second year of the study. In the first days after application, the pesticide migrated deeper than the PERL model took into account, this phenomenon may be due to the preferential migration paths in agro-soddy-podzolic soil. The PEARL model has a parameter 'dispersivity length' – a coefficient that takes into account the vertical heterogeneity of the moisture-conducting pore space. Higher values of this parameter are typical for soils with a high clay content and uneven pore space with preferential migration flows [15]. In previous works [16] it was shown that the dispersivity length can be used to configure the PERL model. In our previous studies [17] the values of dispersivity length were determined for monoliths (d = 10 cm, h = 30 cm) of agrosoddy-podzolic soil in laboratory conditions. But for subsequent modeling in experimental support it is necessary to divide the soil profile into layers, and a length of 30 cm may exceed the horizon thickness, a "scale problem" arises. Therefore, it was decided to conduct another series of filtration column experiments with small monoliths (d=5 cm, h=10 cm), the obtained values of dispersivity length are presented in the Table 2.

Table 2. Laboratory dispersivity lengths for small monoliths (d=5 cm, h=10 cm).

Depth, cm	Filtration rate, cm/day	Dispersivity length, cm
0-10	132	5.0
10-20	57	3.5
20-30	97	7.5
30-40	62	13.0
40-50	55	5.6
50-60	99	5.6

The method of field determination of dispersivity lengths is described in detail in an article with colleagues [18]. In this work, the values shown in Fig. 3 were used.

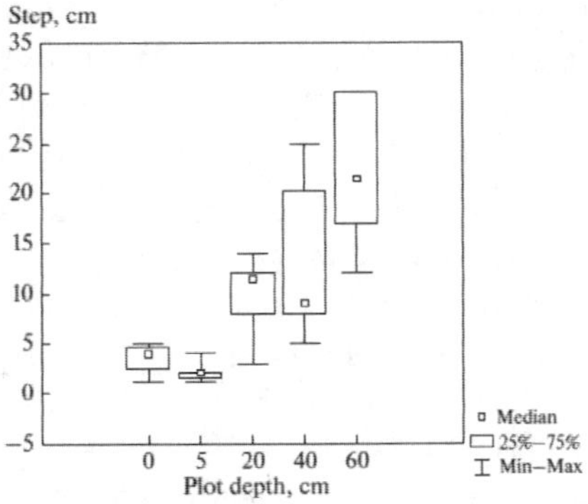

Fig. 3. Statistics of the field determinations of dispersivity length.

Based on a comparison of experimental and predicted results of the dynamics of the pesticide distribution in the soil profile and the analysis of SRMSE (Table 3), it can be stated that the results of modeling the pesticide migration with the dispersivity length=5 cm [6] and with the values determined in laboratory experiments were very close. However, modeling with using field values of dispersivity length better described the behavior of the pesticide, especially in the lower layers, reducing SRMSE (Table 3). Thus, the experimentally determined values of dispersivity length for a particular soil, on which the pesticide is applied, can serve as a tool for configure the PEARL model.

Table 3. SRMSE of predicting the residual pesticide contents in soil at different dispersivity lengths for the second year.

Depth, cm	Dispersivity length		
	5 cm	Laboratory	Field
0-10	0.23	0.23	0.23
10-15	0.75	0.76	0.57
15-20	1.05	1.09	0.80
Soil profile	0.19	0.19	0.19

4 Conclusions

The obtained results allow to assess the significance of the preferential migration flows in the behavior of pesticides in fractured soils. The obtained data (soil experimental support and materials for calibration of the mathematical model PEARL) can be used to create new standard soil and climate scenarios for pesticide leaching models, which used in registration process in the Russian Federation.

References

1. W. Kördel, M. Klein, Pure Appl. Chem., **78** (5), 1081–1090 (2006)
2. T. Katagi, Rev. Environ. Contam. Toxicol., **221**, 1–105 (2013)
3. J. Simunek, N. Jarvis, M.Th. van Genuchten, A.Gardenas, Journal of Hydrology., **272**, 14–35 (2003)

4. *Conclusion on the peer review of the pesticide risk assessment of the active substance cyantraniliprole*, EFSA J., **12**(9):3814 (2014)
5. *World Reference Base for Soil Resources, International soil classification system for naming soils and creating legends for soil maps*. World Soil Resources Reports No. 106. (FAO, Rome, 2014)
6. A. Tiktak, R. F. A. Hendriks, J. J. T. I. Boesten, Pestic. Manage. Sci., 68, 290–302 (2011)
7. F. van den Berg, A. Tiktak, J. J. T. I. Boesten, A.M. A. van der Linden, *PEARL Model for Pesticide Behavior and Emissions in Soil-Plant Systems*, **61**, 134 (Statutory Research Tasks Unit for Nature and Environment, Wageningen, 2016)
8. *The study of the decomposition and migration of cyantraniliprole in the soils of the Russian Federation. Report on research work for 2012* (Center for Ecopesticidal Research «Epicenter», Moscow, 2013)
9. N. Jarvis, J. Koestel, M. Larsbo, Vadose Zone J., **15(12)** (2016)
10. V. N. Kolupaeva, V. S. Gorbatov, I. V. Nyukhina, Vestn. Novosib. Gos. Agrar. Univ., **2 (39)**, 82–91 (2016)
11. A. A. Smetnik, Yu. Ya. Spiridonov, E. V. Shein, *Migration of Pesticides in Soils* (Russian Research Institute of Phytopathology, Moscow, 2005
12. I. Dubus, S. Beulke, C. Brown, Pestic. Manage. Sci., **58**, 745–758 (2002)
13. A. A. Kokoreva, A. B. Umarova, V. S. Gorbatov, Vestn. Orenb. Gos. Univ., **3**, 123–127 (2007)
14. E.V. Shein, M.A. Mazirov, A.A. Korchagin, A.B. Umarova, V.M. Goncharov, C.I. Zinchenko, A.V. Dembovetsky, *Regression analysis in soil science* (VlSU Publishing House, Vladimir, 2016)
15. J. Vanderborght and H. Vereecken, Vadose Zone J., **6**, 29–52 (2007)
16. E. Shein, A. Belik, A. Kokoreva, V. Kolupaeva, P. Pletenev Prediction of Pesticide Migration in Soils: the Role of Experimental Soil Control, Moscow University Soil Science Bulletin, **72(4)**, 185-190 (2017)
17. E. V. Shein, A. A. Kokoreva, V. N. Kolupaeva, A. A. Belik, P.A. Pletenev, Agrokhim. Vestn., **6**, 20–23 (2016)
18. E.V. Shein, A.A. Belik, A.A. Kokoreva, V.N. Kolupaeva, Eurasian Soil Sci. **51(7)**, 797-802 (2018)

E3S Web of Conferences **169**, 01009 (2020) https://doi.org/10.1051/e3sconf/202016901009
APEEM 2020

Risk assessment of ice-melter reagents for urban plants

Yulia Baeva*, *Natalia* Chernykh, *Sofia* Shmerko, and *Varvara* Stepnova

Peoples' Friendship University of Russia (RUDN University), Faculty of Ecology, 6 Miklukho-Maklaya St, Moscow, 117198, Russian Federation

Abstract. The paper assessed the phytotoxicity of the most frequently used ice-melter reagents in Moscow (sand and salt mixture, anti-freeze reagent based on a composition of calcium and sodium chloride, technical salt), snow with varying degrees of contamination of the remnants of ice-melter reagents, as well as soil after the snow cover melting. The toxicity assessment was carried out in the samples taken on the lawns along the roads in nine administrative districts of the city in places with the same anthropogenic load. The test object is oat seeds (*Avena sativa*), as a sensitive biotest showing the most stable and reproducible data compared to seeds from other crops. Salt solutions of reagents with dilution rates of 1, 2, 4, 6, 8 and 10, melted snow and soil extracts were used for biotesting. The phytotoxic effect was determined by comparing the average root length of control and prototype seeds. It was found that all studied ice-melter reagents have a pronounced toxic effect on oat plants even at tenfold dilution. The phytotoxic effect of snow and soil polluted with ice-melter substances, which in turn inherits the chemical composition of snow cover after its melting, was shown.

Introduction

As an important factor in road safety, anti-icing of road pavements also has a negative impact on all components of the urban environment. Current ice-melter reagents - are multi-component chemicals consisting of a set of salts of both organic and inorganic origin [1], most of which contain sodium and calcium chlorides. Sodium chloride and calcium chloride salt solutions formed as a result of ice and snow melting, when they get into the soil, cause salinization of the soil cover [2]. In turn, changes in soil properties can suppress plant growth and development. Thus phytotoxic action of ice-melter reagents is shown both on an organismal level (extinction of roots, twisting of leaves, withering of a plant, etc.), and on cellular level (change of structure of chloroplasts, decrease in the starch content, infringement of process of photosynthesis) [1, 3].

The purpose of this work was to assess the degree of phytotoxicity of various ice-melter reagents, as well as snow and soil contaminated during their application.

* Corresponding author: baeva-yui@rudn.ru

Objects and methods

The objects of research were selected:

1) the most commonly used ice-melter reagents in Moscow (sand and salt mixture, solid multi-component anti-freeze reagent based on a composition of calcium and sodium chloride ($CaCl_2$+NaCl), technical salt (NaCI). General characteristics of ice-melter reagents are presented in Table 1;

2) the snow samples contaminated with ice-melter reagents;

3) the soil after snow melting.

Table 1. General characteristics of ice-melter reagents

Name	Aggregate composition	Component composition	Chemical composition	Operation temperature	Hazard class in terms of impact on the human body
Sand and salt mixture	solid	combined	sand -70% NaCl – 30%	from -12 °C to -15 °C	3 -moderately dangerous. It has no harmful effect on the undamaged skin, but when it hits skin wounds, it impairs their healing
$CaCl_2$+ NaCl	solid	chemical	$CaCl_2$- 20% NaCl – 80%	from -12 °C to -15 °C	3 -moderately dangerous. It has an irritant effect on the mucous membranes of the upper respiratory tract and eyes. Systematically irritates and dries the skin
Technical salt	solid	chemical	NaCl – 93% Ca^{2+} - 0.8% Mg^{2+}- 0.8% K^{+}- 0.9% SO_4^{2-} - 2.2% Fe_2O_3 – 0.1%	from -12 °C to -15 °C	3 -moderately dangerous. It has no harmful effect on the undamaged skin, but when it hits skin wounds, it impairs their healing

Snow and soil samples were taken in nine administrative districts of the city: North-East Administrative District (NEAD), East Administrative District (EAD), South East Administrative District (SEAD), South Administrative District (SAD), South Western Administrative District (SWAD), Western Administrative District (WAD), North West Administrative District (NWAD), Northern Administrative District (NAD), and Central Administrative District (CAD) at the same points on the lawns along roads and sidewalks in places with the same anthropogenic load in December 2018 and April 2019 (Figure 1). Control "conditionally clean" samples were taken in the national park "Losiniy ostrov", where the presence of ice-melter reagents was completely excluded.

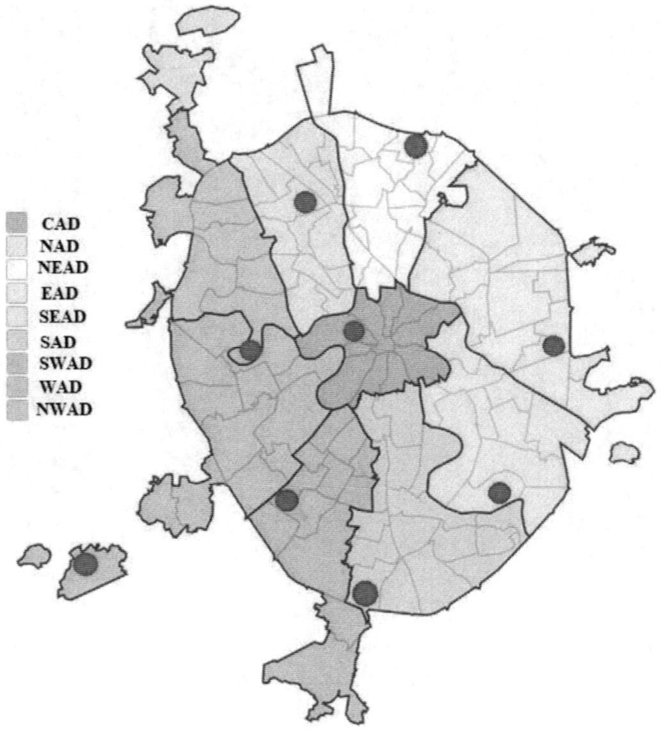

Fig. 1. Sampling map

Phytotoxicity was evaluated using bioassay on the biological effect of saline solutions on oats (*Avena sativa*) of ice-melter reagents salt solutions (at dilution of R=1, 2, 4, 6, 8 and 10), melted snow and soil extraction. Oat seeds used as test plants provide the most stable and reproducible data compared with seeds from other crops (peas, cucumbers, wheat, carrots, etc.) [4]. In Petri dishes with mugs of filter paper 25 oats seeds were placed and 5 ml of the studied solution were added. In the control sample (K) during testing of ice-melter reagents and snow the substrate for germination of seeds was distilled water (pH=6), in the experience with soil samples - "conditionally clean" soil extract. All samples were kept at a temperature of 20-23 °C for 7 days. After this period, the length of the roots of the seedlings in the control and experimental samples was measured, and the root of the maximum length was the object of measurement for each seed.

The toxic effect was determined by comparing the test function of the control and experimental seeds.

Results and discussion

Biotesting of samples of ice-melter reagents currently used in Moscow has shown that all the ice-melter reagents studied in the work have a pronounced toxic effect on oat sprouts even at tenfold dilution of the solution (Fig. 2).

Snow, as the main storage medium in winter, serves as a reliable indicator of atmospheric pollution. However, in contrast to the permanent snow cover, which is formed in parks, public gardens and remote areas of the city during the winter season, in the pollution of snow regularly removed from highways and sidewalks, ice-melter reagents make the main contribution, and the impact of atmospheric deposition is negligible.

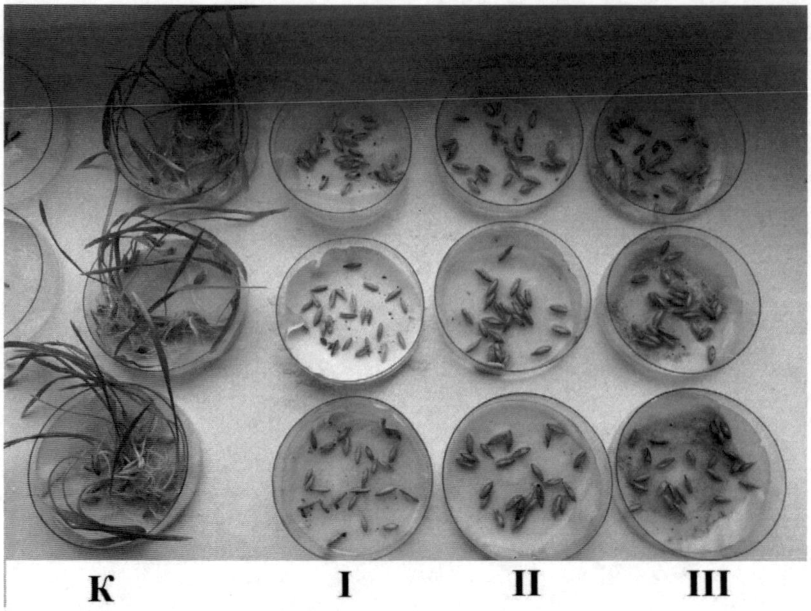

Fig.2. Inhibition of oat seed root growth by solutions of ice-melter reagents (R=10)
(K-control; I- technical salt; II- $CaCl_2$+NaCl; III- sand and salt mixture)

The phytotoxic effect of contaminated ice-melter reagents snow samples can be considered proven in samples taken in the territory of 6 administrative districts of the capital (NEAD, SAD,WAD, NWAD, NAD, CAD) (Table 2). According to the results of biotesting, the most pronounced phytotoxic effect is exhibited by samples taken in the North West Administrative District (55.9%) and the Northern Administrative District (59.7%).

Table 2. Characteristics of the effect of salt solutions of anti-icing agents on oat seeds (snow)

Sample number	Average root length, mm (p<0.05)	Ratio of average length to control, %	Phytoeffect, %	Test reaction
K (control)	68.9 ± 2.9	100.0	0	Standard
I (NEAD)	49.2 ± 3.1	71.3	28.7	Braking effect
II (EAD)	81.4 ± 4.7	118.0	0	Adverse effect is absent
III (SEAD)	69.6 ± 2.8	101.0	0	Adverse effect is absent
IV (SAD)	45.9 ± 1.7	66.5	33.5	Braking effect
V (SWAD)	69.2 ± 5.2	100.3	0	Adverse effect is absent
VI (WAD)	52.6 ±4.3	76.2	23.8	Braking effect
VII (NWAD)	30.4 ± 2.6	44.1	55.9	Braking effect
VIII (NAD)	27.8 ± 3.5	40.2	59.7	BRAKING effect
IX (CAD)	47.1 ± 4.7	68.2	31.8	braking effect

This is probably due to violation of the established norms of distribution and technology of road surface treatment with ice-melter reagents and, as a result, the entry of pollutants into green areas. For example, the density of treatment with solid ice-melter reagents should not exceed 30-50 g/m^2 at air temperature from -4 °C to -8 °C [5]. However, this norm is violated everywhere, and, according to expert estimates, the salt load on the territory of the city in recent years reaches 500 thousand tons [6].

Soil samples taken in various districts of Moscow, contaminated with ice-melter reagents, also have a toxic effect on higher plants. In this case, the maximum values of inhibition of root growth were noted in soil samples on the territory of the Central Administrative District (39.5%) and Southern Administrative District (42.4%) (Table 3).

Table 3. Characteristics of the effect of salt solutions of ice-melter reagents on oat seeds (soil)

Sample number	Average root length, mm (p<0.05)	Ratio of average length to control, %	Phytoeffect, %	Test reaction
K (control)	68.9 ±3.5	100.0	0	Standard
I (NEAD)	57.0 ± 3.6	82.7	17.3	Standard
II (EAD)	78.4 ± 1.9	113.7	0	Adverse effect is absent
III (SEAD)	68.7 ± 3.5	99.6	0.4	Standard
IV (SAD)	39.7 ± 2.5	57.6	42.4	Braking effect
V (SWAD)	53.7 ± 1.8	77.9	22.1	Braking effect
VI (WAD)	79.4 ± 5.8	115.2	0	Adverse effect is absent
VII (NWAD)	43.1 ± 4.7	62.5	37.5	Braking effect
VIII (NAD)	46.0 ± 4.1	66.7	33.3	Braking effect
IX (CAD)	41.7 ±3.9	60.5	39.5	Braking effect

Conclusions

Thus, the use of ice-melter agents in the streets of Moscow has a negative impact on the condition of green areas of roadside territories.

Phytotoxic effect has not only solutions of used ice-melter reagents, but also snow contaminated with salts, as well as the soil, inheriting the chemical composition of the snow cover.

The results obtained indicate the need to take measures to optimize the use of ice-melter reagents, to tighten control over their consumption and compliance with the requirements of their application technology.

References

1. V.A. Korolev, V.N. Sokolov, E.N. Samarin. Inzhenernaya geologiya, **3**, 34-43, (2009)
2. E.M. Nikiforova, N.E. Kosheleva, T.S. Haybrahmanov, Bulletin of Moscow University. Series 5. Geography, **3**, 40-49, (2016)
3. A.G. Starodubov, S.B. Chudakov, *Reports of IV Intern. Congr. waste management*, 17–22, (2005)
4. MR 2.1.7.2297-07, (2007)
5. Ordinance of Moscow City N 05-14-650/1, (2011)
6. D.M. Homyakov, Road power, **47**, 92-96, (2013)

Assessment of heavy metals in modern bottom sediments on the Karadag seashore

Anna Drygval[1*], *Polina* Drygval[2], *Elena* Stanis[2], and *Vladimir* Maltsev[3]

[1]A.O Kovalevsky Institute of Biology of the Southern Seas of RAS, Nakhimov avenue 2, Sevastopol, 299011
[2]Peoples' Friendship University of Russia (RUDN University), Faculty of Ecology, 6 Miklukho-Maklaya St, Moscow, 117198, Russian Federation
[3]T.I. Vyazemsky Karadag Scientific Station – Nature Reserve of RAS – Branch of A.O. Kovalevsky Institute of Biology of the Southern Seas of RAS, Nauki street 24, Feodosia, Kurortnoye, 298188

Abstract. The work presents the results of quantitative assessment of heavy metals (HM) content in bottom sediments of the coastal zone near the Karadag nature Reserve. The data were obtained as a result of field studies in 2016-2018. Estimated CC and Co coefficients and Zc were calculated. It was shown that arsenic concentration limits were exceeded everywhere in the samples. The nickel, lead, copper and zinc limits were found to have been exceeded in some locations. Ecological and geochemical assessment of water and bottom sediments quality in the section of Karadag water area has been made.

1 Introduction

Bottom sediments are an integral component of aquatic ecosystems and largely determine the quality of the water layer. The level of different substances in sediments can provide an objective view of the aquatic ecosystems stagehand and also determine the source of input of a substance into the water area. With negligible surface runoff from the area, most solid runoff accumulates in a fairly narrow coastal zone. Substances dissolved in the surface runoff of continental waters also enter there. The finely dispersed part of the bottom sediments is a good adsorbent and deposit medium for various elements. At the same time, such sediments themselves become sources of secondary pollution of the bottom water layer and substrate on which coastal organisms live [3]. This means that sediments are included in the food chain of bottom organisms. It is known that a feature of heavy metals is their ability to accumulate in bottom sediments, becoming a potential source of secondary contamination of water mass under certain conditions.

The aim of the study is to conduct an environmental-geochemical assessment of the modern sediments quality in a narrow coastal area near the Karadag natural reserve in terms of heavy metals content.

* drygval95@imbr-ras.ru

2 Objects and methods of research

The study area is situated on the shoreline of the Sudak-Karadag lowland [5]. From a geological point of view, the bedding coastal rocks belong to the terrigenous flisch and flishlike sediments of the Triassic-Jurassic period and various volcanogenic rocks represented by rhyolites, rhyolite-dacites, dacites, andesite-dacites, andesite, andesite-basalts, basalts [4]. Quaternary sediments are represented mainly by the eluvia of native rocks.

Expeditionary studies consisted in sampling sediment along the route from Lis'ya bay to Kiik-Atlama Cape. The points of selection are shown in the diagram (Figure 1).

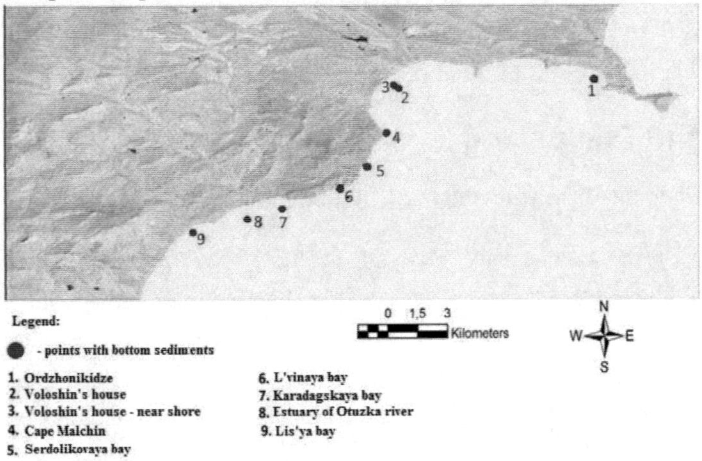

Fig. 1. Sampling points in the narrow coastal area of the Karadag seashore

Expeditionary studies consisted of sampling bottom sediment along the route from Lis'ya Bay to Kiik-Atlama Cape. The procedure of sampling, processing and analysis was in line with the accepted methodology [1].

The total HM content was determined by X-ray fluorescence analysis using a sequential wave-dispersion S4 PIONEER spectrometer. The analysis of bottom sediments was carried out on the basis of the laboratory of chemical and analytical studies of the GIN RAS.

Further assessment of the results obtained was made using several approaches, since at the moment there is no generally accepted methodology for assessing bottom sediments.. One of the approaches is a method based on grading the multiplicity of excess of the assessed substances content in relation to the quality parameters of bottom sediments. In the work such excesses have been calculated using the concentration coefficient (CC) - the ratio of the chemical element content in bottom sediments to the clark of this element. The values of the earth's crust element clarks by A.P.Vinogradov were used [6].

The application of earth's crust element clarks is the most common approach to assessing the environmental-geochemical condition of bottom sediments, since the geological features and presence of different genesis rocks and specific soils in the study areas affect the average content of chemical elements in bottom sediments. In our study, the excess of local concentrations over clarke concentrations will reflect local geochemical features for these elements in the bottom sediments. of the studied area.

The second approach in environmental-geochemical assessment of bottom sediments is related to the use of the regional standard for St. Petersburg [2]. This document is an innovation for Russian practice. The norms and criteria in this document have been established within the framework of Russian-Dutch cooperation (PSO 95/RF/3/1 programme), based on criteria already developed by the Dutch Environmental Protection Agency (DCMR). In general, this standard is based on a comparison of heavy metal content in bottom sediments with APC_0 (this is an approximate permissible concentration below which the bottom sediments are treated as pure). On the basis of these limit concentrations, a coefficient Co (the ratio of the chemical element in bottom sediments to APC_0 for the element) has been calculated.

For further assessment of the contamination level of bottom sediments we used the total pollution index Zc [7], based on comparison with MAC or APC of pollutants in soils:

$$Z_c = \sum\left(\frac{[Me]}{APC_0} - 1\right) \tag{1}$$

where [Me] is the concentration of metals, in mg/kg;

APC_0-an approximate permissible concentration below which the bottom sediments are considered as net [2].

3 Results and Discussion

Ecological and geochemical assessment of bottom sediments have been made on the basis of calculated coefficients Co.

The results of the comparative analysis for heavy metals in bottom sediments are presented in Table 1.

Table 1. Calculated coefficients of heavy metals and arsenic in bottom sediments.

Sample points	Calculated coefficients	Cu	Ni	Pb	Zn	Cr	As
	Clarke, %	0.0047	0.0058	0.0016	0.0083	0.0083	0.00017
	APC_0	0.0035	0.0035	0.0085	0.014	0.01	0.0029
Point 1	*CC*	<0.1	0.2	0.7	0.3	0.7	**9.4**
(2018)	*Co*	<0.1	0.3	0.1	0.2	0.6	0.6
Point 2	*CC*	0.2 / <0.1	0.2 / 0.2	**1.1** / 0.6	0.6 / 0.4	0.4 / 0.6	**18.8** / **16.5**
(2016 / 2018)	*Co*	0.3 / <0.1	0.4 / 0.4	0.2 / 0.1	0.4 / 0.2	0.3 / 0.5	1.1 / 0.9
Point 3	*CC*	0.2 / <0.1	0.4 / 0.1	0.8 / 0.6	0.7 / 0.4	0.7 / 0.6	**19.4** / **14.1**
(2016 / 2018)	*Co*	0.3 / <0.1	0.6 / 0.2	0.2 / 0.1	0.4 / 0.2	0.6 / 0.5	**1.1** / 0.8
Point 4	*CC*	0.4	0.2	0.8	0.8	0.8	**10.0**
(2016)	*Co*	0.5	0.4	0.1	0.5	0.6	0.6
Point 5	*CC*	0.2	0.1	0.6	0.7	0.5	**2.9**
(2018)	*Co*	0.3	0.2	0.1	0.4	0.4	0.2
Point 6	*CC*	0.4 / 0.3	0.1 / 0.2	0.8 / 0.5	0.8 / 0.6	0.8 / 0.5	**4.7** / **5.3**
(2016 / 2018)	*Co*	0.5 / 0.4	0.2 / 0.3	0.1 / 0.1	0.5 / 0.4	0.6 / 0.4	0.3 / 0.3
Point 7	*CC*	0.6 / 0.6	0.2 / 0.7	0.4 / 0.8	0.8 / 0.9	0.6 / 0.9	**4.7** / **22.4**
(2016 / 2018)	*Co*	0.8 / 0.8	0.3 / **1.1**	0.1 / 0.2	0.5 / 0.6	0.5 / 0.8	0.3 / **1.3**
Point 8	*CC*	**1.6** / <0.1	1.0 / 0.6	**1.6** / **1.3**	**1.2** / 0.6	**1.1** / 0.8	**9.4** / **32.9**
(2016 / 2018)	*Co*	**2.1** / <0.1	**1.7** / 0.9	0.3 / 0.2	0.7 / 0.4	0.9 / 0.6	0.6 / **1.9**
Point 9	*CC*	<0.1	0.8	**1.6**	0.7	0.8	**64.7**
(2018)	*Co*	<0.1	**1.4**	0.3	0.4	0.7	**3.8**

Notes: The concentrations of elements above clarke/APC_0 are in **bold;** CC is a concentration clarke, a ratio of chemical element content in bottom sediments to the clarke of this element; Co coefficient is a ratio of chemical element content in bottom sediments to APC_0 of this element.

The table shows the calculated coefficients for the study period 2016 and 2018. At some points were sampled only one year due to technical capacity. However, data at points 2,3,6,7 and 8 provide an estimate of the change in heavy metal content over a period of 2 years.

Concentration clarke values of HM and arsenic show that local geochemical features of bottom sediments are characterized by excessive arsenic concentrations. CC values of arsenic exceed from 2.9 to 64.7. Moreover, the excess concentration of this element gradually increases from the Point 5 (the territory of the reserve) to the periphery - large bays that are not part of the reserve (Points 1, 2, 3 and 9). Temporary variations in arsenic content in bottom sediments are noted: at points 7 and 8 there is a significant increase in the concentration of this element, at point 6 - CC increases slightly from 4.7 to 5.3. Further along the coast in a northeasterly direction, arsenic content in the bottom sediments has decreased over a period of 2 years. In its turn, when comparing arsenic with APC_0, the excess relatively to limit values was found only at point 9 (3.8 times), which is not included in the reserve's water area. According to the calculated indicators it can be concluded that locally in the area under consideration the maximum CC and Co values for arsenic were found at point 9 (Lis'ya Bay). It is the most visited recreational, tourist area, not belonging to the reserve, among all other sampling points. The high content of arsenic in bottom sediments can also be explained by local geochemical features.

In points 2, 8 and 9 an increase in lead concentration (CC value not exceeding 1.6) in bottom sediments has been noted. And the lead content decreased by 2018. Accumulation of copper, nickel, zinc and chromium in the bottom sediments have been observed locally in points 7, 8, and 9 with a tendency to decrease the content of these elements in the bottom sediments, because the calculated indicators generally decreased over 2 years. At the other points the excess of clarke for the lithosphere and the limit value of APC_0 have been not observed in any of the points, there is it has been noted the dispersion of these elements compared with their average content in the Earth's crust.

As part of the assessment of bottom sediments using the methodology described in [7] it has been calculated Zc. The values of the Zc level of bottom sediments have been established for the values exceeding APC_0 for each year under review. The final Zc values are presented in Table 2.

Table 2. Bottom sediment assessment [7]

Sample points	Zc of estimated elements (2016/2018)	Level of HM content in bottom sediments	Level of HM content in water
Point 3 (2016 y.)	0.1	low level	weakly raised level
Point 7 (2018 y.)	1.4	low level	
Point 8 (2016 / 2018 y.)	2.8 / 0.9	low level	
Point 9 (2018 y.)	4.2	low level	

In terms of Zc for heavy metals, bottom sediments have a low level of content, and in the aquatic environment their content is assessed as weakly raised in 2016 and 2018.

4 Conclusions

The geochemical feature of the reserve's territory adjacent to the water area led to the accumulation of a number of heavy metals in the surface parts of the bottom sediments. The research have shown that the levels of heavy metals in the bottom sediments of the Karadag reserve are weak and are not associated with anthropogenic influence.

In accordance with the method adopted in the work using Clarks and limit value of APC_0, the content of heavy metals in the water has weakly raised level.

Such observations will be continued in the future in order to monitor and evaluate the natural potential of the territory, in order to identify the natural background values of the concentration of heavy metals in bottom sediments. But for such purposes a longer observation period is needed.

In this instance, a clearer correlation of the source of elements entering the water from the land will be traced. And one of the most important topic to study in the future is the possibility of evaluating surface deposits in the land-sea complex.

The study was made as part of the IBSS research (registration number: AAAA-A19-119061-190081-9).

References

1. *GOST 17.1.5.01-80 Nature protection. Hydrosphere. General requirements for sampling bottom sediments of water bodies for contamination analysis*
2. *Regional standard: Standards and criteria for assessing the contamination of bottom sediments in water bodies in St. Petersburg*
3. R.V. Galiulin, R.A. Galiulina, *Water: chemistry and ecology*, 3-8 (2011)
4. V.E. Sokolov, *Reserves of USSR. Karadag reserve*, 226-241 (1987)
5. Klyukin A.A., *Exogeodynamics of Crimea*, 320 (2007)
6. A.I. Perelman, *Geochemistry*, 531 (1989)
7. I.I. Kosinova, T.V. Sokolova, VSU Bulletin. Series: Geology, **3**, 113-121 (2015)
8. Warmer H., van Dokkum R, *Water pollution control in the Netherlands. Policy and practice*, 77 (RIZA, Lelystad, 2001).

Heavy metals in the organic soil horizon of the Northern taiga forests resulting from industrial air pollution

Alexander Evdokimov*

Herzen State Pedagogical University of Russia, 48 Moika River Emb., Saint Petersburg 191186, Russia

Abstract. This study focuses on heavy metals contained in organic soil horizons resulting from the industrial air pollution caused by local enterprises. Polymetallic dust containing heavy metals is one of the main by-products in the smelting of non-ferrous metals. We evaluated the contents of Ni, Cu, and Co in the upper soil horizons containing the maximum amount of these pollutants (due to the formation of stable compounds with organic components of these horizons). Sampling was carried out at various distances from the source of pollution at previously established test sites. The study found acid-soluble heavy metals at various distances from the source of pollution and revealed the regularity of their content, both qualitative and quantitative.

1 Introduction

The Kola Peninsula features tundra and forest podzolic soils. These soils are marked by low productivity with little organic matter arriving with litter every year. Organic matter is concentrated in the uppermost part of the soil profile, where the root systems of plants are found. Unfavourable temperature conditions, low ash content of litter and poor soil aeration, adversely affect the microbial activity resulting in insufficient humus content [1]. Another feature of such soils is pronounced gleying. The surface horizons of podzolic soils are depleted in nutrients especially highly-soluble compounds of potassium, phosphorus, and nitrogen. These soils are among the most unsaturated and acidic soils in the world. They are infertile due to their high acidity, low nutrient content, the prevalence of brown humus and poor water physical properties [2].

The vitality of plants growing in the area with industrial air pollution is affected by substances that penetrate the plant not only through its assimilation system, but also through its underground organs along with water [3]. The ability of soil to accumulate these substances is central to our study as it focuses on heavy metals–primarily nickel, copper and cobalt–contained in the soil in soluble and metabolic forms [4, 5].

* Corresponding author: evdokimov89@gmail.com

The study aims to compare the quantitative content of heavy metals (nickel, copper, and cobalt) in the organic soil horizons in common pine communities located in the areas of Northern Taiga affected by industrial air pollution.

2 Materials and methods

The evidence for the study was obtained on the Kola Peninsula in Monchegorsky (buffer and impact zones), Kovdorsky and Olenegorsky districts (non-contaminated area) at previously established test sites located at different distances from the source of pollution (the Severonickel plant) (Fig. 1). These sample plots are located in middle-aged pine forests (*Pinus sylvestris* L.) with disturbance of about 80 years and undergoing regular aerotechnogenic impact. Samples of the upper organic horizon of the Al-Fe podzolic soil were taken from each test area with the subsequent exclusion of undecomposed upper level of forest litter large inclusion, living components (roots) and large mineral components of the parent materials (stones) (in the number of 5 samples for each sample plot) [6]. These samples were used to prepare a soil extract (extraction with a HCl solution of 1.0 n). At the next stage, atomic absorption spectrometry was used to determine the content of acid-soluble forms of Ni, Cu, and Co. The paper presents the average values of the available sample for these sample plots.

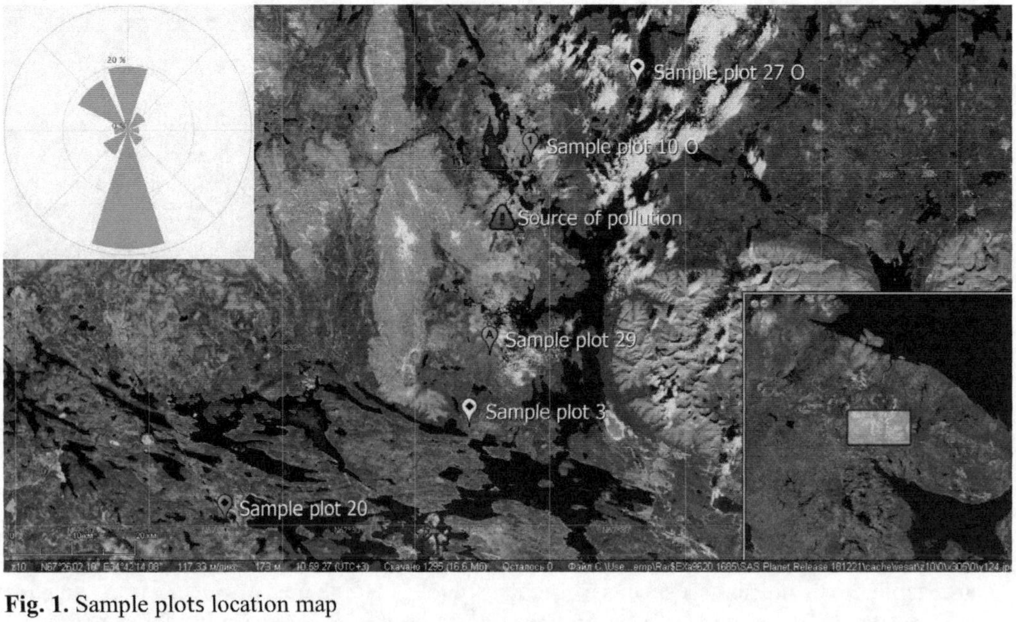

Fig. 1. Sample plots location map

Also, for these sample plots, the technogenic load index (I_I) was calculated, which is the ratio of the current pollution level (total or for each individual pollutant) to the non-contaminated indicator for this region. This index is a generally accepted indicator for assessing the technogenic load of the territories of a given region and is actively used by the staff of the laboratories performing this assessment (including the staff of the Laboratory of Plant Ecology of Komarov Botanical Institute of the Russian Academy of Sciences). This index can be represented as a formula (1):

$$I_I = P/C, \qquad (1)$$

where I_I – technogenic load index; P – pollutant content; C – pollutant content in the non-contaminated area.

 https://doi.org/10.1051/e3sconf/202016901011

The indicator is widely used in assessing the technogenic impact on ecosystems and their components and allows us to determine how many times the pollutant content (in this case, heavy metals) exceeds similar indicators in an unpolluted region. The background values are taken as an indicator of the content of heavy metals of those territories of a given region that are not subject to aerotechnogenic emission of any sources.

3 Results

The study showed that the average content of nickel, copper, and cobalt in non-contaminated areas is 7.5 mg/kg, 16.75 mg/kg, and 1.5 mg/kg respectively. Recent studies [7] report 10 mg /kg of nickel and copper in non-contaminated areas (which corresponds to the "normal" content of heavy metals for soils in this region that do not have the influence of aerotechnogenic emission of these substances). Thus, there is a slight increase in the copper content most likely due to the location of non-contaminated sample plots relative to the prevailing winds (south of the plant). Another possible reason is changes in the plant production practices (the use of various ores, new methods of production and refining, etc.). The cobalt content does not exceed the threshold for non-contaminated areas.

In the buffer zone, the nickel content is about 50 mg/kg, which is 5 times higher than the values in non-contaminated areas. The copper content ranges from 185.5 to 76.9 mg/kg (different test areas), whereas the cobalt values range from 2.72 to 5.63 mg/kg, which also significantly exceeds the values for non-contaminated areas.

The impact zones show much higher concentrations of the three metals reaching as much as 598.5 mg/kg (60 times higher) for nickel, 3,582 mg/kg (almost 360 times higher) for copper, and 28.7 mg/kg (almost 20 times higher) for cobalt (Fig. 2) (Table 1).

Table 1. The values of the content of heavy metals in the organic soil horizon at different distances from the source of pollution

Study area	**Distance from the source of pollution, km**	**Content in the soil, mg / kg**		
		Ni	Cu	Co
Non-contaminated	65	7.80	16.75	1.46
Buffer	35	69.10	185.12	2.72
	25	45.45	76.99	5.63
Impact	15	510.75	2182.15	16.70
	15	434.58	2034.20	11.16
	10	598.52	3582.00	28.68

The general index of technogenic load also varies greatly depending on the approach to the source of pollution. As mentioned earlier, in an area not experiencing an aerotechnogenic load, this indicator is 1. In the buffer zone, the total value of this index (for all three pollutants) is 9.5 (for the trial area south of the pollution source) and 4.7 (for the trial area north of the source of pollution). On average, this figure is 7.1. In the impact zone, this indicator is 91.4 (for the trial area south of the pollution source) and 125.5 (for the trial area north of the pollution source). The average for the impact zone is 108.5 (Fig. 3).

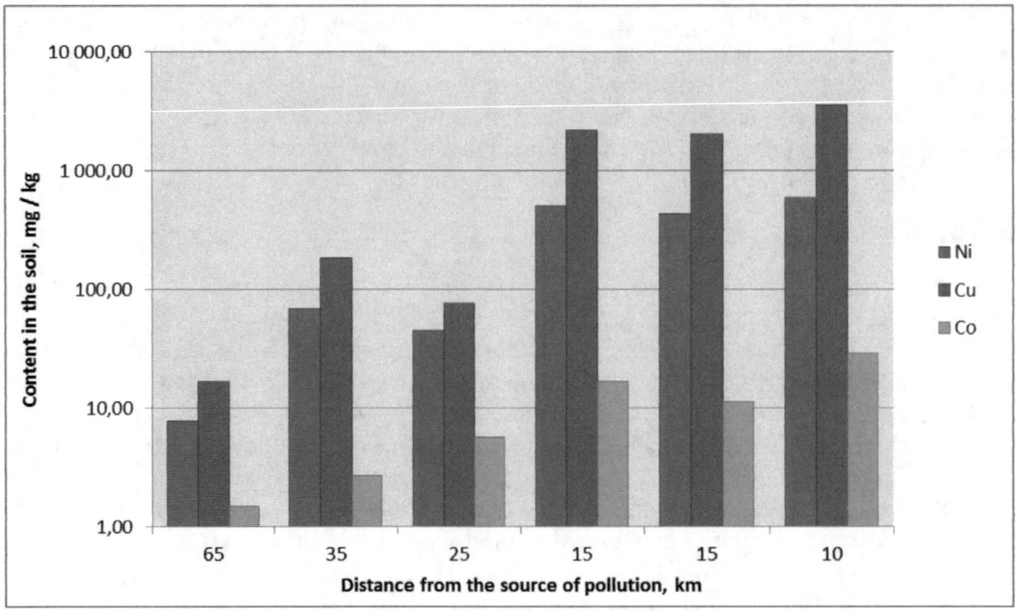

Fig. 2. The content of heavy metals in the organic soil horizon at different distances from the source of pollution (logarithmic scale of ordinates)

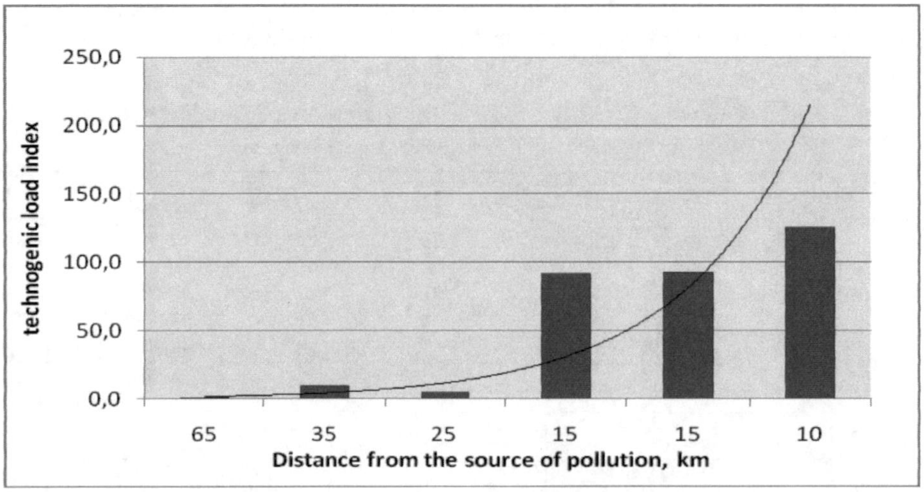

Fig. 3. The technogenic load index of the organic soil horizon at different distances from the source of pollution

As can be seen from the histogram, the content of heavy metals varies progressively depending on the distance to the source of pollution (for this reason the logarithmic scale of ordinates was used). Moreover, this is true for each of the three heavy metals despite a huge differenece in values. Thus, we observe a pattern characteristic of this region as well as other similar places.

An increase in the technogenic load index also occurs progressively in accordance with the exponential curve. On average, in the buffer zone (25-35 km), the content of heavy metals exceeds background values by more than 7 times. For the impact zone (10-15 km), the increase in this indicator is even more noticeable. The content of heavy metals on average for this territory exceeds background values by more than 108 times.

https://doi.org/10.1051/e3sconf/202016901011

4 Conclusion

The content of heavy metals in the organic soil horizon varies progressively (the content of heavy metals in the soil increases exponentially) depending on the proximity to the source of pollution. The content of heavy metals in different components of the community on average exceeds that for the non-contaminated area by 1.5-2 times and 2-15 times in the buffer and impact zones respectively. Regarding the impact zone, it can be assumed that the marked excess of heavy metals (primarily copper) is associated with the accumulation of these substances in the soil. It occurs primarily in organic horizons due to the formation of stable complexes with organic molecules (chilates), which prevent the active leaching of these substances into mineral horizons with precipitation.

There are noticeable differences in the content of each of the three components at any distance from the source of pollution (the content of copper and nickel is on average several times higher than that of cobalt). This can be explained by the nature of the ores used in the enterprise. Namely, the plant focuses on the smelting of copper and nickel. Cobalt is a side component of these ores, which explains such noticeable differences.

References

1. N. P. Belov, A. V. Baranovskaya, *Soils of the Murmansk region.* (JL: Science, 1969)
2. M. A. Glazovskaya, A. N. Gennadiev, *Soil geography with the basics of soil science.* (Moscow State University, Moscow, 1995)
3. I. V. Lyanguzova, Russian Journal of Ecology, **48**, 4 (2017)
4. I. V. Lyanguzova, V.T. Yarmishko, A.S. Evdokimov, A.I. Belyaeva, Plant resources, **54**, 4 (2018)
5. V. S. Barkan, I. V. Lyanguzova, Eurasian Soil Science, **51**, 3 (2018)
6. E. L. Vorobeichik, S. Y. Kaigorodova, Eurasian Soil Science, **50**, 8 (2017)
7. V. T. Yarmishko, I. V. Lyanguzova, M. A. Yarmishko, Agrarian Russia, **2**, (2009)

Assessment of the genotoxicity of water and soil in the places of storage of reserves of old pesticides by Allium-test

Oksana Cherednichenko*, *Serikbay* Nuraliev, *Bakytzhan* Bekmanov, *Anastassiya* Pilugina, *Leila* Dzhansugurova, and *Gulnara* Baigushikova

Institute of General Genetics and Cytology, Almaty, Kazakhstan

Abstract. Pesticides and persistent organic pollutants (POPs) are a serious environmental problem for almost all countries of the world. The toxic and mutagenic activity of samples of natural water and water extracts of soils collected near the storage sites of old and prohibited pesticides located in the villages of Kyzylkairat, Amangeldy, Beskainar, Belbulak, Enbekshi. Territories of comparison - Basshi village (Altyn Emel National Park) and near farmland, where pesticides were used in the past (Taukaraturk village, Almaty region) was investigated. Allium cepa L. was used as a biological test system. The mitotic activity and the frequency of disturbances at different stages of mitosis in onion root meristem were evaluated. Chemical analysis showed that in all soil samples the total content of 24 POPs-Pesticides exceeds the maximum permissible concentration (MPC). It was established that the level of mitotic activity of Allium cepa cells in all samples was slightly reduced compared with the control. In soil samples, the frequency of micronuclei in interphase cells (0.22-3.25%) is higher than in water samples (0.36-0.99%) (control - 0.41%). The revealed increased level of cytogenetic disorders correlates with the total content of organochlorine pesticides.

1 Introduction

The impact that is the pesticides have on the environment is acute and the issue of major concern for many countries. The Republic of Kazakhstan faces the stockpiling due to the large amount of obsolete pesticides which had not yet been disposed since the Soviet times. Still being stored in leaky conditions of the dilapidated warehouses they pollute the surrounding area and water supplies [1].

Persistent organic pesticides are characterized by the following features: they remain in the environment for a long time until their full decomposition; they transmitted over long distances, even to areas far from thousands of kilometers from their emission sources; they are accumulated in the tissues of all living organisms penetrating with food, drinking water or atmospheric air; they poison human and animals, causing wide spectrum of toxic disturbances; they have a mutagenic, teratogenic or carcinogenic effect; they can also

*Corresponding author: cherogen70@mail.ru

concentrate as they move along the trophic chain: "soil - plant - animal - human" locally increasing the accumulation level in the body.

One of the most promising methods for determining the total effect of toxicants on natural environment objects of natural inhabitants is testing of the state of bioindicator organisms, which can be microorganisms, fungi, endemic species of plants and animals. The main condition for using a living organism as an object of bioindication is the possibility of fast fixation of a certain type of disturbances caused by genotoxic agents of the habitat. At the same time, it is important that methods for assessing the state of environmental objects using biotesting methods are rapid and do not require large financial costs. Allium cepa was applied as a biological indication, it is standardized, sensitive and officially recognized method which has been in use widely for more than four decades to assess the quality/pollution of water and water supplies, as well as the ground and the state of environment resulting from the harmful effects from different kinds of pollutants including pesticides [2] (The Allium root tip chromosome aberration assay has been adopted by the International Program on Plant Bioassay (IPPB), for monitoring or testing environmental pollutants [3]).

In this regard, it was necessary to assess the toxic and mutagenic activity of natural water and soil samples collected near the storage sites of old and prohibited pesticides using the Allium cepa model test system. This study is part of a bioecological analysis of long-term environmental pollution in Almaty region by persistent organic pollutants to assess the risk to public health.

2 Materials and methods

The search was implemented for toxicant mutagenic activity of ambient water samples along with the soil-water extracts collected from around the 5 sites of demolished warehouses which are located in the Almaty region, namely the villages Kyzylkairat, Amangeldy, Beskainar, Belbulak and Enbekshias well as from the 2 intercomparison fields: farm lands close to the countryside Taukaraturyk, where pesticides had been used time ago and National Park Altyn-Emel in Basshi village as a control-point area. A clean water was used as an internal control measure.

The sampling of water was carried out in accordance with GOST 17.1.5.05-85 "General requirements for the sampling of surface and sea water" in the spring. Sampling of soil samples was carried out according to the procedure, which meets the requirements of GOST 14.4.4.02.-84., GOST-29269-91, by the method called "envelope" from soil horizons 1 (from 0 to 5 cm deep) and 2 (from 5 to 25 cm deep). For the chemical analysis and evaluation of the genotoxic potential, water and soil samples were taken, each weighing not less than 1 kg. Samples of water were taken in 1l plastic bottle.

Water extracts of soils were prepared as follows: soil samples (30 g) were poured into 150 cm3 of distilled water, stirred for 3 minutes on a mixer and left for 5 minutes to settle (GOST 26423-85).

Conducting quantitative analysis of environmental samples for pesticides and their metabolites (24 pesticides) was carried out with sample preparation of water from natural reservoir sand soil in accordance with GOST 29269-91 Soils. General requirements for analysis. Gas chromatography and mass spectrometry were used to determine pesticides and their metabolites in samples of soil and water.

To study cytogenetic abnormalities using the Allium cepa test object, onion seeds were germinated in Petri dishes at a temperature of 24°C for 72 hours in the test sample. The squash technique for onion root as described [4] was used for the cell division stages investigation. Cells samples were taken from root meristems containing actively growing cells. The mitotic index (MI) and the frequency of disorders in different stages of mitosis were defined in onion root meristem.

3 Results and discussion

When analysing the POPs quantity in natural water around the former pesticides warehouses it was revealed that all investigated samples contain highly toxic metabolite of DDT - DDE. The MPC in water for DDE is 2 mg/dm^3. The residual amount of the metabolite DDE varied in natural water from 2.4 to 5.4 mg/dm^3. It is established that samples of natural water also contain the metabolite 2.4 DDD and 4.4 DDD. Also in some samples, endosulfansulfate in concentration 0.184-1.33 μg/dm^3, dibutylendan 0.325-1.788 μg/dm^3, methoxychlor 0.12 μg/dm^3, hexabromobenzene 0.287 μg/dm^3, deldrin 0.0625-0.1213 μg/dm^3, heptachlorepoxide 0.063 μg/dm^3, chlordan 0.03 μg/dm^3, 2.4 DDD 0.538 μg/dm^3 and 4.4 DDD 0.21-0.365 μg/dm^3 were present.

In the study of soil samples, it has been found that soil around the former warehouses has a multicomponent pollution, i.e. polluted by pesticides and heavy metals. The strongest pollutants of the soil are DDT (0.1624-21.2 mg/kg) and DDE (0.2372-2209 mg/kg). Also collected soil samples contained β-HCH, aldrin, dibutilendan, endosulfansulfate, geptahlorepoksid, deldrin, endrin, chlorobenzilate.

The total content of pesticides in specimens is shown in Fig. 1.

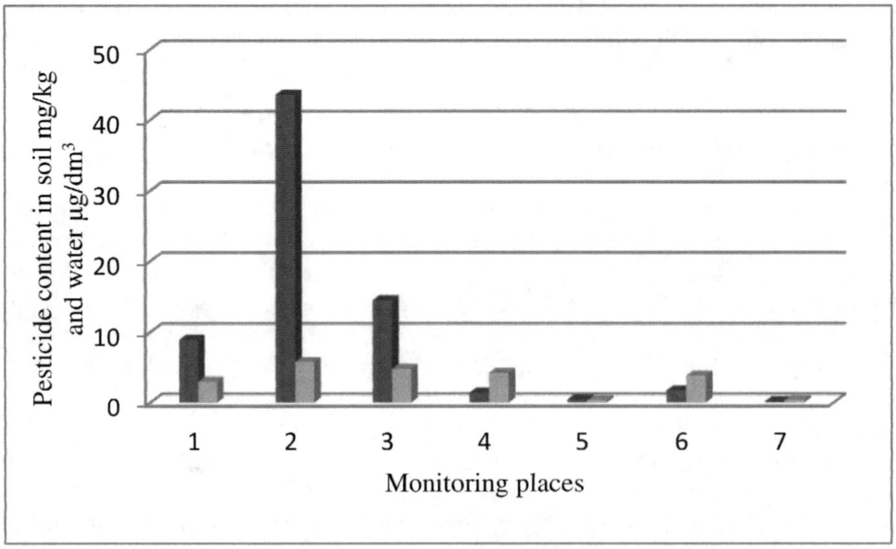

Fig. 1. The total content of organochlorine pesticides in the studied soil and water samples.

1 – Beskainar, 2 – Kyzylkairat, 3 – Amangeldy, 4 – Belbulak, 5 – Enbekshi, 6 – Taukaraturyk, 7 – Basshi

It is established that in all samples the total amount of pesticides in the soil exceeds the MPC. The soil most polluted by pesticides was taken from the territories of former pesticide storage facilities located in the villages Kyzylkairat, Amangeldy and Beskainar. The concentration of pesticides in the soil exceeded the MPC from 60 to 120 MPC. In the village Taukaraturyk, where there are no official data on the location of the former pesticide storage facilities, the concentration of POPs pesticides in the soil exceeded the MAC up to 17 times. The concentration of pesticide POPs in soil samples taken from the territories of former pesticide storage facilities located in the villages of Belbulak and Amangeldy (the burial place of pesticides) also exceeded the MPC in the soil from 9 to 15 times. It should be noted that the ratio of DDT and its derivatives makes it possible to approximately estimate the time of appearance and decomposition of residual concentrations of pesticide contained in soils.

Thus, the ratio (DDE + DDD) / DDT> 1 indicates the "old" use of DDT and its active transformation by microorganisms.

The results obtained from cytogenetic analysis of the apical meristem through Allium test for disorders in interphase cells and during different stages of cell division are presented in table 1. It was identified that the mitotic values derived from Allium test cells in all samples were slightly decreased compared with the control meanings.

Table 1. The frequency of cytogenetic disorders in assessing the genotoxicity of soil and natural water samples using the Allium test.

Variant	Cells viewed	Interphase	Prophase	Metaphase	Anaphase	Telophase	Mitotic index
Beskainar							
Water	3028	0.71	1.48	24.49	9.64	1.89	15.87
Soil	3041	1.38*	-	6.48	4.82	-	12.96
Kyzylkairat							
Water	3546	0.99	0.32	22.45	14.41	1.72	17.38
Soil	3103	3.57*		10.17	3.60	-	15.06
Amangeldy							
Water	3047	0.79	0,86	15.46	21.98	1.89	16.80
Soil	1500	0.87	-	3.22	10.00	-	12.55
Belbulak							
Water	1500	0.92	-	-	11.10	11.10	14.40
Soil	1520	0.93	-	4.50	9.50	-	15.80
Enbekshi							
Water	3532	0.36	0.39	19.19	9.09	-	13.78
Soil	1500	0.22	-	4.54	9.23	-	14.80
Taukaraturyk							
Water	1668	0.72	0.28	13.11	12.82	1.61	19.08
Soil	1530	2.59*	3.33	18.70	12.80		12.80
Basshi							
Water	1500	0.45	-	-	-	-	12.40
Soil	1500	0.46	-	2.00	6.67	-	15.00
Distilled water	1522	0.41	-	5.88	4.17	-	19.00
Note: * - $p \leq 0.01$							

It has been detected that in soil samples the frequency of micronuclei (MN) in interphase cells (0.22-3.57%) is higher than that in water samples (0.36-0.99%) and in some cases it significantly exceeds the reference level. The greatest number of violations was recorded at the key stages of cell division - metaphase and anaphase. Analysis on water samples revealed significantly more violations in different stages of mitosis, suggesting the presence of chemical substances with mitotic modifying and genotoxic effects in water. A particularly high level of disturbances at these stages was revealed when analysing water from the villages of Beskaynar, Kyzylkairat and Amangeldy. According to total estimates, the most polluted soil and water were reported from Kyzylkairat and Beskainar village. It is worthwhile noting that in township Taukaraturyk, where pesticides have not been used for more than 20 years the level of their residues in the soil significantly increases the frequency of cytogenetic disturbances in the interphase and prophase of cell division.

It was evaluated that disorders in interphase nuclei and in prophase were represented by micronuclei, in metaphase - as micronuclei, dropped chromosomes and C-metaphases, in anaphase - chromosome bridges and lagging chromosomes, in telophase - chromosome bridges (Fig. 2).

Based on the results of present cytogenetic studies, the correlation within the chemical determination of pesticides in examined samples was assessed. A direct correlation was established between the frequency of micronuclei in interphase cells and the total content of organochlorine pesticides in natural water samples — 0.96 and soil specimens — 0.74.

Genotoxic substance research is of great importance for the protection of the environment because it enables an insight into the influence of genotoxic substances on organisms. With the Allium-test and related genotoxic tests one can determine the influence of chemical substances on an organism [5].

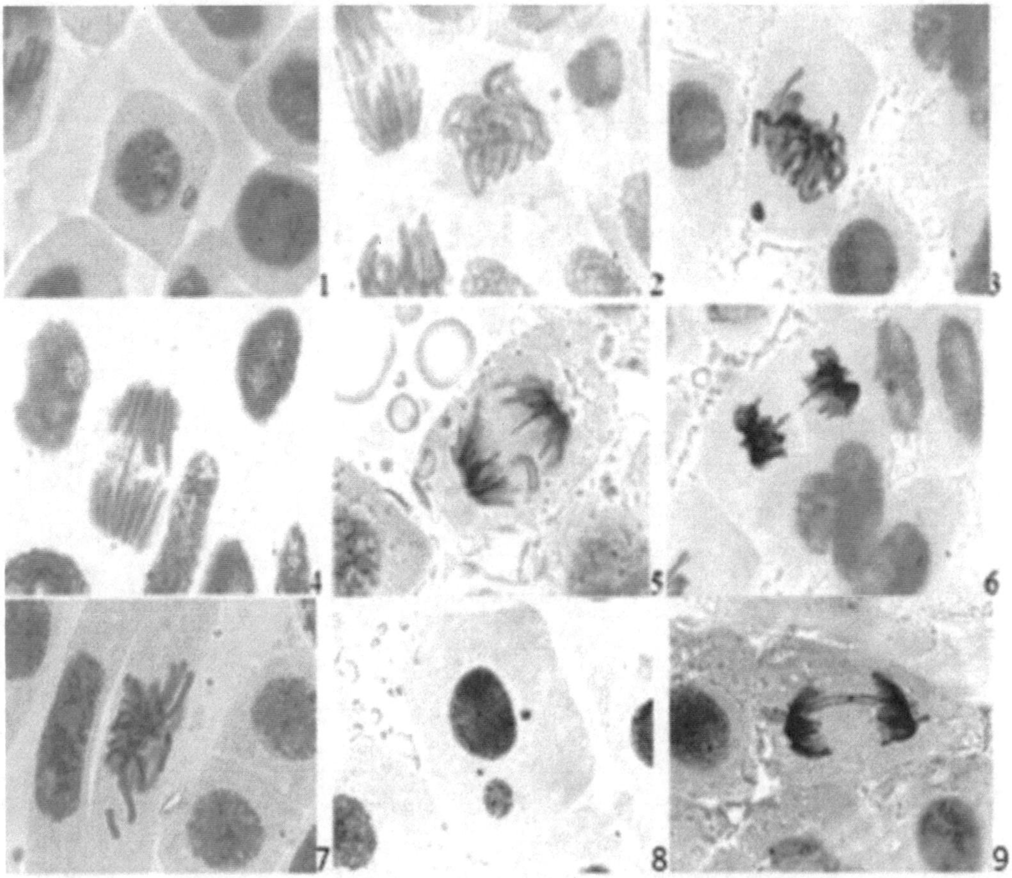

Fig. 2. Types of cytogenetic disorders during the Allium test.

1 - micronucleus in interphase; 2-micronucleus in prophase; 3- micronucleus in metaphase; 4-chromosome bridge in anaphase; 5 – backward chromosomes in anaphase; 6-chromosome bridge in telophase; 7 - C-metaphase with a fallen chromosome; 8 - interphase cell with three micronuclei of different sizes; 9 - anaphase with two bridges.

Pesticides have been widely tested in bioassays with plants and positive results have generally been obtained. In a study to evaluate the negative effects of diuron, a substituted urea herbicide [6] reported cytotoxic and genotoxic effects in A. cepa meristematic cells. The authors showed that diuron causes mitotic alterations by inhibiting the polymerization of the tubulin subunits of microtubules. In a study to evaluate the action of the trifluralin herbicide in A. cepa test system [7], showed the mechanism of MN formation by polyploidization of the genetic material. A study to evaluate the genotoxic effects of the herbicides pentachlorophenol and butachlor by the A. cepa test showed magnification CA values for all the chemicals tested [8]. Also the genotoxicity of roundup [9], atrazine [10], insecticides (cypermethrin, fenvalerate) [11], deltamethrin [12] and many others has been shown. Moreover, many of them are also characterized as clastogenic and aneugenic agents. Thus, the many authors concluded that pesticides can induce cytotoxic and genotoxic effects on the A. cepa test system, which showed to be sensitive to this analysis [13].

Studies of sensitivity and correlation among test systems are fundamental for a more accurate evaluation of the environmental risks, as well as extrapolation of data to other target organisms, e.g. man. Tests to detect the risks to human health are conducted using several test systems which use a wide variety of organisms. Allium-test shows a good correlation with similar tests applied on mammals and other vertebrates [13]. According to Grant, of 148 chemicals evaluated by the Allium test, 76% presented positive results [14]. Fiskesjo reported that the sensitivity of the A. cepa test was virtually the same as the one observed for algae and human lymphocyte test systems [15]. Rank and Nielsen showed a correlation of 82% of the A. cepa test in relation to the carcinogenicity test in rodents [16].

In our study, in addition to the Allium-test system, we also used a test system - cultures of human lymphocytes and mouse-like rodents (micronuclear test and comet-test) that live in the studied villages. All these tests showed a similar degree of pollution of villages, according to a chemical analysis of soils and natural water. A small correlation at 0.50 was shown only between the comet-test in mouse rodents and the Allium-test results in the study of natural water in the studied villages. Some authors believe that this non-correlation may be related to the enzymatic activities of the detoxification process of exogenous substances and DNA repair, which are less relevant in plant cells than in mammals [17].

4 Conclusions

Around the territories of the former warehouses of obsolete and prohibited pesticides, an elevated level of organochlorine compounds remains, significantly exceeding the MPC. The main pollutants are highly toxic compounds - DDT and DDE. Using Allium-test, an increased level of cytogenetic disorders in the root meristematic cells was found, which correlated with the total content of organochlorine pesticides. Nonetheless, we can say that using different test systems is important for a more comprehensive assessment of the different aspects of an agent action, thus obtaining a complete overview of its hazard.

The work was performed under Scientific and Technical Program: No. BR05236379 «Comprehensive assessment of unutilized and banned pesticides impact on genetic status and health of population of Almaty region»

References

1. A. Nurzhanova, S.Kalugin, K. Zhambakin, Environ Sci Pollut Res., **20(4),** 2054-2063 (2013)
2. P. Firbas, T. Amon, J. Bioremed. Biodeg., **4(4),** 2-5 (2013)
3. T.H. Ma, Mutat Res, **426**, 103-106 (1999)
4. K. al-Sabti, Cytobios, **58** 71-78 (1989)
5. P. Firbas, T. Amon, Caryologia, **67(1)**, 25-35 (2014)
6. L.K.S. Chauhan, P.N. Saxena, V. Sundararaman, S.K. Gupta, Pest. Biochem. Physiol., **62**, 152–163 (1998)
7. T.C.C. Fernandes, D.E.C. Mazzeo, M.A. Marin-Morales, Pest. Biochem. Physiol., **88,** 252–259 (2007)
8. B. Ateeq, M. Adul Farrah, M.N. Ali, W. Ahmad, Mutat. Res., **514,**05–113 (2002)

9. J. Rank, A.-G. Jensen, B. Skov, L.H. Pedersen, K. Jensen, Mutat. Res., **300**, 29–36 (1993)
10. P. Bolle, S. Mastrangelo, P. Tucci, M.G. Evandri, Environ. Mol. Mutagen., **43**, 137–141 (2004)
11. L.K.S. Chauhan, P.N. Saxena, S.K. Gupta, Environ. Exp. Bot. **42** 181–189 (1999)
12. L.K.S. Chauhan, S.K. Gupta, Pest. Biochem. Physiol. **82**, 27–35 (2005)
13. D.M. Leme, M.A. Marin-Morales. Mutation Research.**682**, 71–81 (2009)
14. W.F. Grant, Mutat. Res. **99**, 273–291 (1982)
15. G. Fiskesjo, Hereditas. **102,** 99–112 (1985)
16. J. Rank, M.H. Nielsen, Mutat. Res. **312**, 17–24 (1994)
17. D. Feretti, I. Zerbini, E. Ceretti, M. Villlarini, C. Zani, M. Moretti, C. Fatigoni, G.Orizio, F. Donato, S. Monarca, Water Res. **42**, 4075–4082 (2008)

Specificity of accumulation of hydrocarbons in various components of geosystems

Aleksandr Khaustov[1*], and *Margarita* Redina[1]

[1]Peoples' Friendship University of Russia (RUDN University), Faculty of Ecology, 6 Miklukho-Maklaya St, Moscow, 117198, Russian Federation

Abstract. The results of the analysis of thermodynamic characteristics of the processes of hydrocarbon amassment in the components of geochemical systems are presented (waters, soils, plants and other substances forming ecosystem). The need to use an ideology and apparatus for calculating thermodynamics is due to the complexity and diversity of mass transfer in interacting media. The material and energy flows of marker compounds, polycyclic aromatic hydrocarbons (PAHs), are analyzed: Naphthalene (Naph), Acenaphthene (Ace), Fluorene (Flourene), Phenanthrene (Phen), Anthracene (An), Fluoranthene (Flu), Pyrene (Py), benzo[a]anthracene (BaA), Chrysene (Chr), Benzo[b]fluoranthene (BbFlu), Benzo[k]fluoranthene (BkFlu), Benzo[a]pyrene (BaP), Benzo[ghi]perylene (Bghi), Dibenzo[a,h]anthracene (DbA), Indeno[1,2-cd]pyrene (Ip). Estimates are obtained of the intensity of PAH concentration processes in geochemical systems in different states from the point of view of stability, from clearly unstable to metastable. The role of living matter in the formation of stability of geochemical systems is revealed.

Introduction

The proposed study focuses on PAHs – very common in the environment, persistent substances able to cause significant negative effects on organisms. At the same time, an important feature of this group of substances is the ability to mark natural and technogenic processes. Analysis of PAH concentrations reflects the hydrocarbon flows in the environment, allows to identify areas of influence of pollution processes. The specific structure of molecules (2 and more benzene rings) determines the hydrophobicity and lipophilicity of molecules and their behaviour in media.

Ordering the movement of organic molecules dissolved in water solutions in the gravitational field creates conditions for their collective behaviour. Coherent motion occurs even in systems that are "sensitive" to weak fluctuations in the gravitational field. This is the reason for the difference in the concentration gradients of hydrocarbons and, as a result, the variation processes in geochemical systems (GCS), acting from outside, and the development of the system itself, depending on the receipt of energy quanta.

According to Buck et al. (1987), systems with a large number of interacting elements naturally evolve to a critical state in which any (even small) event will lead to the

[*] Corresponding author: khaustov-ap@rudn.ru

destruction of the system or a change of state. This principle is the base of our research on the concentration of PAHs through estimates of measures of heterogeneity (entropy) in GCS is based on. Other thermodynamic characteristics have been also calculated for assessing the direction of processes: changes in Gibbs energy (ΔG) and enthalpy (ΔH). The distribution of substances in octanol and water (log Kow) for PAHs was also used, which allows to judge the selective possibility of their movement with solutions in media.

Research methods and objects

A set of the objects for modelling includes 142 genetically different components of GCS:

- a part of the objects can be considered background (samples of water, soil, silt, brine, various parts of hydrobiont plants on the Kerch Peninsula) [1-3];
- samples from areas with an intense technogenic pressure: soils, vegetation, water and bottom sediments of Aojian river and it's estuary (the area of development of the leather industry in China) [4]; soil samples and needles of various plants from regions with different levels of anthropogenic pressure (zone of influence of the Irkutsk aluminium plant, Melnichnaya Pad settlement, villages Moty, Baklashi, Olha) [5]; samples with different levels of anthropogenic pressures in the Arctic and Antarctic regions [6, 7];
- test soil-plant systems (soil - rhizosphere - root - stem): conditionally clean and irrigated with sewage for more than 40 years [8];
- snow samples from the RUDN campus (the territory of Moscow with variable technogenic loads and under the influence of highways) [9].

The migration and accumulation activity was estimated on the basis of concentration coefficients (CC) – the ratios of the concentrations of individual PAHs in an object to the average concentration in a group of genetically similar objects. The calculation was carried out using hierarchical constructions of data structures and formal classification logic (the concept of division of concepts or the logic of a jump in data). Given the heterogeneity of the media, 5 iterations of classification were adopted. For them, in addition to QC, thermodynamic parameters were calculated. Such a set of methods allows us to consider possible variations of PAH fractions in the system of "signs - objects" within the framework of the ideologies of synergetic and formal kinetics.

Results and discussion

Among the probable reasons for the separation of PAHs into groups (fractionations) are the internal physicochemical (molecular) properties. This is a fundamental conclusion that allows one to interpret many processes of concentration of PAHs in media and the conversion of energy from thermodynamic positions. The identification of the structure of PAH bonds is illustrated by the dendrogram of cluster analysis (Fig. 1). PAHs are distributed into groups as a whole according to their mass (respectively, the number of aromatic rings in the molecule, and hence the shape and properties). Thus, the group of the lightest PAHs included Napa, Ace, Fluorene and Py (the latter has 4 rings and is relatively far from the centre of the group). The remaining group of PAHs can be divided into the heaviest (5-6 rings) and relatively lighter (mainly 4-ring). The mobility and migration of PAHs from water media to other media is also due to the synergistic influence of the media themselves-porosity, temperature and lipophilicity, etc. This is confirmed by correlation coefficients (close relationship of physical and chemical characteristics with PAH concentrations).

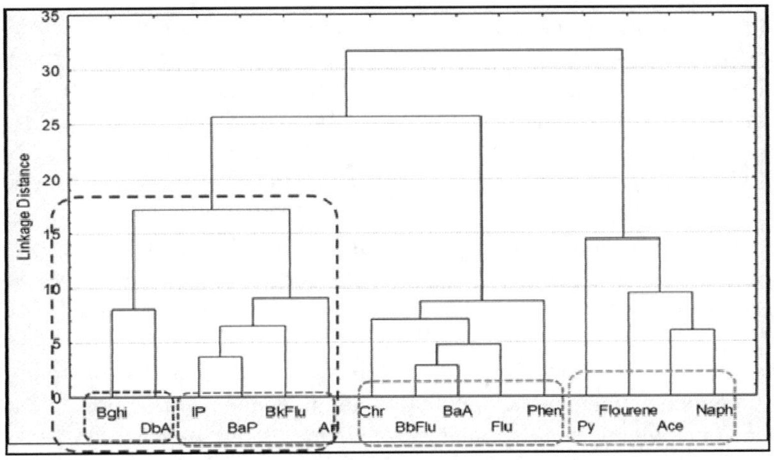

Fig. 1. The results of a cluster analysis of the distribution of PAHs in 142 natural and technogenic objects of various genesis

The distribution of 142 objects on the dendrogram obtained by cluster analysis based on data on concentrations of 15 PAHs is determined by the origin of contamination and the type of objects. There are 2 classes with markedly different values of average ΣPAH (ng/g): 1. Clean and relatively unpolluted environments (ΣPAH 226) and 2. Polluted environments (ΣPAH 3384).

By further iterations, the first class is subdivided into 1.1. Clean (ΣPAH 90.1) and 1.2. Having a tendency to pollution (ΣPAH 810). Further subclasses were divided into smaller groups: from 1.1.1. Clean for these formations (ΣPAH 106.0) to 1.1.1.2.4. Maximum clean (snow outside the pollution zone, ΣPAH 0.47); from 1.2.1. Relatively polluted (ΣPAH 632) to 1.2.2. With severe pollution (ΣPAH 1088).

The second class of contaminated media is divided into: 2.1. With severe pollution (ΣPAH 2430), including 2.1.1 with pronounced pollution (ΣPAH 2564) and 2.1.2 slightly contaminated (ΣPAH 1552) and 2.2. Very polluted (ΣPAH 3384), consisting of 2.2.1. Extremely polluted (ΣPAH 697) and 2.2.2. Very dirty (ΣPAH 5763).

Also, the media with abnormally high Σ PAU = 31,450 μg / kg – soil in the industrial zone of the Bratsk aluminium smelter were identified [8].

The dominant PAHs for all pure formations are Py, Naph, Phen, and Flu. In this pool, PAHs can change their places depending on the prevailing media in the clusters, the role and conditions regarding the press of pollution, as well as the degree of transformation in natural conditions and the influence of other media.

Analysis of specificity of PAHs distribution in the environmental media allowed us to identify some paradox moments.

1. Naph is considered as extremely unstable PAH having naphthidogenic nature, relatively soluble in water. Nevertheless, it is quite firmly fixed in plant communities, rhizosphere, soils, peat, tow. Apparently, this is due to the migration of moisture, permeating the entire cycle of plant development. According to the frequency of occurrence in environments, Naph is the leader (except for snow in the zones of influence of vehicles - it is absent there, due to the high temperature of fuel combustion in engines [9]). Similarly, Naph Phen and Flu are detected, although they are more resistant to environmental influences, and their appearance, in addition to pyrogenic technogenesis, may be due to the generation of organic matter.

2. Extremely low concentrations of PAHs in snow and water, despite the potential for accumulation. Low PAH concentrations in snow are regulated by the time of contact with precipitation. In the rivers most of the PAHs may migrate to the organic suspensions.

Therefore, measurements of their concentrations in water record only the dissolved part of PAHs. In underground waters, PAH concentrations are insignificant due to their vertical migration through rocks (except for hydrocarbon deposits). From the standpoint of thermodynamics, the water-rock interaction can be interpreted as a super-flow system, where the principle is implemented: the more matter and energy entered the system from an external source, the more significant amounts of them are converted with minimal synergetic processes [1].

3. The variety of time intervals during which the maximum accumulation of PAHs is achieved, and the synergy of factors for the formation of the PAH gradient in the media chain. Concentration occurs to a greater extent due to microeffects in each of the media with a relative constancy of macro factors (gravity, etc.). Concentration decreases in the sequence: the needles of cedar, larch, pine, the soil in the influence zone of aluminum production > drains below the ash dump TPP > soil, bottom sediments in the area of impact of the leather industry>soil and sediment in the impact zone exits of underground waters (Crimea) > soils, silts, peat, hards > needles, hydrophytes groundwater sources, the roots of Salicornia, Arctic soil > roots of plants, soils, sludges, polluted snow > water underground springs and mud volcano, surface water, clean soil of Antarctica; snow in Moscow. This sequence indicates the possibility of maximum accumulation of PAHs from the atmosphere in the coniferous trees, even in short periods of time (larch periodically sheds needles). Soils have their own limits of PAH saturation, although their ability to "process" petroleum products is higher than that of other media. Like sediments (silts), they can deposit PAHs up to certain limits. The maximum deposition, in addition to lipophilicity, is controlled by the effective diameter of the mineral matrix. The rhizosphere is a specific subsystem that controls selective PAH inputs from soil to stems; this is a "filter" that regulates mini-mass flows in plants. The spectrum of dependencies of entropy averaged over clusters on the coefficients of PAH concentrations in the considered groups is shown (Fig. 2).

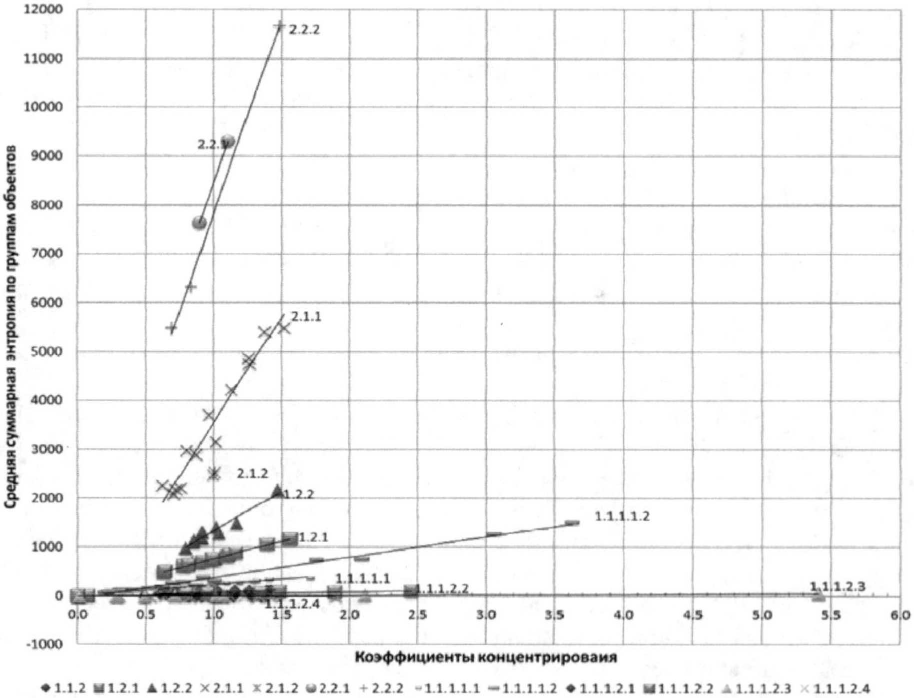

Fig. 2. The dependence of the total entropy on the CC for the selected groups of objects

A clear differentiation of points indicates the abrupt development of dissipative systems, with possible phase transitions at high entropy values. The separation of media with low entropy values and low concentration potential is unambiguous. These are polluted soil and rhizosphere of plants (industrial zone, drainage irrigation-cluster 2.2.1) and cluster 2.2.2 – larch needles from industrial areas of the Irkutsk region.

4. The fourth paradox is the role of biological systems for the transformation of PAHs. The underestimation of the role of living matter as a component of the geological environment (continuously and with minimal energy expenditures fulfilling the barrier role) leads to disruption of the functions of its components. For bottom sediments, it can be argued that they are approaching an altered stable asymptotic state, a weak reaction to external influences. Further, as the entropy decreases in clusters 2.1 1, 2.1.2, and 1.2.1, a regular increase in the concentration of PAHs in media occurs. These are mosses, lichens, rhizosphere, santolin wormwood, soil, and the most polluted bottom sediments of the river. Aojian, soil and plant rhizosphere in an irrigation experiment with industrial waste. For the first cluster, we can talk about the property of the plants themselves with maximum activity to absorb pollutants and natural PAH flows.

5. The fifth paradox: the generation and accumulation of PAHs, reaching maximum concentrations, can occur naturally through evolutionary catalysis. This path in open systems is anti-entropy in nature. Cluster 2.1.2. consists of 2 objects with a high PAH content due to natural concentration in the transitional and lower peat layers. High abilities are known for the generation and accumulation of hydrocarbons by this substance under favourable conditions. Given the close hydrocarbon composition of peat and oil products, these effects can be explained (except for self-generation) by the principle of "like tends to like". From the standpoint of nonequilibrium thermodynamics, the problem of assessing the stability of states as a measure of deviations from equilibrium under various external influences is of particular relevance. The state function of a system or its evolution can be identified based on the values of ΔS and ΔG. Based on empirical data, the values of standardized entropy for the selected clusters were calculated (Fig. 3).

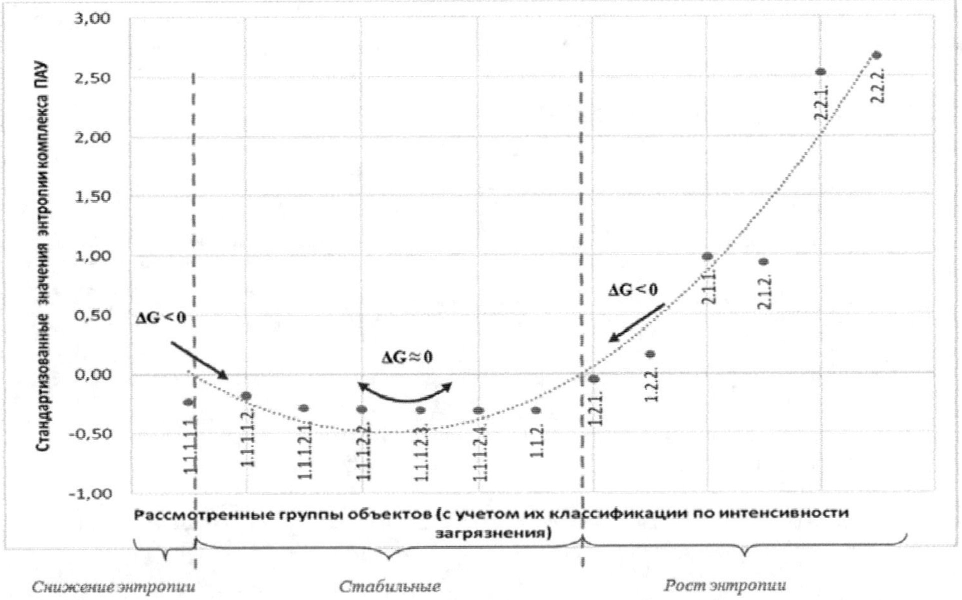

Fig. 3. Changes in the values of standardized entropy of ∑PAH in different genetic types of objects

6. The stability of GCS from the standpoint of thermodynamics is an equilibrium-nonequilibrium state; it is characterized by the centripetal directivity of the flows of matter and energy (Fig. 3).

The general dynamics of the variability of entropy depending on the properties of the media in which PAHs migrate is close to the theoretical model of the kinetics of transformation of the starting materials into reaction products. One can clearly see the tendency to minimize and go to the negative region of clusters from 1.2.1 to subcluster 1.1.1.1.1 (mainly taxa with a low level of PAH contamination in various environments). The values of thermodynamic parameters indicate that a conditional equilibrium is reached in the media when the influx of matter and energy into the system is approximately equal. Coordination of system elements creates an optimal margin of adaptation to external influences. Catabolic processes, biological (they are given the predominant role) and physicochemical (including the ablation of PAHs from a system with moisture streams) are active in all parts of the trophic chain of systems with a pulsed damped mode. This provides a time-delayed evolution of HCS or a steady development trend. Internal fluctuations of systems are balanced by external influences of the supersystem. PAH generation by the system itself is manifested by natural pools of fractions, and their local contribution is suppressed by the cumulative nature of the system itself. The second part of the curve with positive entropy values from the cluster 1.2.2. to the cluster 2.2.2. - heterogeneous systems with a pronounced technogenic influence. Here, PAHs are accumulating (to the maximum) with the participation of self-oscillating and autowave processes. Judging by the increase in energy parameters, the accumulation and destruction of PAHs are abrupt, and the formation of PAH pools at the output is stochastic.

7. The effects of concentration cannot be explained only by the kinetics of substances or their internal molecular properties. Mass flows in media are aimed at equalizing concentration gradients; these processes can be considered as the beginning of the fractionation of substances. They are selective depending on the physical properties of substances and purposefulness. For biota, an important role is played by the time intervals of "life" and the processes of fractionation are more contrasted. The osmotic properties of plants or porous media significantly reduce the role of thermodynamic forces. Mass flows tend to even out gradients and activate feedback between system components, giving an impetus to evolution. Low-gradient formations include clean and relatively unpolluted clusters 1.1.1.1 - 1.2.2; to highly gradient ones are polluted media, clusters 2.1.1 - 2.2.2, and media with abnormally high pollution (Table 1, Fig. 3).

Table 1. Estimates of the significance of thermodynamic parameters in the processes of fractionation and concentration of PAHs

Cluster of objects	Leading factors in the accumulation of PAHs
1.1.2 Aojian River polluted waters (influence of leather industry enterprises)	ΔS>lgKow>ΔHsolid>ΔG
1.2.1 *Technologically loaded :* soils and bottom sediments of the Aojian River	Extremely weak influence ΔS
1.2.2 *Technologically loaded:* plants in the Irkutsk region, relatively clean soil, p. Olha, runoff 2 km below the thermal power plant, Olha river bassin - soils	ΔHgas >ΔS
2.1.1 High PAH content: mosses, lichens, rhizosphere, santolin wormwood, soil, the most polluted bottom sediments of the Aojian River	lgKow>ΔG> ΔHsolid >ΔS> ΔHgas
2.1.2 High PAH content, natural objects - transitional and lower peat	Слабое влияние ΔHgas
2.2.1 *Contaminated soil and rhizosphere (industrial zone, runoff watering)*	No significant connection
2.2.2 *Industrial areas, larch (Irkutsk region)*	ΔHgas

1.1.1.1.1 Hydrophytes of Lake Chokrak and soleros; Arctic soil	ΔG>lgKow>ΔS> ΔHgas > ΔHsolid
1.1.1.1.2 Antarctica soil, Olha river - bottom sediments, silt of Lake Chokrak	ΔG>lgKow> ΔHgas >ΔS> ΔHsolid
1.1.1.2.1 Soils and plant roots in the aquatic system, silts of Lake Chokrak, clean soils of Antarctica	lgKow>ΔG>ΔS> ΔHgas > ΔHsolid
1.1.1.2.2 Water of the Olha River, clean soils of Antarctica, the root of an unpolluted soil system	ΔG ≈ lgKow ≈ ΔHsolid ≈ ΔS> ΔHgas
1.1.1.2.3 Chokrak lake water, soil and plant roots in the aquatic system of Chokrak lake, snow on the RUDN campus, clean soils of Antarctica	ΔG>lgKow> ΔHsolid >ΔS> ΔHgas
1.1.1.2.4 Snow, RUDN campus	No significant connections

For the considered groups, multiple correlations of physicochemical characteristics and PAH concentrations were calculated. The role of these characteristics was evaluated by the number of significant relationships. The host adopted a characteristic that gains most of the significant relationships (correlation coefficient of 0.5 or more). On this basis, "ratings" of the degree of influence of parameters on ∑PAH concentrations were formed (Table 1).

The calculation results indicate a predominant role in all media in the total concentrations of PAHs ΔG and log Kow. Entropy almost never takes the lead except for super-polluted waters in the effluents of the leather industry and technologically loaded soils at the aluminum plant.

In most pure formations, regardless of the genesis, the leading positions change between ΔG and log Kow, which may indicate the manifestation of local equilibrium in irreversible heterogeneous GHS, for example, in peat, needles of plants and some types of soils.

Conclusion

A huge role in the macrokinetics of phase transformations and, consequently, the concentration of migrating substances in nonequilibrium systems is played by aqueous solutions that introduce significant fluctuations in the energy field through the manifestation of mass transfer gradients.

Mass flows of substances in interacting media are aimed at equalizing gradients, reduce the role of thermodynamic forces in nonequilibrium systems. This leads to the formation of feedbacks in the system, according to the degree of influence of which the evolution (degree of self-organization) of the systems can be estimated.

The ratios of the concentration coefficients of PAHs in the media can be considered as the most informative parameter of the effectiveness of the nonequilibrium macro-system within the framework of the "input - transformation (or transformation) - removal of substances" scheme.

The minimum increment of entropy is formed in undisturbed media at the maximum contact of the supply vector and the minimum time intervals, which proves the relative stability of the flows.

References

1. A. P. Khaustov, M. M. Redina., E.V. Yakovleva, *Geoecology. Eng. geology, hydrogeology, geocryology,* **3**, 3-17 (2018) (in Russian).
2. Khaustov, A., &Redina, M. *E3S Web of Conferences*, **98**, 01026. EDP Sciences (2019).
3. Khaustov, A., Redina, M., &Yakovleva, E.. *E3S Web of Conferences,* **98**, 01027. EDP Sciences (2019)
4. Li J., Shang X., Zhao Z. et al. *J. of Hazardous Materials,* **173(1-3)**, 75-81 (2010).

5. I. I Marinayte., A. G. Gorshkov, E. N. Taranenko et al. *Chemistry for Sustainable Development,* **20,** 135-146 (2013).
6. E.V. Abakumov, E.D. Lodygin, D.A. Gabov, V.A. Krylenkov, *Hygiene and sanitation* [Gigiena i sanitariya], **93(1)**, 31-35 (2014) (in Russian).
7. E.V. Abakumov, V.M. Tomashunas, E.D.Lodygin, Gabov, D.N. et al. *Soil Science*, **12**, 1433-1433 (2015) (in Russian).
8. S. Tao, Y.H. Cui, F.L. Xu, B.G. Li et al. *Science of the total Environment,* **320(1)**, 11-24 (2004).
9. A. Khaustov, M. Redina, *Polycyclic Aromatic Compounds*, 1-12, (2019). https://doi.org/10.1080/10406638.2019.1645707

E3S Web of Conferences **169**, 01014 (2020) https://doi.org/10.1051/e3sconf/202016901014
APEEM 2020

Geochemical features of soils from the Mid-Ural ore deposit

Ekaterina Dziuba*, *Dmitriy* Andreev, *Sergey* Buzmakov, and *Yuliya* Khotyanovskaya

Perm State University, Department of Biogeocenology and Nature Protection, Bukireva str., 15, Perm, Russia

Abstract. This paper deals with the contemporary state of soil cover around the Mid-Ural iron ore deposit, focusing on geochemical features. Sample plots were laid based on the technogenic impact in the study area. Soil samples were taken in the zone of the iron ore deposit, in the zones of overburden and gold mining waste dumps influence, and in the rock blasting site. A total of 64 sample plots were laid. Soil samples were sampled incrementally by an envelope method from a depth of 0.00–0.03 m. Soil texture, physicochemical features, and total heavy metal content (V, Mn, Ni, Cu, Zn, As, Cd, and Pb) were defined for the soil samples. As a result, the evaluation of heavy metal soil pollution under the influence of iron ore deposit mining is given. Chemical contamination of soil was assessed by a cumulative chemical contamination index. Geochemical series of elements formed under the influence of iron ore deposit mining, as well as overburden and gold mining are given. Here, geochemical patterns of heavy metal distribution in the soil of iron ore deposit and under the gold mining influence are considered. The analysis of gold mining influence on the elementary composition of soil revealed the anomalous content of Mn and Ni.

1 Introduction.

The Vernadsky law for global dissipation of chemical elements states that all environments have all the elements, but their distribution is uneven and highly irregular [1]. Knowledge of the chemical content in an investigated system or object can provide solutions to various environmental problems. Soil is a very important subject of investigation related to this problem [2-5].

The technogenic transformation of a natural environment is the change of environmental compartments and natural systems caused by industrial activity. The transformation involves changing the biosphere and its components (ecosystems) under the influence of geochemical processes, and comprise human, technical, and technological activities for extracting, concentrating, and relocating minerals and organic compounds [6,7].

The entry of heavy metals (HMs) into the environment is driven by human activity, and the mining industry is one industry that pollutes the environment with heavy metals.

* Corresponding author: aea_eco@mail.ru

Compounds with heavy metals contaminate air, water, and soil, and can enter the bodies of plants and animals [8].

Explored mineral deposits are actual sources of environmental pollution [9]. Metalliferous deposits, which contain set of components controlled in human living environments, pose a potential threat for the environment. These components are extracted incompletely, and the residuals are piled in waste dumps [10].

Soil serves as a good scavenger for HMs, and occupies a significant niche in the cycle of pollutants in the biosphere [11]. It is already known that metals accumulate in soil in a relatively short period, and they are withdrawn from the soil at an extremely slow rate [12,13].

The concentration of HMs is affected by soil properties. Heavy textured soils generally have higher HM concentration, while sandy and sandy loam soils accumulate HMs to a lesser extent. Acid-base soil properties have a significant influence on these characteristics. In acidic conditions, insoluble parts of the HM fraction becomes soluble, resulting in an increased concentration of HMs in acidic soils [14].

The Sverdlovsk region is one of the oldest mining regions in Russia and one of the largest in the Russian Federation in terms of proven and projected ore mineral reserves. This resulted in an intense growth in mining (including gold mining), iron, steel, and non-ferrous metal industries, as well as construction and chemical industries [15].

Mining and smelting waste contains a significant quantity of non-ferrous, ferrous, rare, and trace metals. This waste represents a technogenic material that was piled and accumulated because there were no cost effective and environmentally friendly technologies for its utilization and recycling. During permanent storage of mining waste, several processes occur, such as geochemical transformation with the generation of new technogenic minerals, depletion of valuable metals, removal of elements out of the storage sites, and environmental pollution events [16].

The aim of the research was to reveal special aspects of soil pollution in the Mid-Ural ore deposit. The characteristics of soil cover are given, using the results of a summer field survey with soil type recognition and soil sampling for chemical analysis.

2 Materials and methods.

A field investigation of soil cover was conducted using the methodological recommendations on revelation of degraded and contaminated lands [17] and the instructional guidelines on risk assessment of soil contamination with chemical substances [18]. Typical terrain compartments were selected during a route survey where soil pits down to bedrock were located.

Soil sampling took place for assessment of quality and the level of pollution. Sample plots (20–25 m^2) were laid out according to relief, geomorphological, and landscape features. Soil was sampled incrementally using an envelope method from a depth of 0.0–0.3 m. The samples were then mixed together, giving a composite sample with a weight of at least 1 kg.

Chemical contamination of soil was evaluated using the Total Chemical Pollution Index, which serves as an indicator of harmful health impact. The following indices were applied for the control of soil state: total microelement content (V, Mn, Ni, Cu, Zn, As, Cd, and Pb), pH, total humus, hydrolytic and exchangeable acidity, exchangeable hydrogen and Al, exchangeable base (Ca, Mg, and S-value), degree of base saturation, labile phosphorus, and particle-size distribution.

Calculations related to the local geochemical background were conducted according to a guideline [19-22] as well as other specifications and literature sources [23,24]. For this purpose, the indices of microelement content in soil sampled within the Orefield were used. The fragments in the area with obvious anomalies were excluded from consideration. For instance, the content of an element that is vastly different from the typical value or fragments with technogenic pollution known beforehand would be excluded. The data from the remaining area, which could be characterized as an area with a mainly normal geochemical field were used in an element content sample compilation. This sample was laid on the basis of a local geochemical background evaluation.

A typical element content is defined by an average background content (C_b) or geochemical background. A mean, C, is taken as C_b (normal probability law). For determination of the element content deviation in background samples from an average background value, a standard (mean-square) deviation value is used if the Gaussian law is considered.

For determination of a least value of chemical element content, which could be taken as exceeding background fluctuation limits with some probability, we used an anomaly threshold value. This parameter combines the background level with the standard deviation in case of normal distribution ($C_a = C_b + tS$). The probability that some value of this parameter could be exceeding background fluctuations depends on a t coefficient. We used t =1 in the scope of this research.

General description of soil cover. The mountain group located in the eastern slope of the Ural Mountains in the northern part of the Middle Ural was investigated (**Figure 1**).

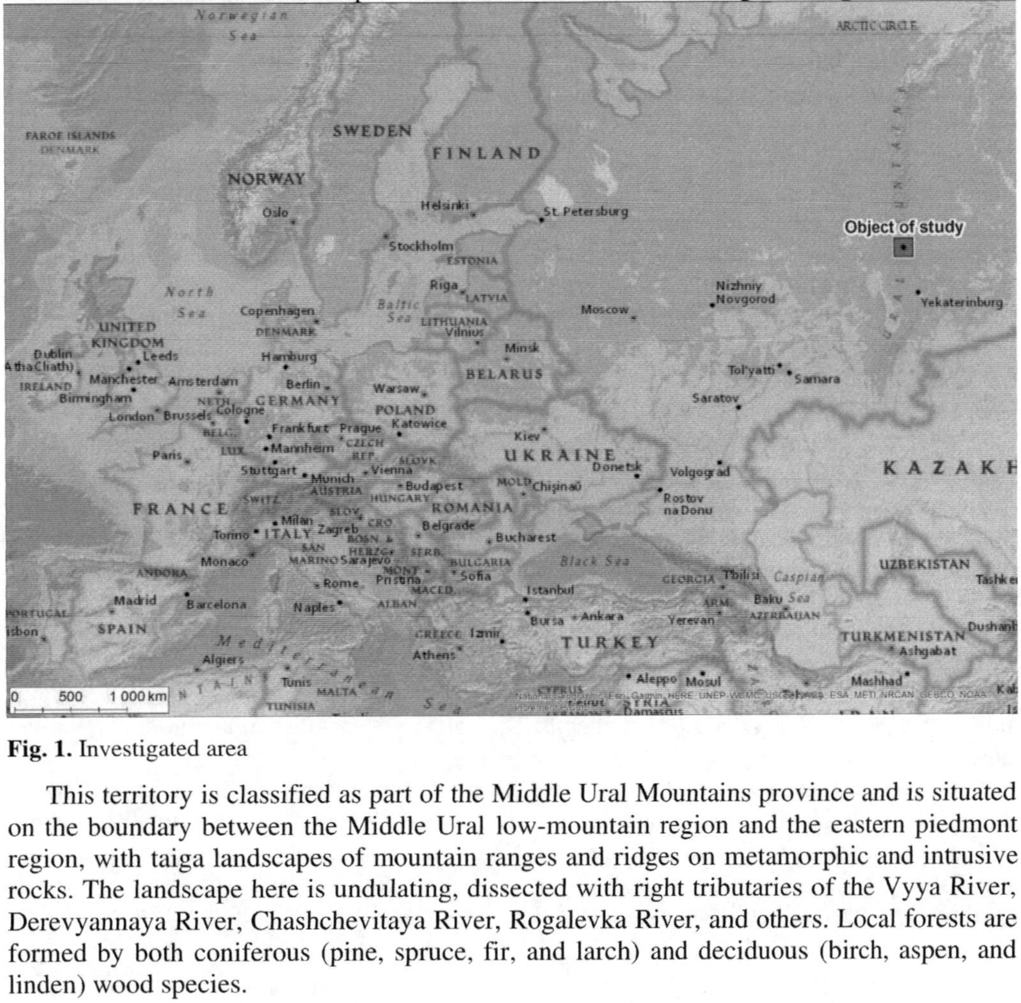

Fig. 1. Investigated area

This territory is classified as part of the Middle Ural Mountains province and is situated on the boundary between the Middle Ural low-mountain region and the eastern piedmont region, with taiga landscapes of mountain ranges and ridges on metamorphic and intrusive rocks. The landscape here is undulating, dissected with right tributaries of the Vyya River, Derevyannaya River, Chashchevitaya River, Rogalevka River, and others. Local forests are formed by both coniferous (pine, spruce, fir, and larch) and deciduous (birch, aspen, and linden) wood species.

The soil cover of the area is characterized by heterogeneity caused by altitudinal zonation, diversity of rocks, and complexity of the landscape. The soils here are formed on eluvial and deluvial deposits underlaid by unstratified rocks that are gravelly and shallow.

The profile of mountain soils is negligibly differentiated and does not have a well-defined podzolic horizon. Accumulation of organic matter occurs in weakly mineralized underlayers and A1 horizon with dark brown humus spots and tongues. Underlying mass represents evenly coloured (usually dark brown or brownish) loam with debris inclusions.

The thickness of the fine-grained stratum in the profile is usually about 50–60 cm. In the upper horizons, the gravel content is about 25%, and it increases down the profile. The thickness of the fine-grained part of the soil profile decreases down to 30–40 cm or becomes fragmentary in steep slopes where rocks lie near the surface with increasing elevation. Upper parts of the slopes are associated with undifferentiated brown mountain forest soils. The mountain forest cambisols have several specific morphological and physicochemical indices that separate them from podzols and calcisols that are prevalent in the middle taiga zones.

3 Results and discussion

The distribution of the sample plots across the area investigated is shown in Fig. 2.

Soil texture. The given soils could be described as light (sandy loam) and heavy (from heavy loams to light clays) types. The increased content of coarse grains (1.0–0.25 mm and 0.25–0.05 mm) down the profile is a specific feature of all mountain soils. Even distribution or some extent of accumulation of sludge (particles less than 0.01 mm) and physical clay (particles less than 0.001 mm) are clearly observed in the upper part of the profile compared to the mother rock as a result of the weakening of erosion and soil forming process with increasing depth. Thus, the soil forming process increases, and mountain relief conditions are accompanied by argillization of the upper horizons. In this regard, eluvial and deluvial profile differentiation, as for podzolic soils, does not occur.

Physical and chemical properties of soils. The soils of the area are acidic. Changes in acidity with depth is determined by the features of the mother rocks. Acidity of acidic rocks changes negligibly across the profile or increases downward. The upper horizons of all mountain soils have high exchangeable acidity, which is caused by hydrogen presence in organics and by exchangeable aluminium in mineral horizons. The soils also have high values of hydrolytic acidity. In upper organic horizons, these values reach 0.5 mg-equivalent/g and decrease dramatically with depth.

Cambisols differ from calcisols and podzols by specific content of exchangeable bases. These soils have an accumulation of exchangeable bases in the upper organic horizons and a steep decrease in their quantity in the mineral stratum. The minimum of exchangeable bases is consisted in soil-forming rock. The degree of base saturation seems to be constant across the profile, which explains the weak profile differentiation.

Forest cambisols are characterized by a high content of organic matter in the upper horizons and a gradual decrease with depth. The soils contain substantial amounts of humus across the whole profile.

Mountain forest saturated cambisols are close to neutral. They are distinguished by the saturation of absorbed bases, which declines with depth. The soils contain notable amounts of humus and have high hydrolytic acidity in the upper horizons. The upper portion contains more sludge and physical clay, while the lower mineral section is predominated by small and medium-sized sand. Acidic cambisol lessivage do not occur in the mountain forest.

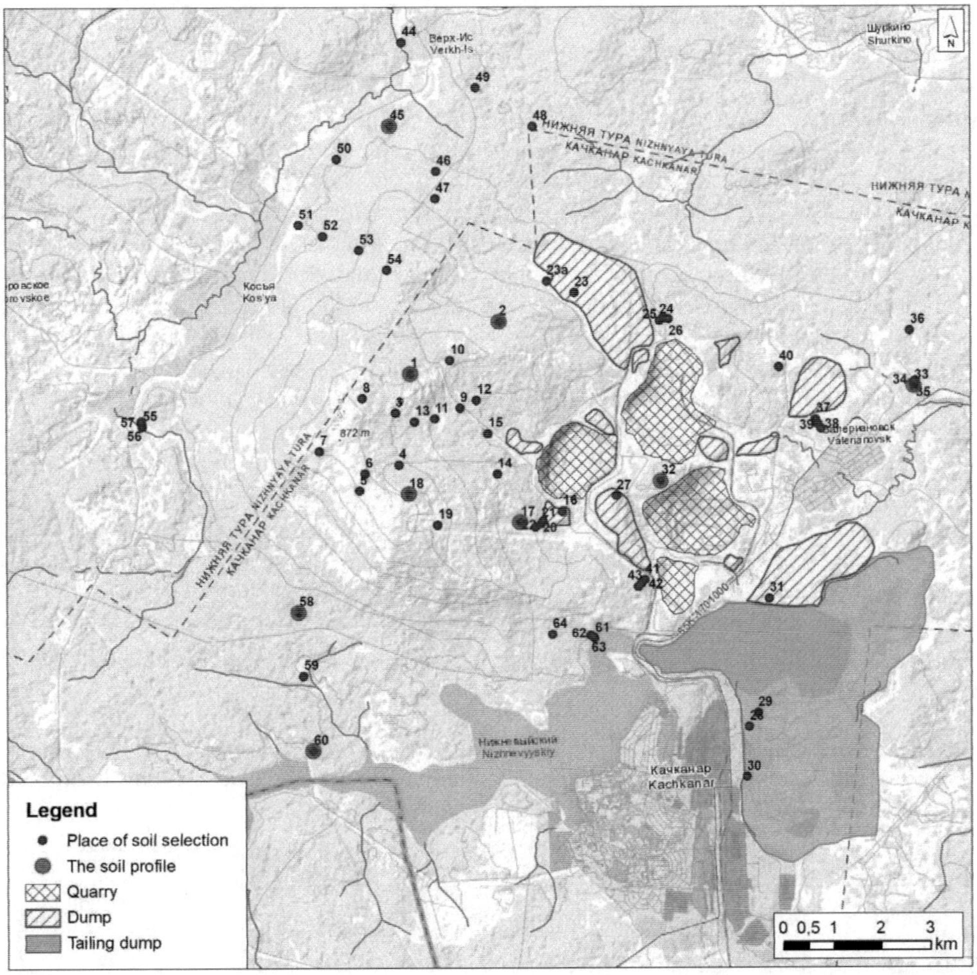

Fig. 2. The distribution of the sample plots (SP)

Low content of labile phosphorous could be noted for all the undisturbed mountain forest cambisols.

Calcisols were determined in the eastern and northern (Verkh-Is River) parts of the area. The soils with podzol features are formed in gentle slopes or poorly drained flat lands that receive additional moisture from water runoff. At heights of 328 m and 214–220 m in dark coniferous forests with moist conditions, gley podzols and calcisols are formed.

Calcisols are specified by high acidity and negligible content of exchangeable bases, accompanied with complete unsaturation of the medium part of the profile. Heavy texture and location within poorly drained lands provide consistent soil moisture, pseudoglazation, and partial accumulation of exchangeable bases relative to the upper part of the profile.

To sum up the data on undisturbed, natural soils of the investigated area, specific aspects of the pedogenic process should be noted. These are the prevalence of cambisols or mountain forest nonpodzolized soils accompanied by limited distribution of podzols. The morphology of the mountain soils is unique—it is negligibly differentiated and highly gravelly. Primitive accumulative soils of the upper parts of the slopes and mountain forest cambisols are formed due to xeromorphous pedogenesis and weathering conditions [25].

The close proximity of underlying dense rocks leads to insignificant soil differentiation and the absence of a clearly visible podzolic horizon. Reduced humidity of the mountain soils is due to water runoff and good rock permeability. These conditions make excessive moistening and reducing conditions impossible and provide for preserving iron in the soil profile. Iron stabilizes soil structure and preserves it from lessivage [26,27]. The intensity of weathering increases from the upper to lower relief components. Increases in moisture and soil depth down the slope lead to podsolization of forest cambisols. Such soils lose sandy fractions while the amount of coarse silt particles (0.05–0.01 mm) increases significantly. The distribution of clay fractions shows podzolic features, with their amounts decreasing in the podzolized horizon and increasing in the enrichment horizon. As the impact of mother rock on the soil disappears, it becomes similar to zonal calcisols [25].

The forest growth conditions could be considered to be quite favorable. The absence of podsolization and the close location of soil-forming rocks, often crystalline, provide considerable amounts of nutrients, while the strong structure of the soils creates good air and water conditions. However, significant earth grades and deforestation can create the conditions for severe erosion and rapid loss of all the soil cover.

Most of the published papers with data for background content of microelements in soil were based only on the research of zonal soils. Besides, the investigated area is characterized as the area of a geochemical anomaly since the large mineral deposits, in particular iron ores with high vanadium content, have been found here.

Based on the above considerations, there is a need for the determination of a local background. The local background is the average modal value of the geochemical index (or indices) of a given geological space (geological feature) and statistically acceptable interval of its variation [23].

The soils of the area were thoroughly sampled for the determination of the background values of microelement content. The results of the calculations are given in Table 1. The data is shown for the elements, with application in the calculation of Total Chemical Pollution Index.

Table 1. Calculations of the geochemical background values in soils

Chemical element	Maximum value (mg/kg)	Minimum value (mg/kg)	Background value (mg/kg)	Standard deviation (mg/kg)	Variation coefficient (%)	Minimum anomaly value (mg/kg)
V	262.2	130.6	219.4	40.6	0.2	259.9
Mn	995.5	339.2	674.4	183.1	0.3	857.5
Ni	54.4	19.1	34.8	10.7	0.3	45.5
Cu	56.9	17.7	29.3	12.4	0.4	41.6
Zn	119.2	52.5	80.3	17.8	0.2	98.1
As	12.0	5.1	7.7	2.0	0.3	9.7
Cd	1.3	0.3	0.7	0.3	0.4	1.0
Pb	64.5	13.3	29.3	17.3	0.6	46.6

The maximum values for the content of the considered elements exceed the background values (Fig. 3). All the studied sample plots in the area of iron ore deposit were combined into 9 groups according to plant communities. For each of these groups, a geochemical series was constructed. We distinguish 5 groups according to the similarity of the geochemical series. Group 1 contains plot soils with mixed geochemical signature, where lithophiles (Mn and V), chalcophiles (Zn and Cu), and siderophiles (Ni) are prevalent. The mixed forest (SP 15, 17) with the series $\frac{Mn}{707} > \frac{V}{353} > \frac{Zn}{83} > \frac{Ni}{58} > \frac{Cu}{54} > \frac{Pb}{25} > \frac{As}{10} > \frac{Cd}{0.7}$ and the sparse spruce-birch forest (SP 19) with $\frac{Mn}{996} > \frac{V}{262} > \frac{Zn}{77} > \frac{Ni}{44} > \frac{Cu}{19} > \frac{Pb}{13} > \frac{As}{5} > \frac{Cd}{0.5}$ series fall under this group.

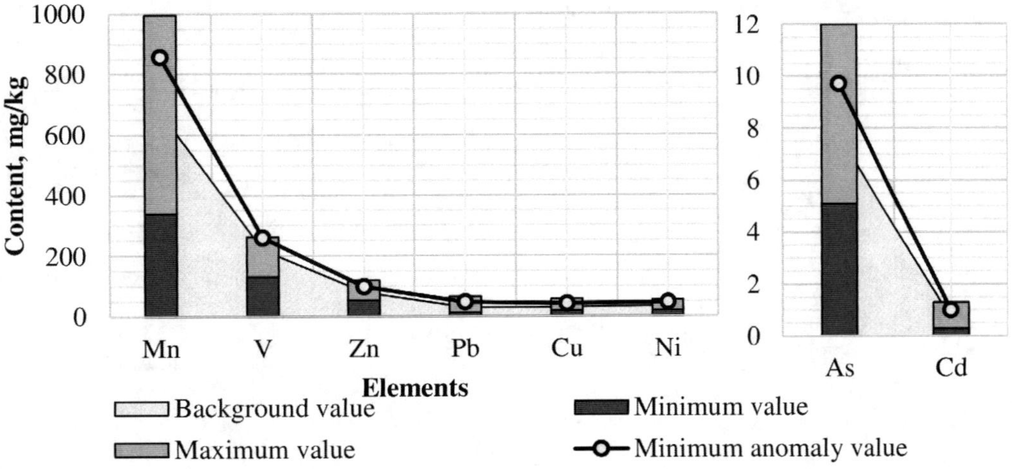

Fig. 3. The geochemical background values in soils

The geochemical series and mixed geochemical signatures of Group 2 and Group 1 are similar, but here we can see a greater content of Pb than Cu. For the spruce-birch krummholz (SP 6, 8) the following series is characterized: $\frac{Mn}{860} > \frac{V}{259} > \frac{Zn}{75} > \frac{Ni}{39} > \frac{Pb}{22} > \frac{Cu}{19} > \frac{As}{6.3} > \frac{Cd}{0.6}$. While the sparse coniferous forest (SP 18) has the following series: $\frac{Mn}{860} > \frac{V}{259} > \frac{Zn}{75} > \frac{Ni}{39} > \frac{Pb}{22} > \frac{Cu}{19} > \frac{As}{6.3} > \frac{Cd}{0.6}$.

Lithophiles (Mn and V) and chalcophiles (Pb and Zn) are prevalent in group 3. It should be noted that Mn content significantly varies from 371 mg/kg to 741 mg/kg. The soils of the spruce forest (SP 4) with the following series: $\frac{Mn}{371} > \frac{V}{131} > \frac{Zn}{98} > \frac{Pb}{56} > \frac{Ni}{39} > \frac{Cu}{31} > \frac{As}{13} > \frac{Cd}{1.3}$ have minimal content of Mn. The content of the elements is higher in the spruce-birch mixed forest (SP 14) with $\frac{Mn}{741} > \frac{V}{244} > \frac{Zn}{132} > \frac{Pb}{63} > \frac{Ni}{47} > \frac{Cu}{40} > \frac{As}{12} > \frac{Cd}{1.6}$ series, and the dark coniferous green moss forest (SP 9, 11–13) with $\frac{Mn}{643} > \frac{V}{209} > \frac{Zn}{92} > \frac{Pb}{44} > \frac{Ni}{32} > \frac{Cu}{31} > \frac{As}{9} > \frac{Cd}{0.9}$ series.

The spruce green moss forest forming group 4 has the following series: $\frac{Mn}{660} > \frac{V}{235} > \frac{Zn}{69} > \frac{Cu}{40} > \frac{Ni}{25} > \frac{Pb}{20} > \frac{As}{7} > \frac{Cd}{1}$. The sparse birch forest (group 5, SP 2), which has the following series: $\frac{V}{237} > \frac{Mn}{229} > \frac{Zn}{53} > \frac{Ni}{19} = \frac{Cu}{19} > \frac{Pb}{16} > \frac{As}{6} > \frac{Cd}{0.3}$, where V content is greatest but not exceeding Mn content in this site`s samples, has shown the biggest difference from other groups.

We calculated the Total Chemical Pollution Index for the territory of the iron ore deposit where pollutants exceed background values (Fig.4). This index was assessed relative to the background content of the elements in Ural soils (Z_c) [25].

As for the background content, the Ural soils demonstrate the following: we can observe excess content in one sampling plot (spruce-birch mixed forest), while some plots` content approach but do not exceed an allowable level (spruce forest, mixed forest, dark coniferous green moss forest).

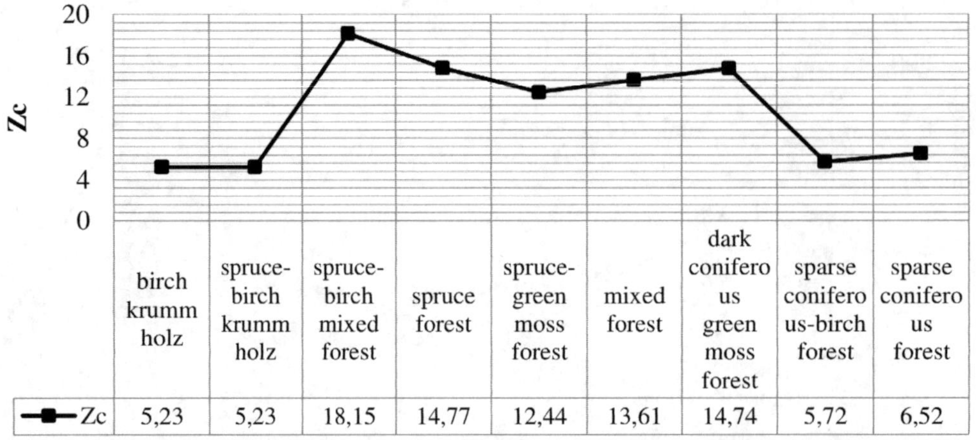

Fig. 4. Total heavy metal chemical pollution of soils from the iron ore deposit

The most important part of the research was the determination of the impact of mining waste dumps on soils. The heavy metal content in soils of sampled plots is shown in Fig. 5–9. The element content values were transformed to one order by raising to the power (the content of V, Ni, Cu, and Zn is equal to X mg/kg×10-1; the content of Mn is X mg/kg×10-2, Cd – X mg/kg×101; the content of As and Pb didn't change).

Fig. 5 shows the content of heavy metals in soils affected by mine waste dump 6 with the following series: $\frac{Mn}{623} > \frac{V}{259} > \frac{Cu}{112} > \frac{Zn}{73} > \frac{Ni}{47} > \frac{Pb}{16} > \frac{As}{6.1} > \frac{Cd}{0.5}$.

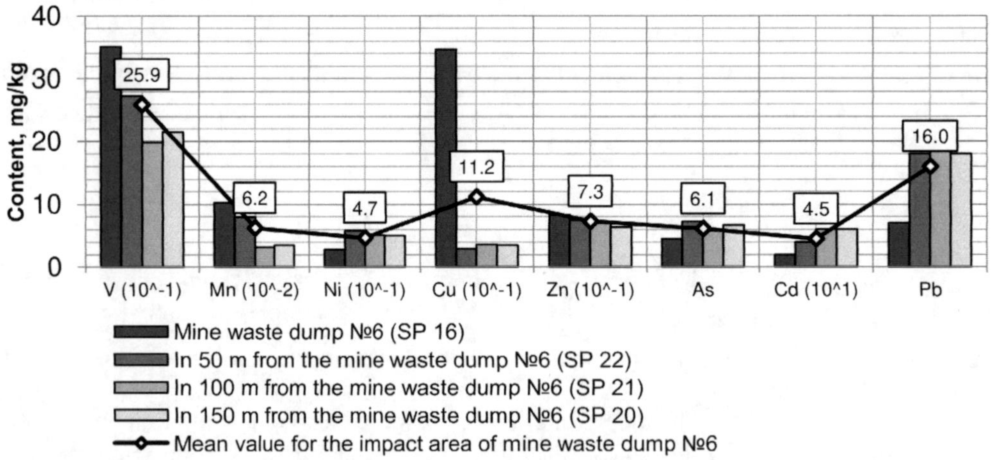

Fig. 5. Heavy metal content in soils affected by mine waste dump Nr. 6

We can observe the decrease in the content of V, Mn, and Zn with increasing distance between sample plots and the mine waste dump (50, 100, and 150 m), as shown in Fig. 3. High content of Cu is noted only in the plot on the territory of the mine waste dump itself. Ni, As, Cd, and Pb tend to increase in content with increasing distance from the dump. We can state, therefore, that the impact of mine waste dump 6 leads to an increase in lithophile (Mn and V) and chalcophile (Zn and Cu) concentrations.

Fig. 6 shows the content of heavy metals in soils affected by mine waste dump 4 with the following series: $\frac{Mn}{623} > \frac{V}{259} > \frac{Cu}{112} > \frac{Zn}{73} > \frac{Ni}{47} > \frac{Pb}{16} > \frac{As}{6.1} > \frac{Cd}{0.5}$.

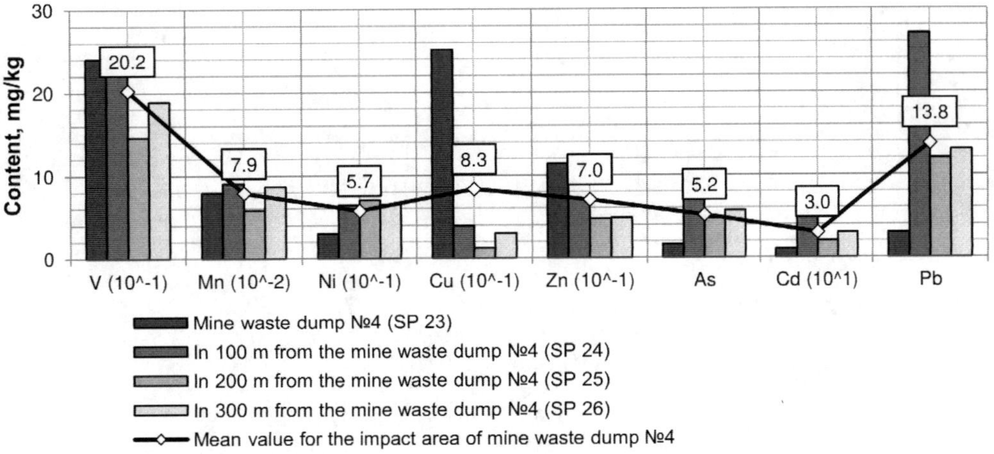

Fig. 6. Heavy metal content in soils affected by mine waste dump Nr. 4

The pattern of heavy metal distribution relative to mine waste dump 4 shown in Fig. 6 is similar to that for dump 6 (Fig. 5). With increasing distance from the mine waste dump, a reduction in V, Mn, Cu, Zn, and Pb content is observed. The greatest difference is seen for Cu.

Fig. 7 shows the data for heavy metal content in several sampling plots, including mine waste dump 1, 7, and the site that is 200 m from the quarry. Geochemical patterns for the soils of the dumps 1 and 7 are different only in the content of Ni and Zn.

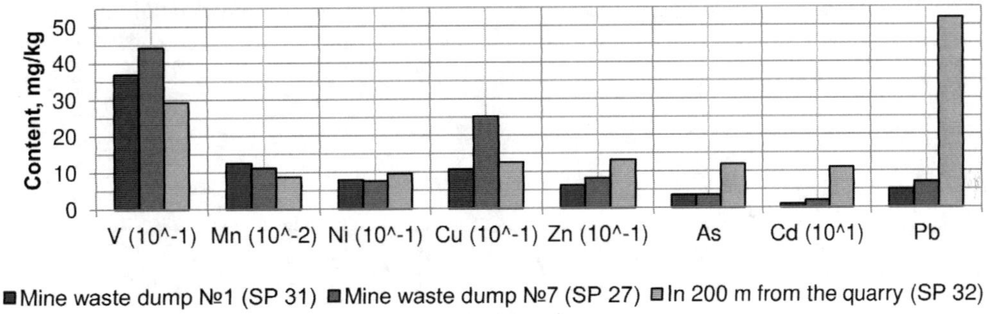

Fig. 7. Heavy metal content in soils affected by mine waste dumps Nr.1 and Nr.7 (soils taken 200 m from the quarry)

The following two geochemical series for the impact area of mine waste dump 1 and for mine waste dump 7 have mixed character in general, with prevalence of lithophiles (Mn and V) and chalcophiles (Zn and Cu): $\frac{Mn}{1253} > \frac{V}{369} > \frac{Cu}{107} > \frac{Ni}{79} > \frac{Zn}{63} > \frac{Pb}{5} > \frac{As}{3.5} > \frac{Cd}{0.1}$ and $\frac{Mn}{1119} > \frac{V}{442} > \frac{Cu}{252} > \frac{Zn}{82} > \frac{Ni}{76} > \frac{Pb}{7} > \frac{As}{3.5} > \frac{Cd}{0.2}$.

The distribution of heavy metals in soils affected by mine waste dump 2 is shown in Fig. 8. As the distance from the mine waste dump increases, a reduction in V, Mn, Ni, and Zn is observed. The content of Cu, As, Cd, and Pb tends to increase as the distance from the mine waste dump increases.

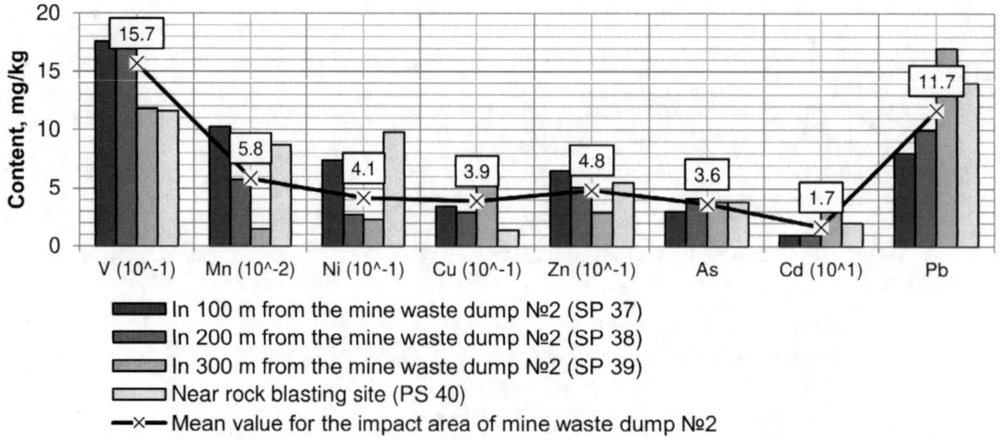

Fig. 8. Heavy metal content in soils affected by mine waste dump Nr.2

The following geochemical series for the impact area of mine waste dump 2 has primarily lithophiles, but their content is significantly less than in the impact areas of other dumps: $\frac{Mn}{583} > \frac{V}{157} > \frac{Zn}{48} > \frac{Ni}{41} > \frac{Cu}{39} > \frac{Pb}{12} > \frac{As}{3.6} > \frac{Cd}{0.2}$. It should also be noted that there is a steep increase in the content of several elements near the rock blasting site. The geochemical series for this area is the following: $\frac{Mn}{871} > \frac{V}{116} > \frac{Ni}{98} > \frac{Zn}{55} > \frac{Cu}{14} = \frac{Pb}{14} > \frac{As}{3.8} > \frac{Cd}{0.2}$. Here, we can see the large content of lithophiles and chalcophiles, in particular Mn, Ni, and Zn.

The southern deposit (Fig. 9) does not affect accumulation of heavy metals in soils in a notable way. The content of all the elements decreases as the distance from the deposit increases. The following series $\frac{Mn}{535} > \frac{V}{176} > \frac{Zn}{49} > \frac{Ni}{41} > \frac{Cu}{18} > \frac{Pb}{13} > \frac{As}{4.1} > \frac{Cd}{0.2}$ for the impact area of the southern deposit is lithophilous, with Mn and V most prevalent: $\frac{Mn}{535} > \frac{V}{176} > \frac{Zn}{49} > \frac{Ni}{41} > \frac{Cu}{18} > \frac{Pb}{13} > \frac{As}{4.1} > \frac{Cd}{0.2}$.

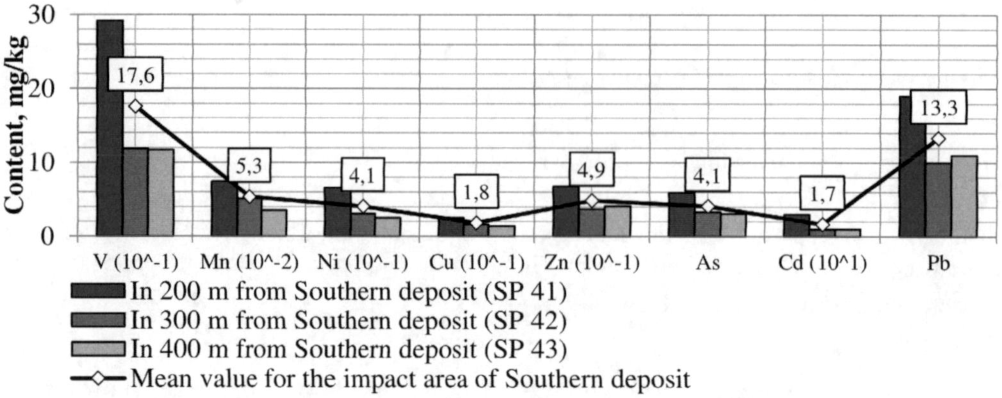

Fig. 9. Heavy metal content in soils affected by of Southern deposit

The total pollution of soils with heavy metals (Fig. 10) exceeds the permissible level (16) in the territory under the influence of the quarry (200 m from the quarry), in the territory of mine waste dumps 6 and 7. We can suggest maximum levels of heavy metal contamination in the sites closest to mine waste dumps, and there is a decrease in those levels when moving away from these dumps.

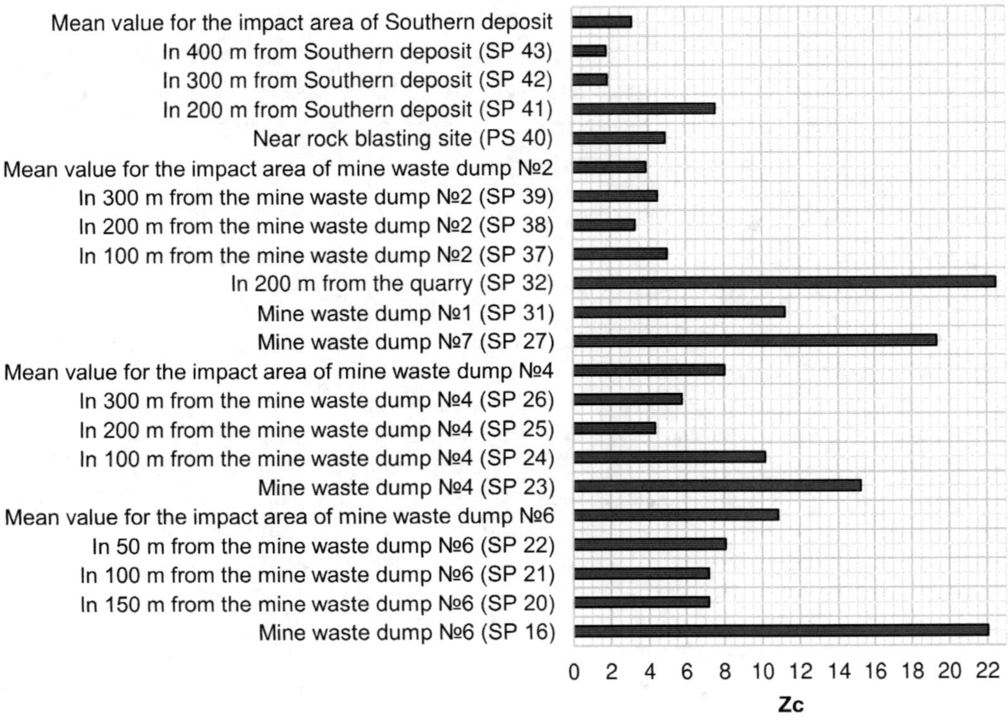

Fig. 10. Total chemical pollution of soils with heavy metals

Fig. 11 shows the content of heavy metals affected due to gold mining. We note a steep increase in heavy metal content compared to the other mine waste dumps. The following average series for the gold mining affected territory has mixed accumulation patterns with prevailing lithophiles and siderophiles: $\frac{Mn}{2624} > \frac{Ni}{650} > \frac{V}{143} > \frac{Zn}{64} > \frac{Cu}{42} > \frac{Pb}{11} > \frac{As}{9} > \frac{Cd}{0.2}$.

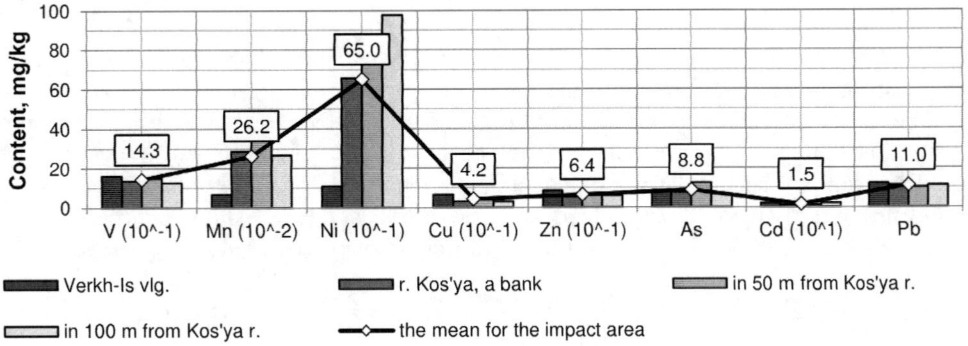

Fig. 11. Heavy metal content in soils affected by gold mining

The largest level of total soil pollution with heavy metals (Fig. 12) was found near the Kosya River (50 and 100 m away). Exceedance of this index was not revealed in the site near. Verkh-Is settlement, while other sites were found to be far beyond that level.

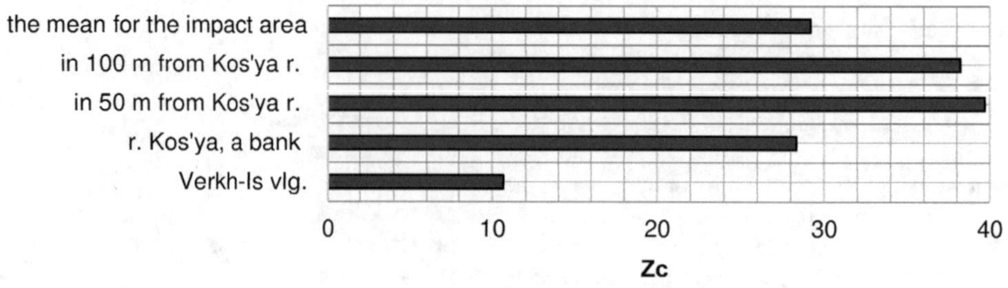

Fig. 12. Total heavy metal chemical pollution of soils affected by gold mining

The analysis of the influence of gold mining on the elementary composition of soil revealed that the content of Mn and Ni increased considerably.

We also discovered moderately hazardous pollution of Ural soils with V, Ni, Zn, and As, which is connected with an increase in their natural levels.

In addition to samples from natural soils, the samples from disturbed lands were also assessed. The largest total pollution index was observed for dumps formed by dredging operations of gold mining in the past (gold mining waste dumps). These dumps are unremediated windrows with heights of 1.5 m. The surface of the windrows is partially covered with grass. A fine-grained soil with 20 cm thickness accumulated in the swale features. Today, these areas are occupied by secondary mixed forests.

Moderately hazardous pollution was also observed in the grounds of the overburden waste dumps 4 and 6, which were formed as the result of mining of the deposit.

We found an insignificant exceedance of the MAC (Minimal Allowable Level; less than 2 MAC) for Mn and of the TAC (tentative allowable concentration; less than 2 TAC) for Cu, Zn, Cd, and Pb in natural soils, which is caused by natural features of the area. The natural background for V and As exceeds the MAC, which led to the exceedance of the regulation values of V in all samples and As in the majority of samples. The natural background is close to the TAC value for Ni, which is the reason why the regulation value was exceeded in a large amount of samples.

4 Conclusion

The mining of iron ore deposits occurs in territories where cambisols are prevalent (mountain and forest cambisols) or in mountain and forest non-podzolized soils with limited distribution of podzols. The morphology of mountain soils is unique—they are negligibly differentiated or differentiated in a way that they cannot be clearly distinguished, and they are highly gravelly. Primitive accumulative soils of the upper parts of the slopes and mountain forest cambisols are formed due to xeromorphous pedogenesis and weathering conditions.

The average geochemical series for all the phytocenosises could be presented as the following:
$\frac{Mn}{676} > \frac{V}{242} > \frac{Zn}{83} > \frac{Ni}{36} > \frac{Cu}{33} > \frac{Pb}{32} > \frac{As}{8} > \frac{Cd}{0.7}$. This series is characterized by high concentrations of V, Zn, Ni, Pb, and As, and quite low concentrations of Mn. Gold mining of the area creates the following mixed geochemical series with high content of Mn and Ni across the territory of the deposit: $\frac{Mn}{2624} > \frac{Ni}{650} > \frac{V}{143} > \frac{Zn}{64} > \frac{Cu}{42} > \frac{Pb}{11} > \frac{As}{9} > \frac{Cd}{0.2}$.

Z_c values in soils around the iron ore deposit are changing from permissible to moderately hazardous. It was revealed that regulation values of V, As, and Ni were exceeded, which could be explained by the increased geochemical background in the area. An increased Cd and Pb content relative to background was observed for Ural soils near the quarry.

References

1. V.I. Vernadsky, *Chemical Structure of the Earth's Biosphere and its Environment* (Nauka, Moscow, 2001) (in Russian)
2. V.A. Alekseenko, *Ecological Geochemistry* (Logos, Moscow, 2000) (in Russian)
3. V.A. Alekseenko, A.V. Suvorinov, V.A. Alekseenko, & A.B. Bofanova, *Metals in the Environment. The Soils of Geochemical Landscapes of Rostov Region* (Logos, Moscow, 2002) (in Russian)
4. V.A. Alekseenko, S.A. Buzmakov & M.S. Panin, *Geochemistry of Environmental: monograph for schools* (Perm state university, Perm, 2013) (in Russian)
5. S.A. Buzmakov, Y.V. Khotyanovskaya, Applied Geochemistry, **113**, 104443 (2020)
6. S.A. Buzmakov Vestn. Perm. Gos. Univ., Biol., **2**, 133–138 (2004). (in Russian)
7. S.A. Buzmakov, Geographical bulletin, **4(23)**, 46–50 (2012). (in Russian)
8. L. Angelovičová, L. Bobuľská & D. Fazekašová, Carpathian J. of Earth and Environmental Sciences, **10(1)**, 193–201 (2015)
9. A. Kelepertsis, A. Argyraki,, & D. Alexakis, Geochemistry: Exploration, Environment, Analysis, **6(4)**, 349–355 (2006)
10. V.I. Golik, O.N. Polukhin, A.N. Petin & V.I. Komachshenko, Mining J., **4**, 91–98 (2013). (in Russian)
11. R.A. Wuana, & F.E. Okieimen, ISRN Ecology, **2011**, 1-21 (2011).
12. Adriano, D.C., *Trace Elements in Terrestrial Environments: Biogeochemistry, Bioavailability and Risks of Metals* (Springer, New York, 2003)
13. G.A. Teplaya, Astrakhan bull. of ecol. education, **1(23),** 182-192 (2013). (in Russian)
14. V.V. Ivanov *The geochemistry of the dispersed elements Ga, Ge, Cd, In and TI in hydrothermal deposits* (Nedra, Moscow, 1966)
15. S.E. Deryagina, O.V. Astaf'eva & A.N. Medvedev, Nature management economics, **1**, 24-30 (2013) (in Russian)
16. V.I. Golik, V.I. Komachshenko & A.V. Polikov, Izvestiya of the Tula State University. Sciences of Earth., **1**, 100–110 (2016) (in Russian)
17. G. A. Vostroknutov, *Temporary methodological guideline on conducting geochemical surveys during geo-ecological works* (Ekaterinburg, 1991) (in Russian)
18. Yu.E. Sayet, B.A. Revich & E.P.Yanin, *Environmental Geochemistry* (Nedra, Moscow, 1990) (in Russian)
19. R.S. Zubareva & V.P. Firsova *On soil characteristics in spruce forests of the mountain band in the Middle Ural region. Soils and hydrologic regime of Ural forests* (UF AN SSSR, Sverdlovsk, 1963) (in Russian)
20. *Methodological recommendation for identification of degraded and polluted lands*, (Russian Federal Environmental Informational Agency, Moscow, 1996) (in Russian)
21. *SanPiN 2.1.7.1287-03. Sanitary and Epidemiological Requirements with Respect to Soil Quality* (Moscow, 2003) (in Russian)
22. *SanPiN 4266-87. Methodological instructions for assessing the degree of danger of soil contamination with chemicals* (Moscow, 1987) (in Russian)
23. S.V. Grigoryan, A.P. Solovrv & M.F.Kuzin, *Instruction on Methods of Geochemical Prospecting of Ore Deposits* (Nedra, Moscow, 1983). (in Russian)
24. B.A. Lebedev, *Agrochemical characteristics of the USSR soils* (Nauka, Moscow, 1964) (in Russian)
25. A.P. Solovov & A.A. Matveev, *Geochemical prospecting methods of ore deposits* (Moscow State University Publ., Moscow, 1985) (in Russian)
26. V. P. Firsova *Soils of Sverdlovsk oblast and their changes under the influence of forestry practices. Institute of plant and animal ecology*, (RAS, Sverdlovsk, 1969. (in Russian)
27. I.S. Urusevskaya, Eurasian Soil Science, **40(11)**, 1145-1157 (2007). (in Russian)

E3S Web of Conferences **169**, 01015 (2020) https://doi.org/10.1051/e3sconf/202016901015
APEEM 2020

Identification of the state of the soil-plant systems on the RUDN-University campus (based on PAH concentrations)

Aleksandr Khaustov[1,*], *Margarita* Redina[1], Zhandos Kenzhin[1], *Dmitry* Gabov[2] and *Evgeniya* Yakovleva[2]

[1]Peoples' Friendship University of Russia (RUDN University), Faculty of Ecology, 6 Miklukho-Maklaya St, Moscow, 117198, Russian Federation
[2]Institute of Biology of Komi Science Centre of the Ural Branch of the Russian Academy of Sciences, 28 Kommunisticheskaya st., 167982 Syktyvkar, Komi Republic, Russia

Abstract. The environmental state of soil-plant systems under the condition of technogenic pressure has been demonstrated on the example of the social significant territory in the city of Moscow – the campus of the RUDN University. The assessment is based on the analyses of the environmental fate of polycyclic aromatic hydrocarbons (PAH) in soil, roots and aerial parts of plants of bluegrass meadow: Naphthalene (Naph), Fluorene (Fluorene), Phenanthrene (Phen), Anthracene (An), Fluoranthene (Flu), Pyrene (Py), Chrysene (Chr), Benzo[a]anthracene (BaA), Benzo[a]pyrene (BaP), Benzo[b]fluoranthene (BbFlu), Benzo[k]fluoranthene (BkFlu), Benzo[ghi]perylene (Bghi), Indeno[1,2,3-cd[pyrene (IP), Dibenzo[a,h]anthracene (DBa). These pollutants are considered as geochemical markers and reflect quite detailed the integral state of soil-plant systems. As a result, the zones with relative prosperous and with critical state are identified. The connection between physicochemical characteristics (log K_{ow}) of pollutants and their accumulation activity (concentration coefficients) is confirmed.

1 Introduction

Urban soil-plant systems are objects of a whole range of technogenic impacts due to the activity, primarily, of industrial and transport pollution sources. Under Moscow conditions, it is transport loads that prevail in the formation of environmental pollution. However, the influence zones of the transport routes depend both on the actual activity of the transport and on the specifics of the polluted systems and the characteristics of the emission components. In this regard, local pollution models are of great scientific interest and are characterized by significant diversity. To identify the zones of influence of technogenic sources, it is necessary to apply the reliable markers that allow to clearly attribute the studied object to a certain type of pollution

Plants play an important role in the accumulation and conversion of pollutants

[*] Corresponding author: khaustov-ap@rudn.ru

(including hydrocarbons) in environmental components. This became the basis for the development of methods and technologies for phytoremediation. The experiments [4] established a positive effect of plants on soil cleaning, but it is believed that the effect of using plants is low with respect to PAHs — a decrease in polyarenes of 0.8–14 times depending on the experiment time, types of vegetation, and concentrations of PAHs introduced. The greatest effect was recorded for pyrene and phenanthrene during two-month observation of remediation by various plants. For all types of phytoremediation, which leads to a natural decrease in PAHs in soils (phytoextraction, phytofiltration, phytovolatilization, rhizoremediation, phytostabilization, phytodegradation), their assimilation (sorption) by vegetation is necessary. The main processes leading to the penetration of hydrocarbons into plant organisms are phytofiltration and rhizomediation, which control the movement (albeit weak due to the high hydrophobicity of hydrocarbons) of solutions in the soil – root – stem system [4]. A significant role is given to the destruction of PAHs in the rhizosphere with the participation of microbial communities, that is, plants themselves are only a part of the general system of PAH decomposition in soils [2, 6]. When predicting these processes, it is necessary to take into account the heterogeneity of hydrocarbons distribution in the soil, the hydrophobicity of pollutants, the growth of vegetation, the possibility of PAHs coming from the atmosphere into plants, as well as the natural generation of PAHs by plant organisms themselves. The role of these factors for soil-plant systems with different levels of technogenic loads can be redistributed. Thus, the property of hydrophobicity of PAH compounds (expressed by the distribution coefficient in the octanol-water system $log\ K_{ow}$) does not always uniquely determine the behavior of PAHs in the environment due to the specificity of technogenic pressure on these media. Let us demonstrate this by the example of soil-plant systems on the campus of the Peoples' Friendship University of Russia.

2 Research methods and objects

The object of research in this article is soil-plant systems on the campus of the Peoples' Friendship University of Russia. The territory is located in a relatively prosperous area of Moscow, however, a major transport highway (Leninsky Prospekt) and streets quite loaded with vehicles are adjacent to it. Since 2017, environmental monitoring of the campus has been organized, covering all environmental components. As of today, about 4000 determinations of pollutants in air, snow cover, soil, vegetation have been carried out at 33 points of the monitoring network; acoustic load levels, electromagnetic fields and radiation background are estimated. These works provide reliable and complete assessments of the state of the environment.

Particular attention is paid to the identification of *marker substances and priority pollutants in the environmental components*. PAHs, which are geochemical markers of natural and man-made processes, can be represented as such compounds [1, 2]. PAH molecules contain 2 or more aromatic rings in their structure, have a relatively high resistance in the environment and low solubility in water, are lipophilic and can cause mutagenic and carcinogenic effects. Analysis of PAH concentrations in environmental objects allows to associate pollution to a certain source: a natural or man-made process that occurs at elevated temperatures or under "normal" conditions. Most often, for the purpose of identifying sources, indicator ratios of PAH concentrations are used: the ratio of 2-3-ring to 4-5-ring sums, or the ratio of isomer concentrations within a single molecular weight.

For the purpose of this study, we analyzed the contents of 14 PAHs in the soils, roots, and aerial parts of plants of meadow bluegrass (Poa praténsis) on campus. The analysis was carried out at the Institute of Biology, Komi Scientific Center, Ural Branch of the Russian Academy of Sciences. A detailed description of the technique is presented in [2].

Previously, on the basis of our earlier observations, we divided the territory into a conditionally clean zone (a "background" zone in a forest park, a recreation zone) and a territory with an increased level of pressure. An analysis of the spatial distribution of PAHs in soils, roots, and aerial parts of plants allows a more detailed assessment of the area of influence of vehicles on the territory. First of all, estimates of PAH migration in the soil – root – stem chain were carried out: - concentration coefficients (CC) were calculated as the ratio of PAH concentration in the "receiving" medium to the concentration in the "giving" one. With QC> 1, PAH transition is claimed. CC calculations were carried based on the PAH analysis data from 33 monitoring points; then estimates of PAH migration activity for various campus areas were obtained. Based on the CC values for the identified individual PAHs, using the cluster analysis procedure (Ward's method, Euclidean distances), the "soil – root" and "root – stem" medium borders were classified. This allows us to compare the activity of PAHs destruction under the influence of the "rhizosphere effect" - to correlate the activity of destruction under the influence of microorganisms living in the rhizosphere to the activity of this process without the participation of microorganisms, by the "forces" of plants alone. As a rule, such affected areas with minimal activity of microbiota are identified in the immediate vicinity of the loaded traffic routes. The differences in these environments are determined by the fact that the number of hydrocarbon-oxidizing bacteria in the rhizosphere is several times higher than in soils without plants.

The self-purification mechanisms of soils are of a staged nature, associated with the seasonal development of vegetation and, accordingly, soils, which are considered as a medium for a bacterial filter that retains and transforms PAHs. According to published data, the bioconcentration of PAHs by vegetation from soils varies from 10^{-6} to 16, and for individual polyarenes it can reach 30–40 [4].

The most important characteristic of PAHs, which determines the specificity of their behavior in relation to organisms, is log K_{ow}. This indicator characterizes not only the hydrophobicity (lipophilicity) of the molecules themselves, but also the ability of substances to bioaccumulate. This property is considered expressed when log Kow> 3; with a value above 4, the substance is classified as of an increased hazard class and a high tendency to bioaccumulation is predicted for such a compound [5]. All PAH compounds (except Naph) have a log K_{ow} of more than 3.2, and starting with 4-ring ones - and higher than 4. In connection with this, the log K_{ow} indicator was given special attention in the study: we assume that the correlation between log Kow and CC substances should testify to the activity of PAH movement processes in soil-plant systems and the levels of technogenic loads on them.

The revealed patterns of PAH distribution were visualized based on the Surfer GIS package.

3 Results and discussion

The groups of borders-barriers in soil-plant systems identified using cluster analysis are presented in Table. 1.

The features of the first group include the active transition of light PAHs (2-3 benzene rings) at extremely low values for the sum of PAHs in general. These are mainly borders in the soil–root system, as well as root–stem borders at the borders of the background or lightly pressured territory, but not in the directly roadside parts.

In cluster 2, PAH migration can be characterized as nonequilibrium technogenic. PAH flows in the soil – root system are weakly expressed, and the stem part is polluted from the atmosphere mainly by multi-ring PAHs due to the combustion products of motor fuel. This is confirmed by calculations of the ratios of kinetic to thermodynamic polyarenes. Point 27 is far enough from the main highway, however, it is located within the car parking of the

RUDN University employees.

Table 1. Characterization of HCB clusters on campus in terms of migration with the accumulation of PAHs

Cluster	Borders	CC of ΣPAH	Individual PAHs with maximum CC
1	**Technogenic migration under the conditions of saturation of environments**	**0.02 - 3.35** **1.09**	**Fluorene, Naph, Phen, An**
1.1	**Active transition is unlikely**	0.02 - 1.09 0.65	Fluorene, Phen, Naph
1.1.1	Soil-roots on the points 10, 11, 16, 17, 21, 22, 25, 26, 29, 31	0.02 - 0.86 0.52	Fluorene, Naph, Phen
1.1.2	Soil-roots on the points 4, 27	0.73 - 1.09 0.87	Phen, (Fluorene, An)
1.2	**Transition is real**	0.63 - 3.35 1.44	Fluorene, Naph, Phen, An, Ip
1.2.1	Soil-roots on the points 7, 12, 24; root-stem on the points 5, 8, 9, 13, 15, 17, 19, 23, 25, 27, 28, 29, 30	0.63 - 1.60 1.10	Fluorene, Phen, Naph, An
1.2.2	Soil-roots on the points 1, 32; root-stem on the points 2, 11, 14, 22	0.63 - 3.35 1.90	Fluorene, Naph, Phen, An, IP, Chr, Flu=Py
2	**Ative technogenic migration**	**0.54 - 3.51** **1.65**	Bghi, An, Fluorene, BkFlu, Chr, Phen, Py, Naph, Flu, BbFlu, BaA (Bap, DbA)
2.1	**Weakly expressed technogenic migration**	0.54 - 3.16 1.51	Bghi, Chr, Phen, An, BkFlu, Fluorene, Flu, Py, BbFlu, Naph, (Bap, DbA)
2.1.1	Soil-roots on the points 5, 6, 15, 18, 19 23	0.69 - 3.16 1.85	Bghi, BbFlu, BkFlu, BaA, Chr, Phen, Flu, An, BaP, Fluorene, DbA=Naph=Py, (IP)
2.1.2	Soil-roots on the points 2, 8, 14, 20, 28, 30, 33	0.54 - 2.96 1.34	Chr, Phen, An, Fluorene, BkFlu, Py, Flu, Naph, (DbA, BaA, BbFlu)
2.2	**Intensive manifested technogenic migration**	1.15 - 3.51 2.58	An, Fluorene, BkFlu, Naph, Py, Chr, Phen, Flu, BbFlu, (BaA, BaP)
2.2.1	Soil-roots on the points 3; 20 root-stem on the points 3, 12	3.09 - 3.51 3.30	An, Fluorene, Naph, Py, Flu, BbFlu, Chr, Phen, BkFlu, (BaP, BaA)
2.2.2	Soil-roots on the points 9, 13 root-stem on the points 1, 4, 6, 7, 10, 16, 18, 20, 21, 24, 26, 31, 33	1.15 - 2.58 1.86	BkFlu, Fluorene, Chr, Phen, Naph= Py, Flu, An (BaP)

The low probability of an active transition in cluster 1.1 indicates the loss of self-cleaning functions of PAHs by the soil layer. Here, the maximum saturation of the rhizosphere with polyarenes is typical, which prevents their entry into the stems. High PAHs toxicity does not allow the microbiota to form favorable conditions for the promotion of nutrients in the stems. Despite regular watering of the highway and the surrounding area, the soils do not manage to form nutrients that can support a stable vegetation cover.

For a real transition (cluster 1.2), CC > 1 is recorded at many points in the transition zone between Leninsky Prospekt and Miklukho-Maklaya street, where vegetation forms flows from the root to the stems and, naturally, from the soil to the roots. These are 2-3 ring PAHs with possible self-generation, but mostly man-made. The cluster covers a forest Park zone with the maximum potential for self-cleaning of soil and vegetation in natural conditions. When the rains fall, certain plant species experience man-made pressure.

Further analysis of the distribution of PAHs in the background territory and in the zone with increased transport load confirmed the correctness of the allocation of this group: regardless of the level of anthropogenic load, it is in the soil-root system that the flows of PAHs are maximally ordered. This border turns out to be weakly permeable for heavy PAHs, and the CC for the total amounts of PAHs is relatively small.

The second group of borders is maximally permeable for a large set of individual PAHs; the CC values are generally higher than in the first group. Most of the borders here are the "root–stem" at points in the zone of direct influence of the roads. Obviously, this is precisely what determines the diversity of the association of migrating PAHs (Table 1): there is contact with air polluted by vehicles. However, it was not possible to detect "heredity" in the intensity of PAHs migration at the observation points. Table 2 shows the numbers of points that are characterized by a complete absence of PAHs transition, minor CC or pronounced transition. Most likely, more factors affect the conditions of PAHs entry and accumulation in the aerial part than in the case of migration in the soil–plant system. At a minimum, it should be noted the probability of introducing PAHs with air flows and the processes of PAHs degradation due to the photolysis. As a result, the distribution of PAHs in the ground parts of plants may be more "blurred".

Table 2. PAH transition activity characteristics
(taking into account the probability of determination error of 30%)

Transition border	**Numbers of points where the transition is…**		
	absent (КК<0.7)	**possible (КК=0.7-1.3)**	**mannifested (КК>1.3)**
Soil – root	8, 9, 13, 14, 20, 28, 30, 33	2, 3, 5, 6, 10, 15, 18, 19, 22, 23, 24, 26, 29	1, 4, 7, 11, 12, 16, 17, 21, 25, 27, 31, 32 CC_{max} = 2.74; $CC_{average}$ =1.72
Root – stem	1, 10	3, 4, 6, 7, 12, 13, 15, 16, 17, 18, 20, 21, 23, 24, 26, 27, 28, 29, 31, 32, 33	2, 5, 8, 9, 11, 14, 19, 22, 25, 30 CC_{max} = 3.51; $CC_{average}$ =2.15

When comparing the activity of polyarene migration in the soil – root and root – stem systems (Table 1), it is impossible to distinguish "unambiguously active" zones of the campus, where both PAH transitions would be equally intense. Of all the points examined, only on the points 11 and 25 significant CCs were noted for both systems (1.5 - 1.9 in both cases: significant, but not maximum CCs). In all other cases, the activity of PAHs migration from soils to the roots and from roots to the aerial parts of plants is different (Fig. 1).

Maps show significantly different conditions for PAH migration in the background and transport-loaded zones. This may be due to the fact that intense pollution has already led to the accumulation of significant amounts of PAHs in all the environmental components. Soils, unlike plants with their seasonality of life processes (especially for annual plants, as in our case), are more prone to PAH deposition. Therefore, the processes of migration and accumulation of PAHs in the soil-root system look more streamlined, and the very ability to take new "portions" of PAHs in vegetation is higher - and due to the greater "capacity" (smaller amounts of PAHs are initially accumulated and there are opportunities to skip larger masses of PAHs across the border), and due to large sources of input (both from the soil through the root systems, and from the atmosphere due to transport).

https://doi.org/10.1051/e3sconf/202016901015

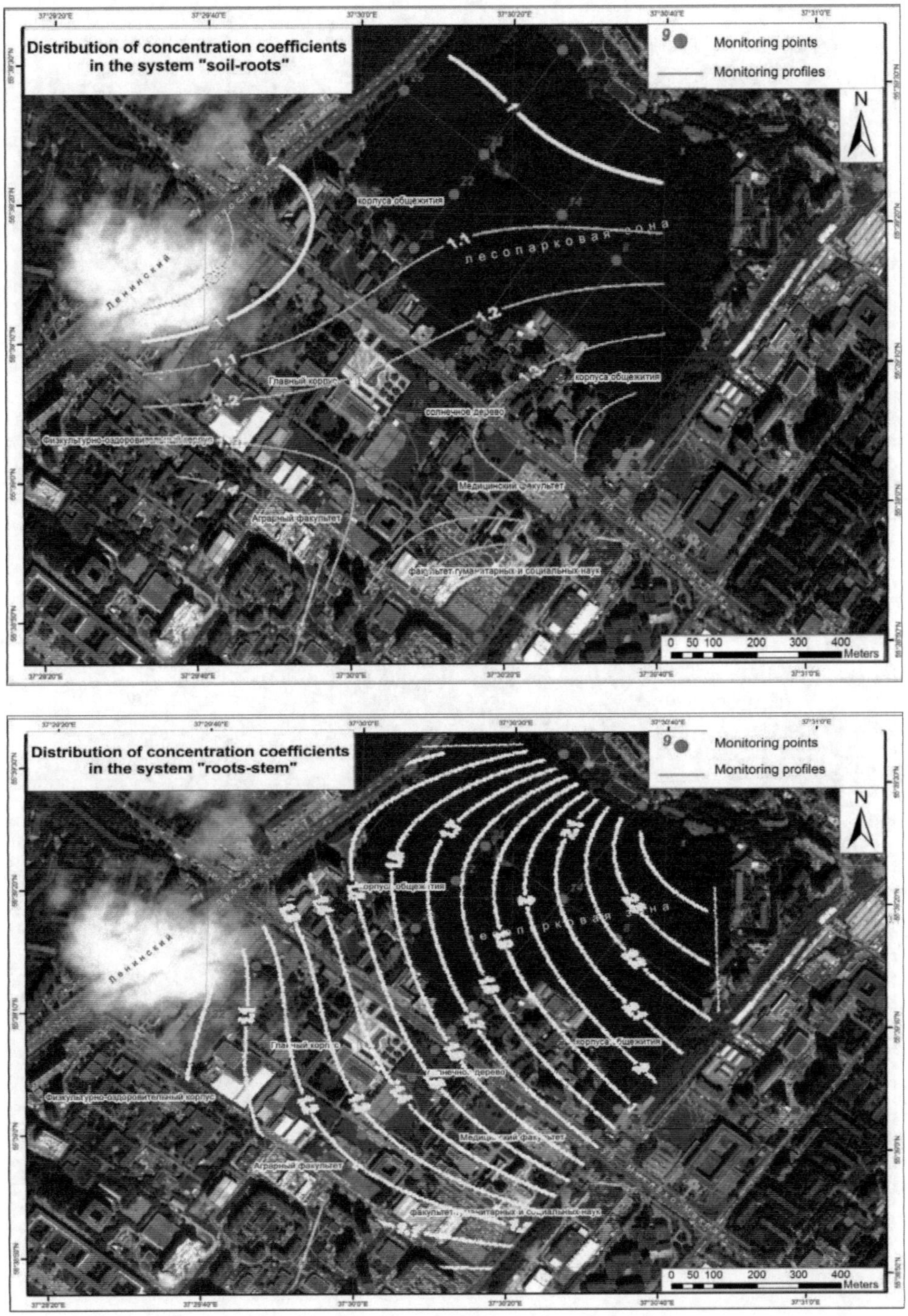

Fig. 1. CC of PAHs in the systems "soil-root" (upper) and "root-stem" (lower)

An analysis of the relationship between log Kow and CC for the “soil – root” and “root – stem” systems (Fig. 2) shows differences in the state of soil – plant systems due to the uneven technogenic loads.

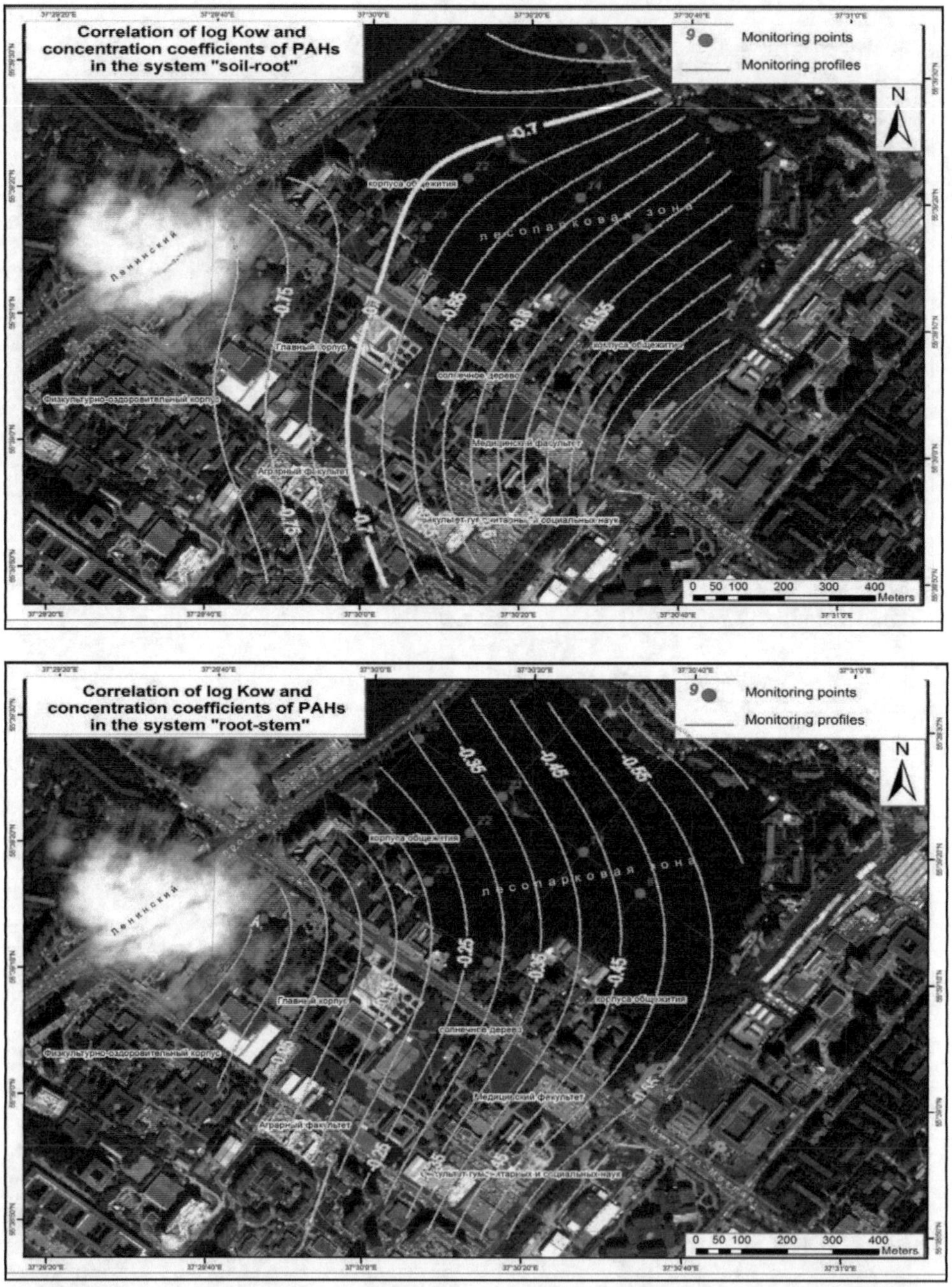

Fig. 2. Correlations between log K_{ow} and CC of PAHs in the "soil-root" (upper) and "root-stem" (lower) systems

As can be seen, in contrast to clearly defined negative relationships for the soil – root system, in the case of the root – stem system, the correlations of log Kow with CC are not so obvious. In some cases, these are insignificant positive connections (vols. 2, 3, 14, 15, 18, 19); but for most points, negative or close to missing points. Such a variety can testify to the significant role of PAHs aerogenic supply to the plant organism. In addition, for points with a positive correlation (14 is the background; 2, 3, 15, 18, 9 are the technogenic load below the maximums), we can assume the natural generation of PAHs.

4 Conclusion

Roots play the role of "filters", creating natural barriers to the separation of PAHs, both in the background and transport-pressured zones. Anthropogenic load plays an unconditional role in the processes of PAH fractionation and determines the initial PAH masses that "claim" to be transferred. However, the composition of associations that penetrate the "soil – root" border and the transition intensity themselves are generally similar: 2-3 ring PAHs migrate relatively easily. QCS for larger molecules may be an order of magnitude smaller or even beyond the scope of significance.

CC in the background zone is clearly higher compared to the CC of polyarenes in the transport loaded zone A possible explanation is that “saturation” of the “host environment” has been achieved and more active PAH transitions do not occur even under conditions of much higher levels of soil contamination compared to the background.

In root – stem systems, patterns are not so obvious in the background territory. You can select the maximum CC coefficients at point 14 (except for BbFlu and heavy PAHs), followed by point 22. In the transport-loaded zone, uniform transitions of the entire PAH Association are characteristic; peaks are at point 23 (An) and 11 (BbFlu).

The obvious dependence of PAH accumulation processes in the root system when received from soils with a log K_{ow} coefficient was confirmed. However, the spatial distribution of the correlation index of this coefficient with the activity of PAHs uptake by roots and subsequent transport to the aerial part indicate that the anthropogenic load has a significant impact on these processes. The correlations are less pronounced in areas with a strong influence of transport pollution. Obviously, the state of the oppressed microbiota in the rhizosphere is also manifested here – it ceases to prevent the entry of PAHs into plant organisms and there is a more active absorption of PAHs by them.

References

1. A. P. Khaustov, M. M. Redina., Geochem. Int., **55(1)**, 98-107 (2017)
2. A. P. Khaustov, M. M. Redina., E.V. Yakovleva, Geoecology. Eng. geology, hydrogeology, geocryology, **3**, 3-17 (2018) (in Russian)
3. A. P. Khaustov, M. M. Redina., E.V. Yakovleva, *E3S Web of Conferences*, **98**, 01027 (EDP Sciences, Les Ulis, 2019)
4. A. A. Oborin, *Oil polluted biocenoses (formation processes, scientific bases of recovery, medical and ecological problems)* (Perm state University, Perm, 2008) (in Russian)
5. *OECD Environment, Health and Safety Publications, Series on Testing and Assessment, No. 27*, (Environment Directorate, Organization for Economic Co-operation and Development, 2001)
6. E. V. Yakovleva, D. N. Gabov, V. A.Beznosikov et al., Polycyclic Aromatic Compounds, **37(2-3)**, 203-218 (2017)

E3S Web of Conferences **169**, 01016 (2020) https://doi.org/10.1051/e3sconf/202016901016
APEEM 2020

Dangerous geodynamic processes of the Eastern Orenburg

Maksim Nesterenko[1,*], *Aleksey* Tsviak[1], *Olga* Kapustina[1], *Aleksey* Nesterenko[1], and *Sergey* Nikiforov[2]

[1]Orenburg Federal Research Center, UB RAS, Orenburg, Russia
[2]NUST MISIS College of Mining, Moscow, Russia

Abstract. The article raises a problem of dangerous geodynamic processes taking place in the eastern part of Orenburg region. The analysis of subsoil geodynamic state of Orenburg region is completed. Effective monitoring of geodynamic processes in the eastern part of Orenburg region using a seismological network is proposed and justified. Seismic events with a magnitude of M_L> 2.0 are confidently recorded on this territory as well as subsidence of land surface reaching 50 cm per year. According to long-term monitoring of seismic events in the eastern part of the Orenburg region, a repetition schedule of seismic events was constructed. It allowed us to make a conclusion of the reason for increased geodynamic activity in this region is high technogenic load on geological environment. To study abnormally high seismic activity in the East Orenburg region, it is necessary to create a seismological network of 3-4 seismic stations in the territory of seismic activity in the Gaysky and Novooorsky districts near the cities of Orsk and Gai, Iriklinskaya State District Power Station and Iriklinskoye Reservoir. This will help to identify seismic activity causes and sources, level of danger, help to develop measures for reducing seismicity and increase the security of specific objects of society, industry and energy.

Introduction

Mining industry development in Russia and the Orenburg region leads to increase in load on geological environment, which leads to geodynamic processes appearance, phenomena in the bowels and on surface and leads to tragic consequences. This problem is especially acute at mining sites. During extraction of solid minerals, the number of dynamic manifestations of rock pressure at mining facilities increases. In majority of regions, there has been a steady increase in the share of large events such as man-made earthquakes, which are accompanied by great damage to mine workings and on surface with negative, including tragic, consequences. Such events are already taking place at mining enterprises in Russia and in the world. For example, a series of earthquakes and catastrophic failures in the Perm Territory in the city of Solikamsk and Berezniki at mine workings of the Verkhnekamsk deposit, in the Sverdlovsk region in areas of the Kizel coal basin development, at the copper ore quarry in Uchaly (Bashkortostan). A strong earthquake with M_L=6.1 with severe consequences was recorded in the Kuzbass region near the Bagatsky section.

Most mining enterprises (including the Perm Territory, Kuzbass, etc.) are monitoring seismic and geodynamic activity of geological environment of a developed mineral deposits. Seismic monitoring involves creation of a network with stationary and mobile seismic

* Corresponding author: geoecol-onc@mail.ru

https://doi.org/10.1051/e3sconf/202016901016

stations and its compaction in identified geodynamically and seismically hazardous areas. Geodynamic monitoring involves highly accurate periodic observations of land surface movement at workings, sides and ledges of quarries and in their vicinity. Geodynamic monitoring only allows us to record facts of dangerous processes on surface, whereas seismological monitoring allows us to continuously identify zones and sections of the bowels which accumulation of tension. Based on integrated seismic activity monitoring of a subsoil and land surface movements, it is possible to predict dangerous geodynamic phenomena for risk reducing of technological and natural-technological disasters

Methods

Orenburg Federal Research Center of the Ural Branch of Russian Academy of Sciences together with Orenburg Region Russian Ministry of Emergencies has been monitoring seismic and geodynamic activity in the western part of Orenburg region since 2005. Seismic events with magnitude M_L>2.0 are confidently recorded in this territory. To clarify parameters of recorded seismic events, data from seismic stations in Bashkortostan and in Perm Territory are used. Monitoring of seismic activity in the western Orenburg region allowed us to obtain a large amount of factual material about its geodynamics. Due to data analysis we concluded that there is a significant technogenic effect on seismic activity of a subsoil of the region. In areas of exploited hydrocarbon fields, density of released seismic energy is three or more times higher than in areas not subject to technogenic impact [1-5]. Fig. 1 shows the registration capabilities of an existing network of seismic stations.

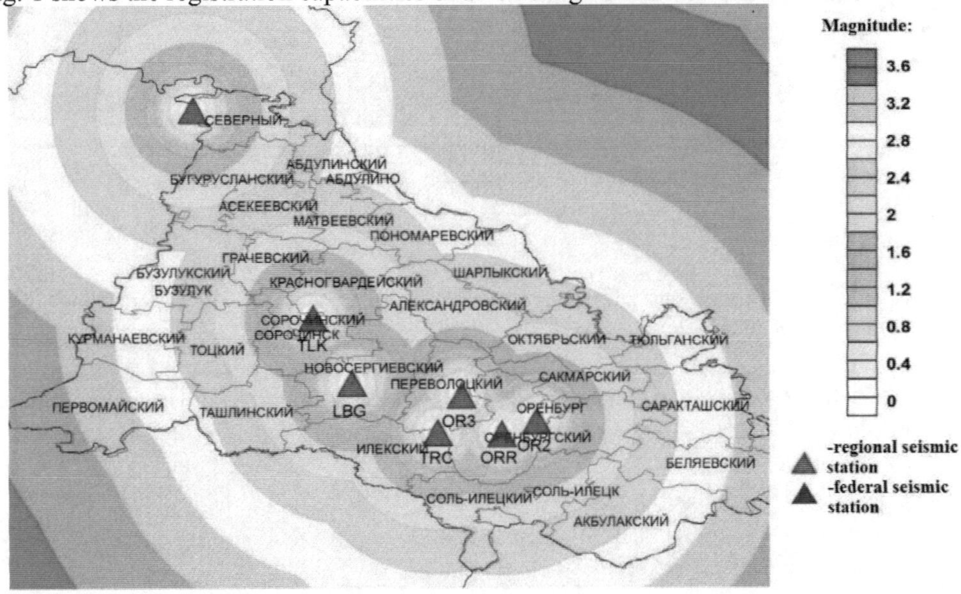

Fig. 1. Registration capabilities of a seismic stations network in the western Orenburg region.

At quarries and mines of Gaisky GOK earth's surface subsidence reaching 50 cm per year is registered [6]. The situation with connection of numerous seismic events with magnitudes M_L 1.5-2.9 is of concern in Geoecology department of Orenburg Federal Research Center. These seismic events were recorded by stations of Orenburg Research Center and Kazakhstan national seismological network, located at a great distance from the area of Gaysky field. This does not allow to determine the source of seismic events, to determine their nature (rock impact, tectonic event, explosion, etc.), to identify stress zones in rock massifs, etc.

 https://doi.org/10.1051/e3sconf/202016901016

Only in October 2019, Kazakhstan National Data Center has recorded more than 10 events that occurred in the eastern part of Orenburg region with magnitudes from 1.5 to 3.2. In fig. 2 events with a magnitude greater than two units that have occurred since beginning of the current year are noted.

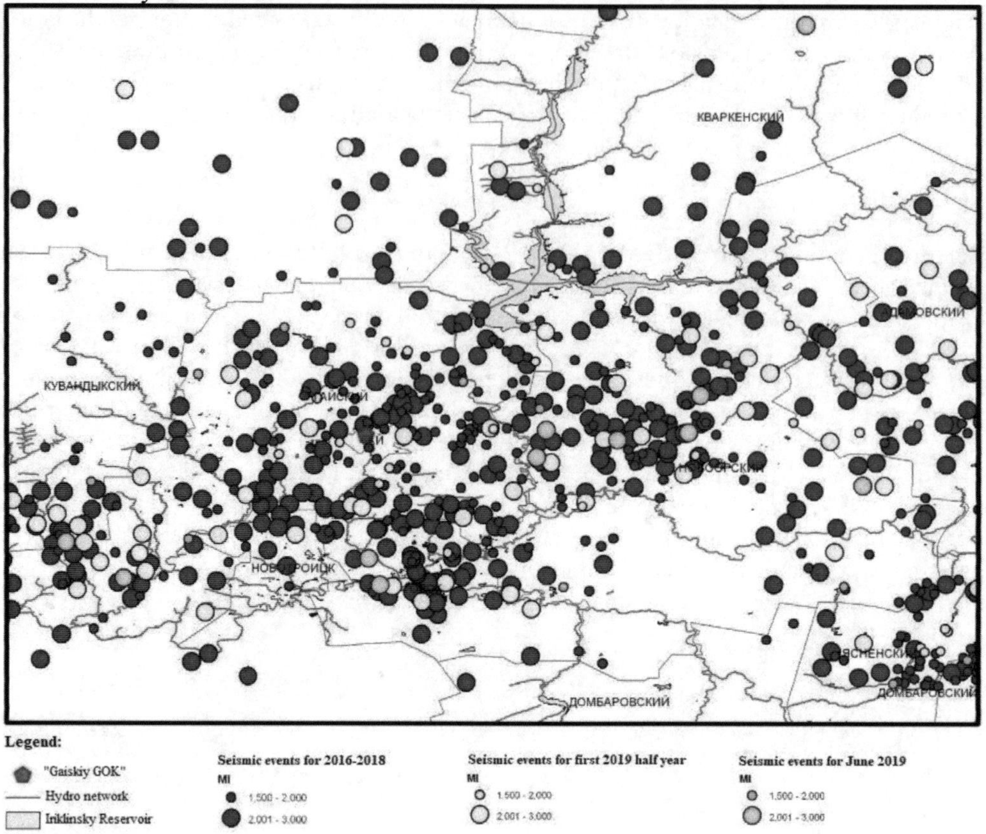

Fig. 2. Seismic events location on a map of the Orenburg region recorded by Kazakhstan National Data Center for January-October 2019

Using well-known methods for assessing a seismic regime for natural and man-made seismic events recorded in the eastern part of Orenburg region by seismic stations belonging to Kazakhstani colleagues, distribution of seismic events depending on a magnitude is compiled.

Results

Results of assessing the seismic regime in the eastern part of the Orenburg region are presented in Table 1.

Table 1. Seismic events distribution in the eastern part of Orenburg region depending on their magnitude.

Magnitude range, *Mb*	(0; 2]	(2; 2.5]	(2.5; 3]	(3; 3.5]	(3.5; +∞)
Number of seismic events, *N*	220	529	153	12	2
Logarithm of seismic event number, *lnN*	5.4	6.27	5.03	2.49	0.69

According to data presented in the table, the plot of seismic events frequency for the period from 2016 to present is constructed (Fig. 3).

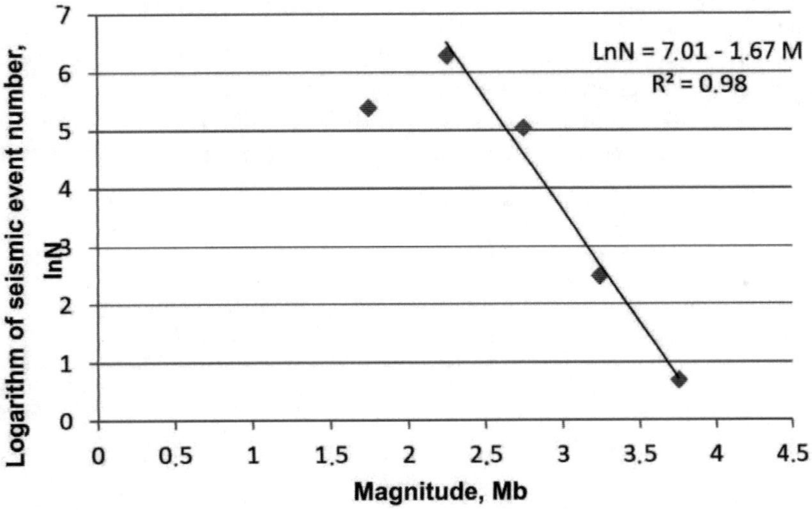

Fig. 3. Seismic events frequency for the period from 2016 to present

Selecting a linear section, we obtain the following dependence of number of seismic events logarithm on magnitude:

$$lgN = 7.01 - 1.67 \cdot M \tag{1}$$

The result obtained can be interpreted as follows. For natural and man-made earthquakes recorded in the territory under consideration, a deviation from the straight line is observed, reflecting a linear relationship between a magnitude and a number of events in the area of a weak seismic events with a magnitude of less than 2.

The slope coefficient of a repeatability plot, which is one of the main indicators of a seismic regime, has an absolute value of b = 1.67, significantly exceeding natural seismicity with the value of b = 0.75, which is characteristic of technogenic and technogenic-induced seismicity. Therefore, the main source of increased seismic activity of a subsoil in the eastern Orenburg region is a technogenic load.

Conclusions

Seismological monitoring is required to study an abnormally high seismic activity in the East Orenburg region, identify its causes and sources, level of danger, develop measures to reduce seismicity and increase protection of specific objects of society, industry and energy. Seismic stations of Kazakhstan National Center are located at a distance of several hundred kilometers from large cities, industrial and energy facilities. It does not allow obtaining sufficient information on a seismic activity of the eastern Orenburg region. A seismological network created in Geoecology department of Orenburg Federal Research Center in the western Orenburg region cannot solve assigned tasks due to remoteness and geological conditions for seismic waves passage through Ural Mountains. To solve them, it is necessary to create a seismological network of 3-4 seismic stations in the territory of seismic activity in Gai and Novoorsk regions near Iriklinskaya State District Power Station and Iriklinsk Reservoir.

References

1. S. Gibowicz, S. Lasocki, Adv. Geophys **44**, 39-180 (2001)
2. J. Suckale, Adv Geophys **51**, 55–106 (2009)
3. A. Tsviak, M. Nesterenk, A. Nesterenko, ISEES (2018) DOI:10.2991/isees-18.2018.100
4. M. Nesterenko, Yu. Nesterenko, OSU Bulletin **9**, 122-127 (2010)
5. Yu. Nesterenko, M. Nesterenko, M. Karpyuk, Risk Analysis Issues **2**, 48-54 (2010)
6. V. Gorbunov, Yu. Kantemirov, Geomatika **2**, 70-76 (2013)

Analysis of dynamic of tropospheric ozone and nitrogen dioxide in the highest region of the Quito city

Cristian Salazar[1,*], *Freddy* Mullo[2]

[1]Peoples' Friendship University of Russia (RUDN University), Faculty of Ecology, 6 Miklukho-Maklaya St, Moscow, 117198, Russian Federation
[2]Peoples' Friendship University of Russia (RUDN University), Agrarian and Technological Institute, 6 Miklukho-Maklaya St, Moscow, 117198, Russian Federation

Abstract. A descriptive study of behavior of daily concentrations of ozone (O_3) and nitrogen dioxide (NO_2) in the period from January 2017 to December 2019 was conducted with the purpose to describe air pollution in the south of the city of Quito. The data collected was processed using the Excel and SPSS v25. The statistical analysis included central tendency values, percentages of violation of the threshold limit value (TLV), maximum values and 25th, 75th and 95th percentiles. The correlation between pollutants was evaluated using the Spearman's Rho coefficient. Multiple linear regression models were made for each pollutant and meteorological variables (wind speed and mean ambient temperature). Inverse moderate correlation was found between the contaminants analyzed. The daily and monthly concentrations of O_3 and NO_2 were lower than the TLV, therefore, the air quality in the period studied with respect to the analyzed pollutants is satisfactory.

1 Introduction

The development of human activities in recent times had caused an increase in emissions of toxic substances into the atmosphere, which led to an increase in the concentration of greenhouse gases and a gradual worsening of air quality.

Different environmental laws around the world regulate the abundance of various air pollutants, given that their presence in the ambient air above certain concentration levels implies an imminent threat to public health and the integrity of goods, materials and even forests and crops [1].

The studies on the concentration of primary pollutants (CO, SOx, NOx) and secondary pollutants (O_3) in the air of Quito in previous years reported low concentrations of carbon monoxide and sulfur dioxide, moderate to high tropospheric ozone concentrations and high concentrations of nitrogen dioxide, which exceeded the concentrations established by the national ambient air quality standards [2-3].

Monitoring of tropospheric ozone and nitrogen dioxide in ambient air is significant for two reasons: due to its impact on public health and direct effect of ozone on the oxidative capacity of the atmosphere.

* Corresponding author: cristian_salazarf@hotmail.com

As for the first reason, the presence of atmospheric pollutants such as nitrogen dioxide and ozone is directly related to a high risk of stroke, heart disease, asthma and lung cancer [4]. Therefore, it is necessary to monitor the levels of these pollutants in the air in order to warn the population if concentration values exceed the permitted values established by the standards of each country. In Ecuador the ambient air quality guideline [5] for ozone is set at the level of 100 $\mu g/m^3$ for daily maximum 8-hour mean and an annual mean concentration for nitrogen dioxide of 40 $\mu g/m^3$.

The second reason for the importance of environmental ozone monitoring is explained from the physical perspective of the mechanisms of secondary pollutants formation. That is, the presence to a greater or lesser degree of ozone in the ambient air directly influences on the formation of OH radicals and at the same time the combined action of these two oxidants impacts on the formation of the aerosol fraction of the fine particulate matter [6].

The mechanism of tropospheric ozone production has been extensively studied by several authors. Ozone is formed when the primary emissions of volatile organic compounds (VOCs) react with the hydroxyl radical (OH) to form hydroperoxyl radicals (HO_2) and other organic radicals, collectively called RO_2. Also, many of the organic compounds are photolysed in daylight and produce radicals. Once RO_2 species are formed, they react with primary NO emissions and form secondary NO_2. Now NO_2 has a photolysis peak of 400 nm, that is, at the beginning of the visible light spectrum. For this reason, daylight acts on primary and secondary NO_2 emissions, with the products of photolysis being atomic oxygen and NO. Finally, atomic oxygen combines with molecular oxygen and forms polluting ozone at the Earth's surface [7].

The distribution of ozone concentrations in mountainous areas can be very heterogeneous and is influenced by the meteorological processes associated with a complex relief, mainly by the regime of slope and valley winds. In these areas an increase in ozone levels is frequently related to altitude and a higher intensity of solar radiation respectively that favors photochemical reactions [8].

The present study focuses its interest specifically on two most problematic pollutants in recent years in the city of Quito: tropospheric ozone and nitrogen dioxide with the objective of describing the status of air pollution in the highest region of the city, in the period from January 2017 to December 2019.

2 Study area

Quito is located in the north of the Republic of Ecuador near the equator. It has a complex topography with heights between 1800 and 4000 meters above sea level. The air quality monitoring network of the city is equipped with eight automatic stations to measure concentrations of carbon monoxide (CO), sulfur dioxide (SO_2), nitrogen dioxide (NO_2), ozone (O_3) and particulate matter.

The Guamani automatic station is located in the south of the municipality (78°33'5" W, 0°19'51" S), in the highest part of the city at 3,600 masl (fig. 1).

All information on air quality is public and can be consulted and downloaded from the official website of the city municipality.

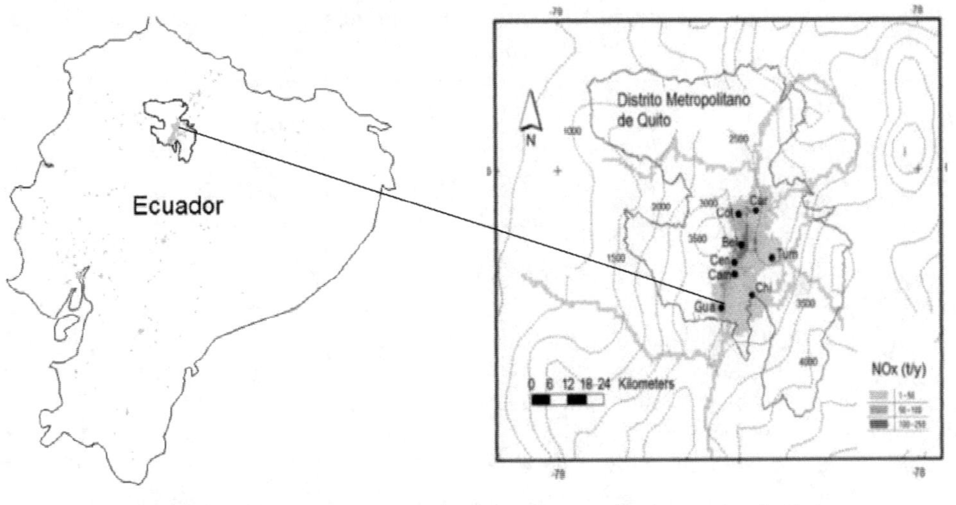

Fig. 1. Location of Guamani (Gua) automatic air quality monitoring station in Quito

3 Research method

The universe of the present study constitutes the daily concentrations of tropospheric ozone and nitrogen dioxide, obtained from Guamani monitoring station for the period between the years 2017-2019.

The Guamani monitoring station is equipped with Thermo Scientific ™ 49C / 49i ozone analyzer that utilizes UV Photometric technology to measure the amount of ozone (EPA equivalent method No. EQOA-0880-047) and a Thermo Scientific ™ 42C / 42i (NO-NO_2-NOx) analyzer that is based on chemiluminescence technology to measure the amount of nitrogen oxides in the air (EPA reference method No. RFNA-1289-074). The location and operation of the monitoring stations in Quito comply with the recommendations of the United States Environmental Protection Agency (US-EPA) (EPA. 40CFR58, Appendix E) and the World Meteorological Organization (WMO, No. 8) [9].

The Kolgomorov-Smirnov test was used to evaluate the goodness of fit of the studied pollutants concentrations distribution. In order to evaluate the relationships between the pollutants, Spearman's Rho coefficient was calculated, given the asymmetric distribution of the majority of data collected. Subsequently, multiple linear regression models were built between each pollutant and meteorological variables (average ambient temperature and wind speed).

4 Results and Discussion

Table 1 shows that for both ozone and nitrogen dioxide, the arithmetic means do not transgress the allowed values, however, in the case of NO_2 at the 95th percentile, the values are close to those established by the Ecuadorian standard. Maximum value of ozone is close to the reference value, and in the case of NO_2 it exceeded more than two times the permitted value, the latter pollutant showed the highest percentage in its frequency of violation of the norm.

Table 1. Summary values and percentages of violation of the threshold limit values (TLV) of the daily means of O_3 and NO_2

Pollutant	Mean (μg/m³)	Percentiles				Maximum concentration (μg/m³)	Percentage that exceeds the TLV (%)
		25	50	75	95		
O_3	27.09	11.78	25.35	39.42	60.43	120.44	0.08
NO_2	16.45	8.04	14.05	22.90	36.40	89.33	2.68

In a study conducted in the period 2005-2012 [10], the arithmetic mean of ozone was higher than that obtained in the present work (33.43 μg/m³) and for nitrogen dioxide it was lower (15.57 μg/m³). The low concentration of nitrogen dioxide in the previously conducted study is explained by the fact that in this period the vehicle technical inspection by the city municipality began, so that the atmospheric pollution by the automotive fleet decreased, moreover, the strongest solar radiation was recorded compared to the other periods [9], as large quantity of nitrogen dioxide reacted to form ozone.

In another study carried out within the period 2013-2016 [11] mean concentrations of ozone 30.31 μg/m³ and nitrogen dioxide 20.35 μg/m³ were obtained, but in comparison with data obtained in the current study these concentrations are higher. This increase in the concentration of nitrogen dioxide is largely due to the fact that in this period a large number of forest fires were recorded compared to other periods [12].

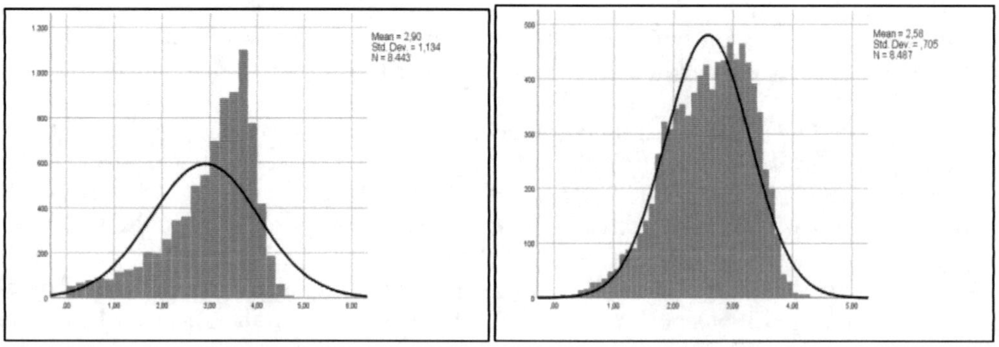

a) b)

Fig. 2. Histogram of the distribution of concentration data a) O_3, b) NO_2

When applying the Kolmogorov-Smirnov test, the napierian logarithm of the variables was determined, and it was found that the studied pollutants do not follow a normal distribution (NO_2: 0.05, $p < 0.05$; O_3: 0.13, $p < 0.05$) (Fig. 2a and 2b).

Although the studied pollutants did not show a normal distribution, multiple linear regression models between their concentrations and the previously described meteorological variables were constructed in order to have an approximation of their interaction.

Regarding NO_2 mean temperature (beta = -0.092, $p < 0.01$) and daily wind speed (beta = -0.542, $p < 0.01$), the highest concentrations were associated with low temperatures and light wind speed. This behavior according to PAHO [13] is due to the fact that low temperatures together with weak winds in a city with a stable climate, such as Quito, cause a thermal inversion, which leads to an accumulation of pollutants.

Tropospheric ozone has a direct and significant relationship with temperature (beta = 0.524, $p < 0.05$) and with wind speed (beta = 0.290, $p < 0.05$). The highest concentrations of ozone were recorded with increasing temperature and increasing wind speeds. This behavior of ozone is due to the fact that the increase in atmospheric temperature associated with the heating of the day, favorably affects the kinetics of the ozone formation reactions [2].

Table 2. Spearman's correlation coefficient between contaminants

Pollutant	NO_2	O_3
NO_2	1.00	-0.58, $p<0.05$
O_3	-0.58, $p<0.05$	1.00

As the pollutants do not follow a normal distribution, the Spearman correlation was used to establish the correlation between them, where nitrogen dioxide and ozone have a moderate negative correlation (see table 2), this is due to the fact that in a city with low traffic flow, for example, the city of Quito, out of two most important ozone precursors (NO_2 and VOC), nitrogen dioxide limits the formation of ozone [14], but in this case without significant differences compared to VOC.

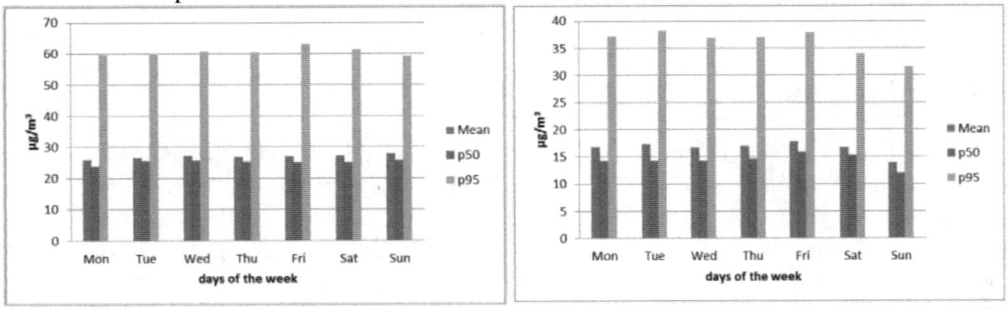

a) b)

Fig. 3. Daily behavior of a) O_3, b) NO_2

Figures 3a, and 3b describe the behavior of the studied pollutants according to the days of the week. According to the mean and the 50th percentile, the weekend effect is evident in the south of the city of Quito, especially on Sundays, where the level of ozone concentration is lower than on the other days of the week, while the concentration of nitrogen dioxide is the highest. On Saturdays there is no aforementioned behavior, since in the south of the city, on this day, commercial activities with the large influx of people are organized that causes greater vehicular flow. The concentrations of the pollutants did not show noticeable differences during the rest days of the week, where the 95th percentile in both pollutants did not exceed the values established by the Ecuadorian air quality standard.

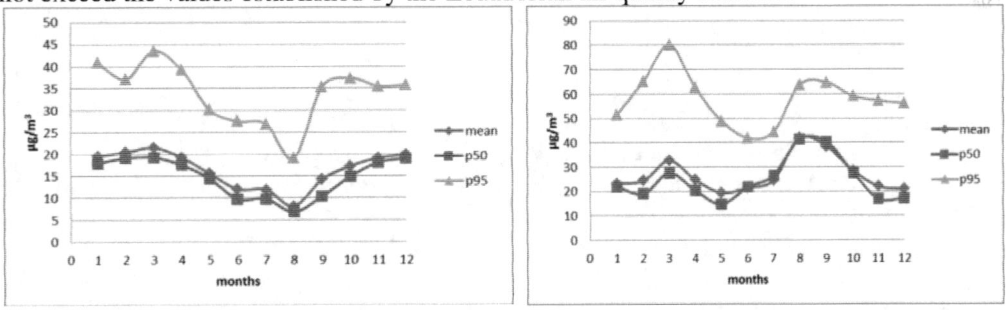

a) b)

Fig. 4. Monthly behavior of a) O_3, b) NO_2

Figure 4a shows maximum values of the mean, median and 95th percentile of ozone concentrations in the following months: March, August and September. The high values in March and September are largely due to the arrival of the equinox, a phenomenon in which the sun rays fall perpendicularly in the equator, as Quito is located in a mountainous area, the city is more prone to the influence of solar radiation favoring the formation of ozone.

When analyzing figure 4b, the monthly behavior of nitrogen dioxide denotes that between February and April the highest concentrations were recorded. These months coincide with the winter season, which is prone to thermal inversions by radiation, making the dispersion of pollutants in the air closer to the Earth's surface less favorable during the night and early morning hours [13]. The lowest concentrations for the mean, median and 95th percentile were recorded from June to August, which corresponds to the period of student vacations, when vehicle flow drastically reduces.

In the city of Quito, the transport sector is one of the most important sources of tropospheric ozone precursors (nitrogen oxides and volatile organic compounds). With the start of metro operation in 2020, it is expected that the concentrations of the pollutants under study will decrease, as it emits far less pollution compared to other types of transport, moreover, the traffic congestion will be reduced.

5. Conclusions

According to the results obtained, it is determined that the air quality in the south of Quito in the period between the years 2017 to 2019 with respect to the analyzed pollutants is satisfactory, this is due to the fact that the arithmetic means of the daily concentrations of ozone and nitrogen dioxide were lower than the maximum permissible concentrations by 72% and 58%, respectively. Moreover, as per the monthly analysis of the pollutants the maximum value of the 95th percentile of the monthly ozone concentration was 20% lower than that established by the Ecuadorian air quality standard, as for the 95th percentile of nitrogen dioxide, it exceeded the allowed value in March by 8%, however, in the rest months it remained below the standard.

The weekend effect is mainly evident on Sundays, where the arithmetic mean of ozone compared to the day with the greatest contamination was 2% higher, and in the case of nitrogen dioxide 19% lower.

The publication has been prepared with the support of the "RUDN University" Program 5-100.

References

1. A. Kurbatova, Y. Mikhaylichenko, A. Dorontsova, *IOP Conference Series: Earth and Environmental Science*, **272(2)**, 022107 (2019)
2. M. Cazorla. ACI, J. **5**, 67-78 (2013)
3. R. Parra, E. Franco, AIR 2016, **207**, 169–180 (2016)
4. O. Roman, M. Prieto, P. Mancilla, REV MED CHILE, **132**, 761-767 (2004)
5. Ministerio del Ambiente de Ecuador, *Norma Ecuatoriana de la calidad del aire* (MAE, Quito, 2011) (in Spanish).
6. J. Seinfeld, P. Spyros, *Atmospheric Chemistry and Physics: From Air Pollution to Climate Change* (Wiley&Sons, 2016).
7. M. Jacobson, *Atmospheric Pollution: History, science and regulation* (Cambridge University Press, 2002).
8. E. Gégo, P. Porter, A. Gilliland, S. Rao. J APPL METEOROL CLIM, **46**, 994 – 1008 (2007).
9. Secretaría del Ambiente. *Informe de la calidad del aire de Quito 2016*. (MDMQ, Quito, 2016) (in Spanish).
10. P. Bazante, *Análisis de la concentración de contaminantes atmosféricos en el DMQ* (USQF, Quito, 2015) (in Spanish).

11. J. Manosalvas, *Análisis temporal multivariante de la contaminación atmosférica dentro del DMQ* (UPM, Madrid, 2017) (in Spanish).
12. K. Ochoa. *Emergencia en Quito: El número de incendios forestales se quintuplicó* (Metro, Quito, 2014) (in Spanish).
13. Organización panamericana de la salud. *Curso básico sobre contaminación del aire y riesgos para la salud* (OPS, México, 1991) (in Spanish).
14. R. Parra, ACI, J. **9**, 104-111 (2017).

Accumulation of heavy metals in aquatic vegetation of the Barguzin River

Valentina G. Shiretorova*, *Svetlava* V. Zhigzhitzhapova, *Elena* P. Dylenova, and *Larisa* D. Radnaeva

Baikal Institute of Nature Management Siberian Branch of Russian Academy of Sciences, 670047, Ulan-Ude, Russia

Abstract. This study is devoted to the analysis of the distribution of heavy metals in the aquatic system of the Barguzin River. This is one of the largest rivers of the Baikal lake system. The element composition of the aerial scion of *Potamogeton pectinatus* L. (syn. Stukenia pectinata (L.) Borner), *Potamogeton perfoliatus* L., *Hippuris vulgaris* L. and *Nymphoides peltata* (S.G. Gmelin) O. Kuntze of the Barguzin River is studied in this research. One of the main results of the study is the establishment of the following sequence of the accumulation of metals in aquatic plants: Mn > Fe > Cu (Zn) > Cr > Ni > Co > Pb > Cd > Hg. *Hippuris vulgaris* L. contains the highest concentration of the studied metals. Thus, this research provides one of the important steps for the development of regional environmental standards and environmental risk assessments.

1 Introduction

Due to the technogenic pollution of surface waters with pollutants, the issue of availability and the quality of fresh water is becoming more acute for humanity. Climate change, accompanied by the increase of average water temperatures, melting glaciers, increase in the number of dry periods and floods, only make matters worse [1]. Inorganic pollutants, including heavy metals and metalloids, such as Cr, Ni, Cu, Zn, Cd, Pb, Hg, As, do not decay, unlike organic pollutants. Heavy metals emitted into the ecosystem, due to significant geo-, bioaccumulation and biomagnification, cause a potential risk to human health and the ecosystem [2].

It is well known that aquatic plants intensively absorb macro- and microelements, mineral and organic substances, accumulate heavy metal ions and radionuclides, and are associated with mineralizers and detoxicants, as well as biofilters of pesticides and oil products [3-5]. They are used as objects of indication and monitoring of water quality, eutrophication and pollution of water bodies, and can be industrial raw materials [6-11]. So, various types of hydrophytes of *Potamogeton* L. and *Myriophyllum* L. genera has been used to assess xenobiotics' emission into water bodies subjected to anthropogenic stress and to determine

* Corresponding author: vshiretorova@rambler.ru

the possible places of pollutant accumulations in ecosystems [5, 12-14]. Data on the microelement composition of aquatic plants can be used to detect both the initial stages of pollutants emission into water bodies and long-term systematic anthropogenic impacts.

This study was devoted to the analysis of the distribution of metals in the aquatic system of the Barguzin River. This is the third largest rivers of the Baikal lake system. Thus, significant attention to the environmental state of this great lake is a reason for the control of each component of the Baikal catchment area. This study contained the element composition of the aerial scion of 2 species of *Potamogeton - Potamogeton pectinatus* L. (syn. Stukenia pectinata (L.) Borner), *Potamogeton perfoliatus* L., *Hippuris vulgaris* L. and *Nymphoides peltata* (S.G. Gmelin) O. Kuntze growing in the Barguzin River.

2 Material and methods

As a result of the intensive use of water resources of the Barguzin basin (agriculture and forestry), there has become a distinguishable trend in the change of water natural composition and the decrease of their quality, especially pronounced in the lower reaches of the river [15]. The distribution of metals in aquatic plants of the Barguzin River has been studied for the first time. The plants were collected near Barguzin village (N 53°36'7,4" E 109°36'20,0"). Plant materials were collected on July 25-27, 2019, without reference to the plant's age. Selected specimens were cleaned from impurities, dried to an air-dry state and grounded. Sample preparation of plant materials was carried out by ashing with concentrated nitric acid in MARS 6 microwave system. The metal content in the samples was determined by atomic absorption spectrometry of an air-acetylene flame on Solaar M6. The mercury content was determined by the cold steam method using a VP-100 hydride attachment to an atomic absorption spectrophotometer. The analysis was conducted in a plant sample with 0.5 g (weigh) of air-dried biomass, taking into account the plant generic assignment, in triplicate.

3 Results and discussion

The results of determination the metal content in the studied aquatic plants are shown in Table 1.

Table 1. The content of metals in aquatic plants of the Barguzin River (mg/kg, air-dry weight)

Metal	**Species**			
	***Potamogeton perfoliatus* L.**	***Hippuris vulgaris* L.**	***Potamogeton pectinatus* L.**	***Nymphoides peltata* (S.G. Gmelin) O. Kuntz)**
Fe	904±64	1829±133	706±46	111.2±1.1
Mn	2004±142	12069±553	930±26	199.1±12.9
Zn	41.7±2.7	51.3±2.8	37.8±0.8	11.4±2.1
Cu	37.7±1.7	51.8±1.5	37.5±0.5	24.2±1.8
Cr	7.7±0.04	12.7±0,2	11.7±0.2	4.9±0.5
Ni	4.22±0.25	11.7±0.4	4.7±0.18	1.5±0.35
Co	3.92±0.28	4.82±0.17	b.d.l.	3.57±0.43
Pb	1.24±0.08	3.68±0.23	0.49±0.19	1.52±0.12
Cd	0.43±0.01	0.70±0.14	0.49±0.09	0.63±0.28
Hg	0.048±0.002	0.101±0.002	0.162±0.003	0.015±0.004

b.d.l.– below detection limit

By the level of metal accumulation the studied species can be arranged in the following sequence:
Potamogeton perfoliatus Mn > Fe > Zn ≥ Cu > Cr > Ni ≥ Co > Pb > Cd > Hg;

Hippuris vulgaris Mn > Fe > Cu ≥ Zn > Cr ≥ Ni > Co > Pb > Cd > Hg;
Potamogeton pectinatus Mn > Fe > Zn ≥ Cu > Cr > Ni > Pb ≥ Cd > Hg > Co;
Nymphoides peltata Mn > Fe > Cu > Zn > Cr > Co > Ni > Pb > Cd > Hg.

The following sequence of metal accumulation in aquatic plants was observed: Mn > Fe > Cu (Zn) > Cr > Ni > Co > Pb > Cd > Hg (exept accumulation in *Potamogeton pectinatus* and *Nymphoides peltata*). The highest concentration among the studied metals was determined in *Hippuris vulgaris* L., the lowest (except As) - in *Nymphoides peltata.* The manganese content exceeds the iron content in *Potamogeton perfoliatus* and *Hippuris vulgaris,* 2 and 6 times, respectively, while in *Potamogeton pectinatus* and *Nymphoides peltata,* it is only 1.2-1.7 times. The concentration of zinc and copper in *Potamogeton* and *Hippuris* was in comparable amounts, while the copper concentration of *Nymphoides* had almost twice much of zinc. Iron and manganese were the most intensely involved in the migration cycles in higher aquatic plants, which is associated with their participation in metabolic processes, since these elements are the part of pigments, vitamins, enzymes and, possibly, during the active growing season, they can overcome anticoncentration barriers in excess quantities accumulate by barrier-free type [16]. The highest contents of nickel and cobalt were investigated in *Hippuris vulgaris*, chromium - also in *Hippuris vulgaris* and *Potamogeton pectinatus*. Nickel content was close in both types of *Potemogeton*, in *Nymphoides peltata* - the lowest. The cobalt content in *Potamogeton perfoliatus* and *Nymphoides peltata* was approximately the same, in *Potamogeton pectinatus* it was below the detection limit (Table 1). According to the literature a high accumulating ability towards zinc, manganese, and iron was observed in *Hippuris vulgaris*. However, the authors referred the high accumulation of manganese in *Hippuris vulgaris* of lakes in Western Siberia to the intense occurrence of photosynthesis, which leads to the oxidation of Mn^{2+} to Mn^{4+} on the cell surface due to the increased pH of the waters of thermokarst lakes [17]. In this case, the more intense course of photosynthesis can be explained by the lower depth growth of *Hippuris vulgaris* compared to *Potamogeton perfoliatus* and *Potamogeton pectinatus*.

Lead, cadmium and mercury are the most toxic and widespread methals. Cadmium ranks first among phytotoxic heavy metals [18]. In living organisms, it can replace zinc in the active centers of metal-containing enzymes, leading to a crude disruption in the functioning of enzymatic processes. In elevated concentrations, cadmium is toxic, especially in combination with other chemical compounds. The mercury content in higher aquatic vegetation can serve as an indicator of water bodies pollution [11]. Since there is no maximum concentration limit of metals in higher aquatic vegetation, it is possible to estimate the pollution of a reservoir by lead, cadmium, and mercury comparing the detected concentrations in other aquatic ecosystems in plants of the same genus.

Lead, cadmium and mercury were found in all studied species of aquatic plants of the Barguzin River. The highest concentration of lead was observed in *Hippuris vulgaris* (3.68 mg / kg), the lowest - in *Potamogeton pectinatus* (0.49 mg / kg). Comparison with literature data showed that such lead content was much lower than indicated in plants of *Potamogeton*, *Hippuris* and *Nymphoides* genera from other reservoirs.

The highest concentration of cadmium was observed in *Hippuris vulgaris* (0.70 mg/kg), then *Nymphoides* (0.63 mg/kg) and the lowest - in both species of *Potamogeton* (0.43-0.49 mg/kg). The indicated cadmium concentrations in *Hippuris* and *Nymphoides* were significantly higher than those indicated in the literature. However, it is necessary to mention that there is only one article where the cadmium content in plants of *Hippuris* and *Nymphoides* genera can be find [19]. The concentration of cadmium in plants of *Potamogeton* genus was higher than that indicated for plants from Siberian water bodies, but significantly lower than from European water bodies [9, 13].

The highest mercury content was found in *Potamogeton pectinatus* (0.162 mg/kg), the lowest - in *Nymphoides peltata* (0.015 mg/kg). The mentioned concentrations of mercury are comparable to the concentration in plants of the Bratsk reservoir with high mercury pollution [11].

4 Conclusions

Thus, the results of determination of the element composition of the aerial scion of Potamogeton pectinatus L. (syn. Stukenia pectinata (L.) Borner), Potamogeton perfoliatus L., Hippuris vulgaris L. and Nymphoides peltata (S.G. Gmelin) O. Kuntze, growing in Barguzin River, showed the following order of metal accumulation in aquatic plants: Mn > Fe > Cu (Zn) > Cr > Ni > Co > Pb > Cd > Hg. The highest concentration among the studied metals was determined in Hippuris vulgaris L.

Thus, the obtained data of the distribution of heavy metals broadens the understanding of the biochemical characteristics of aquatic plants. This research provides one of the important steps for the development of regional environmental standards and environmental risk assessments.

This research was conducted in the framework of the State assignment of Baikal Instutite of Nature Management, and with partial financial support of The Russian Foundation for Basic Research (Project No 17-29-05085).

References

1. R.P. Schwarzenbach, Th. Egli, Th.B. Hofstetter, U. von Gunten, B. Wehrli, Annu Rev Env Resour, **35(1)**, 109-136 (2010).
2. R. Verma, D. Pratima, Recent Res. Sci. Technology, **5**, 98-99 (2013).
3. W. Xing, G. Bai, H. Wu, H. Liu, G. Liu, J Soil, **17**, 2566–2575 (2017).
4. B.B. Bazarova, Water: Chemistry and ecology, **11**, 54-60 (2013).
5. A.P. Kuklin, Water: Chemistry and ecology, **8**, 49-54 (2014).
6. I.V. Dovbnya, *Freshwater hydrobionts and their biology* (Nauka, Leningrad, 1993).
7. N.M. Dayneko, S.F. Timofeev, S.V. Zhad'ko Science and Practice Bulletin, **2**, 99-109 (2017).
8. E.N. Philipenko Bulletin of the Transnistrian University, **2**, 117-123 (2014).
9. J. Li., H. Yu., Ya. Luan, Int. J. Environ. Res. Public Health, **12**, 14958–14973 (2015).
10. D.P. Phillips, L.R.D. Human L., J.B. Adams, Mar. Pollut. Bull., **92,** 227-232 (2015).
11. M.G. Azovsky, M.V. Pastukhov, V.I. Grebenshikova, Water: Chemistry and ecology, **8**, 20-24 (2010).
12. A.A. Fomina, E.I. Tikhomirova, A.I. Korableva, Modern problems of science and education, **6** (2015)
13. C.A. Harguinteguy, M.N. Cofré, A.F.-Cirelli, M.L. Pignata, Microchem. J., **124**, 228-234 (2016).
14. F. Duman, O. Obali, D. Demirezen, Chemosphere, **65**, 2145-2151 (2006).
15. V.V. Dryukker, L.M. Sorokovikova, V.N. Sinyukovich, T.G. Potemkina, I.G. Nikulina, O.A. Molozhavaya, I.V. Korovyakova, Geography and Natural Resources, **4**, 72-78 (1997).
16. M.Yu. Puchkovm V.P. Zvolinsky, V.V. Novikov, A.I. Kochetkova, E.G. Kochetkova, E.G. Loktionova, Basic research, **6**, 392-396 (2013).
17. R.M. Manasypov, O.S. Pokrovsky, L.S. Shirokova, S.N. Kirpotin, N.S. Zinner, The news of Tomsk Polytechnic University, **329**, 50-65 (2018).
18. D.M. Dikieva, I.A. Petrova, *Hydrobiological processes in water bodies (*Nauka, Leningrad, 1983*).*
19. L. G. Vardanyan, B S. Ingole, Environ. Int., **32**, 208-218 (2006).

Geoecological situation in coastal and sea area of the Western Yamal

Roy Shirokov[1,2]

[1] Earth Cryosphere Institute, Tyumen Scientific Centre SB RAS, 86, Malygina str., box 1230, Tyumen, 625000, Russia
[2] State University of Land Use Planning, 125315, 14, Baltiyskaya st, Russia

Abstract. An urgent problem is climate change and marine conditions affecting the formation and evolution of the permafrost zone of the shelf and the continental border of the Russian Arctic. The results of long-term monitoring of the permafrost zone clearly demonstrate the degradation of frozen rocks in both continental and subaquatic conditions of the western sector of the Russian Arctic. The coastal-marine region of Western Yamal is the most vulnerable and therefore model territory with climatic changes. Against the background of the observed climate dynamics, the study of geoecological processes becomes especially relevant when drawing up long-term plans for the development of climate-dependent branches of economic activity.

1 Introduction

Assessment of changes in geo-ecological conditions in the Arctic during climate warming and industrial development in recent decades has advanced several priority problems. Changes in climatic conditions affecting permafrost, sea hydrology, as well as the potential increase in negative anthropogenic pressures associated with growth, oil and gas production in the Arctic and the development of related infrastructure, sharply increase the level of geoecological risks. Climate warming and changing environmental conditions are changing the geo-ecological and land-use conditions.

A set of studies, including climatic and marine hydrological observations, morphology and landscape structure, permafrost dynamics and adverse cryogenic processes, is carried out on the western coast of the Yamal Peninsula near the Marre-Sale polar station. The territory was chosen as a model since all the features of the natural environment are present and manifest here. In the Marre Sale region, the main morphological levels are presented - the third sea terrace, low and high river floodplains, modern low accumulative formations (sea laids). The territory belongs to typical tundra with a full set of characteristic plant communities. Permanent permafrost rocks are widespread. The characteristic features of land use are the combination of the traditional way of life of indigenous peoples and industrial oil and gas development in one territory.

Against the background of the observed climate dynamics, the study of geoecological processes becomes especially relevant when drawing up long-term plans for the development of climate-dependent branches of economic activity.

Changes in climatic conditions affect the socio-economic activity of the regions, living conditions of people and public health; leads to changes in the natural environment and is reflected in the species composition of flora and fauna, river levels, weather conditions, seasonal rhythm, thawing of permafrost, the spread of areas of natural focal diseases, etc. Everything can entail the creation of a fire hazard and the possibility of environmental disasters. Climate change can have both positive and negative consequences.

The coastal-marine region of Western Yamal is the most model and vulnerable area under climatic changes. The geographical position of the seacoast is determined by the position of the border of the sea and the land, called the coastline or water edge. Moving the coastline towards the land causes the destruction and retreat of the coast, the corresponding movement of the entire coastal zone and a decrease in the land area. Moving the coastline in the opposite direction causes an increase in land area and the corresponding movement of the coastal zone.

2 Objects of study

The coastal-marine region of Western Yamal is the most model and vulnerable area under climatic changes. The geographical position of the seacoast is determined by the position of the border of the sea and the land, called the coastline or water edge. Moving the coastline towards the land causes the destruction and retreat of the coast, the corresponding movement of the entire coastal zone and a decrease in the land area. Moving the coastline in the opposite direction causes an increase in land area and the corresponding movement of the coastal zone.

Most of the Yamal Peninsula is occupied by lowland plains and does not rise above 50-60 m above the Kara Sea. Its flat surface is composed of upper Quaternary marine sediments and is dissected by shallow valleys of small tundra rivers starting from the lakes of the interior of the peninsula. The territory is heavily grained. The main role in the geological structure of Western Yamal is played by the Late Quaternary deposits of marine genesis, which form terraces of different heights in the relief [2, 3].

Based on long-term data of meteorological observations, it has been established that, since the 1970s. in the Arctic as a whole and in Western Yamal pronounced warming is observed. It occurs synchronously for the entire region of the Western Arctic, while not monotonously, but with some short-term fluctuations both in the direction of increasing and decreasing average annual temperature. Our data on the trends in average annual air temperature show that over the past 50 years, a moderate scenario of climate warming has been implemented in Western Yamal with an increase in average annual temperatures up to 4.0-4.5 ° C / 100 years [4].

Analysis of climatic data showed that an increase in the duration of the warm period is observed for Western Yamal. Autumn temperature transition through "0" is shifting to ever later dates. In recent years, this date has shifted from October 5 (1980) to October 17 (2016), which is about two weeks over the past 30-35 years. Climate warming is accompanied by an increase in annual precipitation. In the late 1960s and early 1970s. the amount of summer rainfall was 110-130 mm, and by 2015 it had grown to 150-170 mm. According to available observations, an increase in the maximum snow depth is observed: from 16 cm (statistically) in 1960 to 37 cm with a slight decrease in its density. Thus, the insulating role of snow has significantly increased.

It was noted that against the backdrop of global warming of the World Ocean, estimated at about 0.02-0.04 ° C per year, the temperature of the bottom layer of water in the Kara Sea over the past 100 years has been increasing at a rate of from 0.01 ° C to 0.028 ° C per year [5]. This leads to accelerated degradation of subaquatic permafrost, additional methane emission from frozen rocks into the water column and accelerated biogeochemical processes.

An increase in air temperature, an increase in the duration of the ice-free period causes an increase in the average annual water temperature in the shallow area. Based on the results of measuring the thickness of seasonal (fast ice) ice for 1998-2018. we can conclude that climate warming affects the ice cover to the greatest extent. In West Yamal, the maximum thickness of seasonal ice over the past 20 years has decreased (statistically) from 150 to 115 cm. The temperature of permafrost in the continental region rises due to climate warming. From 1978 to 2018, on average in the research area, the average annual temperature increased from -6.8 to -4.6 ° C. Due to climate warming in this zone, there is difficulty in permafrost neoplasm and its stability is violated [5].

Soil temperature plays a very important role in the development of soil processes and vegetation. In recent decades, a distinct trend has been forming in increasing the average annual temperature of the root layer of the soil (to a depth of 1.0 m) from -7.5 to -5.5 ... -6.0 ° C. Thus, more favorable conditions are formed for the development of vegetation cover and biomass growth. Over the period from 1978 to 2018, on average in all dominant landscapes, the depth of seasonal thawing increased by 5-15%. There is a progressive thawing of the upper horizon of frozen rocks and the development of destructive processes, primarily thermokarst.

Climate: Air temperature. Based on long-term data of meteorological observations, it has been established that, since the 1970s. in the Arctic as a whole and in Western Yamal pronounced warming is observed. It occurs synchronously for the entire region of the Western Arctic, while not monotonously, but with some short-term fluctuations both in the direction of increasing and decreasing average annual temperature. In general, the gradient of the increase in average annual temperature for 1970–2018. amounted to about 2.2 ° C, i.e. 0.04 ° C / year. These are high values of the temperature rise gradient. There is no evidence that such rates of climate warming will continue in the future [4].

In recent decades, it is customary to predict climate changes based on the family (ensemble) of climate models [1], each of which has its own set of advantages and disadvantages. Data analysis using a family of models allows to some extent to level out their individual flaws and get more reliable estimates. Our data on the trends in average annual air temperature show that over the past 50 years a moderate but approaching a “tough” scenario of climate warming has been implemented in Western Yamal with an increase in average annual temperatures to 4.0–5.5 ° C / 100 years. The catastrophic scenario provides for a temperature increase of 6 ° C / 100 years or more, an acceptable level that does not cause irreversible changes is considered to be a warming level of 2 ° C / 100 years [1, 5]. An increase in the number of air temperature anomalies during climate warming has not been established. The duration of the warm period. Analysis of climatic data for the Marre-Sale polar station shows that an increase in the duration of the warm period is observed for the Western Yamal.

The transition of the daily temperature through "0" in the spring shifts to ever earlier dates, from June 19 to June 2 (along a linear trend). Autumn temperature transition through "0", on the contrary, is shifting to ever later dates. In recent years, this date has shifted from October 9 (1980) to October 15 (2017). Thus, the duration of the warm period increased by about three weeks over the past 30–35 years.

Precipitation. Western Yamal belongs to the zone of excessive moisture. Climate warming is accompanied by an increase in annual precipitation, while the amount of winter precipitation increases to a slightly lesser extent than summer precipitation. The amount of summer rainfall has grown markedly. In the late 1960s and early 1970s the amount of summer rainfall was 110–130 mm, and by 2015 it had grown to 150–170 mm.

Snow cover. According to available observational data, an increase in the maximum snow cover is observed. From 16 cm (statistically) in 1960, the height of the snow cover increased to 37 cm, with a slight decrease in its density. Thus, the insulating role of snow has significantly increased.

Hydrology of the sea. Bottom water temperature on the high seas. An analysis of the observations showed that against the backdrop of global warming of the World Ocean, estimated at about 0.02–0.04 ° C per year [4], the temperature of the bottom layer of water in the Kara Sea over the past 100 years, depending on the morphological and structural features of the marine bottom and heat transfer conditions, including The flow of warm Atlantic waters through the Kara Gates and Matochkin Shar straits rises at a rate of 0.01 ° C per year (North Taimyr shelf) to 0.028 ° C per year (Baydaratskaya Bay). This leads to accelerated degradation of subaquatic permafrost, additional methane emission from frozen rocks into the water column and accelerated biogeochemical processes.

Water temperature in shallow water. An increase in air temperature, an increase in the length of the ice-free period causes an increase in the average annual water temperature in a shallow area. Based on our observations, the rate of increase in water temperature here is estimated at 0.03–0.035 ° C / year, which exceeds the maximum observed increase in the temperature of the bottom layer of water in the open sea. In other words, the reaction of the shallow zone of the Kara Sea to climate change is more acute than at great depths. Instrumental measurements in shallow water made it possible to establish that rapid changes (hours and the first day) of the bottom water temperature under conditions of prevailing sea waves up to 0.5 m occur only at sea depths less than 6 m. At greater depths, the temperature of the bottom water layer changes due to longer-term changes air temperature - seasonal and interannual changes. Based on general ideas about the laws of mixing of water masses in the sea under wind waves [6] and a predominant wave height of 0.5–1 m, it can be stated that for the Kara Sea, the outer shallow boundary is limited to 6–10 m isobath. It is interesting to note that on at the same depths is the outer boundary of the coastal talik in subaquatic frozen rocks.

Wind waves and storms. An analysis of the available data shows the immutability of wind wave parameters and the frequency and strength of storms over time against a changing climate. The literature contains satellite altimetry data, according to which an insignificant but steady increase in the height of wind waves and the frequency of storms has been observed for 23 years [7]. The growth rates are so small that on a time scale of 10–100 years, the parameters of waves and storms can be considered unchanged.

Seasonal ice cover (fast ice) and freezing zone. Based on the results of measuring the thickness of seasonal (fast ice) ice for 1998–2018. we can conclude that climate warming affects the ice cover to the greatest extent. In West Yamal, the maximum thickness of seasonal ice over the past 20 years has decreased (statistically) from 150 to 115 cm. The width of the freezing band (seasonal habitat of marine biota) along the accumulative coasts has decreased from 150–170 m to 90–120 m. On thermoabrasive shores this the decrease was from 80–130 m to 60–90 m. Even more striking changes are observed in changes in the area of sea ice, but they are not considered here.

Geological environment: Climatic changes do not affect the composition of frozen deposits; therefore, they are not considered here. The temperature of frozen rocks on the third sea terrace. The temperature of permafrost rocks (MMP) in the continental region rises due to climate warming. From 1978 to 2018 the average annual temperature in the study area increased from –6.8 to –4.6 ° C. In all zonal landscapes of a typical tundra, the increase in average annual temperature is approximately the same. In azonal landscapes - swamps and peatlands - the temperature increase is not the same: swamps are characterized by a maximum change in average annual temperature, peatlands - by a minimum. That is, different landscapes have different reactions to climate change.

The temperature of frozen rocks in the transition region depends on the type of lithodynamic processes. On the beaches near thermo-abrasive shores, there is such a high increase in average annual temperature that here we can talk about the transition of the upper stratum of frozen rocks to a seasonally frozen state. Observations in wells located in this type of transition region showed that the average annual temperature varies between –0.9 ... –2.0 ° C, i.e. most often, it is higher than the temperature of phase transitions (–1.4 ° C) [6]. Thus, during the transition from the continental type of cryogenic strata to transit, a sharp jump in the average annual temperature from –4.5 ÷ –6.5 ° C to –0.9 ÷ –2.1 ° C is observed. In the area of marine accumulation, the formation of low accumulative islands is observed, and permafrost is newly formed in syngenetic type. The average annual temperature of rocks increased from –4.4 to –3.5 ° C. Due to climate warming in this zone, there is a difficulty in permafrost neoplasm and its stability is violated [5]. Within the river floodplain, an increase in permafrost temperature led to thawing of permafrost from above and the formation of through-hole taliks with a depth of up to 2.7 m, which had not previously been formed. The average annual temperature of frozen rocks on such landscapes increased to –0.25 ... +0.5 ° C.

Soil temperature plays a very important role in the development of soil processes and vegetation. Our observations have shown that in recent decades a distinct trend has been formed in increasing the average annual temperature of the root layer of the soil (to a depth of 1.0 m) from –7.5 to –5.5 ... –6.0 ° C. Thus, more favorable conditions are formed for the development of vegetation cover and biomass growth.

Depth of seasonal thawing. The depth of seasonal thawing is the most important indicator of the response of permafrost landscapes to climate change and an indicator that determines the development of soil processes and the conditions of root habitat of the vegetation cover. Over the period from 1978 to 2018, on average for all dominant landscapes, the depth of seasonal thawing increased by only 5–15%. On this basis, some researchers consider the depth of seasonal thawing an unreliable indicator of the reaction of permafrost to climate change. MMP, as a cryogenic geosystem, has its own protection mechanisms against progressive thawing based on feedback from external factors. Thus, on the one hand, climate warming should cause an increase in the depth of seasonal thawing, on the other hand, protection mechanisms simultaneously work. As a result, an underestimated reaction of the depth of seasonal thawing to an increase in average summer air temperature is observed. All this relates to natural conditions. For any man-made violations of compensation, there is no additional heat influx, and progressive thawing of the upper horizon of frozen rocks and the development of destructive processes, primarily thermokarst, are observed.

Cryogenic processes. In the coastal-marine region, the main role in the formation of geo-ecological conditions belongs to coastal processes - the destruction of sea coasts or thermal abrasion, the reformation of the adjacent part of the underwater coastal slope, the processes of accumulation of marine sediment and the formation of marine accumulative forms, to a much lesser degree of wind deflation. Based on long-term monitoring of coastal dynamics, it was found that the retreat rate of thermo-abrasive coasts is not directly related to indicators of climate change but is determined by more complex relationships with the characteristics of atmospheric processes - baric circulation processes in the Arctic as a whole. The coastal retreat rate is characterized by extremely high spatial variability. The destruction of seacoasts is not constant over time but varies cyclically with a period of the order of 21–23 years. The average long-term coastal retreat rate in the study area is about 1.7 m / year, varying from 0.4 m / year to 3.2 m / year in different periods. These values should be used in the development of engineering solutions for the industrial development of the coastal-marine region of Western Yamal.

Under natural conditions, the deformations of beaches and the underwater coastal slope are characterized by moderate values - up to 0.7 m, usually even up to 0.10–0.15 m. According to our observations, the deformation of the beaches and the adjoining part of the underwater coastal slope may have a different sign. But carrying out technical work in the abutment zone of the underwater coastal slope, artificial movement or selection of sediments sharply increase the activity of coastal processes. Using the example of the neighboring region of the Kharasey metro station, it is known that the selection of sand deposits in the coastal zone [6] led to a more than twofold increase in the rate of shore destruction [6]. The retreat of the seacoast is accompanied by the development of thermal denudation and linear erosion. Short ravines form along the polygonal vein ice with a speed of movement of the peaks of the ravines up to 10 m / year. Thermal denudation circuses with a length of up to 300 and more meters and a depth of incision towards the coast of up to 200 m are formed in the areas of development of massifs of formation ice. Deflation processes affect the strip of the coast up to 200-400 m wide.

Thus, the width of the risk zone of the coastal-marine region, due to the intensive development of cryogenic processes, is about 400 m from the coastal strip inland [5].

3 Results and Discussion

Analysis of the data obtained, their interpretation makes it possible to assess changes in the geoecological situation of the coastal-marine region of Western Yamal in a changing climate. Climatic changes entail a change in the hydrology of the Kara Sea and key indicators of permafrost conditions. All these characteristics of the environment cause changes in the geoecological conditions of a specific coastal-marine region, which is simultaneously affected by the sea and the continent. Changes occur in all components of the geoecological environment, but here we consider those that can be evaluated based on the data obtained. In most cases, the estimates are of a qualitative nature.

Vegetation cover and soil. Since plant communities in the area of a typical tundra live at the limit of survival, an increase in the average annual and average summer temperatures directly affects the species composition and the growth of bio-productivity of the vegetation cover. The shift in the timing of the transition of air temperature through "0" increases the duration of the growing season. An increase in the amount of summer rainfall in conditions of excess moisture leads to waterlogging of the territory, an increase in the share of bog plant communities and a decrease in the habitat of vegetation of drained tundra. The increase in the thickness of the snow cover favorably affects the tundra shrubs of willow (Salix polaris, S. glauca) and dwarf birch (Betula nana), since the snow cover is a natural protector of the branches of shrubs from frostbite. Over the past 30 years, the height of willow curtains has increased from 20–30 to 40–50 cm. An increase in permafrost and soil cover temperature causes an increase in the root habitat layer and a change in the structure of vegetation adapted to the frozen root substrate.

An increase in air temperature and an extension of the duration of the warm period favorably affects the speed of biogeochemical processes. An increase in the amount of summer precipitation causes waterlogging, waterlogging of the territory and the formation of marsh soils. The proportion of drained soils is declining. An increase in permafrost temperature is also accompanied by an increase in soil temperature. They become “warmer”, the thickness of the soil cover increases. Soils of a typical tundra in the coastal marine region of Western Yamal are subject to wind erosion. The duration and strength of winds during climate warming does not change, so it should be expected that the intensity of wind erosion of soils will not increase.

The traditional way of life of the indigenous population. Climate warming most directly affects the traditional way of life of indigenous peoples, and they adapt to new conditions.

Restructuring of the vegetation cover, shifting the forest border to the north cause a change in the structure of pastureland and a change in traditional nomadic routes. Changes in the terms and duration of the warm period also lead to a shift in the timing of driving reindeer herds, and in some cases to a change in the boundaries of pastures assigned to individual families and households. The increase in the maximum height of the snow cover makes it difficult for the deer to get food, but these difficulties are relatively small.

Climate warming and the resulting increase in the depth of seasonal thawing can cause a deterioration in the sanitary-epidemiological situation. An example is 2016, when, due to the threat of anthrax, quarantine was declared, and appropriate measures were taken throughout the territory of Yamal. In addition, our estimates show that the retreat of the coast leads to the loss of pasture quality land in the Yamalo-Nenets Autonomous Okrug with an area of about 140 hectares annually.

Industrial (oil and gas) development. Climate warming and the resulting changes in geo-ecological conditions have the greatest impact on the principles of industrial development, mainly the oil and gas complex. At the same time, changes in geo-ecological conditions, even without taking into account the technogenic load, must be taken into account when designing and at the stage of construction and operation of industrial facilities. Climate warming improves the climatic and structural characteristics, including an increase in the duration of the warm period for construction work, but at the same time, the terms for using winter roads (winter roads) are reduced by about two weeks. An increase in summer precipitation can complicate industrial development due to flooding and waterlogging, and an increase in the maximum snow cover leads to an increase in snow removal.

The established increase in the temperature of the bottom layer of water by 1–2.8 ° C over 100 years causes the degradation of subaqueous permafrost and the formation of uneven subsidence of the seabed with an amplitude of up to three meters, which complicates the construction of port facilities, sea channels, etc. As the results of studying the topography of the Kara Sea seabed at depths of up to 10 m, the planned size of subsidence areas can reach the first hundred of meters. In the coastal region, due to the reduction in the width of the freezing zone of fast ice and the increase in water temperature in the coastal zone, active degradation and accelerated lowering of the permafrost roof are observed, and as a result, deformations of the foundations of coastal structures are possible and expected. The temperature of frozen rocks increased by 2 ° C from 1978 to 2018. led to the loss of the bearing capacity of the pile foundations by about 20–25% [1,6].

In a generalized form, the influence of environmental factors on the formation of geoecological conditions is given in Table. 1.

Table 1. Change of geoecological conditions of coastal and sea area the western Yamal when warming climate

The environment factors influencing geoecological situation	*Recipients*			
	Plant cover	*Soils*	*Traditional tenor of life of indigenous people*	*Industrial (oil and gas) development*
		Climate		
Air temperature	Change of species composition of vegetation. Shift of borders of bioclimatic zones to the North. Growth of bioproductivity	Accelerati on of biogeoche mical processes	Change of structure of pastures. Change of traditional nomadic routes	Improvement of construction climatic index

Change of duration of the warm period	Increase in duration of the vegetative period		Shift of terms of stage of cervine herds	Increase in duration of the warm period. Reduction of terms of use of winter roads
Summer rainfall	Increase in share of marsh landscapes	Remoistening and bogging		Possibility of flooding of construction objects
Snow	Increase in gain of bushes		Difficulties in getting of forage deer	Increase in volumes of snow removal
Sea hydrology				
Benthonic water temperature in high sea				Formation of sags of sea floor due to degradation of permafrost with reservoir ices
Water temperature in coastal area				Deformations of the bases (technology objects) at the expense of the accelerated degradation of permafrost
Seasonal (pripayny) ice				Narrowing of strip of adfreezing of sea ice and sea floor
Geologic environment				
MMP temperature	Change of structure of plant cover on frozen substrate	Temperature increase of soils		Decrease in bearing capacity of the pile bases
Depth of seasonal thawing	Increase in layer of korneobitaniye	Increase in power of soil covering	Possible deterioration in sanitary and epidemiologic situation	The accelerated cryogenic destruction of headstalls of piles
Cryogenic processes		Wind erosion of soils	Loss of pasturable lands	Development of dangerous processes at technogenic impact on permafrost

4 Conclusions

Based on long-term monitoring (30 years) of coastal dynamics, it was found that the retreat rate of thermo-abrasive coasts is not directly related to indicators of climatic changes, but is determined by more complex relationships with the characteristics of atmospheric processes - baric circulation processes in the Arctic as a whole. The destruction of seacoasts is not constant over time but changes cyclically with a period of the order of 21–23 years. The average multi-year coastal retreat rate in the study area is about 1.7 m / year, varying from 0.4 m / year to 3.2 m / year in different periods. According to our observations, deformations of beaches and the adjoining part of the underwater coastal slope may have the opposite sign [6]. Thus, the width of the risk strip of the coastal-marine region, due to the intensive development of cryogenic processes, is about 400 m from the coastal strip inland.

The current unstable state of the Yamal geosystems is due to the geological, structural and geodynamic features of the territory, the history of its formation and development during the Pleistocene-Holocene time, and current trends in changes in natural and climatic conditions. Due to climate warming in this zone, there is a difficulty in the formation of permafrost and its stability is violated. Soil temperature plays a very important role in the development of soil processes and vegetation [7].

An analysis of the long-term monitoring data allows us to conclude that, within the coastal-marine region of Western Yamal, in a short space in its different natural zones, at the same time, there are reversible reactions (trends) of the components of the natural environment to climatic changes that determine the dynamics of the geo-ecological conditions of these zones.

Estimated climate change can have both positive and negative impacts. To consider the possible consequences of changing geo-ecological conditions of a territory, geographic forecasting, that is, a scientific study of specific prospects for the development of geographical phenomena, is of particular importance.

Overall, the geoecological conditions of the region can be considered quite stable so far, except for cases of industrial development, in which the anthropogenic load can exceed the adaptive capabilities of the natural environment.

The research of conditions and parameters of degradation of permafrost breeds is executed with assistance of the Russian Federal Property Fund, grant No. 18-05-60004, the assessment of change of geoecological conditions is carried out with assistance of grant of the Russian Federal Property Fund 19-35-90049.

References

1. *IPCC. 2018: Summary for Policymakers. In: Global warming of 1.5 °C. An IPCC Special Report on the impacts of global warming of 1.5 °C above pre-industrial levels and related global greenhouse gas emission pathways, in the context of strengthening the global response to the threat of climate change, sustainable development, and efforts to eradicate poverty* (World Meteorological Organization, Geneva, Switzerland, 2018).
2. V.N. Gataulin, *Verkhnechetvertichnye of adjournment of the west bank of the Yamal Peninsula: abstract. yew. ... edging. geol. - the miner. Sciences,* (Leningrad, 1988) (in Russian).
3. V.T. Trofimov, Yu.B. Badu, V.G. Kudryashov, N.G. Firsov, *Yamal Peninsula (engineering-geological sketch)* (Moscow university, Moscow, 1975) (in Russian).

4. J. K. Willis, D. Roemmich, & B. Cornuelle, J. of Geophysical Research: Oceans, **109**, 12 (2004).
5. R. S. Shirokov & A. A. Vasiliev, Natural Resource Management, GIS & Remote Sensing, **1(1)**, 21-27 (2019).
6. A.A. Vasilyev, G.E. Oblogov, I.D. Streletskaya, R.S. Shirokov, Kriosfera Zemli, **12(5)**, 36 (2018) (in Russian).
7. *The report "About strategic estimates of effects of climate changes in the next 10–20 years for the environment and economy of the Union State", considered at meeting of Council of ministers of the Union State on October 28, 2009.* URL: http://212.45.26.46/pub/get-file.aspx? OT=DocumentFile&PN=DocFile&ID=7b95ca54-5132-46e6-b06f-467b294961f6&CT=application/msword&FN=REPORT

Ecotoxicological hazard assessment of deicing chemicals for higher plants

M.D. Fedorchenko[1], *K. Yu.* Mikhaylichenko[1*], *O.V.* Ushakova[2], and *I.A.* Adarchenko[1]

[1]Peoples' Friendship University of Russia (RUDN University), Faculty of Ecology, 6 Miklukho-Maklaya St, Moscow, 117198, Russian Federation
[2]Center of Strategic Planning and Management of Medical and Biological Health Risks of the Russian Federation, 119992, Pogodinskaya ul., 10s1, Moscow, Russia

Abstract. The article presents data on ecotoxicological hazard assessment of different types of deicing chemicals for higher plants. The assessment was carried out according to a specifically developed rapid test on oats seeds. The phytoeffect was determined by comparing the lengths of the seedling roots placed in a solution of deicing chemicals and was calculated by the effect size of root growth inhibition comparing with proof sample. The safest composition of deicing chemicals was determined.

1 Introduction

Currently, many cities use deicing chemicals (hereinafter - PGR) during the cold period to reduce slip effect. However, despite the undoubted benefit, deicing chemicals have immediate as well as delayed effects on the environment and on human health [1,2].

In this regard an experimental study was conducted of the effect of different types of PGR on seed germination oats. The study was conducted from July 1st to August 1st of 2019 at the laboratory for environmental and hygienic assessment of soil and waste of the Center for Strategic Planning and Management of Medical and Biological Health Risks of the Ministry of Health of the Russian Federation with the guidance of Savostikova O.N. [3].

2 Methodology

The phytotoxic properties of PGR were evaluated using the rapid test on oats in accordance with the current method: Guidelines. MR 2.1.7.2297-07: Justification of production and consumption waste hazard class according to phytotoxicity [4]. According to the results of preliminary studies oat seeds was the most stable and gave the most reproducible data compared to seeds of other crops (peas, cucumbers, wheat, carrots, etc.) in this regard they were chosen for the present study [3].

* Corresponding author: mikhaylichenko-kyu@rudn.ru

https://doi.org/10.1051/e3sconf/202016901020

3 Results and discussion

Three types of deicing chemicals were selected for analysis: PGR 1 containing 80% NaCl and 20% $CaCl_2$; PGR 2 containing 72% NaCl, 17% $CaCl_2$, and 11% HCO_2Na — sodium formate; and PGR 3, which includes 35% sodium chloride, 50% $CaCl_2$, and 15% HCO_2Na. A solution with a PGR concentration of 100 g/L was prepared from each sample. Distilled water was used as for proof sample. Also, we used compositions with different PGR obtained by sequential dilution of the initial solution by 10, 100, and 1000 times [5,6].

Seeds were germinated on filter paper in petri dishes for 7 days in dark conditions at a temperature of t = 25 ± 2°C. The phytotoxic effect was evaluated by the magnitude of the inhibition of root growth relative to the proof sample (Phytoeffect,%) [3]. A proof sample, initial solution and all dilutions were analyzed in threes petri dishes for each one [4]. In each cup there were placed 25 dry healthy seeds with germination rate at least 95%.

After that, the root lengths of the germinated seeds were measured, and their average lengths in the proof and experimental samples in each repetition were calculated according to the formula (1):

$$L_{av} = \Sigma L_i / n, \quad (1)$$

where L_i – the length of the maximum root of each seed (ᴍᴍ); n – total number of analyzed seeds [5].

Based on the obtained results, the general average values of root lengths were calculated. Their values are given in table 1.

Table 1. The results of the average root lengths of the proof and experimental samples.

Type of PGR	Sample	Average root lenth, L_{av}, mm			Overall average, L_{ov},mm
		1	2	3	
PGR1	Proof sample	91.19	73.64	74.35	79.70
	Primary solution	0.00	0.00	0.00	0.00
	Delution R 10	21.76	11.61	13.50	15.62
	Delution R 100	63.29	86.04	83.91	77.75
	Delution R 1000	83.61	84.48	85.64	84.58
PGR2	Proof sample	72.04	56.80	68.31	65.72
	Primary solution	0.00	0.00	0.00	0.00
	Delution R 10	5.29	12.13	12.86	10.09
	Delution R 100	76.48	67.12	69.55	71.05
	Delution R 1000	84.13	81.83	86.52	84.16
PRG3	Proof sample	79.96	80.12	75.81	78.63
	Primary solution	0.00	0.00	0.00	0.00
	Delution R 10	11.79	21.25	15.08	16.04
	Delution R 100	85.42	77.21	81.72	81.45
	Delution R 1000	86.72	90.24	94.48	90.48

If the total average length of the seed roots germinated in solutions is greater than (or equal) to the total average length of the seed roots germinated in distilled water, then there is no adverse effect of the PGR solution.

Using the obtained values of the total average root lengths, the values of the inhibition effect for each PGR solution were calculated using the following formula (2):

$$E_{inh} = ((L_p - L_{ex}) / L_p) \times 100 \quad (2)$$

where E_{inh} – inhibition effect (%); L_{ex} – average root length in the experiment (mm); L_p – average root length in proof samples (mm).

Based on the calculation results, a graph of the dependences of the phytoeffect on the dilution of PGR (Fig. 1).

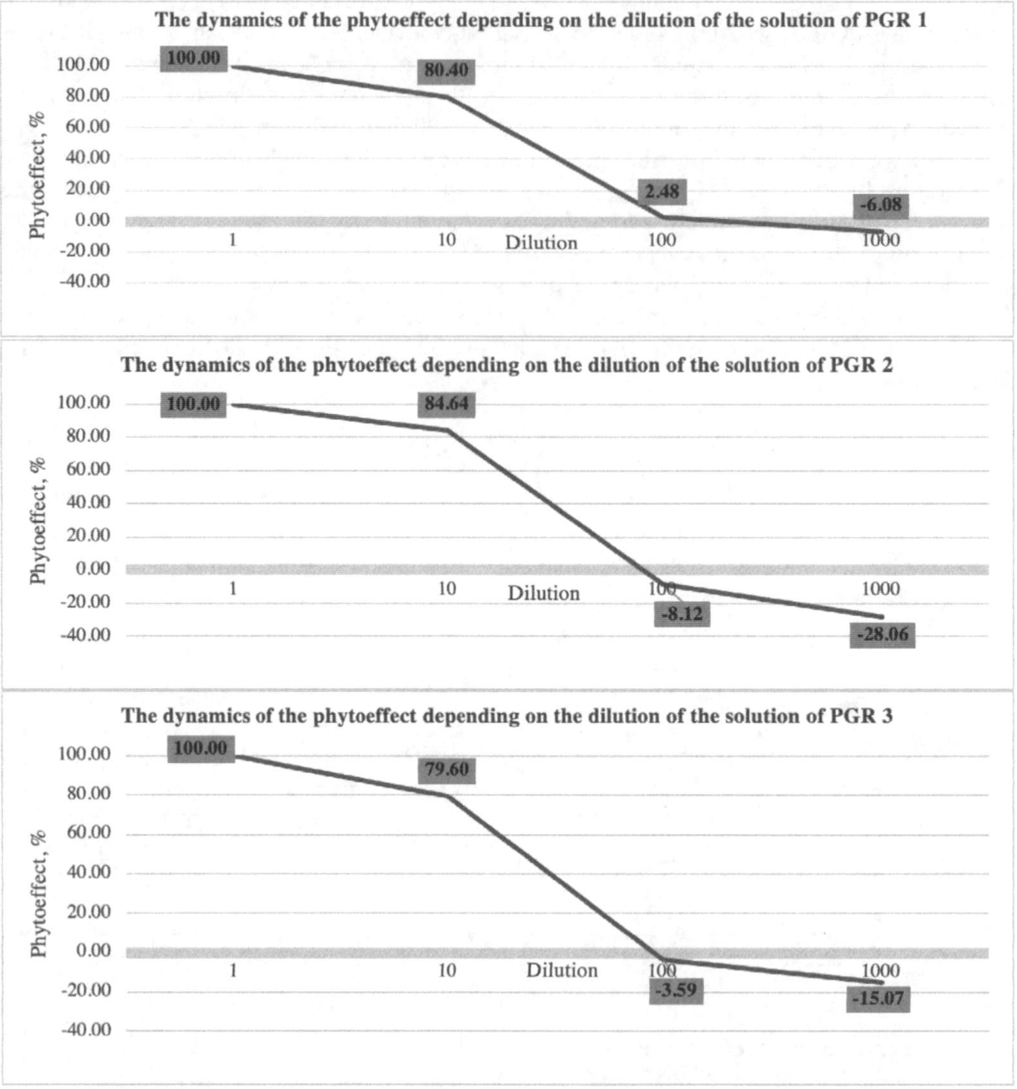

Fig. 1. The dynamics of the phytoeffect depending on the dilution of the solution of PGR.

As figure shows the complete absence of a negative phytoeffect is observed only in diluting the PGR solution more than 100 times.

After conducting a regression analysis of the graph, we simulated a straightforward relationship between dilutions of PGR solutions and the magnitude of the toxic effect (dependence of LgR on E_{inh}). As a result, the equations of the regression model were derived, based on which a coefficient was obtained corresponding to each phytoeffect value and a regression coefficient for each type of PGR. According to formula (3), threshold dilutions of LimR (dilutions causing a phytoeffect at the level of 20%. The phytotoxicity threshold assumes that dilutions of the PGR solution exceeding its value will be safe for the growth and development of higher plants) were calculated:

$$LgR = (-m \times E_{inh}) + b \tag{3}$$

Table 2. The threshold dilutions of the studied deicing chemicals.

Type of PGR	Threshold dilution $LimR_{20}$	Threshold concentration $LimC_{20}$, g/L
PGR1	113	0.9
PGR2	67	1.5
PGR3	85	1.2

Therefore, as a result of the studies, it was found that all analyzed samples of PGR are highly phytotoxic. Thus, for the purpose of recommending the safe use of the studied deicing chemicals, it should be noted that dilution of the initial solutions (100 g/l) 113 times (0.9 g/l) for PGR 1, 67 times (1.5 g/l) for PGR 2 and 85 times (1.2 g/l) for PGR 3 are safe for seed germination of higher plants. For all dilution values lesser than indicated, phytotoxicity should be expected [5]. PGR 2 has the smallest phytotoxicity.

4 Conclusion

As a result of the rapid test on oat seeds and subsequent calculations, it was found that all analyzed samples of PGR have high phytotoxicity.

The adverse effect of PGR terminates only if the initial solution (100 g/l) is subjected to strong dilution. The largest dilution of 100 g/l solutions is required for PGR1 (113 times), the smallest is for PGR2 (67 times).

References

1. S.B. Chudakova, *Dis. ... cand. medic. Sciences, Toxicological and hygienic assessment of the degree of danger of anti-ice reagents*, 202, (2006) (in Russian).
2. A.I. Kurbatova, K.Y. Mikhaylichenko, A.Y. Dorontsova, *IOP Conf. Ser.: Earth Environ. Sci.*, **272**, 7, (2019).
3. M.A.Vodyanova, O.V. Ushakova, L.G. Donerjan, I.S. Evseeva. Modern problems of science and education, **5**, 10, (2018) (in Russian).
4. *Guidelines. MR 2.1.7.2297-07. Justification of production and consumption waste hazard class according to phytotoxicity*, 15, (2018) (in Russian).
5. V.A. Korolev, A.K. Gornyakov. Engineering survey, **12**, 9, (2018) (in Russian).
6. A.V. Sbitnev, M.A. Vodianova, I.A. Kriatov, L.G. Donerian, I.S. Evseeva, O.V. Ushakova, D. I. Ushakov, I.S. Matveeva, O.M. Rodionova, Hygiene and Sanitation, **95(8)**, 5, (2016) (in Russian).

Using remote sensing for monitoring the dynamic of soil organic carbon concentration in Lake Valencia basin, Venezuela, based on Landsat 8 data

Israel Jose Arias Govín[1*], *Elena V.* Stanis[1], *Elena N.* Latushkina[1], and *Aigul* Ospanova[1]

[1]Peoples' Friendship University of Russia (RUDN University), Faculty of Ecology, 6 Miklukho-Maklaya St, Moscow, 117198, Russian Federation

Abstract. Maintaining or increasing SOC concentration is fundamental for reducing the effects of global warming and increasing soil productivity. In this paper, a method based on Landsat 8 OLI products was developed for qualitatively monitoring in the Lake Valencia basin (Venezuela) the dynamic of SOC concentration between the years 2013 to 2018. The developed method uses the Green (B3), NIR (B5) and SW1 (B6) bands of Landsat 8 OLI sensor for detecting changes in the spectral signatures of bare soils that indicate possible variations in their concentrations of SOC. It was found that for the study period, the Lake Valencia basin soils do not present spectral features of significant variation in SOC concentration. An area of 8.61Km^2 (0.3% of the study area) was identified as a zone with a possible reduction of SOC concentration. In case of insufficient data for developing remote sensing based predictive models, the proposed method allows qualitatively monitoring and categorizing the dynamic of SOC concentration and identifying areas with spectral features of a possible variation in SOC concentration.

1 Introduction

Soil organic carbon (SOC) represents 69.8% of organic C in the global carbon cycle in the biosphere [1]. The total amount of organic C stored in soils has been estimated by different methods [2, 3] and it is considered to be close to 1500 Pg at 1 m depth [1, 4]. This means more C than the content between the atmosphere and the terrestrial vegetation [5] representing the largest world reserve of terrestrial C [6, 7]. Therefore, C sequestration in soils could contribute significantly to reducing global warming levels [8]. On the other hand, the increase in SOC directly influences the improvement of the structure, nutrient retention, physical and chemical properties of soils, and consequently an increase of soil fertility [9].

Due to the importance of the SOC, the FAO technical report states that *keeping the pool of COS in balance or increasing the SOC content towards the optimum level for the local environment can contribute to achieving the Sustainable Development Goals* [10]. For

* Israel Jose Arias Govin: israelgovin@gmail.com

achieving this, it is essential to develop profitable tools and methods applicable to different types of soil allowing regular monitoring and mapping of SOC variations. The above represents a challenge because the spatial distribution of the SOC is not uniform across large surfaces and at different depths of the ground [10]. Besides, variations in the concentration and spatial distribution of SOC in soils are controlled partially by spatially and temporally variable factors such as environmental conditions, vegetation cover, land use, and soil edaphic characteristics [11-13].

Until the date, there are no standardized methods for determining the concentration of SOC [10]. Most of the traditional methods based on the oxidation of C involve high costs in time and resources for collecting, preparing, and processing the soil sample [14]. Satellite remote sensing techniques have been used for developing predictive models of SOC concentration in different types of soil [15-18]. The results obtained in these investigations demonstrate that using satellite remote sensors for estimating the SOC concentration is feasible. Nevertheless, for developing statistical models it is essential to calibrate them with data analysed in laboratory conditions. This is because the spectral response of the soils is affected not only by SOC concentration but also by its chemical and physical characteristics [19]. The extrapolation of predictive models to different areas for which they were developed would imply considerable errors in the prediction of SOC concentration [20].

In this study case, due to insufficient data for developing a statistical model for estimating SOC concentration in the Lake Valencia basin (Venezuela), it will be developed a method based on Landsat 8 OLI products for qualitatively analysing the dynamic of SOC concentration between the years 2013 to 2018 in the study area.

2 Study area

Lake Valencia is the most important natural endorheic lake of South America [21] with an approximate surface of 344 km^2. The basin of Lake Valencia covers an area of approximately 3000 km^2. Figure 1 shows the location of the study area and land use classification.

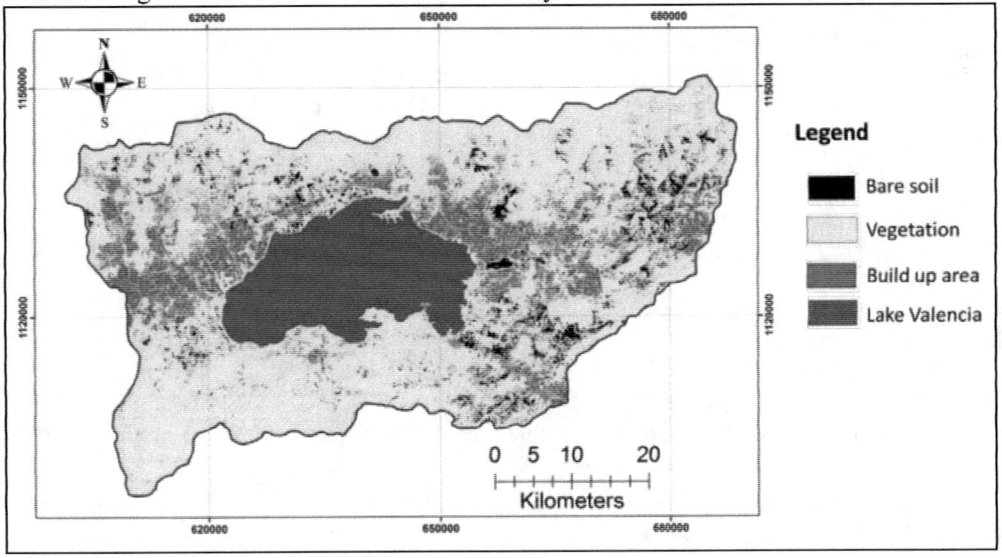

Fig. 1. Location of the study area and land use classification

The predominant soil type belongs to the category of inceptisols, with medium texture. Mollisols soils are medium textured with high organic matter content. In the northwestern region of the basin, oxisol soils with high iron concentrations are scarcely found [22-24]. Figure 3 shows the soil types distribution in the study area.

3 Materials and methods

Several studies have characterized the spectral behaviour of soils [14, 19, 20, 25, 26]. The results of these investigations agree that soils with higher SOC concentration have lower reflectance values in their spectral signature compared to the reflectance values of the spectral signature of soils with lower SOC concentration. Given the above considerations, for the specific same sample of soil, an increase in the reflectance of its spectral signature could be largely associated with a decrease in their SOC concentration.

Figure 2 shows the spectral profile of 5 types of soils with different SOC concentration based on the studies by Shi et al. [19], and modified according to the spectral bands of the Landsat 8 OLI sensor. Figure 2 shows that the decrease in SOC concentration is characterized by a significant increase in the reflectances of the bands Green (561nm), NIR (865nm) and SW1 (1609nm), the bands corresponding to Blue (483nm), Red (655nm) and SW2 (2201nm) do not have a relevant impact on the variation of the spectral signature associated with an increase or decrease in the SOC concentration. Following this, the bands Green (B3), NIR (B5) and SW1 (B6) were selected for developing the method applied. The method developed consists of 4 stages: a) multispectral image pre-processing; b) selection of bare soil points; c) determination of the spectral signatures variations at the selected points; and d) interpolation and obtaining the distribution map of the SOC concentration variation during the study period.

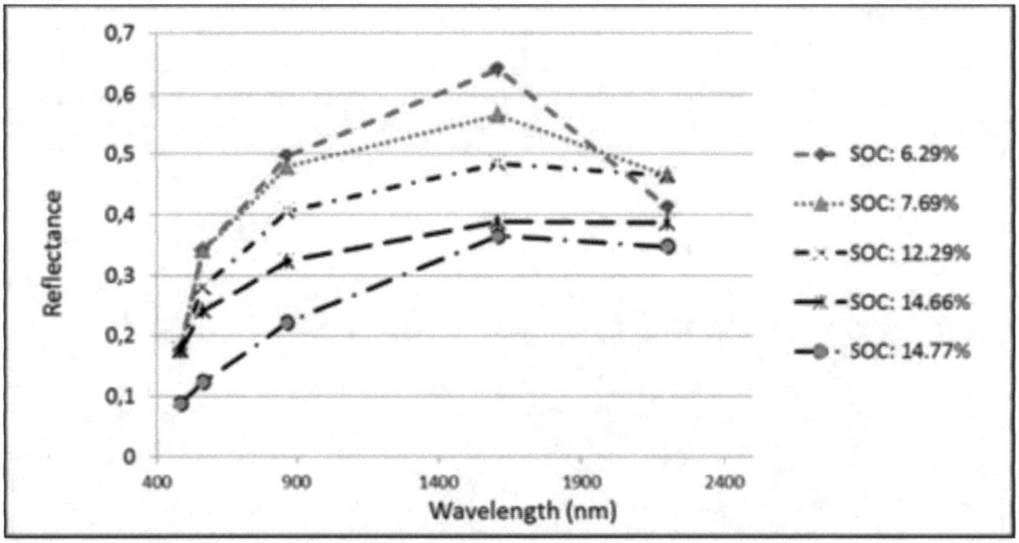

Fig. 2. Spectral profiles of soils with different SOC concentration adapted to the spectral bands of the Landsat 8 OLI sensor. Modified from Shi et al. [19]

In the first stage, four images of the Landsat 8 OLI sensor were used. Two images with the lowest cloud coverage percentage were selected for each year of study (2013 and 2018) to cover the area of the basin. The images correspond to the month of November (dry season) to guarantee the lowest level of humidity in the soils. The top of atmosphere (TOA) reflectance values were obtained according to the procedures described in the Landsat 8 user manual [27]. Atmospheric correction for obtaining the top of canopy (TOC) reflectance values, was performed by applying the DOS method developed by Chavez [28].

In the second stage, the coincident points of bare soils in the two years studied were selected. Initially, to identify areas of bare soil without vegetation, water, and cloudiness, a mask was created by applying the NDSI (Normalized Difference Soil Index) [29] in both scenes using Equation 1. In Equation 1, NDSI is the Normalized Difference Soil Index, B5 and B6 are the TOC reflectance raster of the NIR and SW1 bands. Then, the classification of the NDSI rasters obtained was performed and polygons were generated for bare soil areas and build-up areas. Points were created randomly within the corresponding polygons to the bare soil area. A total of 2475 coincident points for both years were selected. Figure 3 shows the spatial distribution of the selected bare soil points.

$$NDSI = (B6 - B5) / (B6 + B5) \quad (1)$$

In the third stage, the variations of the spectral profiles were obtained at the selected points. For this, a mask was created by applying Equation 2 using the TOC reflectances bands B3, B5, and B6. The pixels of the raster obtained have values 1 (one) or 0 (zero). Pixels with a value of 1 correspond to the points whose spectral signatures showed an increase in reflectance between 2013 and 2018. A total of 163 points that are shown in Figure 3 were identified.

$$(((B3_{2013}-B3_{2018})<0)+ ((B5_{2013}-B5_{2018})<0)+ ((B6_{2013}-B6_{2018})<0)) / 3 \quad (2)$$

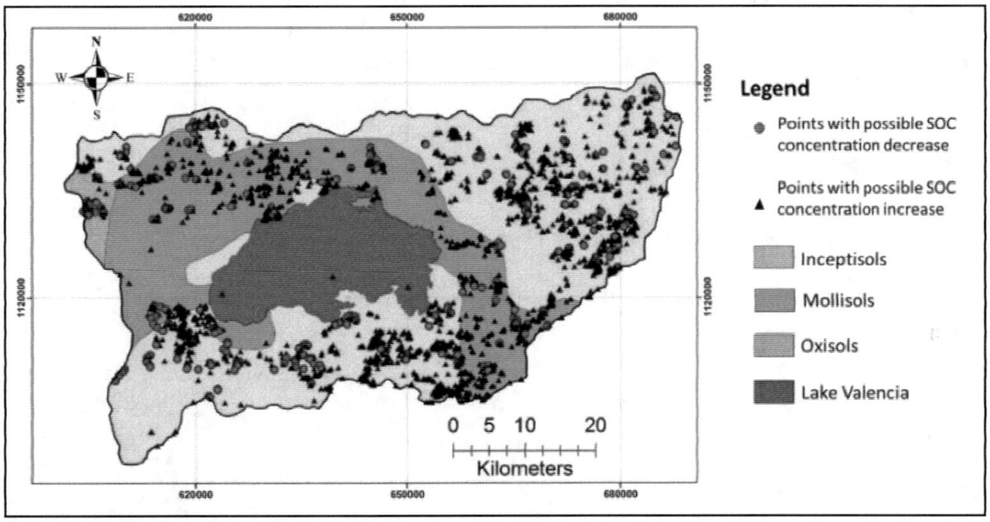

Fig. 3. Soil types and bare soil points distribution

In the fourth stage, the method applied for classifying the SOC concentration variation was based on the comparison of the area under the reflectance curve between the wavelengths 0.561 µm (B3) and 1,609 µm (B6) for each point between the years 2013 and 2018. For determining the area under the curve, the model presented in Equation 4 was developed, where A_{i-j} is the area under the reflectance curve for each point "i" in the year "j" and B3, B5 and B6 the TOC reflectance rasters of the Green, NIR and SW1 bands respectively. Equation 4 was obtained from Equation 3 using the wavelength values of the corresponding bands for the Landsat 8 sensor. In Equation 3, L_{B3}, L_{B5}, and L_{B6} are the wavelengths of bands B3, B5, and B6 respectively. Two rasters corresponding to the values of A_{i-j} of every point were obtained, one for the year 2013 (A_{i-2013}) and the other for the year 2018 (A_{i-2018}).

$$A_{i-j} = 0.5*((L_{B5} - L_{B3})*(B5+B3) + (L_{B6} - L_{B5})*(B6+B5)) \quad (3)$$

$$A_{i-j} = 0.152*(B3) + 0.524*(B5) + 0.372*(B6) \quad (4)$$

Six (6) categories were established to qualitatively classify the variation of the SOC concentration according to the difference of the areas under the reflectance curve for the same point between 2013 and 2018 using Equation 5. In Equation 5, V_{SOC} is the grade of variation of the SOC concentration, $A_{i\text{-}2013}$, and $A_{i\text{-}2018}$ are the raster previously obtained. The categories used were the following: a) $V_{SOC} < -0.2$, severe decrease; b) $-0.2 \leq V_{SOC} < -0.1$, moderate decrease; c) $-0.1 \leq V_{SOC} < 0$, no decrease or slight decrease; d) $0 \leq V_{SOC} < 0.1$, no increase or slight increase; e) $0.1 \leq V_{SOC} \leq 0.2$, moderate increase; and f) $V_{SOC} > 0.2$, high increase. As a result, the raster corresponding to the variation of the SOC concentration for each point selected in stage 2 was obtained. Finally, after obtaining the V_{SOC} values for each point, the IDW interpolation was performed to obtain the distribution and categorization map of the SOC concentration dynamic in the study area between the years 2013 and 2018.

$$V_{SOC} = (A_{i\text{-}2013} / A_{i\text{-}2018}) - 1 \tag{5}$$

4 Results and discussion

For validating the results of the method developed in the study area, the variation of the spectral signatures of 1610 points (65% of the bare soil points) was checked. The validation confirmed the spectral behaviors of possible decrease or increase of the SOC concentration. Figure 4 shows the spectral signatures of six characteristic points within the zones of the SOC concentration variation categories established: a) severe decrease; b) moderate decrease; c) no decrease or slight decrease; d) no increase or slight increase; e) moderate increase; and f) high increase.

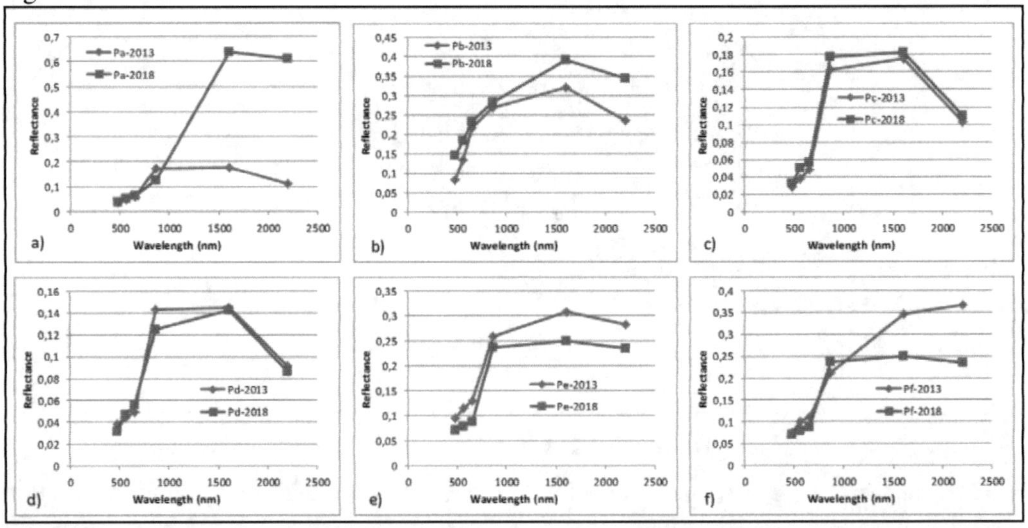

Fig. 4. Spectral signatures of characteristic points for every SOC variation categories established

Figure 5 shows the distribution map of the SOC concentration variations in the study area. The results show that for the study period, the soils of the Lake Valencia basin do not preset relevant signs of SOC concentration decrease. 96.57% of the total area (2919.86 km^2) of the basin did not present variations in their soils spectral signatures that indicate significant increases or decreases in the concentration of SOC. The 3.13% of the study area (91.59 km^2) corresponds to zones with possible increases of SOC concentration and only 0.3% (8.61Km^2) showed signs of a possible reduction of SOC concentration.

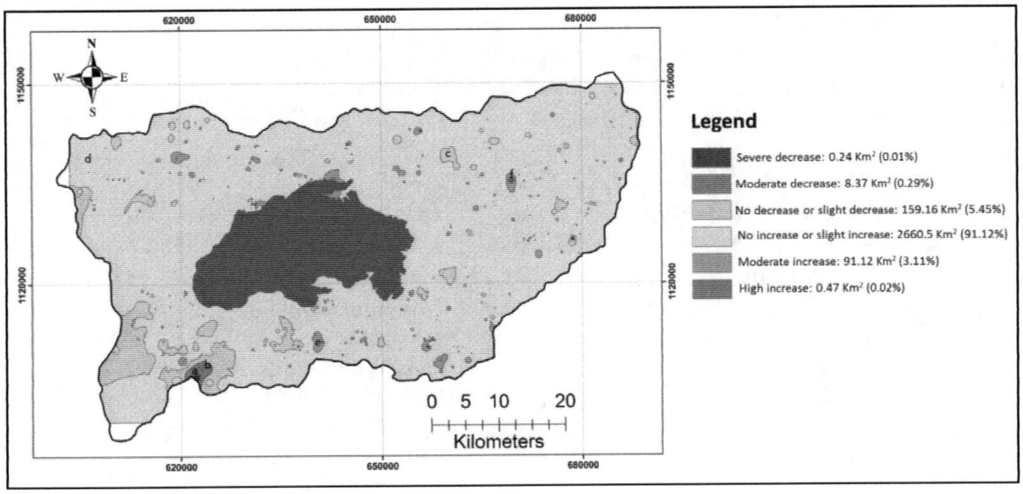

Fig. 5. Distribution map of the SOC concentration variation in the study area from 2013 to 2018

5 Conclusions

The developed method is based on the relationship between the spectral behavior of soils and its concentration of SOC. It was found that using Green (B3), NIR (B5) and SW1 (B6) bands of the Landsat 8 OLI sensor is feasible for detecting changes in the spectral signatures of bare soils that indicate possible variations in the concentration of SOC. The proposed method allows qualitatively monitoring the temporal and spatial dynamics of the SOC and identifying the areas with signs of possible decrease or increase in SOC as well as to categorize the degree of variation. The NDSI allows discriminating areas of vegetation, water bodies, cloudiness, and shadows, highlighting areas of bare soil and build-up areas. Its use is recommended together with visual review for a more accurate selection of points corresponding to bare soil. Adaptation of the method for using with other remote sensors should consider the adjustment of Equation 3 according to the wavelength values for the Green, NIR and SW 1 bands of the sensor to be used. The main limitation of the method is that it is not able to quantitatively estimate the SOC concentration.

References

1. FAO. *Soil carbon sequestration for improved land management*. **96**, p.58, (2001).
2. W. M. Post, et al. Nature. **298**, 156-159, (1982)
3. R. S. Swift. J. Soil Sci., **166**, 858-871, (2001)
4. W. H. Schlesinger. Nature, **348**, 232-234 (1990)
5. FAO-GTIS. *Status of the World's Soil Resources*, (2015)
6. E. Jobbágy, R. Jackson. Ecol. Appl., **10,** (2000).
7. J. Scharlemann, E. Tanner, R. Hiederer, V. Kapos. Carbon Manag., **5**, 81–91, (2014)
8. T.W. Crowther et al. Nature, **540**, 104–108, (2016)
9. T. Ontl, L. Schulte. Nat. Educ. Knowl., **3**, 35-45, (2012)
10. FAO. *Soil Organic Carbon: the hidden potential*, (2017)
11. J. Bouma. J. Environ. Qual., **26(1)**, 26-31, (1997)
12. D. Karlen, et al. S. Sc. Soc. Am. J., **61(1)**, 4-10, (1997)
13. Y. Guo, R. Amundson, P. Gong, Q. Yu. Soil Sci. Soc. Am. J., **70**, 590–600 (2006)
14. A. Gholizadeh, et al. Appl. Spectrosc., **67(12)**,1349-1362, (2013)
15. D. Rogge, et al. Remote Sens. Environ, **205**, 1–17, (2018)

16. F. Castaldi, et al. Remote Sensing, **10 (2)**, 153, (2018)
17. T. Angelopoulou, et al. Remote Sensing, **11(6)**, 676, (2019)
18. F. Castaldi, et al. Remote Sensing., **11**, 2121, (2019)
19. Z. Shi, et al. Science China. Earth Sciences, **57(7)**, 1671–1680, (2014)
20. H. Bartholomeus, et al. Geoderma, **145**, 28-36, (2008)
21. M. W. Binford. Ecol. Monogr., **52(3)**, 307-333, (1982)
22. INE. *Informe Geoambiental, Estado Carabobo*. INE, Rep. B. Venezuela, (2011)
23. INE. *Informe Geoambiental, Estado Aragua*. INE, Rep. B. Venezuela, (2011)
24. L. Mogollón, J. Comerna. *Suelos de Venezuela*. PDVSA, (1995)
25. E. Stoner, M. Baumgardner. Tech. Rep., Purdue Univ., (1982).
26. T. Henderson, et al. Soil Sci. Soc. A. J., **53**, 1778-1784, (1989).
27. USGS. *LANDSAT 8 (L8) Data users handbook*. Version 3.0 - LSDS-1574, (2018)
28. J. Chavez. Remote Sens. Environ., **24**, 459-479, (1988)
29. A. Rogers, M.S., Kearney. Int. J. of Rem. Sens., **25(12)**, 2317-2335, (2004)

Study of cyantraniliprole sorption in soils of Russia

Victoria Kolupaeva*, *Inna* Nyukhina, and *Aleksandra* Belik
All-Russia Research Institute of Phytopathology, 143050, Bolshie Vyazemy, Russia

Abstract. The sorption of the new insecticide cyantraniliprole was studied by the batch equilibrium method. The study of quantitative characteristics of cyantraniliprole sorption by the soils of different types showed that this process is well described by linear isotherms. According to obtained values of K_{oc}, cyantraniliprole belongs to medium-mobility active substances of pesticides. The sorption value of cyantraniliprole was closely correlated with the content of organic carbon and clay. The regression equation of the dependence of K_d on soil organic carbon content was obtained.

1 Introduction

Cyantraniliprole is a new broad-spectrum insecticide. It is approved for use in Argentina, Canada, the USA, Great Britain and several EU countries. In the Russian Federation, cyantraniliprole was registered in 2019 as the active substance of insecticides recommended for pest control in plantings of cabbage, tomatoes, cucumbers and onions [1]. Laboratory filtration experiments showed high mobility of cyantraniliprole [2, 3]. In the lysimeter experiment cyantraniliprole was found in the leachate 2 weeks after its application (after heavy showers), and also was found in most samples during the experiment [4].

Pesticide sorption by soil is considered the process that has the greatest impact on pesticide behavior and fate in the environment [5, 6]. Sorption is a result of the interaction of pesticides with soil colloids, non-colloidal organic matter of the soil and soil solution, and it can be reversible and irreversible. The extent of pesticide sorption to soil is expressed by the distribution coefficients between the aqueous and soil phases (K_d). The distribution coefficient value varies widely with soil type and soil organic matter content, for this reason the sorption coefficient normalized to soil organic carbon content (K_{oc}) is more often used [7].

Sorption indexes are necessary to assess the potential pesticide mobility in soil, and they are also widely used as input parameters to mathematical models of the pesticide behavior in soils. The accuracy of the prediction of pesticide migration in soil largely depends on the quality of input sorption data. Since in further studies it is planned to use the data of the sorption experiments for mathematical modeling of the behavior of cyantraniliprole in soils of Russia, the main goal of this work was to study the cyantraniliprole sorption in soils of the different climatic zones of the Russian Federation - Luvic Chernozem (Kursk region), Halpic Kastanozem (Saratov region) and Albic Glossic Retisol (Moscow, large lysimeter of Moscow State University) [9] and obtain quantitative characteristics of cyantraniliprole sorption.

* Corresponding author: v.kolupaeva@vniif.ru, amulanya@gmail.com

2 Objects and Methods

2.1 Pesticide

Cyantraniliprole is a second-generation insecticide from class of anthranilamides, which is used on a wide range of crops and is effective against sucking pests and thrips. In terms of resistance it refers to low-persistent - persistent active substances: DT_{50} = 8.7 - 91.9 days ($DT_{50 geomean\ lab.}$ = 34.4 days), in terms of migration ability it refers to medium-mobility active substances: K_{oc} = 157-367 ($K_{oc\ average}$ = 241 ml/g) [8].

2.2 Soils

The properties of soils are shown in Table 1.

Table 1. Physical and chemical properties of investigated soils.

Soil	Horizon	Depths, cm	$C_{org.}$, %	pH_w	Particle size, %		
					Sand	Silt	Clay
Luvic Chernozem	A	0-24	4.9	6.9	14.1	35.6	50.4
Halpic Kastanozem	A	0-18	2.5	7.1	51.1	21.1	27.8
Albic Glossic Retisol	A	0-36	2.2	6.5	1.5	87.9	10.6
	E-EB	36-51	0.7	6.1	1.0	91.5	7.5

2.3 Study of sorption kinetics for one concentration of cyantraniliprole in one soil

The sorption study was carried out in accordance with the OECD methodology [10].

A preliminary experiment was conducted to study the sorption kinetics of cyantraniliprole in order to determine the time period for establishing equilibrium in the soil – solution system. At the first, 10 g of air-dried soil samples (Retisol, arable horizon) were sifted through a 1 mm sieve and placed in flasks, then 20 ml of 0.01M solution of $CaCl_2$ were added and flasks were shaken for 4 hours before the experiment. After that, 5 ml of the test compound solution was added into the flasks to bring the final volume to 25 cm^3 and to obtain cyantraniliprole concentration of 0.1 μg/ml. Flasks were shaken for 16 hours.

Sampling was conducted after 0.1, 2, 4, 8, 12, 16 hours. The study was carried out in 2 replicates. The selected solutions were centrifuged at a speed of 4000 rpm for 15 min, equilibrium content of cyantraniliprole in the liquid phase was determined by high performance liquid chromatography (HPLC).

The solutions were prepared from an analytical standard containing 99.8% cyantraniliprole.

2.4 Study of cyantraniliprole sorption

At the first, 10 g of air-dried soil samples were sifted through a 1 mm sieve and placed in flasks, then 20 ml of 0.01M solution of $CaCl_2$ were added and flasks were shaken for 4 hours before the experiment. After that, 5 ml of the test compound solution was added into the flasks to bring the final volume to 25 cm^3 and to obtain cyantraniliprole concentration of 0.5; 0.2; 0.1; 0.05 and 0.02 μg/ml. The flasks were placed on a mechanical shaker; samples were taken for analysis after 16 hours. The study was carried out in 2 replicates.

2.5 Determination of cyantraniliprole in soil

Quantitative determination of cyantraniliprole content in water and soil was carried out in accordance with the methodological guidelines [11]. Detection limits of cyantraniliprole were 0.5 µg/L and 2.5 µg/kg for water and soil respectively.

2.6 Construction of isotherms and calculation of sorption coefficients

The dependence of the sorption value of cyantraniliprole on its equilibrium concentration in solution was approximated by the equation of linear sorption isotherm:

$$K_d = q/C_e \quad (1)$$

where K_d – sorption coefficients, q – concentration of pesticide adsorbed on soil (µg/g), a C_e – equilibrium concentration of pesticide in solution (µg/ml).

Since the soil organic matter plays a dominant role in the pesticide sorption, the sorption coefficient is often normalized by the content of soil organic carbon and K_{oc} coefficient is used for a comparative assessment of the sorption ability of soils, regardless of the content of organic matter in them [7]. K_{oc} value was calculated as:

$$K_{oc} = (K_d\ x\ 100)\ /\ C_{org} \quad (2)$$

where C_{org} – content of soil organic carbon (%).

Moreover, the relationship between the concentration of adsorbed cyantraniliprole and its concentration in an aqueous solution was approximated by the Freundlich equation:

$$q = K_f\ x\ C^{1/n} \quad (3)$$

where K_f and $1/n$ are the Freundlich coefficient and the Freundlich exponent, which indicate the affinity of the soil for the pesticide and the degree of linearity between the amounts adsorbed and the solution concentration, respectively.

3 Results and discussion

The preliminary experiment on determining the time to reach equilibrium in the soil-solution system for cyantraniliprole showed (Fig. 1) that after 8 hours of the experiment, the sorption rate decreases significantly, and 16 hours are enough to achieve the equilibrium of cyantraniliprole between solid and liquid phases.

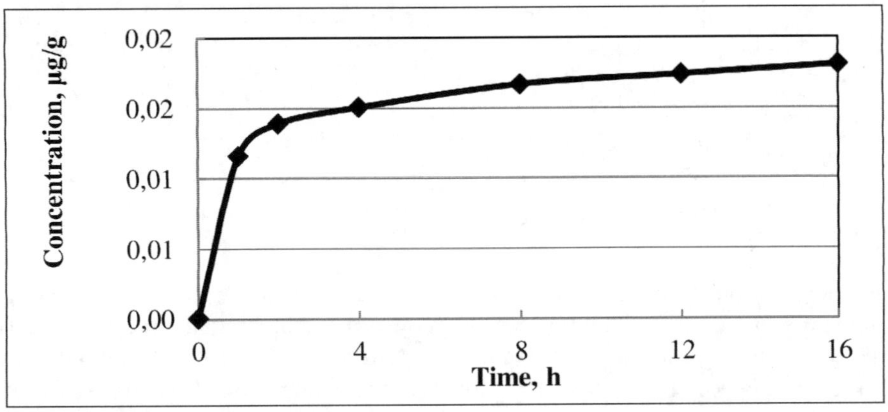

Fig. 1. Sorption kinetics of cyantraniliprole in soil.

The sorption of cyantraniliprole is well described by linear isotherms (Fig. 2), which allows to use a simple and universal index, the distribution coefficient (K_d). The values of K_d, as well as Freundlich coefficient (K_f) and Freundlich exponent (1/n) are shown in Table 2. The values of determination coefficient R^2 indicate that the accuracy of the description of the obtained data is higher when using the linear equation. For linear equation all values of R^2 are above 0.99.

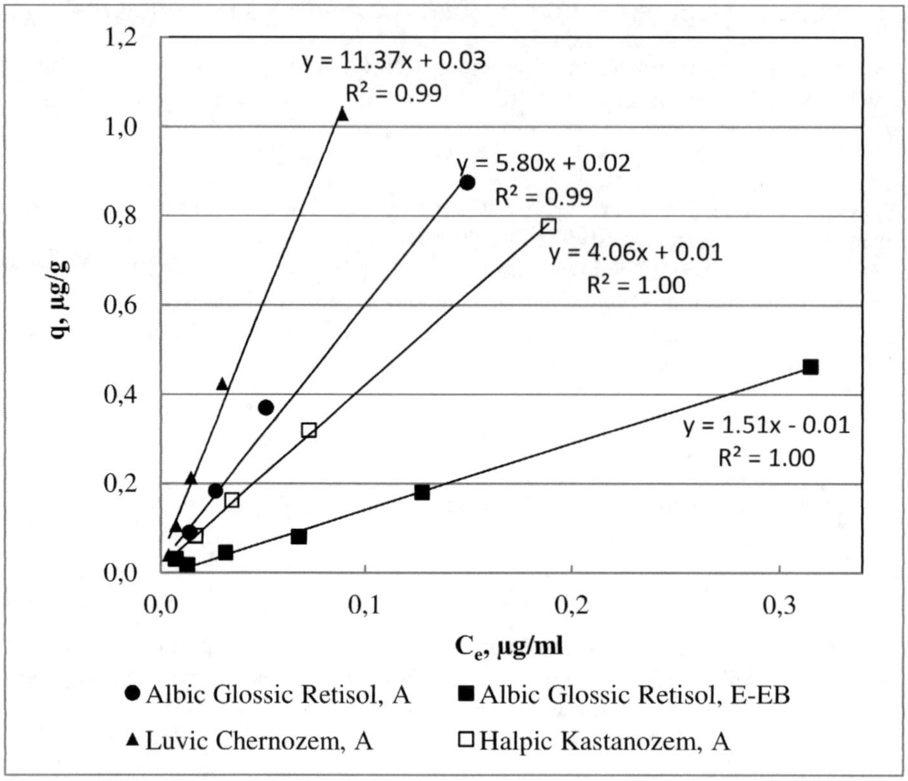

Fig. 2. Isotherms of cyantraniliprole sorption.

Table 2. Sorption parameters of cyantraniliprole for Freundlich equation and linear sorption model.

Soil, horizon	K_d*	R^2	K_{oc}**	K_f***	1/n****	R^2
Halpic Kastanozem, A	4.06±0.22	0.999	185	4.07±1.35	0.97±0.09	0.998
Luvic Chernozem, A	11.37±1.78	0.993	258	13.80±2.88	1.01±0.24	0.984
Albic Glossic Retisol, A	5.80±0.31	0.992	387	7.94±1.45	1.07±0.08	0.991
Albic Glossic Retisol, E-EB	1.49±0.11	0.998	229	3.02±9.33	1.41±0.59	0.916

* Linear adsorption coefficient with 95% confidence interval;
** linear adsorption coefficient normalized for organic carbon with 95% confidence interval;
*** Freundlich adsorption coefficient with 95% confidence interval;
**** Freundlich exponent with 95% confidence interval

The values of the sorption coefficient varied from 1.49 in the mixed horizon E-EB of Albic Glossic Retisol to 11.37 in the arable layer of Luvic Chernozem. Sorption coefficients normalized to soil organic carbon content (K_{oc}) ranged from 185 (in the arable horizon of Halpic Kastanozem) to 387 ml/g (in the arable layer of Albic Glossic Retisol), which allowed cyantraniliprole to be classified as a medium-mobile pesticides [12]. These data generally correspond to the results of cyantraniliprole registration research conducted in EU (K_d = 2.51 - 7.14, K_{oc} = 157 - 367 ml/g) [13], (K_{oc} = 251 ml/g) [2].

All obtained K_d values are significant at a 0.01% significance level.

The highest value of K_{oc} was obtained for the arable horizon of Albic Glossic Retisol, this suggests that, in addition to organic matter, mineral part of soil is also involved in the sorption of the pesticide.

To determine the factors affecting the cyantraniliprole sorption, the regression equations were obtained (determination coefficients R^2 are given in Table 3).

Table 3. Determination coefficients of regression relationship of sorption coefficient K_d to soil properties.

$C_{org.}$, %	Clay, %	pH	Silt, %	R^2
x				0.94
	x			0.73
		x		0.01
			x	0.20
x	x			0.97

According to the Table 3, the sorption coefficient most depends on the content of organic matter (R^2 = 0.94) and clay (R^2 = 0.73) in the soil. The regression equation of the dependence of the sorption coefficient on the soil organic carbon content has the following form (R^2 = 0,94):

$$Kd = 2{,}23 \times Corg - 0{,}33 \quad (4)$$

4 Conclusions

1. The study of quantitative characteristics of cyantraniliprole sorption by the different soils showed that this process is well described by linear isotherms (R^2=0.99 for all soils).
2. According to obtained values of K_{oc}, cyantraniliprole belongs to medium-mobile pesticides.
3. The sorption coefficient of cyantraniliprole was closely correlated with the content of organic carbon and clay.

References

1. Gosudarstvennyj katalog pesticidov i agrohimikatov URL: http://mcx.ru/ministry/departments/departament-rastenievodstva-mekhanizatsii-khimizatsii-i-zashchity-rasteniy/industry-information/info-gosudarstvennaya-usluga-po-gosudarstvennoy-registratsii-pestitsidov-i-agrokhimikatov/ (in Russian)

2. N. Vela, G. Pérez-Lucas, M.J. Navarro, I. Garrido, J. Fenoll, S. Navarro, Bull. Environ. Contam. Toxicol., **99(4)**, 465 (2017)

3. E. Shein, A. Belik, A. Kokoreva, V. Kolupaeva, P. Pletenev, Moscow Univ. Soil Sci. Bull., **72(4)**, 185 (2017)

4. V. Kolupaeva, A. Kokoreva, A. Belik, P. Pletenev, Open Agriculture, **4**, 599 (2019)

5. R. Green, S. Karickhoff, *Pesticides in the Soil Environment: Processes, Impacts, and Modeling*. Soil Science Society of America Book Series 2. Soil Science Society of America, Madison, WI, 79 (1990)

6. Z. Gerstl, Crop Prot, **19**, 643 (2000)

7. D. I. Gustafson, Environ. Toxico.l Chem. **8**, 339 (1989)

8. PPDB - Pesticide Properties Database URL: **https://sitem.herts.ac.uk/aeru/ppdb/en/**

9. *IUSS Working Group WRB. 2015. World Reference Base for Soil Resources 2014, update 2015 International soil classification system for naming soils and creating legends for soil maps. World Soil Resources Reports No. 106.* (FAO, Rome, 2015)

10. *Adsorption - Desorption Using a Batch Equilibrium Method. OECD guideline for the testing of chemicals. Test No. 106.* (US EPA, 2000)

11. *Opredelenie ostatochnyh kolichestv ciantraniliprola v vode, pochve, kapuste, plodah tomata i tomatnom soke, v luke pere i luke repke metodom vysokoeffektivnoj zhidkostnoj hromatografii Metodicheskie ukazaniya*, 32 (Federal'nyj centr gigieny i epidemiologii Rospotrebnadzora, Moscow, 2015) (In Russian)

12. Metody ocenki ekologicheskoj opasnosti pesticidov pri ih registracii (Rukovodstvo po klassifikaciyam ekologicheskoj opasnosti pesticidov) (VNIIF, 2010) URL: **http://penreg.ru/pesticide-ecological-evaluation/opasnost.html** (in Russian)

13. Conclusion on the peer review of the pesticide risk assessment of the active substance cyantraniliprole, EFSA Journal, **12**, 9 (2014)

Distribution of heavy metals in anthropogenic soils: on the example of the industrial zone of Moscow

Elena Latushkina[1] and *Anna* Oleynik[2]

[1]Russian State Social University, 4/1, Wilhelma Pika St., Moscow, 129226 Russia
[2]Peoples' Friendship University of Russia (RUDN University), Faculty of Ecology, 6 Miklukho-Maklaya St, Moscow, 117198, Russian Federation

Abstract. The article presents the results of the assessment of the distribution of heavy metals content in anthropogenic transformed soils of the industrial zone of Moscow. Excesses of norms are shown, geochemical associations are obtained. The work proved to exceed the maximum permissible concentrations of cadmium, lead, copper, nickel, zinc, arsenic for eleven test sites per 13.95 hectares of industrial zone in 57% of the obtained empirical data on the content of pollutants in the soil. In the investigated area, the contents of lead, copper, nickel and zinc are not uniformly distributed and are within wide range. No excess of the norm was recorded for mercury. The isolation of two geochemical associations of man-made genesis is justified. One association includes cadmium lead, copper, and zinc with the system-forming element copper. The other consists of nickel and mercury.

1 Introduction

Heavy metals participation in biogeochemical cycles, their toxicity, the ability to form geochemical associations, accumulate and transform the qualities of environmental components together with insufficient study of their behaviour in the biosphere determine the relevance of studies on the distribution assessment of heavy metals in environmental elements. In works of many authors it has been proved: "Soils are the depositing medium for heavy metals" [1, 2, 3, 4]. In cities, especially in industrial zones, special, anthropogenic changed soils have been formed characterized by increased content of ecotoxicants [3].

Industrial areas where economic activities have ceased are usually in need of recultivation. Depending on the level of soil contamination, the algorithm, methods and technologies for the restoration of industrial areas will depend.

Contamination of soils by heavy metals in industrial areas of urban agglomerations has a negative impact on public health. This is due to the wind transfer of dust from the soil surface and its ingress into the human respiratory system and on the skin surface [7]. Heavy metals have been proven to have toxic, allergic, carcinogenic, gonadotropic effects. Among the elements with toxic effects on human health, cadmium, copper, arsenic, nickel, mercury, lead, zinc are accepted [8].

2 Materials and methods

The study was carried out during the summer period of 2019 on the territory of the former industrial zone located on a plot of 13.95 hectares at the intersection of the Enthusiasts Highway and the Nizhny Novgorod trestle. There has been no economic activity in this territory for several years.

Soil samples were collected in accordance with the principles of uniformity of placement and availability of physical accessibility of test sites. A total of 11 sites were designated for soil sampling, with 11 average samples taken. Sampling was performed in accordance with the requirements of state standards [9, 10].

Each sample was signed and delivered to the accredited laboratory of ANO "Independent Institute of Expertise and Standardization" to determine the content of lead, zinc, copper, nickel, cadmium, arsenic, and mercury by atomic absorption spectrometry and spectrophotometry.

The content of lead, zinc, copper, nickel, and cadmium was revealed by using an atomic and absorbing spectrometer of AAS-30. Arsenic was measured spectrophotometrically on the SF-26 spectrophotometer. The mercury content was determined by the flame-free atomic absorption method on the "Julia-2" mercury analyzer.

The analysis of the obtained empirical data was carried out using the index method, the descriptive mathematical statistics (arithmetic mean, minimum and maximum value, range), correlation analysis, correlation arm method.

3 Results and discussion

Exceeding of the maximum permissible concentration (MPC) of heavy metals in the soils in of the test areas is given in Table 1.

Table 1. Excess of MPC of heavy metals in soils

Sample number	Chemical element						
	Cd	Pb	Cu	Ni	Zn	As	Hg
1	1.0	2.6	10.7	9.7	4.2	0.7	0.01
2	2.1	3.7	8.5	4.9	3.0	0.5	<0.01
3	0.7	0.8	4.3	4.9	1.5	0.5	<0.01
4	0.8	1.0	4.8	4.1	1.6	0.4	<0.01
5	0.5	0.1	16.1	2.8	1.4	0.6	<0.01
6	0.7	0.3	3.5	2.8	2.3	0.8	<0.01
7	0.8	2.7	14.6	9.3	6.0	1.0	<0.01
8	0.3	6.2	13.4	2.8	5.3	0.7	<0.01
9	1.6	11.9	23.3	5.0	6.6	0.5	<0.01
10	1.5	10.6	29.3	7.5	12.6	1.6	<0.01
11	0.5	13.7	196.3	9.8	252.0	0.7	0.01

It can be seen from in Table 1 that over 57 % of all values obtained exceeds the norm. All samples have increased copper, nickel, and zinc content; seven samples show the excess of MPC for lead; three samples for cadmium, and one for arsenic.

Table 2 shows that the content of lead, copper, nickel, and zinc varies widely enough. This indicates uneven scattering of the pollutants on the studied site.

Table 2. Average, minimum, maximum values of heavy metals content in anthropogenic soils and their range of variation, mg/kg

Indicator	Cd	Pb	Cu	Ni	Zn	As	Hg
Average value	0.480	146.13	88.6	23.13	619.8	1.47	0.04
Minimum value	0.169	2.06	10.36	11.2	31.17	0.829	<0.02
Maximum value	0.783	411.3	589	39.24	5797.00	3.22	0.04
Amplitude	0.614	409.24	578.64	28.04	5765.83	2.39	0.02

To identify geochemical associations, correlation coefficients were calculated using the K. Pearson formula (Table 3), and correlation pleiades were constructed (Figure 1).

Table 3. Correlation coefficients between indicators heavy metal content in anthropogenic soils

Chemical element	Cd	Pb	Cu	Ni	Zn	As	Hg
Cd	1.0						
Pb	0.2	1.0					
Cu	0.7	0.7	1.0				
Ni	0.1	0.4	0.5	1.0			
Zn	- 0.3	0.6	0.9	0.5	1.0		
As	0.1	0.3	0.1	0.4	- 0.1	1.0	
Hg	- 0.1	0.1	0.3	0.7	0.4	-0.1	1.0

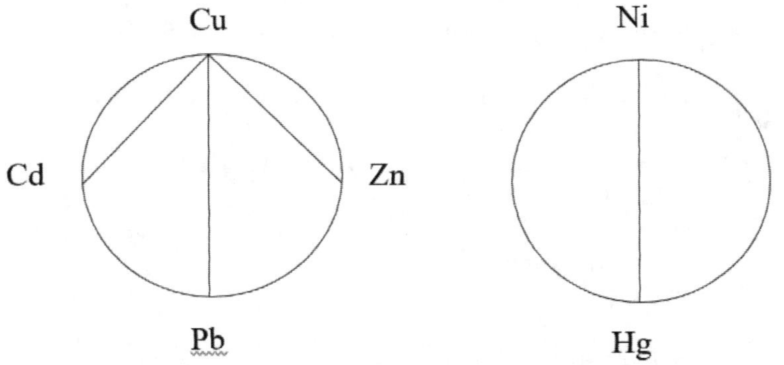

Fig. 1. **Geochemical associations of heavy metals in anthropogenic soils**

The geochemical association with the binding element (copper), which unites cadmium, lead, and zinc into a single group, is characterized by a high close relationship between copper and zinc (0.9). This may be due to the use in production brass or parts of equipment made of this alloy in the production. It should be noted that lead brass is used in mechanical engineering for making fasteners and gears. The high value of the correlation coefficient between cadmium and copper may mean the use in production of cadmium bronze, which is necessary for the manufacture of collector plates, contact wires of electrified transport. The presence of lead in the soil is due to its presence in the exhaust gases of road transport passing near the research site on the roads of the city.

The geochemical association of nickel and mercury with the close relationship of elements, which is 0.7 indicates the joint application of these chemical elements in the analysed industrial zone. It can be using or amalgam obtaining required for electrical engineering.

The revealed peculiarities of distribution of heavy metals in the decommissioned industrial zone located in the immediate vicinity of roads and residential buildings indicate the need to carry out recultivation measures.

Practical recommendations for the environmentally sound use of a territory with a high content of heavy metals in anthropogenic soils should be:

- Soil removal and cleaning;
- Creation of a reclamation layer, replacement or dilution of the contaminated soil layer;
- Control of chemical element ratios in soil;
- Planting of herbs or roll lawn flooring;
- Planting of plants resistant to increased content of heavy metals.

4 Conclusion

In the territory of the former industrial zone where there is no economic activity for a long time heavy metals are:

- not uniformly distributed in the soil, variations in lead, copper, nickel and zinc are within wide limits, the standards for cadmium, lead, copper, nickel, zinc and arsenic are exceeded. At all points the run-up exceeding the maximum permissible concentration is recorded by copper, nickel, and zinc;
- Form two geochemical associations. The first is represented by cadmium, lead and zinc with the system-forming element copper. The second is nickel and mercury. The genesis of these associations is man-made.

5 References

1. N.A. Kirpichnikov, N.A Chernykh, I.N. Chernykh, Agrochemistry, **2**, 93-101 (1993) (in Russian)
2. E.V. Stanis, E.N. Latushkina, N.V. Marsheva, E.A. Parakhina, Problems of regional ecology, **2**, 109-112 (2018) (In Russian)
3. G.A. Warm, Astrakhan J. of Environmental Education, **1 (23)**, 182-192 (2013)
4. D.V. Vlasov, Geophysical Research, Abstracts, **16** (2014).
5. N.S. Kasimov, D.V. Vlasov, N.E. Kosseleva, E.M. Nikiforova, *Geochemistry of landscapes of Eastern Moscow* (APR, Moscow, 2016)
6. I.F., Medvedev, S.S. Zhelyagin, *Heavy metals in ecosystems* ("Racurs" Publ., Saratov.) (2017) (In Russian)
7. I.V. Wise, Hygiene and sanitation, **4**, 32-37 (2008)
8. M.I. Chubirko, Population Health and Habitat, **1**, 19-20 (2008)
9. GOST 17.4.3.01-2017 Nature Protection. Soils. General sampling requirements (in Russian)
10. GOST 17.4.4.02-2017 Nature Protection. Soils. Methods of sampling and preparation of samples for chemical, bacteriological, helminthological analysis (In Russian)

Content of heavy metals in Haplic Chernozem under conditions of agrogenesis

Anna Medvedeva[1*], *Olga* Buryukova[1], *Alexey* Kucherenko[1], *Yaroslav* Ilchenko[1], *Victor* Chaplygin[1], *Vishnu* D. Rajput[1], and *Ridvan* Kizilkaya[2]

[1] Southern Federal University, Stachki Ave, 194/1, Rostov-on-Don, 344090, Russia
[2]Ondokuz Mayis University, 55139, Samsun, Turkey

Abstract. The paper presents the results of a long-term study on the content and distribution of heavy metals in Haplic Chernozem of agrocenoses in southern zone of Rostov Region. The impact of various agricultural technologies - traditional (using moldboard ploughing) and resource-saving (minimum and No-till) on accumulation and mobility of Pb, Zn, Cu in Haplic Chernozem in winter wheat cultivation is considered. The total content of Pb, Zn, Cu in the winter wheat areas corresponds to the background level and does not exceed permissible concentrations. Low mobility of elements (Pb - 2.0-3.0%; Zn and Cu up to 1.0% as a percentage of their total forms) is a regional soil feature. In addition, the low availability of Zn and Cu in soils is due to their removal with the crops in the absence of micronutrient fertilizer replenishment. The significant role of soil organic matter in the accumulation of both total and mobile forms of heavy metals in soil has been confirmed. The intensity of the cultivation methods used has a significant effect on the total content of Zn, Cu and on the mobility of Pb, Zn, Cu in Haplic Chernozem.

1 Introduction

Soil is one of the most important natural resources. Anthropogenic pressure on soils is often accompanied by their degradation, including pollution by heavy metals. Among the many anthropogenic factors affecting the change in the content of heavy metals in the soil, crop production technologies play a significant role. Intensive farming contributes to a change in the content of metal compounds in the soil, which can negatively affect crop quality, both due to its decline caused by insufficient micronutrition and due to contamination of products with highly dangerous substances [1]. The content and composition of heavy metals in agricultural soils is largely determined by the source material (parent rock). But their entry from other sources, including organic and mineral fertilizers, chemical ameliorants can be significant [2]. Widespread application of advanced resource-saving technologies based on minimal tillage is a pressing issue. However, it is the least studied under the soil and climatic conditions of Rostov Region. The introduction of resource-saving technologies is possible with high farming standards, sufficient supply of fertilizers and pesticides. To determine the content of heavy metals in soils taking into account their regional characteristics and the

* Corresponding author: medvedeva.estelior@yandex.ru

intensity of modern agricultural technologies is necessary for assessing and predicting the availability of macro- and microelements to plants, and environmental monitoring.

2 Materials and methods

Field research studies have been carried out for 3 years (2014 - 2017) on the premises S. M. Kirov JSC in Peschanokopsky district of Rostov Region where minimum tillage technology has been used since 2000, No-till has been applied since 2008. In terms of natural and economic division, this territory belongs to the Southern Haplic Chernozems (Ciscaucasia) with a semi-arid climate, moderately hot summers and moderately cold winters. According to the World Reference Base for Soil Resources (WRB), this type of chernozem belongs to Calcic Chernozems [3].

Nine soil profiles were laid in the winter wheat areas: 4 - using No-till (Semeato TDNG-420 made in Brazil); 4 – minimal cultivation to a depth of 10-12 cm (heavy disk harrow-3); 1 – moldboard ploughing (ploughing to a depth of 25-27 cm, lift-type disc plough-4-35). Soil samples from two soil profiles in the virgin area were compared.

To determine the mobile forms of Zn, Cu, Pb in the soil acetate–ammonium buffer solution (pH 4.8) was used followed by atomic absorption spectrometry (soil to solution ratio 1:10) [4]. The total content of elements was determined by X-ray fluorescent (XRF) scanning spectrometer SPECTROSCANMAX-GV.

Mathematical processing of the results was carried out by correlation analysis and ANOVA methods using the STATISTICA 10 software package.

3 Results and discussion

It was established that lead accumulated in the humus horizon (Ad, A), decreased down the profile accumulating in the lower horizons - Bca and C (Figure 1a). The organophilicity of Pb is well known. According to Morin [5], the content of lead in the silt fraction of contaminated humic soil (Corg = 6.4%) in northern France reached 80%. In arable soil with a low content (Corg = 1.5%), the fraction of Pb-humates was lower, it constituted only 35%. The accumulation of total Pb content in the lower horizons of the soil profile was due to its binding by carbonates. Zn and Cu were characterized by a uniform distribution over the entire soil profile (Figure 2a and Figure 3a).

The total Zn content in the upper humus horizon(Ad) of Haplic Chernozem ranged from 71.8 to 88.3 mg/kg, Cu - 41.6 - 54.2 mg/kg, Pb - 25.3 - 31.9 mg/kg, which corresponded to the background level in this region [6, 7].

Ecotoxicological assessment of the soil collected from winter wheat areas and virgin land showed no excess of the approximately permissible concentrations of the studied elements [8].

More reliable information on the content and impact of trace elements on the ecological status of agrocenoses was provided by their mobile forms. The content of mobile compounds of the elements under study can be represented in the form of a decreasing series: Pb> Zn> Cu. The percentage of mobile elements was Pb - 2.0-3.0%; Zn and Cu up to 1.0% of the percentage of their total content. According to agrochemical standards, soil is characterized by a low availability of mobile compounds of Zn (<2.0 mg/kg) and Cu (<0.2 mg/kg) [9]. This is probably due to active absorption by plants and indicates insufficient use of zinc and copper-based fertilizers.

The low mobility of Cu and Zn in the soils of Rostov Region is due to the presence of carbonates, their highly dispersed micellar carbonates and a slightly alkaline reaction of the medium [6, 7].

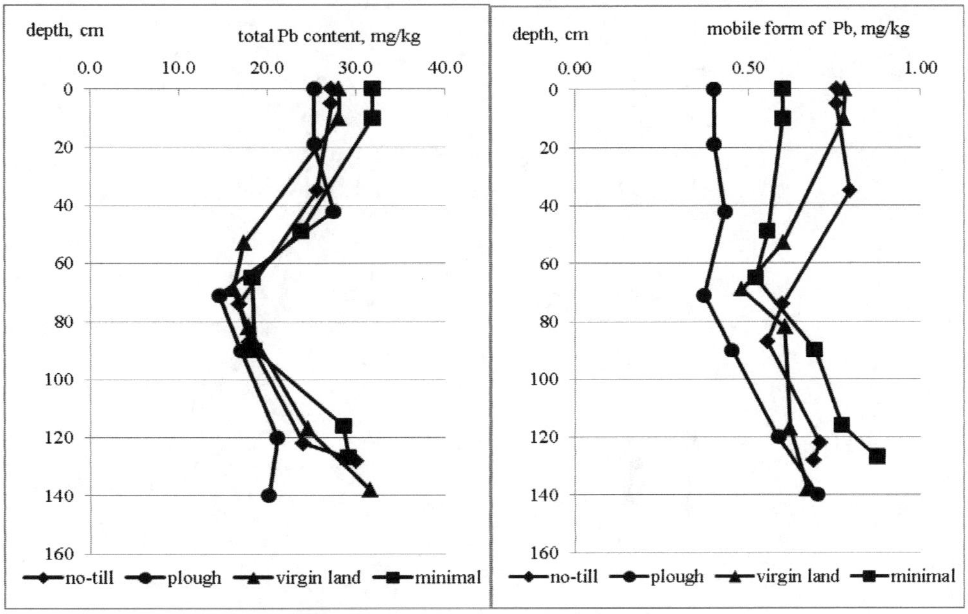

a) b)

Fig. 1. Profile distribution of lead in Haplic Chernozem (a) total, b) mobile), mg/kg.

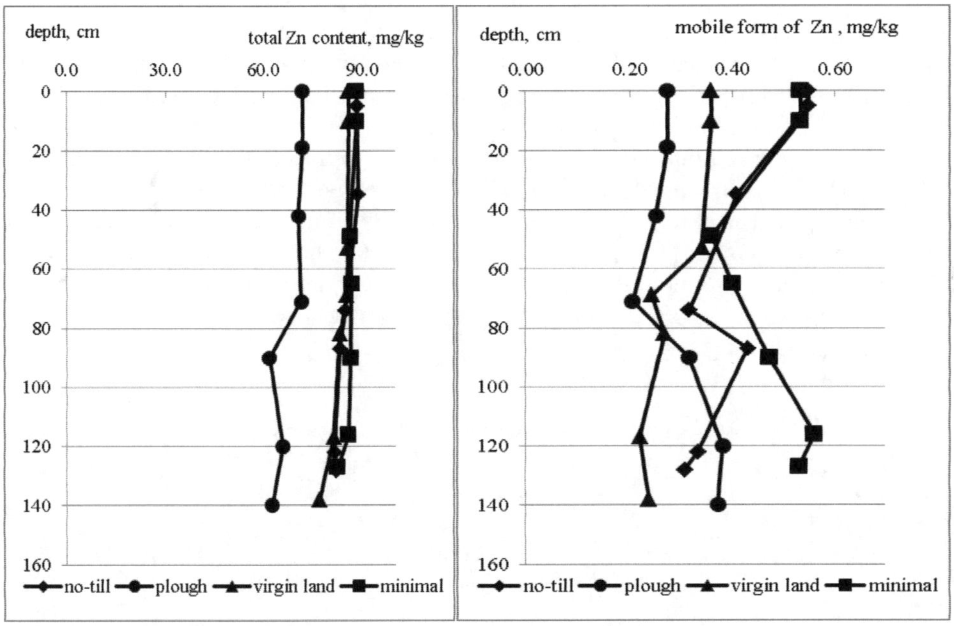

a) b)

Fig. 2. Profile distribution of zinc in Haplic Chernozem (a) total, b) mobile), mg/kg.

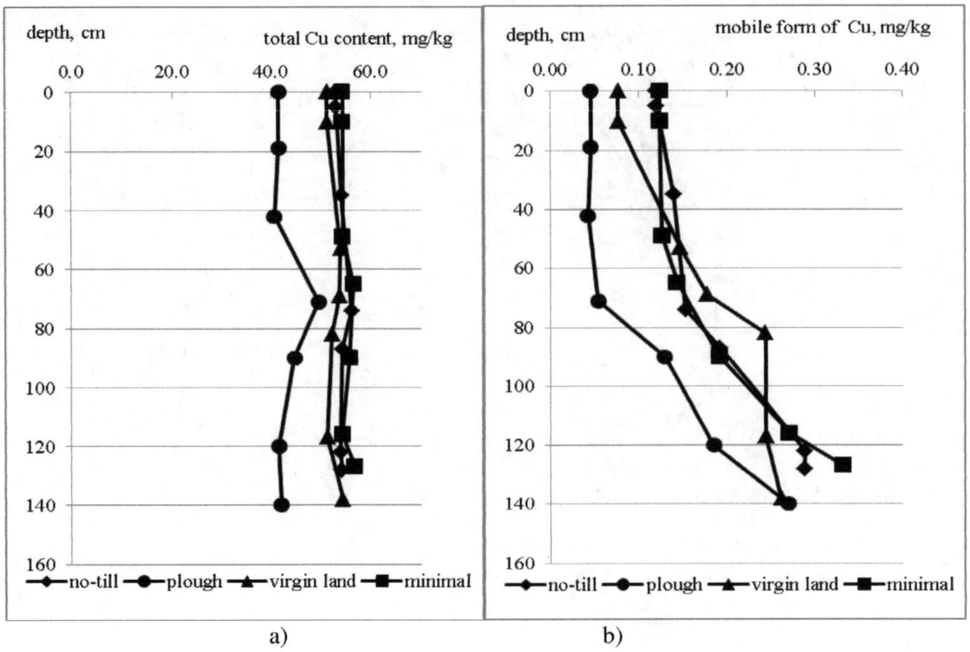

Fig. 3. Profile distribution of copper in Haplic Chernozem (a) total, b) mobile), mg/kg.

Soil is the major source for plants to intake Zn and Cu. The deficiency of these microelements deteriorates the quality of crop production, including the ratio between macro- and microelements [10]. These changes often cause animal and human diseases [11].

The use of resource-saving technologies, including No-till, with sufficiently large doses of mineral fertilizers and intensive care of crops increased the content of mobile Zn and Cu compounds in the soil, but the degree of availability remained the same.

ANOVA of the data obtained indicated a statistically significant effect of the cultivation method on the total content of Cu and Zn. Changes in total Pb content were insignificant. However, analysis of the mobile Pb content both on average in the soil profile and in the humus horizon (Ad-AB) showed a statistically significant effect of the intensity of the studied cultivation methods (Figure 4).

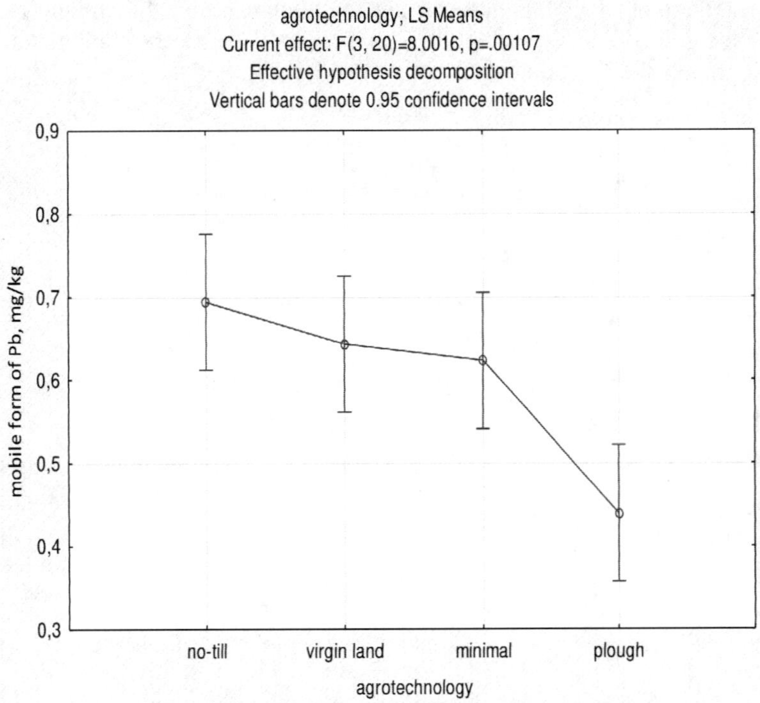

a)

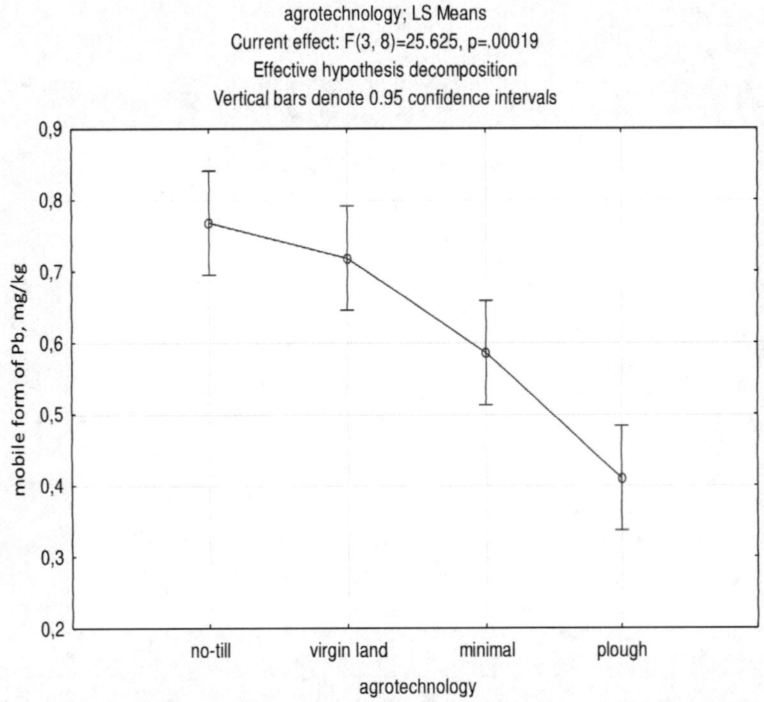

b)

Fig. 4. The influence of a cultivation method on the content of mobile lead compounds in Haplic Chernozem (a) - on average in profile, (b) - and in the humus horizon (Ad-AB).

A significant effect of the intensity of the methods used on the content of mobile Zn compounds was revealed over the entire soil profile, and on the content of mobile Cu compounds - only in the soil humus horizon (Ad-AB) (Figure 5).

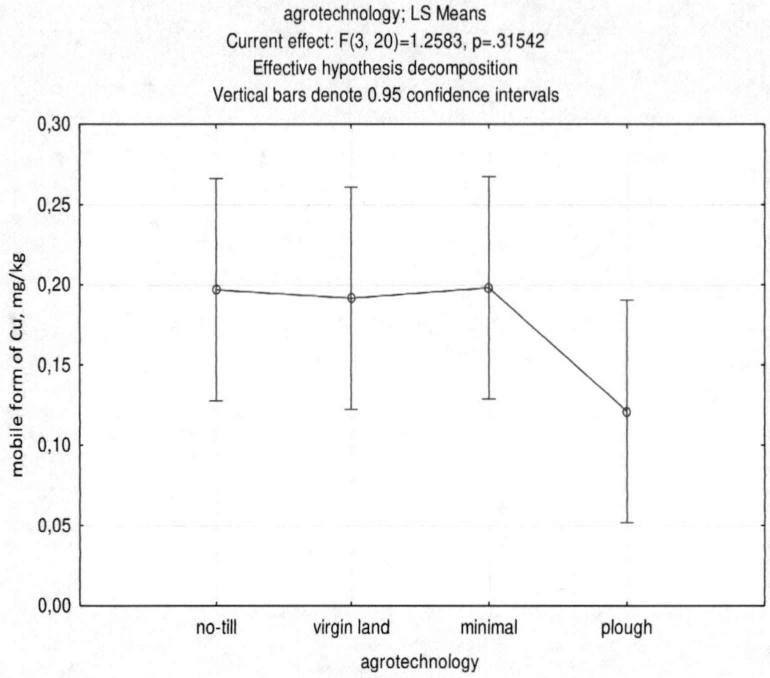

a)

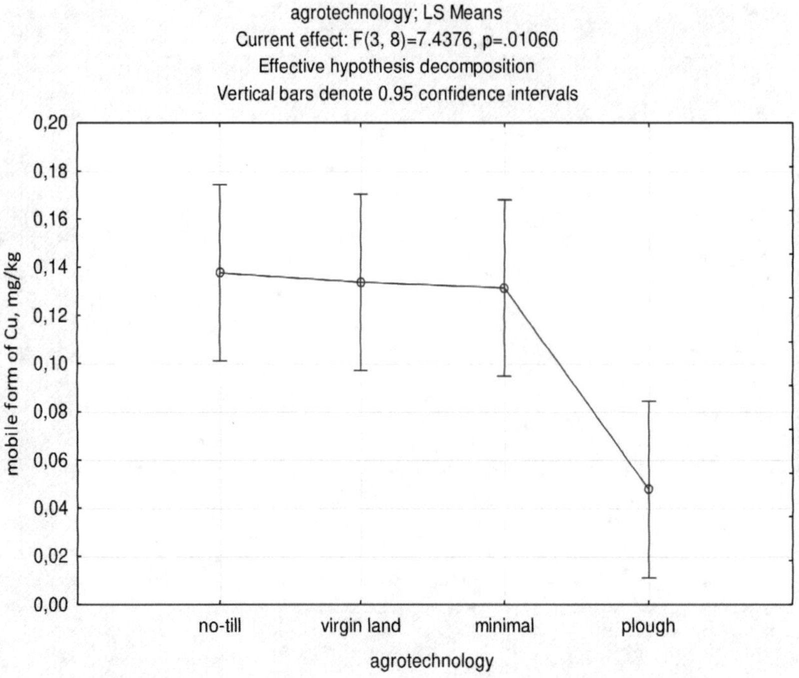

b)

Fig. 5. The influence of a cultivation method on the content of mobile copper compounds in Haplic Chernozem (a) - on average in profile, (b) - and in the humus horizon (Ad-AB).

Soil organic matter (SOM) is one of the main components of the soil affecting the accumulation of heavy metals. A direct average dependence of the total Pb content (r = 0.56 at a confidence level of 0.95%) on the humus content was revealed (Figure 6a). A similar dependence (r = 0.57) was revealed for total Zn. Feedback has been found between the mobile Cu compounds and the humus content (Figure 6b).

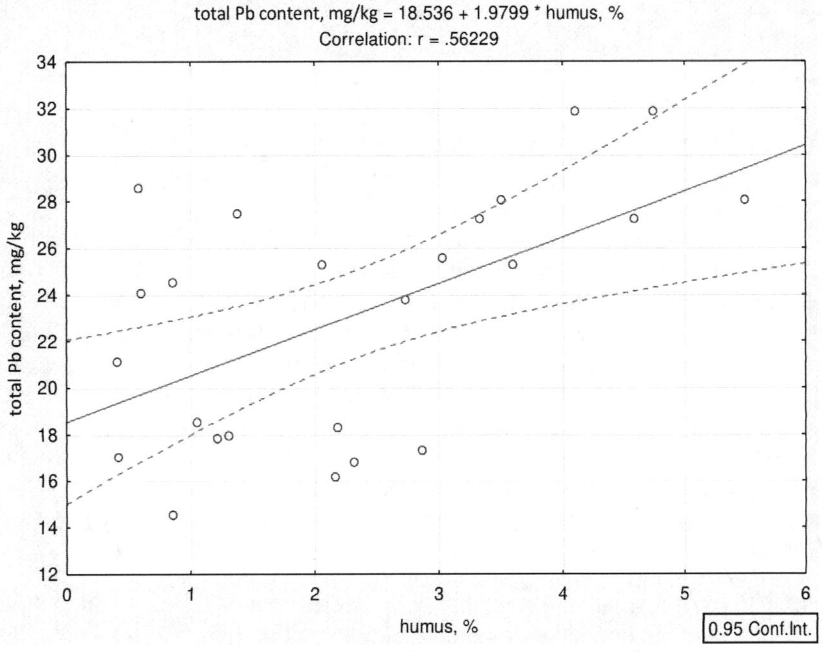

a)

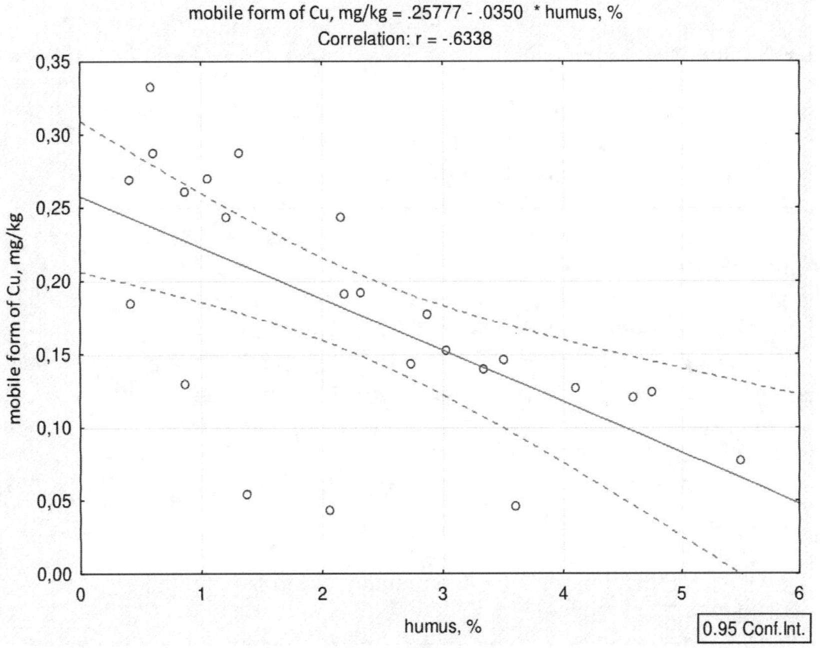

b)

Fig. 6. Dependence of the accumulation of lead (a) and copper (b) in Haplic Chernozem on humus content.

Cu is known to have affinity with the organic complex of the soil. Our previous studies have shown that the use of resource-saving technologies, including No-till, increased the humus content compared to ploughing [12]. According to Moreiraa [13], an increase in SOM concentration with No-till causes the decrease in the availability of Cu to plants. Therefore, when increasing the soil organic material in the fields it is necessary to apply copper-containing fertilizers.

4 Conclusion

The content and distribution throughout the profile of heavy metals (Pb, Zn, Cu) in Haplic Chernozem of agrocenoses in the southern zone of Rostov Region is due to regional soil characteristics and the intensity of cultivation methods.

Comparing the obtained results with current approximate permissible concentrations (APC), it should be noted that the content of heavy metals (Pb, Zn, Cu) when using both traditional and resource-saving technologies cannot cause any pathological changes or anomalies of biological processes or lead to their accumulation in agricultural plants.

SOM is one of the key factors determining the mobility of Pb, Zn, Cu in Haplic Chernozem. An increase in the content of SOM when using resource-saving technologies, including No-till, can reduce the availability of trace elements (Cu and Zn) to plants. To preserve soil fertility and improve the crop quality, introduction of both macro- and microelements is necessary when growing crops.

The research was supported by the project of the Ministry of Science and Higher Education of the Russian Federation within the framework of the state task of the Southern Federal University for 2020-2022 (№ FENW-2020-0028)

The research was financial supported by the project of the Ministry of Science and Higher Education of the Russian Federation within the framework of the state task (Southern Federal University) and the grant from the President of Russian Federation, No. MK-2818.2019.5.

5 References

1. O.A. Biryukova, I.I. Elnikov, V.S. Kryshchenko, *Rapid diagnostics of plant nutrition*. (Publishing house of SFU, Rostov/D, 2010) (In Russian)
2. B.J. Alloway, Environmental Pollution, **22**, 11-50 (2012)
3. *World reference base for soil resources 2014. International soil classification system for naming soils and creating legends for soil maps. World Soil Resources Reports No. 106* (FAO, Rome, 2014)
4. *Methodological Recommendations for the Determination of Heavy Metals in farmland soils and crop plants* (TsINAO, Moscow, 1992) (in Russian).
5. G. Morin, J.D. Ostergren, F. Juillot, P. Ildefonse, G. Calas, J.E. Brown, XAFS determination of the chemical form of lead in smelter contaminated soils and mine tailings: Importance of adsorption process. Am. Mineral., **84**, 420–434 (1999)
6. E.V. Agafonov, *Trace elements-HM in the research of the Department of Agricultural Chemistry*, 262 (DonSAU. Publishing house LLC "Printing complex ESMA-PRINT", r.p. Kamenolomni, 2012) (In Russian)
7. T.V. Bauer, T.M. Minkina, S.S. Mandzhieva, V.A. Chaplygin, D.G. Nevidomskaya, S.N. Sushkova, Scientific journal of the Russian Research Institute of Land Reclamation, **4(20)**, 186–199 (2015) (In Russian)
8. *Approximately permissible concentration (APC) of chemicals in the soil: Hygienic standards*, 10 (Federal Center for Hygiene and Epidemiology of Rospotrebnadzor, Moscow, 2009) (In Russian)

9. L.M. Derzhavina, D.S. Bulgakova, *Methodological Guidelines for comprehensive monitoring of soil fertility in agricultural lands*, 240 (Federal State Budgetary Institution "Rosinformagroteh", Moscow, 2003) (In Russian)
10. N.P. Bityutsky, *Microelements of higher plants* (Publishing House St. Petersburg University, St. Petersburg, 2011) (In Russian)
11. A.V. Rocky, *Human microelements (diagnosis and treatment)* (Scientific world, Moscow, 1999) (In Russian)
12. A.M. Medvedeva, O.A. Biryukova, Y.I. Ilchenko, A.V. Kucherenko, E.V. Kuchmenko, Successes in modern natural sciences, **1**, 29-34 (2018) (In Russian)
13. A. Moreiraa, L.A. Cardoso Moraes, G. Schroth, Soil & Tillage Research, **193**, 133–141 (2019).

Metodological aspects in the studying of soil particle size distribution under contamination and after reclamation

Inna Zamulina*, *Marina* Burachevskaya, *Saglara* Mandzhieva, *Tatiana* Bauer, *Anatolii* Barakhov, and *Mahmoud* Mazarji
Southern Federal University, prosp. Stachki, 194/1, Rostov-on-Don, 344090, Russian Federation

Abstract. In this study, a model experiment was performed under artificial copper contamination of Haplic Chernozem soil. Following the introduction of granular activated carbon (GAC) and biocharin contaminated soil, data on the particle size distribution were obtained. The particle size distribution of Haplic Chernozem was determined by two different methods namely pipette method (PM) and laser diffraction method (LDM). For the LDM, changes in the composition of fractions of more than 0.01 mm were more obvious, while for PM, the main differences were of characteristic for the content of fine fractions less than 0.01 mm.

1 Introduction

Soil pollution originated by industrial companies is a worldwide problem. The content of heavy metals (HMs) in soils is one of the indicators of anthropogenic load [1]. Controlling of these priority pollutants are of great concern [2]. The adsorption and desorption behavior of HMs in the soil media is an important factor that affects their distribution in solid and liquid phases [3]. HMs quickly and firmly adsorbed from solutions by surface soil particles [4]. Moreover, the adsorption of HMs by the soil significantly alters the physical and chemical properties of soils and their corresponding components. HM cations would alternatively affect the specific surface of soils, the microstructure of clay minerals, the thermodynamic state of moisture in clay minerals, and the mineralogical composition of the clay fraction [5]. Consequently, the structure of the soil is altered while its density increases, the overall porosity and water permeability decrease, and water and air conditions deteriorate.

The use of carbon sorbents is one of the most effective and inexpensive methods for soil reclamation, especially for those subjected to technogenic pollution. The introduction of carbon sorbents would alternate eventually the physical properties of the soil. The effect of sorbent additives applied to the soil may vary depending on one case to another. Low-density sorbents can reduce the density of soil. Another possible effect is in the case of low mechanical strength of the sorbents [6]. The use of organic matter appears to favor soil

* Corresponding author: ivzamulina@sfedu.ru

water retention capacity through an increase in soil porosity and the mean pore size diameter [7].

Carbon sorbents additions to soil may enhance organo-mineral interactions *via* adsorption and/or ligand exchange reactions [8]. These adsorbents can also provide improvement and long-lasting environmental effects for mycorrhizal growth [9]. Besides, this can result in support of soil structure building processes [10].

The particle-size composition of soils is one of the most important basic properties of soils. It governs the ability to determine many other physical and chemical characteristics of the soil, as well [11]. Soil particle-size fractions, i.e., the percentage compositions of clay, silt, and sand, are widely used to classify soil texture and also estimate some hydraulic properties. Standard methods for grain-size analysis are based on sedimentation rates for the fine fractions and screening of the coarse fractions. The classical standard technique established using a pipette method (PM) for soil particle size analysis. Over recent decades, various new methods for grain-size analyses have been developed. The example of the most common method is laser diffractometry (LDM) [12].

In this regard, the characteristics of changes in the particle size distribution applied in cupper polluted Haplic Chernozem alongside introducing carbon sorbents were carried out by two methods of PM and LDM. The results would allow assessing the effect of carbon sorbents on the physical properties of the soil and especially the particle size distribution.

2 Materials and Methods

The upper layer (0–20 cm) of Haplic Chernozem from the specially protected natural area "Persianovskaya steppe" was taken. The sampling site was located far from possible sources of pollution. The soils for the experiment were dried and passed through a sieve with the opening of 2 mm. 2-liter vegetation vessels with a closed drainage system were used for conducting model vegetation experiment. The vessels were poured with 2 kg of soil. Afterward, Cu was subjected in the form of metal oxide (in a dry state) at two different dose levels of 300 and 2000 mg/kg and thoroughly mixed. After 2 months of incubation, biochar and granular activated carbon (GAC) were added to contaminated soil at doses of 1% and 2.5%. These doses were selected according to our previous results of the phytotest. The soil was incubated at a steady water content of 60% of the total field capacity. The temperature was maintained in a range between 20-22°C during the span of the experiment. After 4 months of incubation, two-row spring barley (Hordeum sativum distichum) of the Ratnik variety was seeded. Watering was carried out every other day in order to maintain the corresponding humidity at field moisture capacity (60%). Following the completion of the plant's sampling in the phase of full ripeness, soil samples were taken from the entire volume of the vegetation vessel along with the root system.

The particle size distribution was determined by classical PM and the LDM methods followed by pyrophosphate sample preparation.

3 Results and Discussion

The results of the study of particle size distribution using PM method showed that according to the classification of N.A. Kachinsky Haplic Chernozem corresponded to a heavy loamy in all experimental variants. The content of the sum of fractions less than 0.01 mm in the control was 48.1%, varying within the range of 6.8% (44.4-51.2%) (Fig. 1). In this case, the taxonomic unit of particle size distribution remained unchanged. Variation of the fraction less than 0.001 mm was 15.6%. The maximum fraction content of less than 0.001 mm was noted in the variant with the introduction of 2000 mg/kg Cu (31.4%) into the soil. The minimum was noted in the variants with the addition of GAC at a dose of 2.5% and combined introduction of Cu at a dose of 300 mg/kg + GAC 2.5% which accounted for 16.8% and 15.8%, respectively.

The main changes in the quantitative content were revealed for small fractions of less than 0.001 mm, 0.001-0.005 mm, 0.005-0.01 mm.

The particles are sorted by shape and size in the process of formation of natural undisturbed soils. On the contrary, contaminated soils are often formed by a mixture of various materials The particle size distribution, under contamination with HMs, can reveal a change due to the chemical reactions between the mineral particles of the soil and the pollutant [5]. Changes in the distribution of particle size fractions are possible due to micro-aggregation. When HMs solutions enter the soil, the thickness of the double electric layer decreases. Hence, with the same amount of energy, the particles come together and become more condensed [13].

The differences in experimental variants can be due to two reasons: 1) the features of the determination method and 2) the influence of Cu and sorbents. The first reason is the use of weighted average density values when calculating the sedimentation rate using the Stokes formula. Particles with different values of density and diameter can potentially fall into one fractional sample. For example, particles of biochar and GAC with a low apparent density (not less than 0.37 g/cm^3) felt into a fraction of 0.001-0.005 mm. So, in the variants with the addition of 2.5% GAC and 2.5% biochar, the content of this fraction increases by 7.2% and 5%, respectively (Fig. 1). The second reason is associated with the behavior of the introduced HM in the soil. The absorption of the Cu^{2+} ion is largely carried out by the ion exchange mechanism [14]. The substitution of Ca^{2+} and Fe^{2+} ions for Cu^{2+} can increase the yield of finely dispersed fractions due to the formation of less stable organometallic complexes and partial destruction of the mineral organic compounds [15]. The maximum content of the silt fraction was established in the variant with the introduction of 2000 mg/kg Cu - 31.4%.

The introduction of carbon sorbents eliminated the effect of Cu, bringing the data on the content of fractions closer to the control. However, due to different doses and also the adsorption characteristics of carbon sorbents, their effect was observed to be different (Fig. 1).

For LDM, the content of the sum of fractions in the Haplic Chernozem less than 0.01 mm was lower than in the PM - 30.6-42.5%. The same result was observed for the amount of fraction less than 0.001 mm (Fig. 1). The difference in the results obtained by the two methods has been discussed in the literature [11, 16, 17].

Granulometric composition, as a multicomponent system, can be characterized using fundamental physical characteristics. In this regard, entropy (H) [18] was calculated by the Shannon formula:

$H = - \sum (x_i / G) \log x_i / G$, equation (1)

where x_i is the content of the fraction, G is the sum of all fractions.

The calculations showed a decrease in the average values of entropy relative to the control in both LDM and PM methods. It was found that the value of this indicator in PM was higher (1.96) than LDM (1.78).

Maximum entropy characterizes the uniform distribution of all fractions. A decrease in entropy indicates a high differentiation of particle size distribution when a predominance of one or more fractions is observed [18].

Significant correlation coefficients were established between the entropy and the content of fine sand (0,05-0,25 mm) ($r = 0.79$) in LDM and coarse silt (0,01-0,05 mm) ($r = -0.54$) and medium silt (0,005-0,01 mm) ($r = -0.59$) in PM.

 https://doi.org/10.1051/e3sconf/202016901025

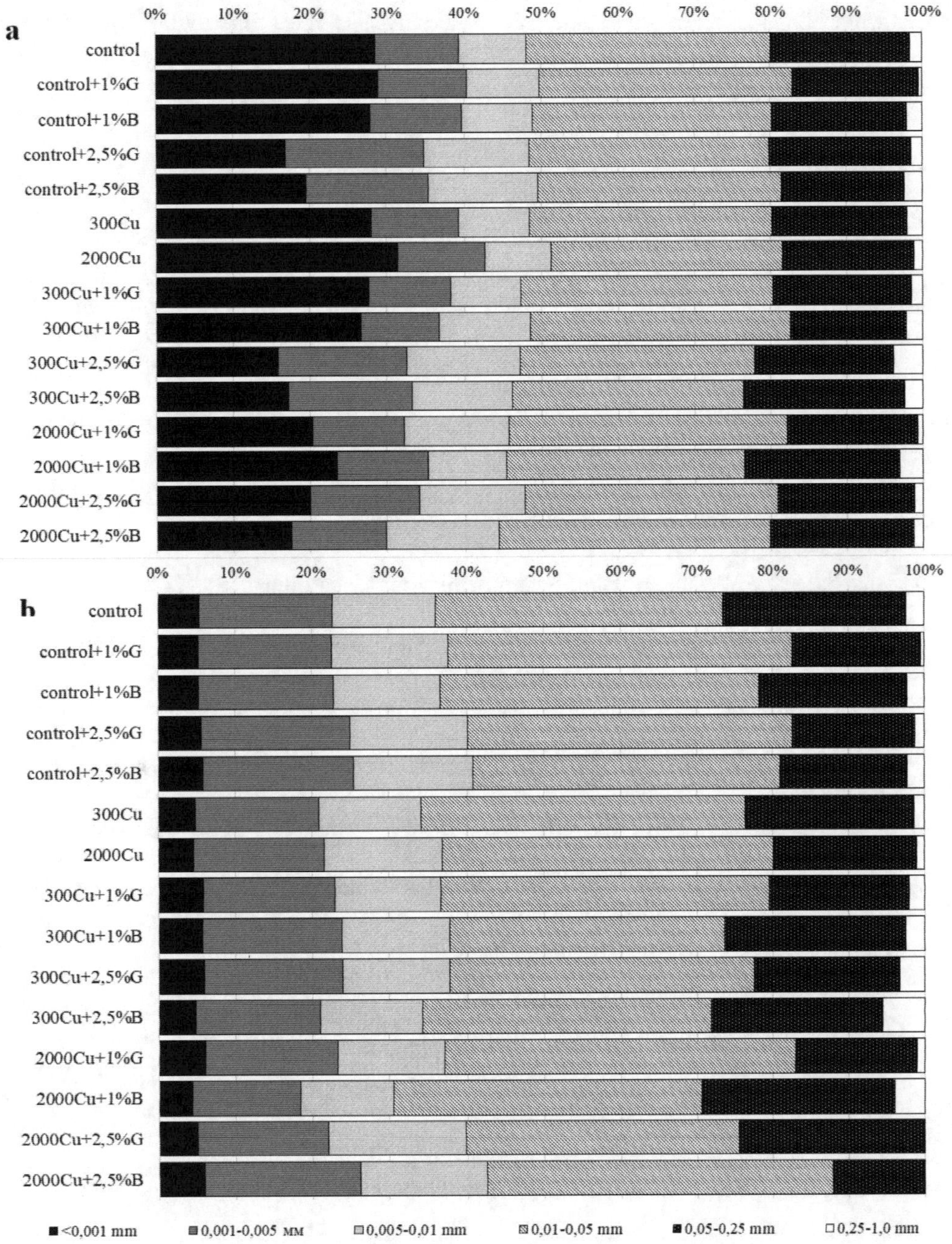

Fig. 1. Particle-size distribution of Haplic Chernozem contaminated with 300 and 2000 mg/kg Cu and adding different doses of biochar (1% B and 2.5% B) and GAC (1% G and 2.5% G): a –data obtained by the pipette method (PM), b - data obtained by laser diffraction method (LDM)

The main difference in the experimental variants falls on fractions whose size exceeds 0.01 mm (i.e. 0.05-0.25 mm, and 0.01-0.05 mm). The principle of LDM is based on the scattering of electromagnetic waves. The angle of diffraction of the laser beam is inversely proportional to the size of the particles. The results obtained by this method reflect the true particle size distribution. The introduction of sorbents as new components of the solid phase, which have different sizes, played a major role in the redistribution of fractional composition. At the same time, the number of fractions less than 0.001 mm did not show difference according to the experimental variants.

4 Conclusions

By using two different methods of PM and LDM, dissimilar data have been obtained in terms of the particle size distribution. In the case of LDM, changes occurred in the group of fractions larger than 0.01 mm. For PM, the change was observed for the group of a fraction of less than 0.01 mm. This observation could be due to differences existed in the principles of the used methods. Using the LDM method gives a real picture of the particle size distribution when sorbents are added.

This study was carried out with the financial support of the Russian Science Foundation in the framework of research project № 19-74-00085.

References

1. X. Zhao, B. Gao, D. Xu, L. Gao, S. Yin, Environ. Sci. Pollut. Res., **24**, 20844–20858, (2017)
2. N. Barsova, G. Motuzova, K. Kolchanova, A. Stepanov, M. Karpukhin, T. Minkina, S. Mandzhieva, Chemistry and Ecology, **35:1**, 86-101, (2019)
3. X. Fei, G. Christakos, R. Xiao, Z. Ren, Y. Liu, X. Lv, Sci. Total Environ., **661**, 168–177, (2019)
4. T.M. Minkina, G.V. Motuzova, O.G. Nazarenko, Eur. Soil Sc., **39**, 720–726, (2006)
5. T.M. Minkina, D.L. Pinskii, I.V. Zamulina, D.G. Nevidomskaya, C. Gülser, S.S. Mandzhieva, T.V. Bauer, I.V. Morozov, S.N. Sushkova, R. Kizilkaya, J. Soils Sediments, **18(6)**, 2418-2430, (2018)
6. F. Verheijen, S. Jeffery, A.C. Bastos, M. Van Der Velde, I. Diafas, C. Parsons, *Biochar application to soils: a critical scientific review of effects on soil properties, processes and functions*, (Joint Research Centre, Ispra, 2009)
7. G. Baiamonte, C. De Pasquale, V. Marsala, G. Cimò, G. Alonzo, G. Crescimanno, P. Conte, J Soils Sediments, **15,** 816–824, (2015)
8. S.D. Joseph, M. Camps-Arbestain, Y. Lin, P. Munroe, C.H. Chia, J. Hook, L. van Zwieten, S. Kimber, A. Cowie, B.P. Singh, J. Lehmann, N. Foidl, R.J. Smernik, and J.E. Amonett, Aust. J. Soil Res., **48**, 501–515 (2010)
9. D. Fischer, B. Glaser *Synergisms Between Compost and Biochar for Sustainable Soil Amelioration*, (INTECH Open Access Publisher, 2012)
10. L.D. Burrell, F. Zehetner, N. Rampazzo, B. Wimmer, G. Soja, Geoderma, **282**, 96-102 (2016)
11. G.N. Fedotov, E.V. Shein, V.I. Putlyaev, T.A. Arkhangel'skaya, A.V. Eliseev, E.Yu. Milanovskii, Eur. Soil Sci., **40**, 281-288, (2007)
12. Bittelli M., M.C. Andrenelli, G. Simonetti, S. Pellegrini, G. Artioli, I. Piccoli and F. Morari, Soil and Tillage Research, **185**, 36-46, (2019)
13. J.C. Santamarina, K.A. Klein, M. Fam *Soils and Waves: Particulate Materials Behavior, Characterization and Process Monitoring* (J. Wiley & Sons, New York, 2001)
14. A.A. Ponizovskii, T.A. Studenikina, E.V. Mironenko, Eur. Soil Sci.., **32(7)**, 766-775, (1999)

15. Motuzova G.V., Makarychev I.P., Dergham H.M., Stepanov A.A., Barsova N.U. *Soil organic matter and their interactions with metals: processes, factors, ecological significance* (Nova Science Publishers, New York, 2012)
16. A.V. Yudina, D.S. Fomin, A.D. Kotelnikova, E.Y. Milanovskii, Eur. Soil Sci., **51(11)**, 1326-1347, (2018)
17. G. Eshel, G.J. Levy, U. Mingelgrin, M.J. Singer, Soil Sci. Soc. Am. J., **68**, 736-743, (2004)
18. Yu. N. Vodianitskiy, Sov. Soil Sci., **11**, 105-110, (1986)

E3S Web of Conferences **169**, 02001 (2020) https://doi.org/10.1051/e3sconf/202016902001
APEEM 2020

Interdisciplinary integration on sustainable development issues

Dmitry Sergeyevich Ermakov[1,*], and *Alexander Sergeyevich* Ermakov[2]

[1]Peoples' Friendship University of Russia (RUDN University), Department of Psychology and Pedagogy, 117198, 6 Miklukho-Maklaya Street, Moscow, Russia
[2]Lomonosov Moscow State University, Department of Embriology, 119991, 1, b. 12 Lenin Hills, Moscow, Russia

Abstract. In order to achieve sustainable development, we need a deep understanding of both social and humanitarian processes and relationships in living and non- living nature. The development of scientific research areas that are related to sustainability are very important. Unfortunately, sustainable development is not among scientific and technical priorities in Russian Federation. This study analyzes the dynamics of PhD and Doctoral dissertations on sustainable development performed in various areas of science in the USSR and Russia in 1987–2019. The majority of dissertations (1 322 in total) were in the area of economics (85.4 %), while significantly fewer were in philosophy (2.9 %), technical (2.3%), and geographical (2.2 %) sciences. There are almost no dissertations in the fields of biological and chemical sciences, pharmaceutics and medicine. Our data indicate a significant imbalance and economic bias in the study of sustainable development issues. In this regard, it can be assumed that interdisciplinarity, which can be implemented in the framework of convergent and transdisciplinary approaches, is one of important methodological problem of sustainable development. Examples of relevant interdisciplinary projects are considered.

1 Introduction

In regard with the adoption of the sustainable development goals (2015) [1], the challenges of practical sustainability are even more important now than three decades ago, when they were formulated, since they go beyond purely environmental, economic and social problems and affect the survival of humanity [2]. In relation to this, it is necessary for science to meet the needs not only of the international community, countries and governments, but also of numerous other actors. In our opinion, sustainable development as an innovative strategy for the development of civilization, which takes into account the balance of three components of the environment: nature, society and economy, is one of the important fundamental problems and requires special understanding from the point of view of history, philosophy, methodology, organization, and ethics of science. The decisions of all UN conferences and summits on environment and development (Stockholm, 1972; Rio

* Corresponding author: ermakov-ds@rudn.ru

de Janeiro, 1992, 2012; Johannesburg, 2002) emphasize an importance of scientific research. The concept of the transition of the Russian Federation to sustainable development (1996) also declares the leading importance of science.

At the same time, sustainable development in our country is not among the priority areas for development of science and technology, according to the list of critical technologies of the Russian Federation. Nomenclature of scientific specialties [3] does not assume levels of candidates of science (PhD) and doctors of science (DSc) in this field.

2 Research Methods

Since 2011 we conduct scientometric research [4], which is based on content analysis of the catalog of the electronic library of dissertations by the Russian State Library (*diss.rsl.ru*). This catalog highlights dissertations for the degree of Candidate of Science (approximately equivalent to the western PhD) and Doctor of Science (approximately equivalent to DSc), the subject of which includes the words "sustainable development" (taking into account morphology). Dissertations are distributed by branches of science in accordance with the nomenclature of scientific specialties, according to the approval of the Ministry of Education and Science [3]. This method is well known and widely used, and, particularly, in the field of sustainable development [5]. It should be noted that the results of a similar study [6] were published without the reference to the original source.

3 Findings

In total, since 1987 till 2019, 1 322 dissertation studies on the problem of sustainable development were completed, including 1 115 PhD and 207 DSC, in the Russian Federation (and in the Union of Soviet Socialist Republics until 1992) (Figure 1).

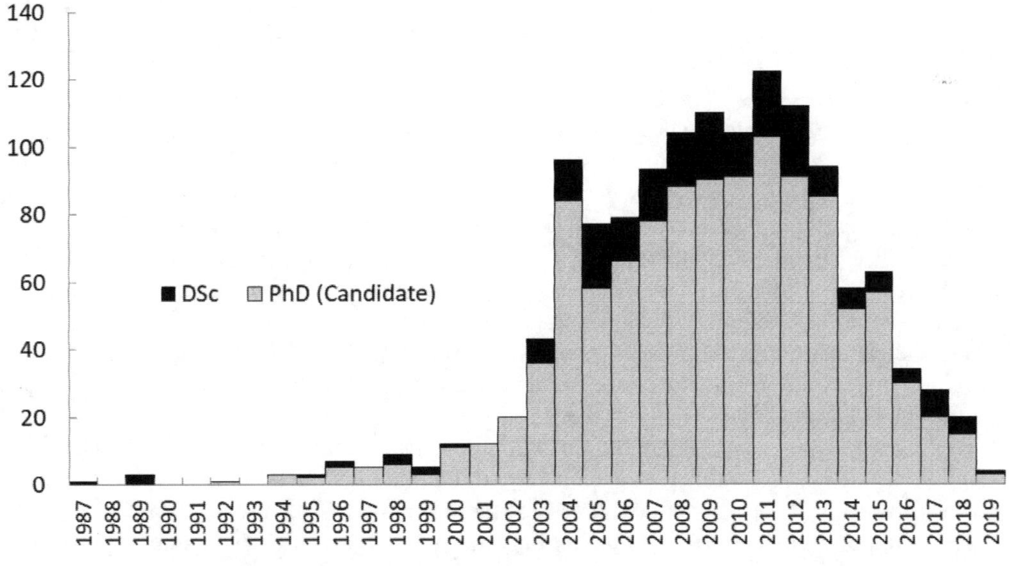

Fig. 1. Dynamics of dissertation researches on the problem of sustainable development.

The largest number of dissertations (more than 100 per year) occurred in 2009–2012. Subsequent decline after 2012 can be explained by organizational conditions (restructuring of the network of dissertation councils, increasing requirements for both the level of thesis themselves and the thesis councils in which they are defended [6]).

However, in our opinion, the reasons for this are of a methodological nature: first, a decrease in interest in the studied problems, and secondly, the mismatch between the disciplinary structure of traditional science and the integrative nature of sustainable development as a socio-ecological and economic phenomenon.

As we can see from Table 1, the vast majority of the works that were considered in this research, were performed in the area of economics (85.4 %), followed with a huge gap by philosophical (2.9%), technical (2.3%) and geographical (2.2%) sciences. There are separate dissertations on political and pedagogical (1.3 %), sociological (1.1 %) sciences, law (0.6%), biological and psychological (0.5 %) sciences, architecture (0.5 %). Very few studies were carried out in the field of physics and mathematics (0.3%), cultural studies, agriculture, geology and mineralogy (0.2% each), and also (0.1% each) in the fields of historical, pharmaceutical and philological sciences. The ideas of sustainable development are not considered at all in the aspects of veterinary medicine, theology and chemistry

Table 1. Distribution of dissertation research by branch of science.

Branch of science	Number of dissertations	Branch of science	Number of dissertations
agricultural science	3	pedagogical science	17
architecture	6	pharmaceutical science	1
biological science	12	philological science	1
chemical science	–	philosophical science	38
culturology	2	physical and mathematical sciences	4
economics	1129	political science	17
geographical science	29	psychological science	6
geological and mineralogical sciences	3	technical science	30
historical science	1	theology (since 2015 г.)	–
juridical science	8	veterinary science	–
medical science	–	*total*	*1322*

It is obvious that social and humanitarian aspects prevail. Understanding of sustainable development issues from the point of view of natural sciences is less common, as the scientific justification of the relevant technologies. It should be noted that the ideas of sustainable development have a long history. However, today the "concept of sustainable development" is a conglomerate of heterogeneous philosophical, environmental, sociological, economic, technological, etc. ideas that have no unity: dozens of definitions, principles, criteria, concepts, strategies, lists of indicators.

At the same time, a number of scientists criticize the formulation of the question of sustainability of development: development means change, and sustainability, on the contrary, means stability [7].

In this regard, we can assume that an important methodological problem of sustainable development is interdisciplinarity, which can be implemented within the framework of two approaches: convergent (includes blurring the boundaries between separate scientific disciplines, between scientific and technological knowledge, which opens up opportunities for creating objects in all spheres of human activity that are close in function and purpose to natural ones) [8] and transdisciplinary (a way of considering a subject when solving complex problems of nature and society outside the framework of traditional scientific disciplines, simultaneously at several levels) [9]. Examples of practical application of these approaches are the projects implemented with the participation of the authors: "Valaam island is our common home", "Living village". To solve the problems of sustainable development of the Valaam archipelago on the Ladoga lake, Republic of Karelia, Russia, and its population, an interdisciplinary team of experts, including specialists in the fields of geography and ecology, economics, law, management, philosophy, psychology and conflict studies, philology and arts, sociology, pedagogy, etc., worked at that area. Taking into account the principles of system analysis, this team studied the processes of crisis and mechanisms of local community development, the conditions for overcoming the crisis and sustainable self-organization, based on social partnership, dialogue and responsible leadership, joint decision-making and collective practices like disposal of solid municipal waste, energy efficiency, development of infrastructure for culture, education and health of the population, tourism, formation of children's craft and environmental skills, strengthening of local self-government [10].

The "Living Village" project, which is implemented in the Ustyansky district of the Arkhangelsk region, provides not only interdisciplinary integration, but also work at different scales, at different levels: mental and physical, theoretical and practical, global and local. Its goal is to reproduce the traditional way of life and types of rural labor of the indigenous peoples of the North of the Russian Federation. Transformation of the rural area is one of the key priorities of sustainable development. At the same time, it is important to teach the villagers to live in their natural and cultural environment, not just preserving it, but conducting economic activities in a symbiotic landscape! In this regard, a promising innovation (paradoxically) is the revival of the traditional way of life, which provided the person and the community with the necessary products in conditions of limited and hard-to-reach natural resources, based on the principles of self-organization, intentionality, common thinking, ideas, values, etc.

The project defines the conditions for sustainable rural development, based on the annual economic cycle of the northern village, including harmonious interpersonal relationships, the relationship of generations, the participation of elders in the education of young, self-government, revival of traditional types and modern types of labor and employment, participation of residents in settlement development, full use of available renewable resources, conservation of natural and cultural heritage [11, 12].

4 Conclusion

Despite the popularity of sustainable development, there is a risk of turning "sustainable development" into a fashionable rhetorical cliché. The term is used to describe a goal that is certainly desirable, but is defined so broadly, from fairly accurate to very vague interpretations, that it can lead to misunderstandings [13]. It is obvious that sustainable development is an iterative process which involves many disciplines and points of view, requiring linking the production of new scientific knowledge with practical solutions to specific problems, including the creation of technological innovations [14].

Since the beginning of the discussion initiated by the publication of the report "Our common future" (1987), many attempts have been made to define the science of sustainability [15]. However, based on the actual results, it is doubtful that we have made progress: many problems (climate change, biodiversity loss, poverty and discrimination, etc.) remain unresolved, or their scale and consequences are even increased. Sustainable development issues are complex and should be integrated into both traditional disciplines and integrative trans-disciplinary research programs, the results of which would contribute to the improvement of the environment, society and human condition [16].

The reported study was funded by Russian Foundation for Basic Research according to the research project № 19-013-00722 "Education for sustainable development in action".

References

1. *Transforming our world: the 2030 Agenda for sustainable development*. URL: http://www.un.org/ga/search/view_doc.asp?symbol= A/RES/70/ 1&Lang=E (2015)
2. E. Kumi, A. A. Arhin, T. Yeboah, Env. Dev. Sust., **16**, 539–554 (2014)
3. Nomenclature of scientific specialties. URL: http://docs.cntd.ru/document/542610966 (2017).
4. *Scientific and methodological support of environmental education for sustainable development.* URL: http://disus.ru/konferencii/364554-1-nauchniy-otchet-grantu-prezidenta-rossiyskoy-federacii-dlya-gosudarstvennoy-podderzhki-molodih-rossiyskih-uchenih-doktorov.php (2011).
5. *Sustainability science in a global landscape* (Amsterdam, Elsevier, 2015)
6. D. V. Vlasov, Vestn. Mosk. Univ. Geogr., **1**, 16–23 (2017)
7. A. S. Ermakov, D. S. Ermakov, Chem. Life, **11**, 24–29 (2012)
8. *Converging technologies for improving human performance. Nanotechnology, biotechnology, information technology and cognitive science* (Dordrecht, Netherlands, Kluwer Academic Publishers, 2003)
9. D. J. Lang, A. Wiek, M. Bergmann, M. Stauffacher, P. Martens, P. Moll, M. Swilling, C. J. Thomas, Sust. Sci., **7**, 25–43 (2012)
10. D. Yermakov, D. Slavinsky, S. Chernikova, *Environmental ethics: the power of ethics for sustainable development* (Vilnius, M. Romeris University, 2010)
11. A. Kulyasova, Vestnik Commis. Rus. Fed. UNESCO (2015)
12. A. A. Kulyasova, A. S. Ermakov, D. S. Ermakov, *Psychology of extreme professions* (Northern (Arctic) Federal University, Arkhangelsk, 2019)
13. J. A. Dixon, L. A. Fallon, Soc. Nat. Res., **2**, 73–84 (1989)
14. G. Wuelser, C. Pohl, Sust. Sci., **11**, 789–800 (2016)
15. T. R. Miller, Sust. Sci., **8**, 279–293 (2013)
16. R. J. Baumgartner, J. Korhonen, Sust. Dev., **18-2**, 71–75 (2010)

Contaminated sites investigation. Objectives and methods

Ioan Bica*,
Technical University of Civil Engineering Bucharest, Blv Lacul Tei, no. 122-124, Bucharest 020396, Romania

Abstract. The past human activities, mainly that related to industrial development caused in many cases a strong contamination of soil, subsoil environment and of groundwater. This type of pollution is a concern because it poses risks to human health and to the ecosystem. More than this, such areas may not be used for new development, requiring solutions for remediation. The management of these sites consists of three main activities: characterization, remediation and, finally, redevelopment. The paper presents a case study dedicated to the first step of contaminated sites management, respectively characterization. This phase is very important, a good characterization could ensure a performant solution for the second step – remediation. Two new techniques for site characterization are presented, as technical principles, but also as performances obtained for the mentioned case study.

1 Introduction

Industrial activities, municipal and industrial waste disposal, mining activity and other human developments during the past industrial revolution all over the world generated the contamination of soil, subsoil and groundwater. In Europe (EU) there are over 2 million of potential contaminated sites and 150.000 contaminated sites. It is estimated that until 2025 the number of contaminated sites will rise by 50% [1]. The concern of the management of these areas results from the consequences that pollution puts on the future use of these areas, on one hand and from the risks on human health and the ecosystem as a whole on the other hand.

As it is well known, the main stages of contaminated sites management are (i) characterization of contaminated sites, (ii) remediation and (iii) redevelopment of the area (Fig. 1). The difficulty of remediation of contaminated areas is generated by a series of aspects, which relate to their specificity: (a) accessibility of the underground environment; (b) complexity of the underground environment, formed by the interlocking of three phases: liquid, gaseous, solid; (c) the physical, chemical and biological mechanisms that act on the pollutants reaching in the underground environment [2].

Many useful and innovative technologies for contaminated site characterization and clean-up as well as methods to support decision making processes exist, but they are only rarely applied in their full potential.

* Corresponding author: ioan.bica@utcb.ro

Characterization Remediation Redevelopment

Figure 1. Stages of contaminated sites management.

Establishing and achieving the remediation levels is a difficult problem because the quality norms specific to the future use of the area, respectively the norms established by the national regulations could be changed in time, on one hand, and on the other hand, the technical performance of the available remediation technologies can't achieve very restrictive goals. The efficiency of remedial technologies is dependent on the difficulties arising from combining the characteristics of the contaminated area (physical, geological, hydrogeological etc.), with the properties of the contaminants (physical, chemical and biological properties).

To achieve decontamination levels based on human health protection is not always feasible – the costs and the time necessary for this goal could be very high. Based on this fact, the future use of contaminated sites must be designed considering the level of remediation that may be achieved in a technical – economical judgment. Or, in other words, the level of remediation could be fixed based on the requirements of a designed activity that has been planned for a specific contaminated site, meaning that it must not be subjected to human health-based criteria on every occasion.

2 Methods and procedures

2.1. Conceptual model

The analysis of the remediation possibilities will consider the characterization of the area realized by geological, hydrogeological investigations, sampling for analytical determinations, assessing the dynamic of contaminant in the subsurface etc. All these will allow to obtain the necessary information to evaluate the behavior of the contaminant, respectively to evaluate the possibilities of remedying or minimizing the risks posed by the contaminated area. These coherently assembled data help to establish a qualitative and quantitative representation of the characteristics of the area, forming the so-called *conceptual model* (Fig. 2).

This model must provide the physical characteristics of the area, the type of chemicals founded in the subsurface and their concentrations, and more importantly it must define the main characteristics of water and pollutants motion (Table 1).

The conceptual model provides an image of the structure of the underground environment, its important characteristics for pollutant migration and its behavior, the characteristics of the pollutants found underground and any other aspects that can help to understand the temporal and spatial evolution of the pollutant front and the risks that it puts on the receivers.

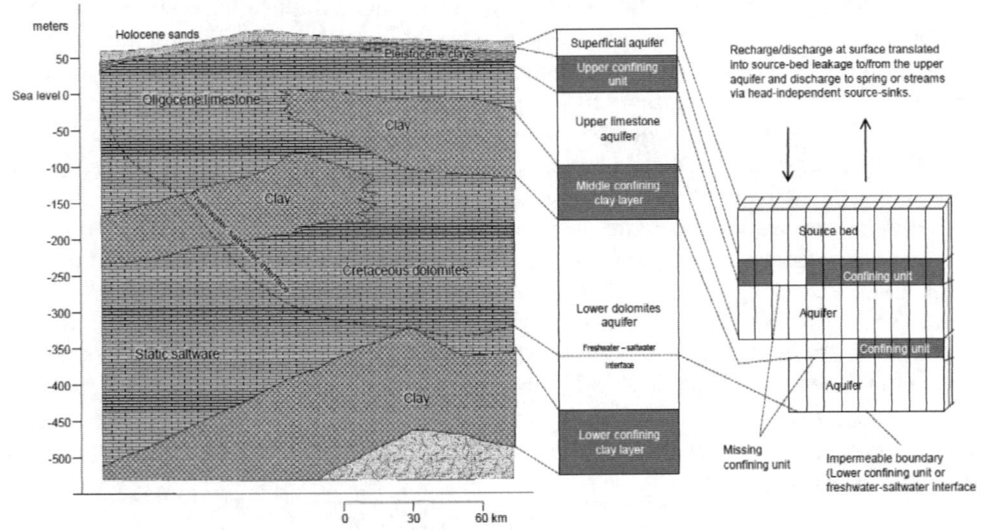

Figure 2. The construction of the conceptual model [3].

The conceptual model offers a three-dimensional image of the conditions of a contaminated site, illustrating the distribution, the mechanisms of contamination behavior, the routes of exposure, the migration routes, and the potential specific receptors [9].

Table 1. Type of data for Conceptual model [2]

1	**General information**	Wells position, type of groundwater bodies, environmental receptors position
2	**Geology and hydrogeology**	Local and regional geology, physical properties, stratigraphy, groundwater level, hydraulic conductivity, isotropy, porosity, groundwater recharge/discharge, etc.
3	**Contamination source**	Position and history of the contamination source (activities). Position of underground contamination sources.
4	**Contamination distribution and transport**	Underground spatial distribution of contaminants, contamination mass, contamination processes and transport, geochemical properties of the rocks, etc.

The conceptual model is a dynamic, flexible working tool that will be improved as new characterization data for the contaminated area are available. The conceptual model will be validated by simulating, with the help of mathematical models, the processes carried out underground. The calibration of the mathematical model on situations determined by in situ measurements offers arguments in accepting its conformity with the real situation.

2.2. Investigation of contaminated sites

The methods of investigating the subsurface environment have developed over time and have reached significant performances. However, the investigation of the contaminated subsurface environment is a relatively new activity and imposes procedures, criteria, and objectives different from the investigation carried out for other purposes. As a result, its transposition to the contaminated environment without adjustment and adaptation to its specifics can lead to disturbance of the subsurface environment, to the further complications of the behavior of the pollutant in the subsurface, respectively difficulties in applying the remediation solution.

The establishment of the investigation method, appropriate to a site, is done by quantifying the advantages and limitations dictated by the specificity of the investigated site as well as by the type of data required for its optimal characterization. In principle, the data obtained should ensure an adequate risk assessment, in order to allow the identification and establishment of the potential and the optimal solution for remediation (Fig. 3).

Attempting to achieve perfection in characterizing the contaminated area can be detrimental because performing too many drills for sampling will disrupt the structure of the natural environment, thus causing changes in the behavior of pollutants in this area. In these conditions, the studies of the characterization of the polluted area must strike a balance between the desire to obtain all the information necessary to design an efficient remediation system and the need to avoid the wide disturbance of the subsurface environment through the characterization investigations [2].

From those multitude site investigation methods, only two were used for this study, *Direct Push* and *Phytoscreening*.

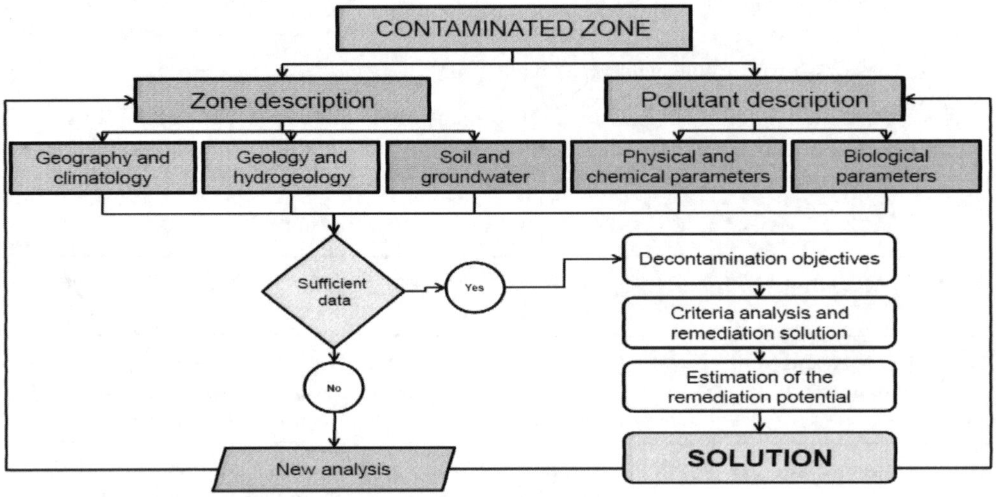

Figure 3. Flow chart of the contaminated zone characterization.

2.2.1. Direct Push

Direct Push [2,3] is a hydraulic-powered subsurface investigation method where both static force and percussion are used to advance sampling and logging tools into the subsurface. No rotation is needed to advance and there is no drilling debris on the surface (Fig. 4).

DP-techniques are multifunctional investigation methods able to provide a detailed characterization of the underground. The DP-based techniques can be applied in the unsaturated/saturated zone and are especially efficient in unconsolidated clayey, sandy and gravely deposits. They offer the possibility of continuously recording investigation relevant parameter, sampling of the soil, soil gas and groundwater and installing monitoring wells.

One important role in the spatial distribution of the contaminants is played by the hydro-stratigraphic features of the site. Retrieving valuable information about the subsurface can be done in real-time using DP-methods through different specialized sensors [3].

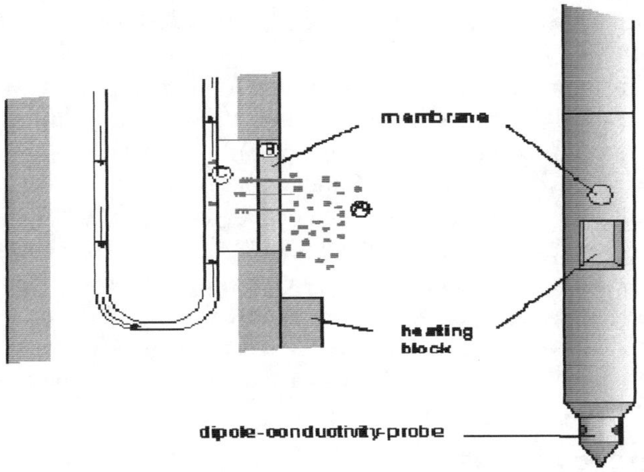

Figure 4. Direct Push method

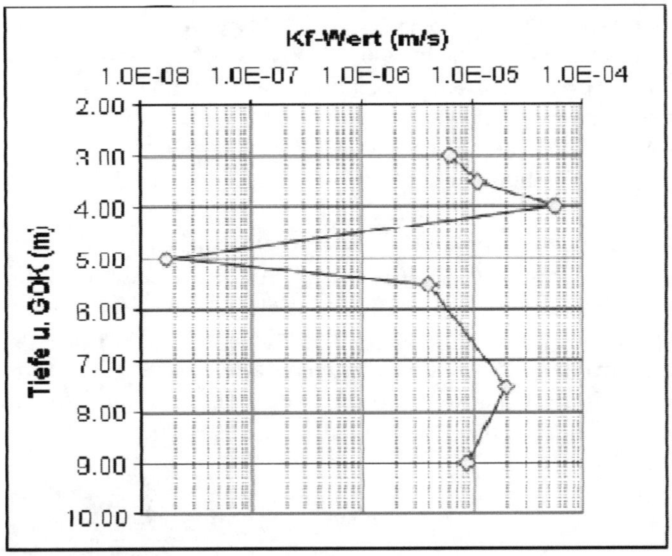

Figure 5. The result of hydraulic conductivity test

The *Electrical conductivity* (EC) profiling method provides on-site, through a direct measure of the electrical conductivity of the underground, an indirect information of the subsurface lithological characteristics. The EC technique is an excellent tool for rapid delineation of the underground heterogeneity. Through this method, vertical profiles of the electrical conductivity are derived in order to map thin layers [3].

In situ testing of *hydraulic conductivity* can be performed by injecting water under a specified pressure (4-5 bar) into the aquifer. The water flows into the layers in an easier or heavier way, depending on the hydraulic properties of the medium. The interpretation provides in a preliminary stage a relative profile of hydraulic conductivity. By means of several slug tests, the results are site-specific and translated into absolute values of hydraulic conductivity (Fig. 5.).

Online profiling of contaminant distribution (MIP) - Membrane Interface Probe is a rapid screening tool for locating volatile organic compounds in the subsurface. MIP is a powerful logging device able to detect volatile organic compounds and record semiquantitative data in a continuous depth profile. MIP collects real-time, vertically continuous data on the distribution of VOCs as well as an electrical conductance (EC) log that is indicative of the crossed lithology.

The in-situ chemical instrumentation is based on a probe that is connected to laboratory instrumentation by a capillary tube system and control cable. The online analysis takes place using a MIP-System connected to a gas chromatograph - GC. The combination of three GC detectors allows distinguishing between different types of VOC's.

Discrete groundwater sampling - The groundwater sampler has a 36 mm outer diameter (OD) and will be driven to the sampling depth. When probes are placed at desired sampling depth, a stainless-steel filter will be exposed by retracting the sheath. The groundwater that flows through the filter is pumped according to low-flow method procedures. Low-flow methods minimize the volatilization of VOC's. The field parameter (ORP, O2, pH, temperature, and electrical conductivity) are measured on site by using a flow-through cell.

2.2.2. Phytoscreening with tree cores

The sampling of plants represents an economical and non-invasive method that complements conventional methods of investigating pollutants in the groundwater and soil (leachate) [3]. The fundamental principle thereby is that the solved contaminants are absorbed through water by the roots. Inside the plant the substances move upwards with the sap current into the trunk or stem and then into the leaves [3,4].

Tree coring is a semi-quantitative screening tool which can be applied initially for location and mapping of pollution. The method is well suited for large sites and to locate (single) hot spots. However, the results should be confirmed by quantitative screening methods such as groundwater sampling or soil sampling [4,5].

Using an increment borer (Fig. 6), tree cores can then be taken from the trunks of trees and analyzed in the laboratory for pollutants. The measured contents can then provide us with insights into the subsurface contamination of a particular site. The scientific literature refers to this semi-quantitative procedure as *phytoscreening*. The suitability of *phytoscreening* for investigation or monitoring purposes could already be undoubtedly proven for some pollutants, particularly those in the group of chlorinated solvents.

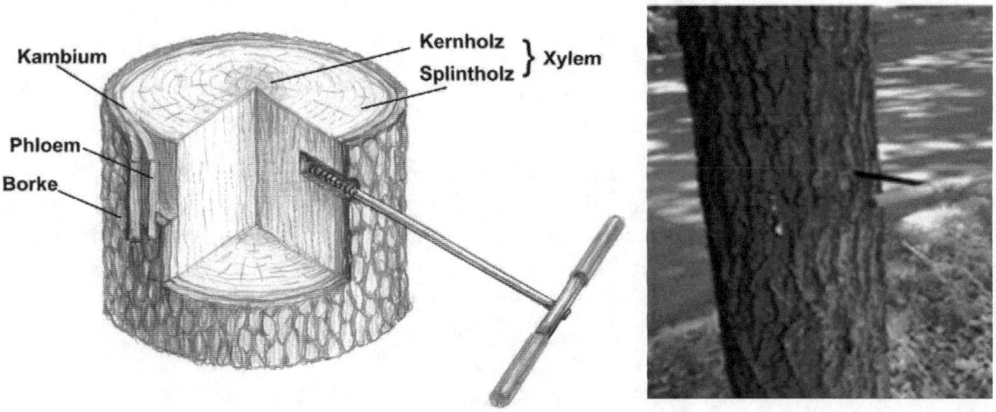

Figure 6. Phytoscreening with tree cores.

The method is recommended by its many advantages [5]: fast, easy, and only little effort is needed to screen a large test site; cheaper than others; suitable at sites where application of conventional site characterization methods is limited/problematic due to the sensitivity of a site; certain site features (moor, rocks, forest), location of the site (inner cities, risk of pipes or cables) or due to former/current activities at the site; minimal impact on the site and the environment; allows a fast overview of the spreading of the pollutants and can this way serve as initial screening method; root systems can be larger than the crown of the tree [6].

3 Case study. Investigations and results

The case study refers to a former industrial activity, dating back to 1884, with alternating stops and re-commissioning, and periods of development or activity restraints. The integrated steel mill included tall furnaces, coke ovens, classic ovens, electric ovens for sheets, wires or steel profiles. Part of the chemical-coke sector was located in the West side, a storage area of the resulted tar from this section being identified. Starting with 1999, no more steel activities were performed, but only demolishing processes.

From the lithologic point of view, the area is characterized by the presence of gravel until 10 m depth, covering a carbonate (karstic) environment. The main aquifer is represented by a karstic groundwater body located under the gravel layer. The groundwater table of the superficial aquifer is located at about 5 m depth.

The 51 percussion holes drilled down up to 2 m depth served for a good geotechnical description of subsurface and for reliable soil gas measurements as well as for the investigation of heavy metals presence and distribution in top soil (0-20 cm) and in depth (1-2 m). High concentrations of ionizable gas were measured in vicinity of remnants of tanks. Generally low soil gas values were detected throughout rest of area. Higher values of heavy metals were measured throughout the investigated area (Zn, As, Cu, Ni, Pb, Cd) but vertically limited mostly to topsoil (0-20 cm). In depth (1-2 m) generally low values are present [8].

By the aid of DP-techniques (Fig.7), the following have been obtained: derived aquifer geometry (bottom, top and thickness), hydraulic properties distribution maps (based on lithological information, (Fig. 8) and grain size distribution values on 18 samples, hydrogeological map (groundwater level) - based on groundwater and surface water measurements. All these were after that used for the configuration of conceptual model of the area [3,8].

Figure 7. DP investigation boreholes location.

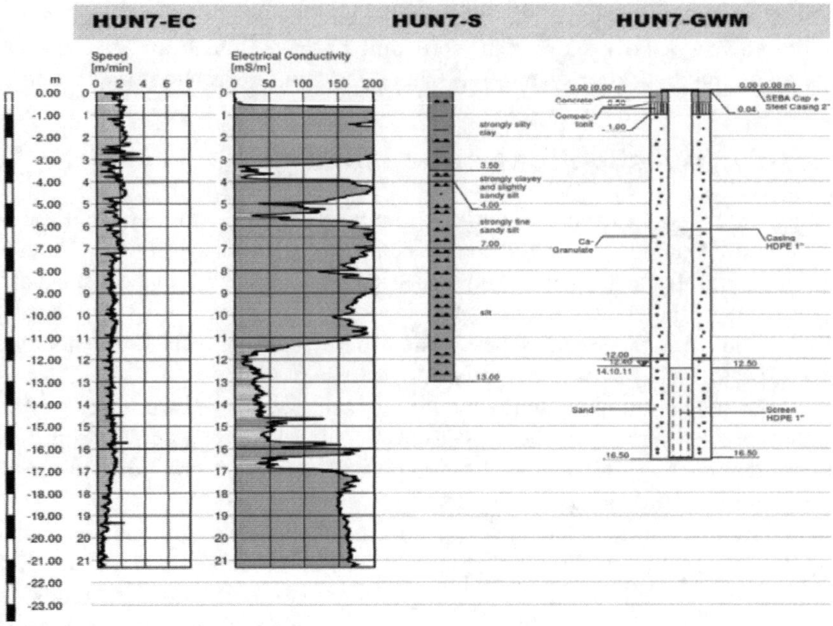

Figure 8. Typical hydrogeological features

Based on same technology, 3D representation of the structures of the geology is obtained. It is useful for understanding the stratigraphy, the potential continuity of the layers and the contaminants' behavior (Fig. 9). The soil samples allow the representation of the distribution of the contaminants in subsurface zone (Fig. 10).

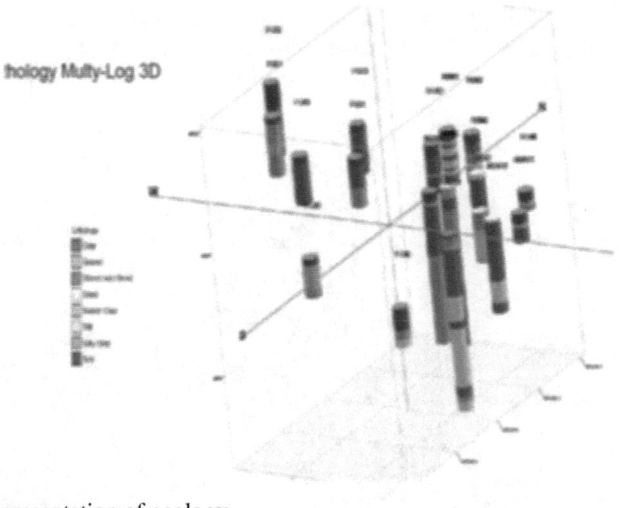

Figure 9. Spatial representation of geology

 https://doi.org/10.1051/e3sconf/202016902002

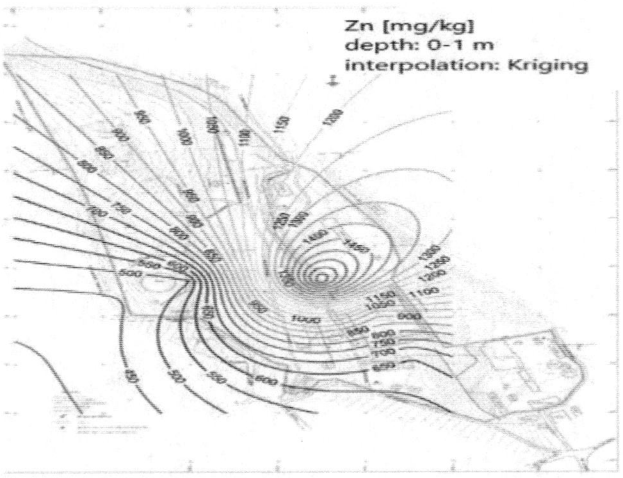

Figure 10. Spatial distribution of contaminant: Zn.

The major contamination issue for the area are the heavy metals [8]. Also, other contaminants can be found in the soil and in the groundwater in significant concentrations, as for example naphthalene, toluene, benzene (Table 2).

Table 2. Vertical distribution of main contaminants (mg/kg) [3]

Depth (m)	Oil products	Arsenic	Lead	Toluene	Benzene
2.8	970	7.62	16.00	0.09	0.23
4.8	2048	4.00	10.30	5.36	12.90
7.5	903	4.00	7.80	0.23	0.76
9.2	3923	10.70	15.20	1.53	7.23
11.5	3061	4.00	6.59	0.54	0.49
12.8	727	7.12	9.17	0.75	2.72
14.7	1629	10.90	6.81	1.83	11.00

Table 2 shows a high distribution of contaminants in depth. This could be explained by lithological affinities (organic compounds - benzene, toluene - hydrophobic accumulation in pelitic rocks). Metals (lead) very immobile. Once introduced in the soil matrix Pb is very difficult to remove. However, this industrial platform, during its 100 years of operation, was the subject of many reconstructions, excavation and earth fillings and mixing it from different levels; sometimes excavations for deep foundations etc.

The tree coring method was tested as a bioindicator for both heavy metals and BTEXN compounds. The sampling for heavy metals was done without knowing much about the heavy metal level at the site beforehand. The samples were therefore collected all over the site and allocated to subareas (Fig. 11). The average concentrations of each subarea are shown in Fig.12. Most of the average concentrations of the test site are above the average concentrations of the reference samples [5,8].

For BTEXN compounds, tree coring as a phytoscreening method was tested in the initial phase in order to answer the question if the compound can be detected or not in wood. BTEXN compounds have been detected in 4 out of 21 samples, all in the high-risk area. For benzene, even different levels were observed. This indicates that the trees can take up BTEXN compounds from the subsurface and transport them to the stem [4.5].

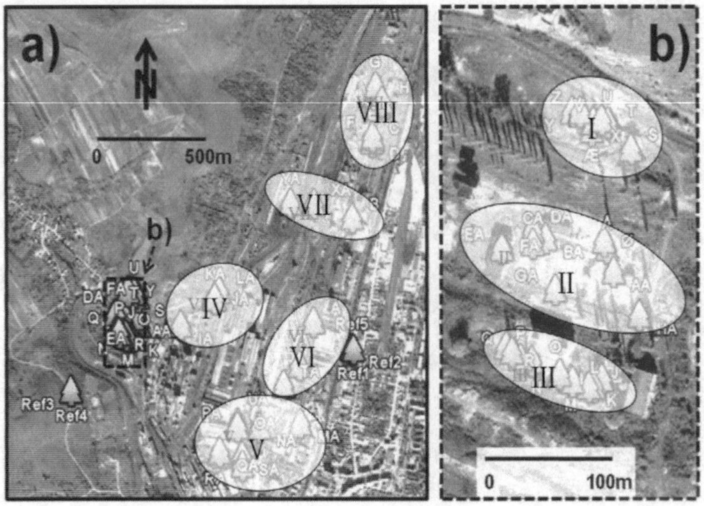

Figure 11. Map of the test site showing the subareas used to evaluate the pollution of heavy metals. **b)** is a zoom into **a).**

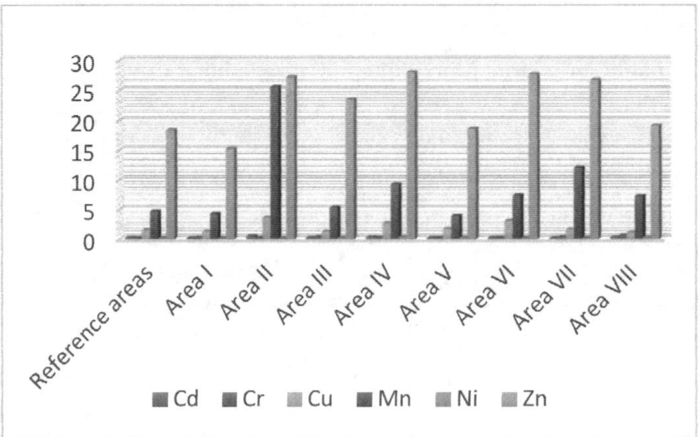

Figure 12. Average concentrations measured in the trees from each subarea.

4 Conclusions

Characterization of contaminated sites is one of the most difficult activities, because the local data must be extrapolated to have a clear image of the whole subsurface area. The conceptual model can be used to have a picture of the subsurface area, and to understand its behavior. It can also be used in modeling the trend of contaminated plume and to test the efficiency of different remediation solutions. Nondestructive (noninvasive) methods of investigation are preferred because they do not disturb the underground zone, but they can fail to provide very accurate data. DP technology and phytoscreening are one the best novel methods for this purpose.

Based on the results of this investigation, the main conclusion is that the redevelopment of the platform area could be divided in three sectors: first one will be used for light industrial activities, the second one is proposed for recreation (vegetation and forests plantation, where phytoremediation may be activated), and last one could be used for residential development (after using soil washing as remediation solution).

The author acknowledges the BRIGAID EU project for technical, scientific, and financial support.

References

1. *EEA, The European Environment. State and outlook.* (European Environmental Agency, Copenhagen, 2005)
2. I. Bica. *Remediation of contaminated sites.* (Orizonturi Universitare, Timisoara, 2014)
3. I. Bica, D., Gaitanaru, *Solving Environmental Problems Through the Understanding of Risk and Environmental Processes: International Workshop* (Institute of International Health, Michigan State University, Technical University of Civil Engineering, Bucharest, 2014)
4. M. Algreen, *Guideline for application of Tree Coring as an initial screening tool for typical pollutants in the subsurface*, (2013)
5. M. Algreen, St., Trapp, *Guideline for application of Tree Coring as an initial screening tool for typical pollutants in the subsurface* (TIMBRE, 2014)
6. M.C. Dobson, A.J. Moffat. Waste Manag. and Res., **13**, 579-600 (1995)
7. N. Fatin-Rouge, *Feasibility of specific soil washing with recycled solutions (SSWRS) for the removal of selected metals and organic compounds from contaminated soil and mathematical model to assess pollutant transfer by SSWRS* (TIMBRE, 2013)
8. E. Martac, *Comparative study of site investigation approaches and potential in situ remediation techniques: model-assisted evaluation of advantages and uncertainties* (TIMBRE, 2014)
9. *US-Environmental Protection Agency, Innovative Treatment Technologies: Annual Status Report* (Eighth Edition), 542-R96-010 (EPA, 1996)

E3S Web of Conferences **169**, 02003 (2020) https://doi.org/10.1051/e3sconf/202016902003
APEEM 2020

Methodology of the geoecological functional zoning of territories subjected to anthropogenic disturbance due to waste disposal

Stanislav Dubrova[1*] , *Sónia* Chelinho[2] , *Pavel* Zelenkovskiy[3] , and *Ivan* Podlipskiy[1]

[1]Department of Geology and Geoecology, Herzen State Pedagogical University of Russia, 199034, Moika river emb., 48, block 5, St. Petersburg, Russia
[2]Centre for Functional Ecology - Science for People & the Planet (CFE) Department of Life Sciences, University of Coimbra, 3000-456, Calçada Martim de Freitas, Coimbra, Portugal
[3]Department of Environmental Geology, Head of Department, Professor, Institute of Earth Sciences, St. Petersburg State University, 199034, Universitetskaya emb., 7/9, St. Petersburg, Russia

Abstract. In the present contribution, the geochemical relationships of territorial units (landscapes), namely sewage sludge disposal sites and historical municipal solid waste landfills, that can be considered as chemical elements migration models and as dynamic components of ecological-geological systems (landfill site - adjacent area - recreation area)., is described. Since the study of pollutants migration in multicomponent systems is extremely time-consuming, the development of theoretical methods, able to simulate these processes, is of particular relevance. Therefore, in the present study, a geochemical model for substance migration in anthropogenically modified geoecological bodies and their surrounding was conceived, making use of complex mathematical processes and data analysis methodologies (parallel use of factor, cluster, canonical and discriminant analysis tools), and considering both geochemical specialization/composition of parent rocks and land use (functional geoecological zoning).

1 Introduction

With the increasing number of megacities, the constant expansion of their borders, and the often chaotic, unreasonable land-use planning, even those sites with potential recreational use suffer from enormous anthropogenic pressure. Geoecological assessments on the influence of various anthropogenic objects on the environment often reflect the level of negative impacts only and do not take into account relationships with other objects, e.g. those within the boundaries of the technosphere and those interacting with the upper horizons of the lithosphere. Moreover, the dimensions of sanitary protection zones do not

* Corresponding author: dubrova.stansilav@gmail.com

take into account these features. In order to preserve ecosystems and assure the rational planning of urban development, it is necessary to pay special attention to the functional purpose and land use, which is the basis of zoning and urban management. The development of migration patterns in the matter flows (man-made and natural), considering objects and territories together, is the only method for justifying the state of ecosystems "downstream". It is fundamental to use the standard amount of geochemical information used in geoecological and engineering-ecological surveys for the characterization of geochemical communication, as a "flow" of substances migration, i.e., as a dynamic component of geoecological systems with no need for additional and time-consuming research. The main goal of the present study was to present an updated complex methodology for geo-ecological functional zoning of territories and their environmental impact assessment using a particular case-study, consisting of territories subjected to different levels of anthropogenic impact. The proposed approach allows the optimal ratio of use and/or protection measures to be set for a particular city site - the basic unit of a city territorial administration, acting as an essential tool for metropolis land resources management.

2 Materials and methods

The present study includes "exceptional objects" (geoecological anthropogenically modified objects of different genesis) with distinct eco-geochemical characteristics [5], located in "exceptional conditions" - within the catchment area of the Lakhta spill, flowing into the Gulf of Finland (Fig. 1). Representative sites of three different functional geoecological zones in St. Petersburg were selected, on the basis of close geomorphological, geological and hydrogeological conditions, but with distinct levels of anthropogenic impact, namely:

1. Industrial functional zone (Historical Municipal Solid Waste Landfill (MSWL) "Novoselki" (food and packaging waste, packaging, waste paper, household and office garbage) and Sewage Sludge Landfill "Severny" (SSL) (solid fraction of waste water, consisting of organic, mineral substances and a complex of microorganisms (excess activated sludge), hereafter designated as "Industrial (MSWL+SSL)". It consists of territories occupied by industrial enterprises, landfills as well as their correspondent sanitary protection zones and also transport infrastructures. 212 soil samples (53 sampling points times four horizons - 0-0.2 m, 0.2-1.0 m, 1.0-2.0 m, 2.0-3.0 m);

2. Residential functional zone (adjacent to the landfills area (MSWL+SSL), hereafter designated as "Residential". This zone is a transition between industrial and recreational areas and it is used for residential and business development purposes. (102 sampling points times four horizons);

3. Recreational functional zone (Novoorlovskaya Special Economic Zone (SEZ), a part of Novoorlovsky Forest Park), hereafter designated as "Recreational". This area was the reference for the geochemical background and it is a green area, mainly used for recreational purposes 235 soil samples (112 sampling points times four horizons).

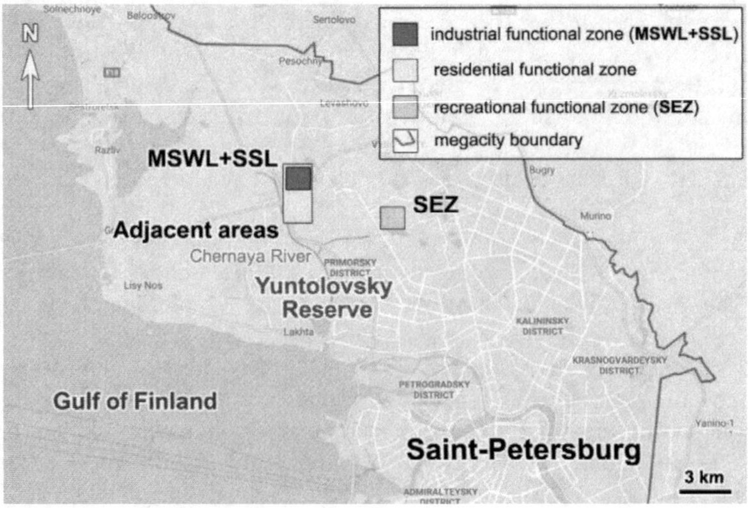

Fig. 1. Schematic map of the study areas

To assess the contamination of soils analysis for the presence of two organic compounds (petroleum and polycyclic aromatic hydrocarbon (PAH) - benzo (a) pyrene) and 10 heavy metals (mercury, lead, arsenic, cadmium, zinc, nickel, cobalt, chromium, copper, manganese) were performed. The analysis of the three matrices (soil) were determined by the Research Park of SPbU "Environmental Safety Observatory" by atomic absorption spectrometry (AAS) for heavy metals [4]. The determination of benzo (a) pyrene in the soil was carried out by high performance liquid chromatography using a chromatograph with fluorimetric detection [3]. Concentrations of petroleum were determined by the fluorimetric method on the "Fluorat-02" liquid analyzer.

To evaluate the degree of contamination of the study areas, the measured concentrations of both organic and metal contaminants were compared with the maximum permissible concentration (MPC) or approximate allowable concentration (APC). Another benchmark to assess and compare the levels of chemical pollution is the "background" content. The assessment of the danger of soil contamination for public health, due to metals were also carried out using the total pollution (Zc) indicator. According to the available standards, the soils were included in category 1 (permissible), 2 (moderately dangerous), 3 (dangerous) or 4 (extremely hazardous pollution) whenever the Zc is, respectively, lower than 16, >16<32, >32<128 or higher than 128.

3 Results and discussion

3.1 Geo-environmental assessment

The actual assessment of the levels of chemical pollution of the territories of domestic and industrial waste landfills was based on a comparison of the existing pollution with the regional and local background geochemical levels for the objects, as well as maximum permissible concentrations (MPC). On the territory of MSWL, the concentrations of pollutants in all the studied horizons remain at a high level, exceeding the MPC for individual elements up to 50 times: Pb - 1580 mg / kg, As - 80 mg / kg, Cd - 24 mg / kg, Zn - 1000 mg / kg, Ni - 300 mg / kg, Cr - 10 000 mg / kg, benzo (a) pyrene - 0.27 mg / kg. According to the degree of chemical pollution, the soils of the territory of the Novoselki landfill, in all horizons , are considered as "extremely dangerous": the content of organic compounds of the first hazard class exceeds 5 MPC, the total pollution indicator taking into account toxicity (Zc) of pollutants takes values over 128 (Fig. 2). On the territory of the storage site of SSL, Zc values are significantly lower than those fixed at MSWL and range from 3.3 to 820 in the upper levels of 0.0–0.2 m and 0.2–1.0 m: Pb - 111 mg / kg, As - 20 mg / kg, Cd - 50 mg / kg, Zn - 1000 mg / kg, Ni - 105 mg / kg, Cr - 206 mg / kg, benzo (a) pyrene - 0.07 mg / kg. Also, at a depth of 2.0-3.0 m, the maximum permissible concentrations of As were exceeded 2 times, Cd - 9 times, Zn - 3 times. The pollution halo is located in the horizons at 1m deep, in the "old" north-western part of the silt sites.

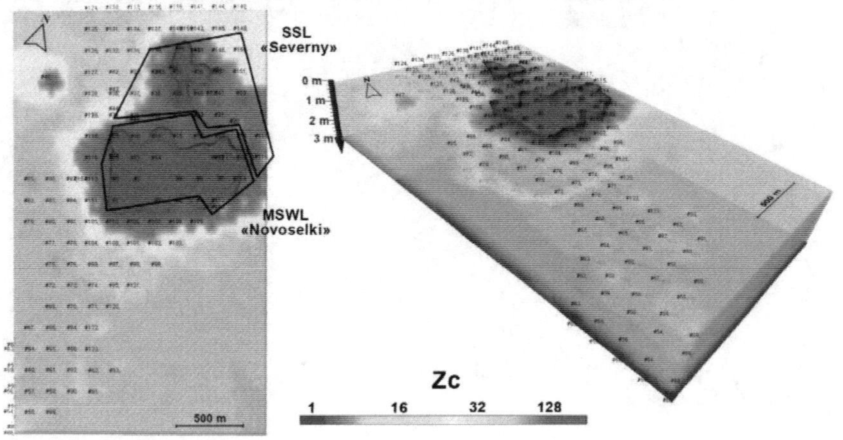

Fig. 2. 2D and 3D schemes of the distribution of soil Z_c values in the territory of MSWL and SSL relative to the regional background.

3.2 Multivariate statistical analysis

In order to perform the geochemical portraits of the functional zones under study, as well as the subsequent creation of geochemical models, a canonical analysis was carried out. This approach enables the reduction of the initial geochemical data and highlight the most significant variables for each of the geoecological objects [1,2]. Historical MSW landfills, the most contrasting geoecological objects from a geochemical point of view, are ideally suited for modeling the flow of anthropogenic load. Forest park recreation zones under similar natural conditions are the best standards for modeling the geochemical background. An essential assumption of the method of geochemical modeling is that the identification signs of the industrial functional zone fully, accurately and consistently reflect the anthropogenic load. Such assumption can be further compensated with the creation of a geochemical portrait database, including various geoecological objects under different environmental conditions, that is, by maximizing the distinctive features of the evaluation set (Table 1).

Table 1. Formatting sections, subsections and subsubsections.

Study Areas	Significant variables	Non-significant variables
MSWL "Novoselki"	Ni	Hg, Pb, As, Cd, Zn, Co, Cr, Cu, Mn, Benz, Petrol
SSL " Severny"	As, Pb, Co	Cd, Zn, Ni, Cu, Mn, Benz, Petrol
Adjacent areas	Pb, Cr, Hg, Ni, Petrol, Zn	Cu, Benz
SEZ "Novoorlovskaya"	Cu, Zn, Pb, Mn	Hg, Cd, Ni, Co, Cr

For geochemical monitoring studies, you can include an explanation of the percentage scale, that divides the set of probabilities into the following classes: 0-30% - the probability of anthropogenic load is absent (geochemical background for the object of study); 31-50% -

the probability of anthropogenic load is possible; 51-70% - the probability of anthropogenic load is insignificant; 71-100% - the probability of anthropogenic load is present (Fig. 3).

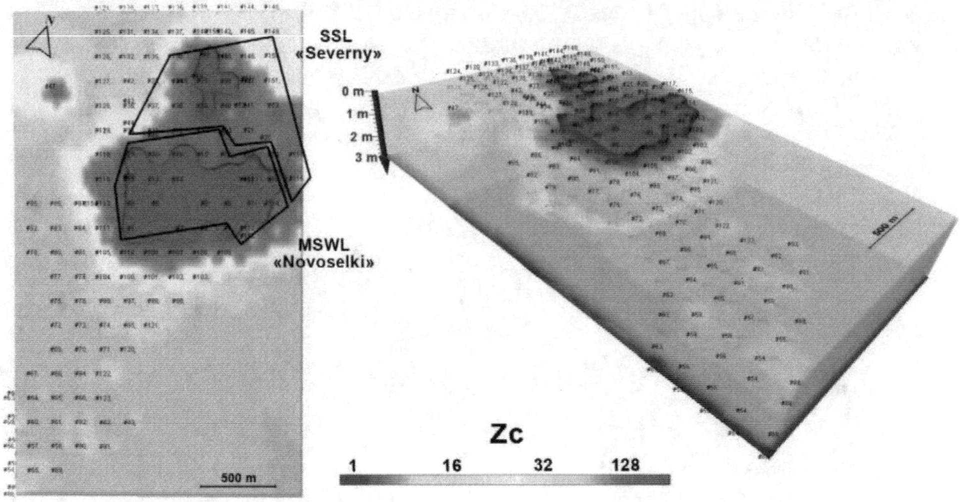

Fig. 3. 2D and 3D schemes of the probability distribution of anthropogenic load on the residential functional area of the first group study areas

The likelihood of anthropogenic load from the "Novosleki" landfill is observed along the bypass trenches of the landfill, along the channel of the Chernaya River, in the south-west direction by lowering the relief and located at 2 meters deep. At Northeast of the sewage sludge landfill, in the area of peat bogs, the estimates also point for anthropogenic load.

4 Conclusions

In the present study, within the framework of the functional zoning of the city's territories, detailed geochemical series of pollutants and three-dimensional geochemical maps of the study areas have been created. A significant improvement was the use of a multi-stage method for processing the high volume of geochemical data obtained from engineering and environmental surveys [6]. Indicator pollutants for each of anthropogenically modified geoecological objects and paragenetic associations for functional zones were identified. The presented methodological approach can be recommended for cadastral work. The simple and universal characteristics of the methodology make it possible to compile both small-scale maps of megalopolises and municipal areas, as well as large-scale maps of individual groups of sites. The necessary information database is based on a standard set of assessment criteria (sampling depth, scale of the testing grid) adopted in engineering and environmental surveys for construction and interstate standards.

References

1. R.E. Dashko, Y.A. Karpova, J. Min. Inst. (2014)
2. S.V. Dubrova, I.I. Podlipskiy, V.V. Kurilenko, W. Siabato, Environ. Pollut., **197**, (2015) http://dx.doi.org/10.1016/j.envpol.2014.12.013
3. J.P. Durand, N. Petroff, J. Chromatogr. A., **190**, 1 (1980) https://doi.org/10.1016/S0021-9673(00)85514-4

4. A. Maurya, L. Kesharwani, M.K. Mishra, IJRASET, **6**, VI (2018) https://doi:10.22214/ijraset.2018.6173
5. N. Mora-Naranjo, J.A Meima., A. Haarstrick, D.C. Hempel, Waste Manage., **24**, 763-773, (2004) https://doi.org/10.1016/j.wasman.2004.04.006
6. M. Vorenhout, N.M. VanStraalen, H.J.P. Eijsackers, Environ. Toxicol. Chem., **19**, (2000) https://doi.org/10.1002/etc.5620190901

Improving radio-ecological monitoring of potentially radon-hazardous territories

Tatiana Lashchenova[1,3], *Lidiya* Karl[1,2*], *Albert* Marenny[2], and *Sergey* Kiselev[3]
[1]Peoples' Friendship University of Russia (RUDN University), Faculty of Ecology, 6 Miklukho-Maklaya St, Moscow, 117198, Russian Federation
[2]State Research and Technical Center of Radiation-Chemical Safety and Hygiene FMBA Russia, 123182, Moscow, Russian Federation
[3]A.I. Burnasyan Federal Medical Biophysical Center FMBA, 123098, Moscow, Russian Federation

Abstract. Radioecological monitoring of radon-hazardous territories in residential and public buildings of the towns of Lermontov (Russia, Stavropol Region) and Baley (Russia, Transbaikal Region) made it possible to reveal that the volume activity of radon in premises on 1 floor of buildings reliably characterizes the radon inflow from the soil surface. This control parameter, which can be used to estimate the radon danger of the territory, to make decisions on the sanitary condition of the building, to estimate the potential danger to the population or personnel, to calculate the potential dose rates. Improvement of radio-ecological monitoring in potentially radon-hazardous areas is aimed at monitoring the volume activity of radon on 1 floor of buildings, the excess of which is higher than the normal values is the basis for making managerial decisions to protect the population or staff.

1 Introduction

According to the UN Scientific Committee, the main contribution to population radiation exposure is from natural sources of ionizing radiation. The main contribution to the population exposure dose from 50 to 90 % is made by natural radioactive gas radon (^{222}Rn) and its decay products. Epidemiological studies have shown that ^{222}Rn, as a potential lung cancer initiator, is in second place after smoking [1,2]. The International Agency for Research on Cancer (IARC) classified ^{222}Rn as carcinogen of group 1.

During the last decade, the scientific institutes of the Federal Medical and Biological Agency of Russia (State Research and Technical Center of Radiation-Chemical Safety and Hygiene and A.I. Burnasyan Federal Medical Biophysical Center) conducted monitoring studies on the content of ^{222}Rn in residential and public buildings. These studies allowed to specify the regions where the problem is particularly acute (Altai Territory, Trans Baikal region, North Caucasus, Urals, etc.).

These areas are located in close proximity to uranium, thorium, gold and other mining sites. In buildings on such territories, due to an unfavourable combination of geological,

* Corresponding author: Lidiakarl@yandex.ru

social and other factors, the concentration of ^{222}Rn is tens of times higher than permitted values [3].

The purpose of this study is to improve radio-ecological monitoring of potentially radon-hazardous territories.

Studies have shown that there are a number of problems in conducting monitoring studies in potentially radon areas. These problems are connected with analytical and methodical support of control of ^{222}Rn content in air, with complex analysis of received data, the volume of which needs correct interpretation for making management decisions, and other problems. There are also problems in the organization of monitoring in potentially radon-hazardous territories, due to financial indifference in the works of owners of territories and objects located on them. In this regard, the relevance of this work is undoubted, it is of great theoretical and practical importance.

Volumetric activity control of ^{222}Rn in residential and public buildings is an important control parameter, and in potentially radon-hazardous areas it is a required one [4,5]. There are no regulatory documents, which would oblige officials responsible for this territory, to conduct this type of monitoring on a permanent basis. In this direction, it is necessary to implement a set of measures aimed at improving radioecological monitoring in potentially radon-hazardous areas.

Fluctuations of ^{222}Rn in buildings depend on a number of factors: the content of natural radionuclides in the soil under the building, the features of the construction of the building and the ventilation condition; the type of building materials used in construction, the geological conditions in the survey area.

Any anthropogenic human activity on the territory can create or change existing inflow ways, increasing the ^{222}Rn output into the atmosphere, including in residential and public, as well as industrial buildings. These routes of entry on each territory should be identified, hazard assessment should be carried out; if necessary, control and continuous monitoring should be organized; protection of the population should be carried out through preventive (preventive) or corrective actions.

At present, the basic principles of radiation safety of the public and staff are defined in ICRP documents (Publication 103, Publication 126) and IAEA publications (SSG-32). They are based on establishing reference levels and applying the principle of optimization in the adoption and implementation of appropriate protective measures. In the Russian Federation, the basic requirements for ensuring radiation safety are set out in the basic regulatory documents NRB-99/2009 and OSPORB-99/2010, which are currently being reviewed in the light of international requirements. [6,7]

2 Materials, methods, methodology for control and assessment

The most reliable and promising methods for measuring the volumetric activity of ^{222}Rn are integral methods based on the use of track detectors to register heavy charged particles, such as alpha-particles, the separation of which occurs at the collapse of the nucleus of the ^{222}Rn atom. The use of track detectors makes it possible to consider fluctuations in the excretion of ^{222}Rn within a day, on this basis to reduce uncertainties in the calculation of the annual average dose to the population due to natural irradiation. The sensitivity of these methods is sufficient and meets the needs of monitoring studies.

The volumetric activity of ^{222}Rn in residential and public premises has been standardised since 1989 in accordance with the requirements of the item. 5.3.3. NRB-99/2009 by the parameter of equilibrium equivalent concentration (EEC_{Rn}). considering products of decay of subsidiary radionuclides (EEC_{Rn} + 4.64 EEC_{Tn}), the value of which should not exceed 200 Bq/m^3 in operated residential premises [6,7]. Average annual EEC_{Rn}

value is calculated from measured EEC_{Rn} with an exposure of at least 2 months performed in cold and warm seasons.

Measurement of volumetric activity of ^{222}Rn by the integral track method in industrial, residential and public premises is carried out by the method, described in the following link [8]. Fundamentals of the method, the principle and method of work with the equipment, as well as the issues of the formation of samples of premises during the examination of settlements and placement of the exposimeters in the examined premises are discussed in detail in the publication [9]. Measurement of ^{222}Rn volumetric activity during screening studies on large territories is developed by the authors [10-16].

For buildings that are already in operation, EEC_{Rn} + 4.64 EEC_{Tn} should not exceed 200 Bq/m^3. If this value is exceeded, radon protection measures should be taken to reduce radon inflow into buildings [6,7].

When constructing new residential and public buildings in the air of premises EEC_{Rn} + 4.64 EEC_{Tn} should not exceed 100 Bq/m^3, and the effective dose rate of gamma radiation should not exceed the dose rate in the open area by more than 0.2 μSv/h. [3]

Radio-ecological and sanitary-hygienic assessment of industrial, residential and public premises is conducted according to the requirements of the following documents [17-21].

3 Results and discussion

In potentially radon-prone areas, average regional data were obtained for screening studies. Details at the district level are not always conducted, and not everywhere, so protection of the population in such conditions cannot be guaranteed.

At the control the content of ^{222}Rn problems of analytical and methodical support are connected with the increase of measurement reliability and creation of methodical support, which are now actively developed with the participation of authors, and will be realized in the form of analytical methods and methodical documents for practical sanitary-hygienic application [10].

While monitoring within the framework of regional programs of ^{222}Rn maintenance in residential and public buildings on potentially radon-hazardous territories one has to face the problems of processing a large volume of accumulated data, with a complex analysis of the received data, which are now being implemented using GIS-technologies [10]. FMBA of Russia is developing an information and analytical system for all types of natural sources of ionizing radiation. Application of GIS-technologies allows to carry out the complex analysis of the received data, to visualize the received results which allow to prepare visual maps on territorial distribution of the increased emission of ^{222}Rn flows, and to give offers for acceptance of administrative decisions in territory, to optimize regional monitoring programs.

As part of the improvement of radio-ecological monitoring of potentially radon-hazardous territories, monitoring was carried out in the well-known radon-hazardous territories in the towns of Lermontov, Stavropol Krai and Baley, Transbaikal Krai, which are located territorially in different regions with different geological structures. For comparison, the territories in Chita were taken as background territory in relation to Baley.

The study of regularities of ^{222}Rn volume activity distribution according to EEC_{Rn} in the buildings of Baley, Chita and Lermontov towns (5239 measurements) and revealed that distribution of values in all premises for all three territories is lognormal, the median was used as an average value. In the background area for the Transbaikal region in Chita, EEC_{Rn} values are significantly lower than 200 Bq/m^3, the median is 27 Bq/m^3.

In Baley and Lermontov, the average geometric and median values do not exceed 200 Bq/m^3, and are 109 Bq/m^3 and 126 Bq/m^3 respectively. At the same time, in Baley, EEC_{Rn} values exceed the standard value of 200 Bq/m^3 in 30% of the total number of measurements; and in Lermontov, the excess is 35% of the total volume of measurements, respectively. Maximum values in Baley and Lermontov exceed 200 Bq/m^3 by 10 times.

Statistical processing of data showed that premises with EEC_{Rn} excess above 200 Bq/m^3 are mainly located on the first floors of buildings.

Our study of the patterns of distribution and accumulation of ^{222}Rn in buildings, showed that the main source of income to the premises on the first floors is the soil under the building. The first floors of buildings experience the largest anthropogenic burden, the magnitude of which depends on the content of natural radionuclides in the soil under the building and violations associated with the peculiarities of the construction of buildings, ventilation and other violations.

The measurements were made and it was found that the gas content of ^{222}Rn in the premises on the ground floors in Baley exceeds 28 %, and in Lermontov 35 % of the total number of measurements. We carried out statistical processing of the received data on the ground floors in Baley and Lermontov. The results are presented in the table 1.

Table 1. Parameters of distribution of EEC_{Rn} values exceeding 200 Bq/m^3 on the first floors of buildings in Baley and Lermontov towns..

Distribution parameter	Baley	Lermontov
N	254	694
Arithmetic mean, Bq/m^3	505	479
Geometric mean, Bq/m^3	421	427
Standard geometrical deviation	1,8	1,6
Median, Bq/m^3	365	422
Max, Bq/m^3	2087	1980
Excesses, % of total excesses	64	60

According to the table, the number of premises with excesses from the total number of measurements on the ground floors are: in Baley town - 64%, in Lermontov town - 60%. Median on the first floors is - 365 Bq/m^3(Baley), and 422 Bq/m^3(Lermontov), that is above 200 Bq/m^3 almost in 2 times. Maximum EEC_{Rn} values are 2087 Bq/m^3 in Baley and 1980 Bq/m^3 in Lermontov.

In order to find out what influences increased EEC_{Rn} values in premises on the first floors, a study was carried out to identify the dependence of radon volume activity on various factors and revealed the following. A multivariate data analysis was carried out in the program "Statistica", for the analysis of data array the method of multivariate classification cluster analysis was used, which allows to identify homogeneous data groups on radon volume activity. All data were divided into groups (clusters) by the method of single linage, using the calculation of similarity method Manhattan distances.

It was found out that the main criteria for assessing the potential radon danger of the territory is the volume activity of radon on 1 floor of buildings, which reliably characterizes the flow of radon from the soil surface, regardless of the source of inflow. These are control parameters by which one can estimate the radon hazard of the territory, make decisions about the sanitary condition of the building, estimate the potential danger for the population or staff, calculate the potential dose rates.

4 Conclusions

Improvement of radio-ecological monitoring in potentially radon-hazardous territories is aimed at control of radon volume activity on 1 floor of buildings, exceeding of which is higher than standard values is the basis for making managerial decisions to protect the population or personnel.

For optimization of radio-ecological monitoring it is necessary to introduce in practice modern analytical methods allowing to control the content of ^{222}Rn in residential and public buildings on large territories; to carry out development of methodical documents for sanitary and hygienic application; development of information-analytical systems with application of GIS-technologies for acceptance for optimization of regional monitoring programs of territories on the radiation factor.

References

1. *UNSCEAR 2006 Report: Annexe E: Source-to-effects assessment for radon in homes and workplaces* (*UNSCEAR,* 2009)
2. *UNSCEAR 1988 Report: Annexe E: Sources, effects and risks of ionizing radiation (UNSCEAR, 1988)*
3. A. Marenny, *Nuclear and Radiation Safety of Russia*, 36-63 (2002)
4. S. Kiselev, A. Marenny, V. Romanov, Radiation hygiene, **12(2)**, 94-102 (2019)
5. M. Zhukovsky, I. Yarmoshenko, S. Kiselev, ANRI (2011)
6. *NRB-99/2009.Standards of radiation safety* (2009)
7. *OSPORB-99/2010. Basic sanitary rules to ensure radiation safety* (2010)
8. *MVI 2.6.1.003-99. Measurement method of volume activity by integral track method in production, residential and public premises*
9. A. Marenny et al. ANRI, **2(81)** 15-26 (2015)
10. A. Marenny, S. Kiselev, Radiation hygiene, **12(S2)**, 97-108 (2019)
11. S. Kiselev, I. Stamat, A. Marenny, L. Il'in, Hygiene and sanitation **97(2)**, 101-110 (2018)
12. A. Marenny, V.Romanov, V. Astafurov et al., Radiation Hygiene, **8(1)**, 23-29 (2015)
13. M. Zhukovsky, I. Yarmoshenko, Proceedings of the Institute of mathematics and mechanics of the Ural branch of the RAS, **231**, 9 (1997)
14. I. Korenkov, O. Polsky, A. Sobolev, *Central Institute for Advanced Training of Doctors*, 252. (1993)
15. D. Shakin, T. Lashchenova, Proceedings of the Institute of mathematics and mechanics of the Ural branch of the RAS, (Kaliningrad, 2014)
16. S. Kiselev, M. Zhukovsky, I. Stamat, I. Yarmoshenko, *Radon. From basic research to regulatory practice* (Moscow,(2016)
17. *SanPiN 2.6.1.2800-10. Radiation safety requirements for exposure of population to natural sources of ionizing radiation*
18. *MU 2.6.1.2397-08. Assessment of doses to population groups exposed to increased exposure due to natural sources of ionizing radiation; Guidelines*
19. *MU 2.6.1. 037 – 2015. Determination of average annual values of radon ERE in the air of premises by measurements of different duration*
20. *MU 2.6.1.1088-02. Evaluation of individual effective doses to the population due to natural sources of ionizing radiation*
21. *MU 2.6.1.2838-11. Radiation control and sanitary-epidemiological assessment of residential, public and industrial buildings and facilities after their construction, overhaul, reconstruction on radiation safety indicators*

Possibilities of using sentinel-2 satellite data to estimate crop cover and LAI of winter wheat at different stages of development

Feddy Mullo[1,*], *Elena* Prudnikova[2,3]

[1]Peoples' Friendship University of Russia (RUDN University), Faculty of Ecology, 6 Miklukho-Maklaya St, Moscow, 117198, Russian Federation
[2]Peoples' Friendship University of Russia (RUDN University), Agroengineering Department, Agrarian-Technological Institute, Faculty of Ecology, 6 Miklukho-Maklaya St, Moscow, 117198, Russian Federation
[3]V.V. Dokuchaev Soil Science Institute, Pyzhevskiy pereulok, 7, Moscow, 119017, Russia

Abstract. The study was conducted on a test plot located in the Yasnogorodsky district of the Tula Region; With a camera with fisheye lens photographs were taken at different points of the plot, each point with a geographical reference. Subsequently, the possible relationship between the information extracted from the classification of the photographs using the CAN-EYE software (LAI, percentage of vegetation) and the vegetation indices (Ratio, NDVI, SAVI and EVI) calculated with spectral values obtained from the different channels of Sentinel-2 (B2, B3, B4, B5, B6, B7, B8, B8a) were evaluated. Finally, best regression models obtained for each phase of the winter wheat development were used to create the maps of LAI and percentage of vegetation. According to our results, Sentinel-2 can be successfully used to map LAI in the studied region at the shooting stage of winter wheat development with accuracy of 85 %. At other stages and for percentage of vegetation the accuracy of the models was below 50 %.

1 Introduction

Remote sensing of the Earth from space using digital methods and image processing techniques is widely used to monitor environmental changes; and in recent decades it has been widely used in agriculture (Martinez, 2005). In vegetation studies actual information on canopy chlorophyll content, in addition to properties like leaf area index, biomass and fraction of absorbed photosynthetically active radiation, is important in understanding plant functioning and status (Clevers and Gitelson, 2013).

Photosynthesis is one of the key processes in plants that is responsible for the energy and carbon balance. Actual chlorophyll estimates are important for applications such as precision agriculture, since chlorophyll is the main plant constituent determining the reflectance in the visible region of the spectrum, optical remote sensing techniques have great potential in providing information on canopy chlorophyll.

* Corresponding author: mfreddy.28fm@gmail.com

Leaf area index (LAI) is a dimensionless variable defined by Watson (1947) as the area on one side of the photosynthetic leaf tissue per unit area of soil. LAI is a key biophysical variable fundamental in natural vegetation and agricultural land monitoring and modelling studies. LAI is the main determinant of the processes of photosynthesis and the evapotranspiration of crops.

Satellite-derived LAI maps have been used in many earth observation studies including evapotranspiration (Allen *et al.*, 1998 and Duchemin *et al.*, 2006), net ecosystem production (Barr, 2004) and rainfall interception by forests (Dietz *et al.*, 2006), as well as in yield estimation and in downstream services for crop monitoring (Busetto, 2017). In the last few years, non-destructive methods have been developed to determine the percentage of vegetation cover and LAI. There are few studies that validate LAI estimates in crops with data taken *in situ* (Sandmann, 2013).

Sentinel-2 provides images with high spatial, spectral and temporal resolution. It covers the visible and near-infrared (VNIR) and the shortwave-infrared (SWIR) spectral region in 13 bands, incorporating two new spectral bands in the so-called red-edge region, which are very important for retrieval of chlorophyll content (Delegido *et al.*, 2011, Gitelson *et al.*, 2005). Sentinel-2 has, an increased sensitivity towards LAIgreen, while the shortwave infrared bands are sensitive to cellulose and lignin (dry matter) absorption (Delegido *et al.*, 2015).

Thus, the objective of the study was to analyze the possibilities of using Sentinel-2 data to assess crop cover and LAI of winter wheat at different stages of development on the basis of ground-truth measurements.

2 Research methods

Test plot is located in Yasnogorodsky district of the Tula Region. During the year 2018 three field trips were made to the test plot: on 27.04.2018 (tillering stage), 1.06.2018 (shooting stage), 6.07.2018 (milky ripeness). During each trip photos of crops with a fish-eye lens camera were made: at 22 points at tillering stage, at 14 points at shooting and milky ripeness stage. At each point 5 photos were taken.

For the analysis of the photos, we used CAN_EYE software. As a result, for each sample point at each stage of development we determined percentage of vegetation (percentage of surface covered by vegetation) and LAI.

The EarthExplorer portal of the Geological Survey of the United States (USGS) was used to download Sentinel-2 images corresponding to the dates closest to the field trips (Table 1). For each of the development phases, Level-1C images were obtained: for the first stage - for April 27, 2018; for the second stage - June 1, for the last stage - July 6. The atmospheric correction of the images was performed with the Sen2Cor tool using the Sentinel Application Platform (SNAP).

Table 1. Bands of Sentinel-2 used in the study (Source: Sentinel Online*)

Bands	Resolution (m)	Central wavelength (nm)
B2 – Blue	10	490
B3 – Green	10	560
B4 – Red	10	665
B5 – Vegetation red edge	20	705
B6 – Vegetation red edge	20	740
B7 – Vegetation red edge	20	783
B8 – NIR	10	842
B8a – Narrow NIR	20	865

*https://earth.esa.int/web/sentinel/user-guides/sentinel-2- msi/resolutions/spatial

With the spectral values obtained from the different bands, we proceeded to calculate vegetatation indices such as: Ratio (R_{NIR} / R_{red}), NDVI ((NIR-Rred) / (NIR + Rred)), SAVI ((R_{NIR}- R_{red}) (1 + L) / (R_{NIR} + R_{red} + L)) and EVI ((2.5 (R_{NIR}-R_{red}) / (R_{NIR} + 6R_{red}-7.5R_{blue} + 1)) These indices are widely used to determine the values of biomass, LAI and plant cover (Jerry *et al.*, 2010). 25 (referred to as R1 to R25) possible combinations were obtained for the RATIO index, for Red NDVI - 21 combinations (referred to as RED1 to RED21), assessment of crop phenology during the growing season has been assessed using a comparison of the NDVI with a visible atmospherically resistant index expressed either as Green NDVI ((R_{NIR}-R_{green}) / (R_{NIR} + R_{green})) (referred to as Green1 to Green5) or Red Edge NDVI ((R_{NIR}-$R_{red\ edge}$) / (R_{NIR} + $R_{red\ edge}$)) (referred to as Edge1 to Edge4) (Vina *et al.*, 2004), and SAVI and EVI- 5 combinations respectively.

Then, regression analysis was performed in SPSS to determine the relationships between the parameters obtained from the analysis of photos (LAI_3rings, LAI_4rings, LAI_5rings, percentage of vegetation) and vegetation indices, calculated from Sentinel-2. The best models with highest adjusted R^2 and statistically significant variables were used in ILWIS 3.3 Academic to create maps of LAI and percentage of vegetation for each development stage of winter wheat.

3 Results and Discussion.

Tillering stage: For the tested regression models for percentage of vegetation, the highest accuracy was below 35% (adjusted R^2 = 0.337) The model includes R13, RED13, EVI2 and EVI5 as predictor variables (Fig. 1 a).

The best model was obtained for LAI_5rings and it explaines 36.7% (adjusted R^2 = 0.367) of the changes of the dependent variable. It includes the following vegetation indicies: R12, R13, R24 and RED21 (Fig. 1 b).

Shooting stage: Best regression model for vegetation percentage explaines 43.8% of the changes (adjusted R^2 = 0.438) in this parameter and includes such predictor variables as R7, R11, R16 and R19 (Fig. 1 c).

Best model obtained for the variable LAI_3rings with predictor variables R7, R11, R14, R19, R20 and R22 explains 85.4% of the variation of LAI on the test plot (Fig. 1 d).

Milky ripeness stage: Best regression model for vegetation percentage variable explains 43.2% of variations in dependent variable (adjusted R^2 = 0.432) with predictor variables R11 and R20 (Fig. 1 e).

Best obtained regression model explains 44.2% of the variation of LAI (LAI_3rings) with predictors variables R6, R13, R24, R4 and SAVI2 (Fig. 1 f).

For the variable LAI at all phases of winter wheat development, a range of 0 to 3 was considered, due to the results obtained in the classification of photographs using CAN-EYE. The map for variable LAI_5rings at the phase of tillering (Fig. 1 b) and LAI_3rings at the shooting phase (Fig. 1 d) indicate a greater presence of values in a range of 0.0004 to 1.20, these values agree with those obtained in the classification of photographs. In the phase of milky ripeness, LAI values are higher (1.23 to 2.99) as indicated by the map of the variable LAI_3rings (Fig. 1 f) which corresponds to the results obtained by CAN-EYE.

In the tillering stage of winter wheat development (Fig. 1 a, b), low area of surface was covered by vegetation according to the vegetation percentage map, due to the presence of bare soil that has not been occupied by the crop yet. This affected modelling and resulted in low prediction ability of acquired models.

 https://doi.org/10.1051/e3sconf/202016902005

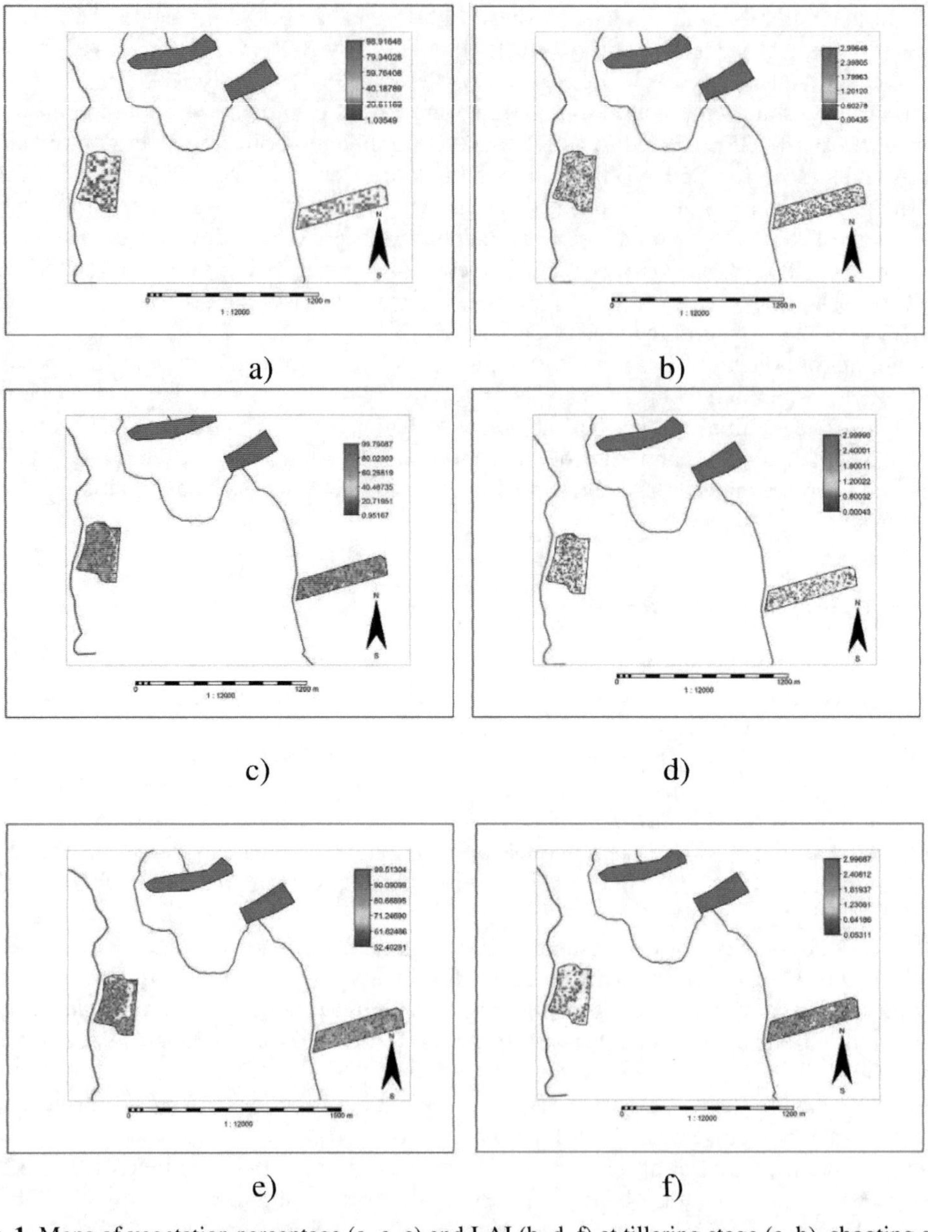

Fig. 1. Maps of vegetation percentage (a, c, e) and LAI (b, d, f) at tillering stage (a, b), shooting stage (c, d) and milky ripeness (e, f) (white color inside test fields corresponds to the areas where models fail (produce negative or too high values))

According to the map for the vegetation percentage (Fig. 1 d), at the shooting phase of winter wheat, greater area of surface was covered by vegetation (60.25 - 99.79%), since in this period the crop reaches its highest vegetative development before ripening.

During the milky ripeness stage, the vegetation percentage varies from 71.24 to 99.51% (Fig. 1 e), because the crop is at the final phase of its cycle.

4 Conclusion

To sum up, according to the results obtained on the basis of ground-truth data, at each stage of winter wheat development the prediction accuracy of the models for determining the percentage of vegetation stayed at 33-43% with the highest accuracy at the stages of milky ripeness and shooting. As to LAI, the highest accuracy (85 %) was obtained for shooting stage when green biomass of winter wheat was the highest. At other stages the accuracy was below 50 %. So, Sentinel-2 data can be successfully used to map LAI at shooting stage of winter wheat. However, this applies only for studied regions and winter wheat.

References

1. R. Allen, L.S. Pereira, D. Raes, M. Smith, *Crop Evapotranspiration-Guidelines for Computing Crop Water Requirements-FAO Irrigation and Drainage Paper*, 56, (1998)
2. A.G. Barr, T.A. Black, E.H. Hogg, N. Kljun, K. Morgenstern, Z. Nesic, Agric. For. Meteorol. **126**, 237–255 (2004)
3. L. Busetto, S. Casteleyn, C. Granell, M. Pepe, M. Barbieri, M. Campos-Taberner, R. Casa, F. Collivignarelli, R. Confalonieri, A. Crema., *et al.*, J. Sel. Top. Appl. Earth Observ. Remote Sens., **10**, 5423–5441 (2017)
4. J. Dietz, D. Hölscher, C. Leuschner, For. Ecol. Manag., **237**, 170–178 (2006)
5. B. Duchemin, R. Hadria, S. Erraki, G. Boulet, P. Maisongrande, A. Chehbouni, R. Escadafal, J. Ezzahar, J.C.B. Hoedjes, M.H. Kharrou., *et al.*, Agric. Water Manag. **79**, 1–27 (2006)
6. M. Sandmann, J. Graefe, C. Feller, Scientia horticulturae, **156**, 113-120 (2013)
7. M. Weiss, F. Baret, *CAN_EYE V6.4.91 USER MANUAL* (INRA, 2017). URL: https://www6.paca.inra.fr/can-eye/Download
8. M. Martinez, *REMOTE PERCEPTION "Fundamentals of Space Remote Sensing"*, (NATIONAL WATER COMISSION, 2005) URL: https://es.scribd.com/document/365840214/Fundamentos-de-Teledeteccion-Espacial (Last login 12.09.2019). (In Spanish)
9. J. Delegido, J. Verrelst, L. Alonso, J. Moreno, Sensors, **11**, 7063-7081 (2011)
10. A.A. Gitelson, A. Viña, V. Ciganda, D. C. Rundquist, T. J. Arkebauer, Geophysical Research Letters, **32** , L08403, (2005)
11. J.G.P.W. Clevers, A.A. Gitelson, International Journal of Applied Earth Observation and Geoinformation, **23**, 344-351 (2012)
12. Delegido J., Verrelst J., Rivera J.P., Ruiz-Verdú A., Moreno J. Brown, Int. J. Appl. Earth Obs. Geoinf, **35**, 350–358 (2015)

E3S Web of Conferences **169**, 02006 (2020) https://doi.org/10.1051/e3sconf/202016902006
APEEM 2020

Fast screening of some flavonoids content in raw plant materials: opportunities of ^{1}H NMR spectroscopy

Gennady Kalabin[1,*] , *Vasilii* Vasil'ev[1] , *Vasilii* Ivlev[1] , and *Vasilii* Babkin[2]

[1]Shared Research and Educational Center, Peoples' Friendship University of Russia (RUDN University), 6 Miklukho-Maklaya Street, Moscow, 117198, Russian Federation
[2]Institute of Chemistry SB RAS, 1 Favorsky street, Irkutsk, 664033, Russian Federation

Abstract. Environmental monitoring and assessment of the prospects for extracting biologically active substances (BAS) from various types of plant biomass requires the development of simple and fast methods for measuring their content in raw materials. A new approach for measuring the content of various flavonoids groups in plant raw material using ^{1}H NMR spectroscopy has been developed, which allows to characterize its resource capabilities and study the effects on their composition different environmental factors without complex sample preparation and standard samples.

1 Introduction

The concept of effective environmental management is not limited by the main environmental imperative - the conservation of environmental biodiversity. Its economic component is the rational consumption of natural raw materials. Trees and higher plants are important renewable resources of the biosphere, the processing methods of which should be non-waste and produce not only cellulose, construction and composite materials. Scientifically-based their use involves obtaining a wide range of food, medical and technical products. Assessment of the content of biologically active substances in biomass should be based on environmentally friendly quantitative analytical methods free from complex sample preparation, derivatization, and the need for standard samples. Taking into account the multicomponent nature of the metabolite extracts, NMR spectroscopy meets these requirements due to sufficient sensitivity, informativity and standardless quantitative measurements.

Among the extractive polyphenolic compounds of plant biomass, a special place is occupied by flavonoids, of which more than 10,000 are identified. We used ^{1}H NMR spectroscopy to develop a universal technique for assessing the content of flavonoids with unique biological activity in significantly different objects: taxifolin **1** in Siberian larch wood (*Larix sibirica*), flavonolignans **2-8** in the seeds of Milk Thistle (*Silybum marianum*), group of flavonol glycosides (FG) **9** in the leaves of a Ginkgo biloba (*Ginkgo biloba*). The

* Corresponding author: kalabinga@mail.ru

aim of the work is to study the possibility for determination of the above biologically active substances content in the studied extracts without using of standard samples for the purpose of eco-monitoring and assessment of their resource potential for drugs creation.

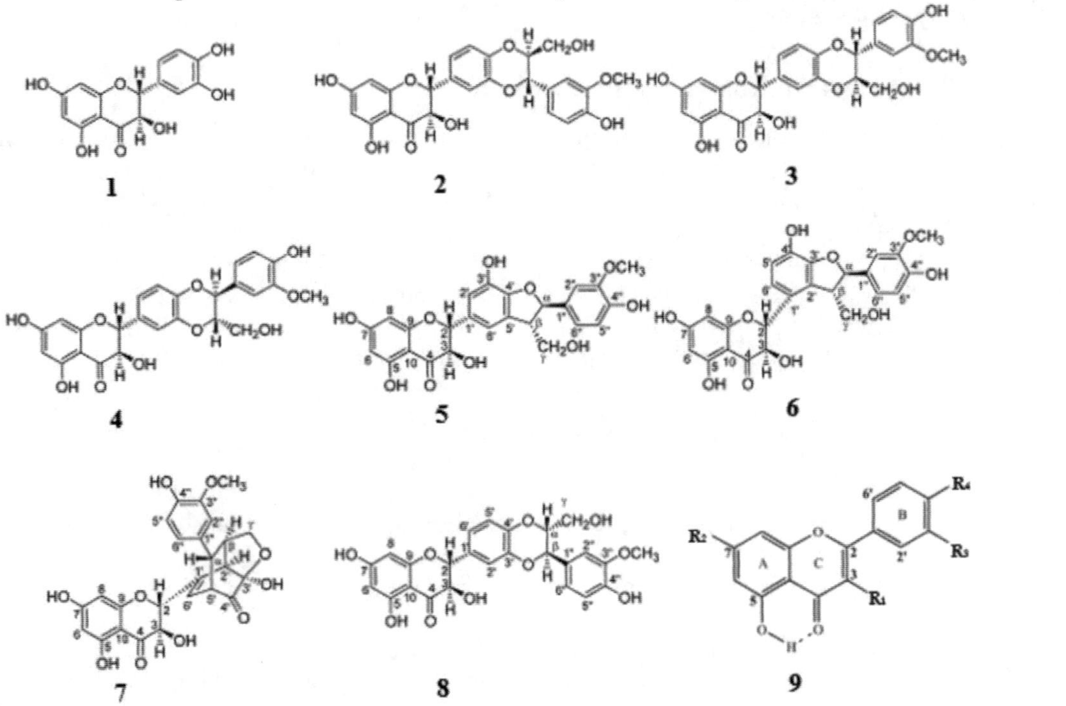

Fig. 1. Structural formulas of the main biologically active substances of the considered extracts: 1. Taxifolin; 2. Silybin A; 3. Silybin B; 4. Isosilybine A; 5. Isosilybine B; 6. Silicritin; 7. Isosilicritin; 8. Silidianin; 9. FG of Ginkgo biloba.

Taxifolin. Russia's coniferous forests contain about half of the world's timber reserves, and the largest area (≈260 million ha) is occupied by larch (≈37%) [1], which UNESCO identified as a symbol of Russia. A unique phenomenon of Siberian larch (*Larix sibirica*) is a high content of **1** in its wood (up to 4%), which is an order of magnitude higher than its content in any other types of plant materials. Due to its ability to inhibit lipid peroxidation, **1** is recognized as a stronger natural antioxidant.

Flavonolignans. *Silybum marianum* is an annual/biennial plant of the Asteraceae family, now growing and cultivated worldwide. It has been used for centuries in Mediterranean medicine mainly to treat liver, kidney, spleen, and gallbladder diseases. Silymarin is a well-established *Silybum marianum* seeds standardized dry extract containing mainly flavolignans (more than 65%) as well as polymeric and oxidized polyphenolic compounds. The main silymarin flavonolignas are **2-8** and taxifolin (Figure 1), where the flavonoid part (taxifolin) of the molecule is fused with a lignan. Content of the amount of **2-8** in raw materials is usually more 2% [2].

Flavonol glycosides. Preparations based on extract from the leaves of *Ginkgo biloba* are the most popular plant nootropics in the world. Its medicinal properties were known more than 2,000 years ago and are used now for treatment central nervous system diseases, such as dementia and Alzheimer's disease.

The extract from ginkgo leaves contains a wide variety of FG (more 70 are classified and more 30 are isolated) [3]. FG of ginkgo has from one to three carbohydrate residues and their acylated derivatives which aglycones are predominantly quercetin, kaempferol and isoramnetin, content of which is minimal. The content of **9** is more than 0.5% in raw materials.

2 Materials and Methods

2.1 Plant material, solvents and standards

Finely ground absolutely dry samples of *Larix sibirica* wood (LLC "Scientific and Production Association "Wood Chemistry", Irkutsk, Russia), *Silybum marianum* seeds (CJSC "Pharmacenter VILAR", Moscow, Russia) and *Ginkgo biloba* leaves (CJSC "Evalar", Biysk, Russia), as well as commercial samples of extracts from this raw material.

A dimethylsulfoxide-d_6 (DMSO-d_6) solvent (≥99.9%, Sigma-Aldrich, CAS Number 2206-27-1) was used. Taxifolin (USP Reference Standard, CAS Number 480-18-2), silybin (USP Reference Standard, CAS Number 22888-70-6) and diosmin (USP Reference Standard, CAS Number 520-27-4) standards have also been used.

2.2 Sample preparation and experiments

50,0±0,5 mg of the raw material powder were placed in eppendorf and 1 ml of DMSO-d_6, containing 0,1% of diosmin standard, was added. Eppendorfs were shaken for 10 minutes on Vortex, 10 minutes were sonicated and placed into a centrifuge (14,000 rpm) for 5 minutes. The supernatant was transferred to a standard 5 mm diameter NMR tube and the ^{1}H NMR spectrum was recorded.

The NMR spectra were acquired at 298 K with a JEOL JNM ECA 600 NMR spectrometer (JEOL, Japan) with an operating frequency of 600 MHz for protons under the following conditions: flip angle - 45°, acquisition time – 3.5 s, the relaxation delay - 4 s, spectra width - 18 ppm, spectrum offset - 7 ppm, free induction decay (FID) data points - 32 kbyte, the number of scans - 64, the time for each spectrum was about 8 minutes. Data acquisition and processing were performed with Delta IV. The phase and baseline correction were done automatically, the integration was done manually. RSD is less 2%.

3 Results and Discussion

Earlier during the screening of flavonoids content in dietary supplements, we found [5] that ^{1}H NMR spectra in a wide range of concentrations (1-50 mg / ml) contain narrow proton signals of a hydroxyl group in position 5 (5-OH) in the range from 11.8 to 13.5 ppm. The protons of 5-OH form a strong intramolecular hydrogen bond with the oxygen of the neighboring carbonyl group, so these nuclei are deshielded and their signals appear in DMSO-d_6 in the above range depending on the presence of a multiple bond in position 2-3 and the nature of the substituents. For example, for flavones - 12.80-13.50 ppm, flavonols - 12.40-12.65 ppm, flavonones - 12.0-12.20 ppm, flavononols - 11.80-11.96 ppm. [4,5]. Using of the 5-OH groups proton signals, it is possible to identify the FG in the extracts of plant origin and also to determine their content, because signals of aromatic and aliphatic protons do not appear in this region of the spectrum. As follows from Figure 1, all flavonoids **1–9** are characterized by the presence of 5-OH.

Biologically active substances of plant materials are usually extracted solvents with low boiling point (water, methanol, ethanol, acetone, ethyl acetate), which are subsequently removed. The presence of a solvent is possible for quantitative estimates of the **1–9** content from ^{1}H NMR spectra, if 5-OH proton signals with a chemical shift >11 ppm are used. DMSO is an environmentally friendly non-toxic organic solvent with the highest dissolving ability for flavonoids. The high basicity of DMSO contributes to the narrowing of the signals of 5-OH protons in the spectrum due to the slowdown of proton exchange. Diosmin was chosen as an internal standard for quantitative measurements of the **1–9** contents, the 5-OH protons T_1 relaxation time of which is less than or equal to the analyte protons T_1. The 5-OH protons T_1 of diosmin and **1-9** does not exceed 2 seconds in DMSO-d_6, which significantly reduces spectrum acquiring time. The 5-OH proton signal of diosmin is quite narrow, located in the same spectral region (12.95 ppm) with the 5-OH proton signals of analytes (Figure 2).

Validation of the new methodology in terms of linearity, limit of detection (LOD) and limit of quantification (LOQ) were carried out for individual solutions of **1** or **2+3** at concentrations from 0.01 to 40 mg/ml. The linear correlation for **1** is described by the equation: y=0.9447x+0.2973, n=7, r=0.9986. LOD (S/N=3) was found to be 0.01 mg/ml for **1** (0.02 for **2+3**) and LOQ (S/N=10) was 0.03 mg/ml for **1** (0.05 mg/ml for **2+3**). The obtained results give the possibility of using the methodology for another plant materials containing flavonoids in smaller quantities than in the biomass of *Larix sibirica* wood or *Silybum marianum* seeds.

To assess the variations in the resource potential of *Larix sibirica* as a source **1**, its content in samples from different parts of wood (root, trunk) of one living tree of mature age (100-120 years) in the vicinity of Irkutsk city was monitored during the year. It was found that the content of **1** in the root biomass is about 50% of the sum of extractive substances in DMSO-d6 and varies from 0.8% in spring to 3.4% in autumn. The trunk biomass always contains **1** 2-3 times less, which is consistent with the results [1]. Comparison of the DMSO-d6 effectiveness as extractant **1** with the commonly used solvents (methanol/water, acetone, ethyl acetate) showed the completeness of extraction. The results of the ^{1}H NMR and HPLC-UV methods [6] are consistent, although the latter is needs more time, requires standard samples and has errors 4-5 times higher.

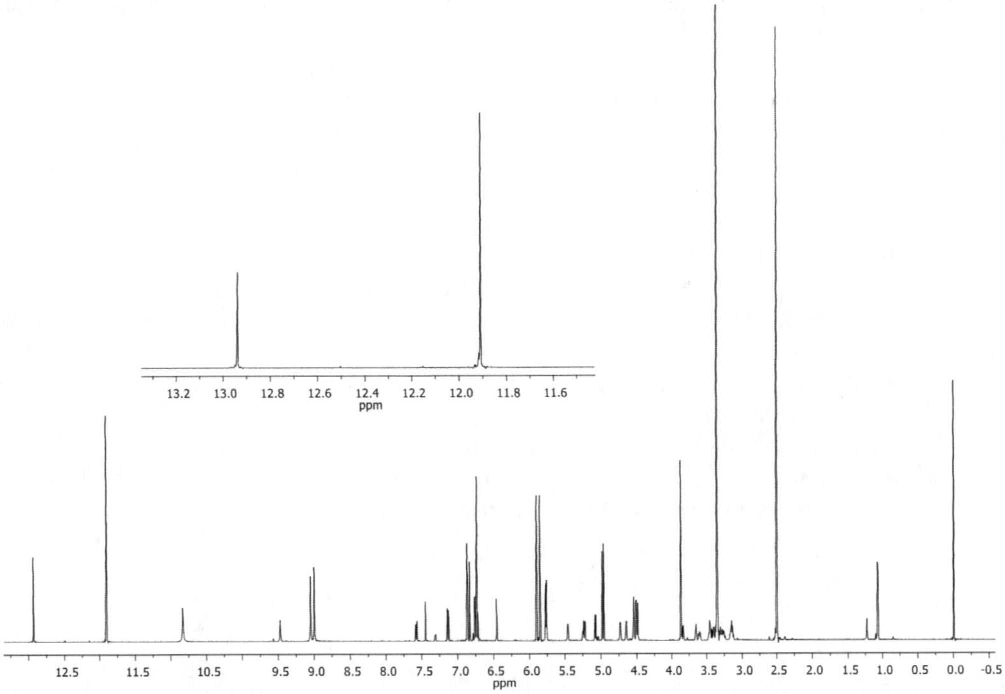

Fig. 2. ^{1}H NMR Spectrum of *Larix sibirica* wood extract in DMSO-d_6.

^{1}H NMR spectrum of flavonolignans of *Silybum marianum* seeds contain overlapping signals of 5-OH protons of components **2-8** in the range 11.9-12.0 ppm (Figure 3A), which is free of signals from other components. Their area corresponds to their total content and allows to calculate it using the known molecular weight of **2-8**, which is equal to 482.4 g/mol. Proposed method allows to determine the suitability of using biomass, extracted from various parts of *Silybum marianum* at different periods and geographical locations of the raw materials. The quality of *Silybum marianum* also depends on the content of its oxidation products, which contain the C_2-C_3 double bond. These components have the 5-OH protons signals in the region of 12.1 - 12.5 ppm (Figure 3A). The proposed method for resource estimation of the most effective biologically active flavonolignans content in *Silybum marianum* significantly exceeds in simplicity, laboriousness, and accuracy the known method, based on HPLC-UV and recommended for quantitative determination of flavonolignans both in extracts and pharmaceutical preparations based on them [6].

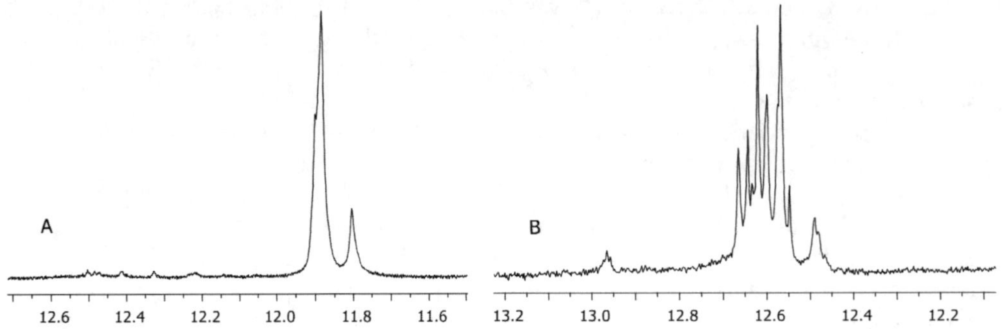

Fig. 3. Enlarged regions of ^{1}H NMR Spectra of A - *Silybum marianum* seeds extract and B - *Ginkgo biloba* leaves extract in DMSO-d_6.

According to [6], the FG content in the leaves of *Ginkgo biloba* is carried out by HPLC-UV after acid hydrolysis of the sample and recalculation of the flavonoid aglycones content to the mass of FG. The stage of sample preparation of our method does not include hydrolysis and, therefore, allows to characterize the native FG in the leaves of *Ginkgo biloba*, so their quantitative determination is direct. Figure 3B illustrates the ^{1}H NMR spectrum region of *Ginkgo biloba* extract, containing 5-OH proton signals in the 12.50-12.65 ppm region, which proves the presence of **9**. According to this Figure, analyzed sample contains at least ten dominant FG. The spectrum also has small signals at 12.47, which indicates the presence of a small amount of aglycons. The *Ginkgo biloba* extract contains flavonoids with different sizes of glycosidic part (1-3 carbohydrate residues), and, accordingly, molecular weight. The average ratio of the molecular weight of FG and their aglycones is close to 2.5 [3], and the value of the average molecular weight is about 759 g/mol. Quantitative determination of FG in different *Ginkgo biloba* samples (I-VI) was carried out by comparing the integrals of their 5-OH proton signals with the 5-OH proton signal of diosmin. The NMR results were compared with the results obtained on HPLC-UV to determine the accuracy. The results of the FG content determining by ^{1}H NMR and by HPLC in the analyzed different *Ginkgo biloba* samples are presented in Table 1.

Table 1. Results of FG determination using ^{1}H NMR spectroscopy and HPLC-UV (n=5, P=0,95)

Sample	NMR	HPLC-UV
	C, mg/ml	C, mg/ml
I	0.19±0.02	0.18±0.02
II	0.32±0.01	0.33±0.04
III	0.19±0.02	0.18±0.05
IV	0.19±0.04	0.19±0.02
V	0.41±0.04	0.35±0.04
VI	0.22±0.02	0.18±0.02

The reproducibility of the NMR measurement results is less 10%. The limit of detection (LOD) of the method with a signal to noise ratio (S/N) = 3 is 0.04 mg/ml, and the limit of quantitation (LOQ) (at S/N = 10) is 0.1 mg/ml. The developed ^{1}H NMR method for the FG determination in different *Ginkgo biloba* raw material is sensitive, linear and reproducible.

4 Conclusion

A new approach for the rapid measurement of taxifolin in *Larix sibirica* wood, flavonolignans in *Silybum marianum* seeds, flavonol glycosides in *Ginkgo biloba* leaves by ^{1}H NMR spectroscopy is proposed. NMR spectroscopy has showed significant advantages, such as no need for standard samples and simple sample preparation, over other methods of analysis. According to the obtained results, it was determined that the use of deshielded 5-OH protons signals of flavonoids is convenient for their identification and quantification in plant materials. Validation characteristics indicate the possibility of applying of the proposed methodology for assessing plant resource potential. This approach was also tested on other plant materials containing flavonoids, such as a roots of *Scutellaria baicalensis* and leaves of *Vitex agnus-castus*, which expands its applicability.

The publication has been prepared with the support of the "RUDN University Program 5-100"

References

1. V.A. Babkin, L.A. Ostrouhova, N.N. Trofimova, *Biomassa listvennicy: ot himicheskogo sostava do innovacionnyh produktov* (SO RAN Publ., Novosibirsk, 2011) (in Russian).
2. A. Cheilari, S. Sturm, D. Intelmann, C. Seger, H. Stuppner, J. Agric. Food Chem., **64**, 1618–1626 (2016).
3. X.G. Liu, S.Q. Wu, P. Li, H.Yang, J. Pharm. Biomed. Anal.,**113**, 212-225, (2015).
4. V.I. Sheichenko, O.P. Sheichenko, V.V. Anufrieva, O.N. Tolkachev, K.M. Dyumaev, T.A. Sokol'skaya. Pharm. Chem. J., **50 (2)**, 83-89 (2016).
5. V.G. Vasil'ev, A.S. Prokop'ev, G.A. Kalabin, Russian Journal of Bioorganic Chemistry, **43(7)**, 86–92 (2017) (in Russian).
6. V.A. Tutel'yan, K.I. Eller, Metody analiza minornyh biologicheski aktivnyh veshchestv pishchi. («Dinastiya» Publ., Moscow, 2010) (in Russian).

E3S Web of Conferences **169**, 02007 (2020) https://doi.org/10.1051/e3sconf/202016902007
APEEM 2020

Impact of black yeasts on the durability of polyurethane foam in tropical conditions in Vietnam

Hoai Nguyen Thu[1*], *Hong* Do Thi Thu[1], *Cuong* Ngo Cao[1], *Thinh* Do Tat[2], and *V.N.* Sereda[3]

[1]Institute of biotechnology, Vietnam Russia Tropical Center, 63 Nguyen Van Huyen Str., Vietnam
[2]Institute of ecology, Vietnam Russia Tropical Center, 63 Nguyen Van Huyen Str., Vietnam
[3]The Russian Academy of Sciences, Moscow, Russia

Abstract. Polyurethane (PU) is a material that is widely used in industries. However, tropical climatic conditions in Vietnam often make the durability of PU lower than temperate climatic conditions. One of the causes of this problem is the biodegradation of microorganisms. In this study, biological characteristics of black yeast on PU is evaluated, thereby, initially assessing the impact of black yeast on the durability of PU in tropical conditions. Ten strains of black yeast were isolated from used PU foam samples in Vietnam. These black yeast strains were tested for enzyme activity including: urease activity, protease activity, polyurethanase activity. There were two strains selected to identify based on sequence analysis of the ITS1/ITS2 region, they belong to genus *Aureobasidium*. VN1Y3 was strain selected for biodegradability testing of PU foam samples. After 28 days of testing, the PU foam samples that have been infected with black yeasts have a change compared to the control in terms of color, surface, tensile strength, elongation. Research results have showed the impact of black yeast on the durability of PU.

1 Introduction

Polyurethane (PU) is a synthetic polymer that is formed by condensation polymerization reaction between polyisocyanate and polyol having intramolecular urethane bonds. Urethanes are derivatives of carbamic acids which exist only in the form of their esters [1]. In tropical climatic conditions in Vietnam, PU often has a short durability due to many reasons such as temperature, humidity, radiation and biodegradation of microorganisms,... Biodegradation is a chemical degradation of polymers by the action of microorganisms, such as bacteria, yeasts and fungi. Microorganisms degrade the polymers like polyurethane by using it as a substrate for their growth. J. R. Crabbe *et al.* found four species of fungi including *Curvularia senegalensis*, *Fusarium solani*, *Aureobasidium pullulans*, and *Cladosporium sp.* They were isolated based on their ability to utilize a colloidal polyester PU (Impranil DLN™) as the sole carbon and energy source [2]. Halim *et al.* tested the growth of several species of bacteria on PU military aircraft paint. The investigators

* Corresponding author: hongdt1009@gmail.com

isolated four species of bacteria including: *Acinetobacter calcoaceticus*, *Pseudomonas sp.*, *Pseudomonas cepacia*, and *Arthrobacter globiformis* . All species were capable of utilizing the polyurethane paint as a sole carbon and energy source with the exception of *P. cepacia* [3]. In addition, many authors argue that the biodegradation of PU is due to the activity of several extracellular enzymes, including polyurethanase [1-3]. In this paper, we described the biological characteristics of black yeast isolated from used PU foam samples and initially demonstrated their impact on the durability of PU.

2 Materials and methods

2.1 Materials

The materials that were used in this study included:

Samples: Used PU foam samples were collected in aircraft fuel compartment in Vietnam.

Media: Hansen medium [4], mineral medium with Impranil DLN™ (6 g/l) [1].

2.2 Methods

The study used the following methods:

Enrichment and isolation method: used PU foam samples are cut into small pieces of size 1.5 x 1.5 (cm) and enriched in mineral medium with Impranil DLN™ (6 g/l) at temperature 28°C and shaking 250 rpm for 7 days. Enrichment culture at the third inoculation was continued incubating on Hansen medium at temperature 28°C for 5 days to isolate black yeast colonies.

Determination of yeast morphology: Black yeasts were cultivated and differentiated by their macro and micro-morphologies according to Kurtzman *et al.*[5]

Determination of extracellular enzyme activity: the experiment was carried out by agar diffusion method according to Bauer *et al.*[4]

DNA extraction: the experiment was carried out using Zymo Research Kit (USA).

PCR: PCR is established using the ITS1/ITS4 primers and the corresponding heating programs.

Sequence analysis: The sequencing results were compared to related data in Genbank by the BLAST search on NCBI.

Biodegradation of PU test: according to ASTM G21-96 [6] and GOST 9049-91 [7].

Scanning electron microscopy: SEM was used to characterize surface modifications of polyurethane samples.

3 Results and discussions

3.1 Isolation results

Ten black yeast strains were isolated from used PU foam sample in Vietnam (Table 1). They had a variety of colonial and cell morphologies with typical characteristics of black yeast group, such as porous colonies, diverse colors, turning black when old (after about 7 days of culture), rod or oval cells, cells in chain or multiple faults.

Table 1. Colony and cell of black yeast strains

№	Strains	Morphology	№	Strains	Morphology
1	VN1Y1	Rounded, yellow colony. Rod-shaped, sequined cells	6	VN2Y1	Milky white colony. Rod-shaped cells, many faults.
2	VN1Y2	Colony round, slightly glossy, turning black when old. Cells spherical, budding.	7	VN2Y2	Creamy white colony, turning black when old. Rod-shaped cells, many faults.
3	VN1Y3	Milky white colony, wrinkled, turning black when old. Oval cell and rod, budding.	8	VN2Y3	Yellow colony, the center has the mind. Rod cells, small, many faults
4	VN1Y4	Colony round, wrinkled, turning black when old. Cells spherical, budding.	9	VN2Y4	White colony, hard surface. Rod cells, small, many faults.
5	VN1Y5	Creamy white colony, wrinkled, turning black when old. Oval cells, sequined.	10	VN3H3	White colony, hard surface, turning black when old. Rod cells, small, many faults.

The morphology of colonies and cells of 10 strains of black yeast have complex differentiation. Many researchers believe that the complexity of the black yeast's life cycle (changes in colony color and cell shape that can occur during 3 weeks of culture) have led to grouping and identifying become less accurate. Therefore, to confirm the results, more biochemical tests and sequencing are needed [8]. In this study, an evaluation of enzyme activity and molecular identification was performed.

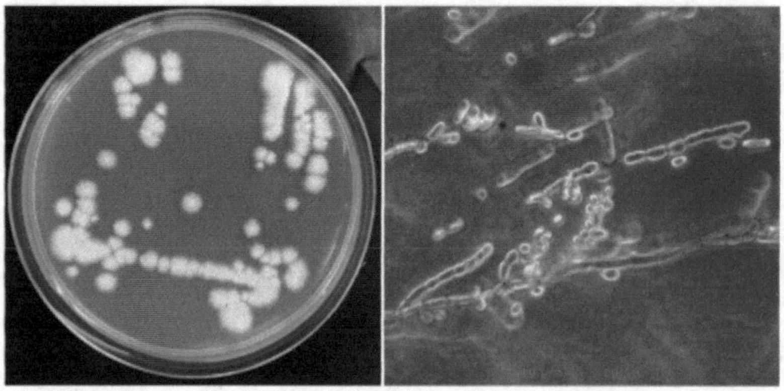

Fig.1. Colony (left) and cell (right) of strain VN1Y4

3.2 Determination of enzyme activity

The results showed that all strains had extracellular enzyme activity. In which, VN1Y3, VN1Y4 and VN2Y1 were the three strains with the highest polyurethanase activity, strain VN2Y1 and strain VN2Y4 had the highest protease activity, the urease activity of black yeast strains in this study was relatively low (except strain VN1Y3). VN1Y3 and VN1Y4 were the best enzyme activity strains, thus they continued to be identified by ITS1/ITS2 region sequences (Table 2).

Table 2. Enzyme activity of black yeast strains

№	Strains	PU	Pr	Ure	№	Strains	PU	Pr	Ure
1	VN1Y1	-	++	+	6	VN2Y1	++	++++	-
2	VN1Y2	+	+	+	7	VN2Y2	+	+	-
3	VN1Y3	++	++	++	8	VN2Y3	-	-	+
4	VN1Y4	++	+++	+	9	VN2Y4	+	++++	-
5	VN1Y5	+	+++	+	10	VN3Y3	+	++	-

([+] positive, [-] negative; PU - polyurethanase, Pr - protease, Ure - urease)

Extracellular enzyme activity of microorganisms isolated from PU foam has been mentioned in many previous studies [9,10]. However, only a few studies have evaluated the polyurethanase activity of black yeast. Studies have also shown the role of polyurethanase in the use of PU materials by yeasts as carbon source for their growth [10]. Therefore, the research results can be considered as the first step to assess the impact of black yeast on the quality of PU during the use of this product.

3.3 Molecular identification

The sequence of ITS1/ITS2 of two black yeast strains (VN1Y3, VN1Y4) was sequenced based on primers ITS1 (5'-TCCGTAGGTGAACCTGCGG-3') and ITS4 (5'-TCCTCCGCTTATTGATATGC-3'). Phylogenetic tree was established by comparing the sequencing results on Genbank using the BLAST search on NCBI (Fig. 2).

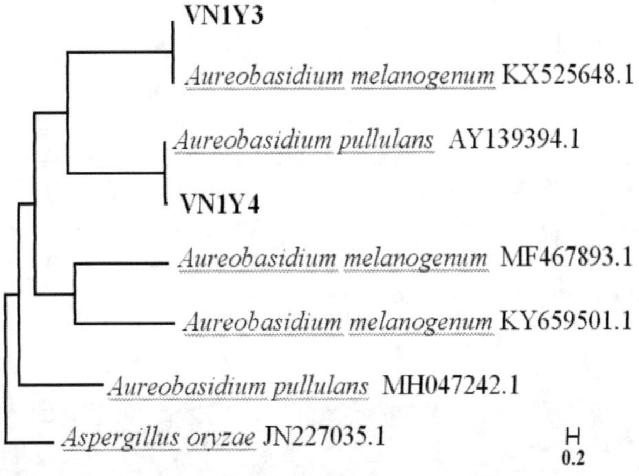

Fig.2 Phylogenetic tree of yeast, based on the ITS1/ITS2 region sequences

VN1Y3 and VN1Y4 a were closely related to species in the following order: *A. melanogenum*, *A. pullulans*. Combined with the results of analyzing the morphology of colonies, the morphology of cells and checking the extracellular enzyme activity of the black yeast strains in the above section (section 3.1 and 3.2), the results of identification by sequencing was completely reliable. These results were similar to the results of J.R. Crabbe *et al.* about the existence of yeast belonging to genus *Aureobasiditum* on PU foam [2].

3.4 Biodegradation of PU test results

VN1Y3 was the black yeast strain selected for biodegradation testing because it had high extracellular enzyme activity for all enzymes studied and determined to be closely related to species *A. melanogenum* (black yeast species was assessed as being capable of producing polymerases [11]). After 28 days of testing with strain VN1Y3, PU foam samples were changed in color, surface, tensile strength and elongation compared to the control PU foam.

3.4.1 Colour of PU

After drying, the PU foam sample tested with strain VN1Y3 has changed its color from light yellow to dark yellow (Fig. 3). The cause of this problem may be due to the yeast adhesion on the surface of PU and the PU material has been modified due to the growth of yeast. To clarify this problem, PU foam surfaces continue to be observed by scanning electron microscopes.

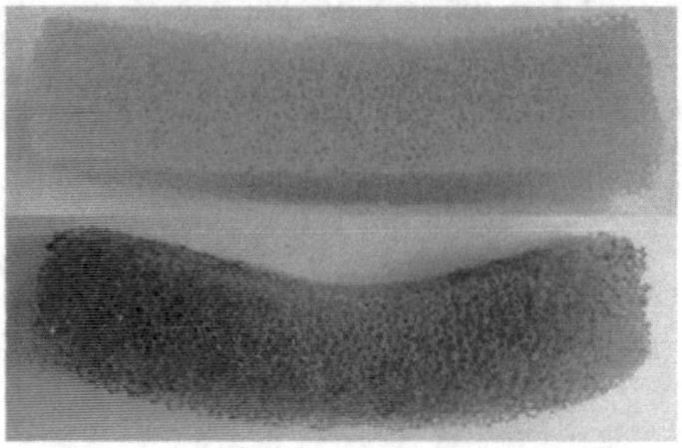

Fig.3 Colour of PU foam with VN1Y3 (below) and control (above)

3.4.2 Surface of PU

The surface of the PU foam after testing was observed by scanning electron microscopy at magnification 500x (Fig.4). Fig 4 showed the adhesion of yeast biomass on PU foam surfaces. Compared with the control under the same test conditions, it can be seen that the PU foam samples tested with VN1Y3 appear to have plaque on the PU surface, their fiber systems form dense networks on the foam surface. In addition, at higher magnifications, pits and dimples can be observed while the control PU foam had a smooth and flat surface after 28 days of testing.

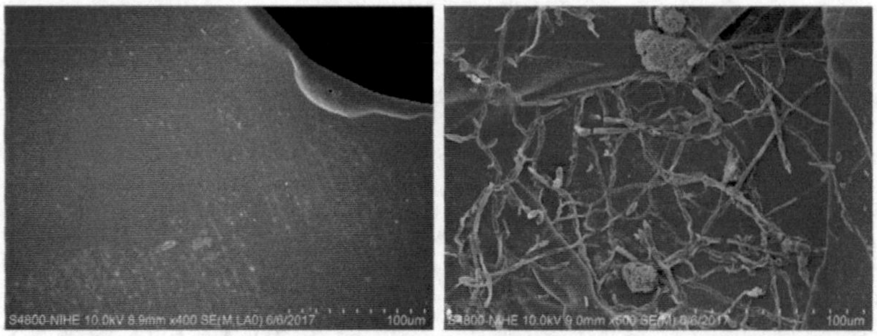

Fig. 4 Surface of control (left) and PU foam with VN1Y3 (right)

The results of this study are completely similar to those of T. H. Gary and A. Nakkabi *et al.* on the growth of microorganisms on PU. A. Nakkabi *et al.* observed a much net growth of the bacterium *Bacillus safensis* on the polyurethane surface after 7 days of testing [1]. T. H. Gary also suggested that microorganisms have used PU as a source of carbon and energy for their growth [10].

3.4.3 Tensile strength and elongation of PU

Many studies suggest that the extracellular enzymes of microorganisms can cut the chemical bonds of PU, thereby reducing their durability [3,10,12]. In this study, tensile strength and elongation of foam samples were checked by specialized equipment. The results showed that the presence of yeast VN1Y3 caused the tensile strength of PU foam sample was 4,9 N/cm^2 corresponding to 38,9% compared to the control (12,6 N/cm^2). The elongation was 175% while the control reaches 335%. The results demonstrated the obvious impact of black yeast VN1Y3 on the durability of PU foam.

4 Conclusions

Ten strains of black yeast were isolated from used PU foam samples with biological characteristics that showed their significant impact on the durability of PU foam, especially the ability to produce polyurethanase. After 28 days of testing with black yeast VN1Y3, PU foam was changed in color, tensile strength and elongation compared to the control PU foam. In which, tensile strength decreases to 38.9%. The strain VN1Y3 is closely related to species *Aureobasiditum melanogenum*.

This research has been supported by Project T 1.3 from The Scientific committee of Vietnam – Russia.

References

1. A. Nakkabi, M. Sadiki, S. Ibnssouda, M. Fahim, Int. J. of Recent Advances in Multidisc. Res. **02**, 4 (2015)
2. J.R. Crabbe , J.R.Campbell, L.Thompson, S.L. Walz, W.W. Schultz, Int. Biodeterior. Biodegrad. **33**, 11 (1994)
3. E.S. Halim, W.M. Mahmoud, E.M. Davis, R.W. Coughlin, Int. Biodeterior. Biodegrad. **37**, 11 (1996)
4. W. Bauer, W.M.M. Kirby, J.C. Sherris, Microbiol. and Medic., **45**, 7 (1966)
5. C.P. Kurtzman, J.W. Fell, T. Boekhout, V. Robert, *The yeasts: a taxonomic study* (Elsevier, San Diego, 2011)
6. ASTM G21-13, *Standard practice for determining resistance of synthetic polymeric materials to fungi* (ASTM International, 2013)
7. GOST 9.049.91, *Polymer materials and their components*, Methods of laboratory tests for mould resistance to the effect of mold fungi.
8. S. Seyedmousavi, M. G. Netea, J. W. Mouton, W. J. G. Melchers, P. E. Verweij, G. S. Hooge, ASM journal, **17** (2014)
9. R. Gautam, A.S. Bassi, E.K. Yanful, E. Cullen, International Biodeterioration & Biodegradation, **60**, 5 (2007)
10. T. H. Gary, Recent Developments in Polymer Recycling, 24 (2011)
11. C. Gostincar, R. A. Ohm, T. Kogej, S. Sonjak, M. Turk, J. Zajc, P. Zalar, M. Grube, H. Sun, J. Han, A. Sharma, J. Chiniquy, C. Y. Ngan, A. Lipzen, K. Barry, I. V. Grigoriev, N. G. Cimerman, BMC Genomics, **14**, 29 (2014)
12. R. S. Gisele, S. C. Armando, B. C. Francine, A. Eliane, L. O. Rodrigo, Polymer Degradation and Stability, **95**, 9 (2010)

E3S Web of Conferences **169**, 02008 (2020) https://doi.org/10.1051/e3sconf/202016902008
APEEM 2020

Dewatering as a primary treatment of faecal sludge in individual residential sector (a technologies review)

Alena Basamykina[1] *, *Marianna* Kharlamova[1], and *Sharon Yeukai* Mada[2]

[1]Peoples' Friendship University of Russia (RUDN University), Faculty of Ecology, 6 Miklukho-Maklaya St, Moscow, 117198, Russian Federation
[2] The Catholic University of Zimbabwe, Faculty of Humanities and Social Sciences, 18443 Cranborne Ave, Harare, Zimbabwe

Abstract. Dewatering of faecal sludge (FS) is the most efficient method of pretreatment of bioorganic human waste for minimizing the faecal sludge volume. This process may be considered as one of management tool in service chain of Faecal Sludge Management (FSM). For this purpose technological features of methods, the advantages, disadvantages and possibilities of the use have been analysed. The article provides a review of the most efficiently methods of faecal sludge dewatering generated in the individual residential sector (unplanted drying beds, planted drying beds, solar drying, mechanical dewatering, geotubes) and their analysis in order to select the best available technology in accordance with the selection criteria. We used the analytic hierarchy process to determine the most multi-operated and effective processing method of faecal sludge dewatering. The final ranking for each dewatering technology was based on the global weights of 10 criteria (dewatering efficiency, operation time, process automation, required space, climate dependent, capital expenditures, operating costs, energy efficiency and environmental impact). The results identify dewatering method with the highest priority result based on priority eigenvalues of criteria.

1 Introduction

Faecal sludge (FS) generated in the individual residential sector is a natural product of human activity. It is highly concentrated mixture of natural organic matter of various compositions and is very different from municipal wastewater. Due to the high variability of the organic matter composition, the design of the processing system should not be based on standard characteristics, but on the results of laboratory tests obtained for each case [1,2,3,4].

Nowadays, there is much tension around the issue of collection, storage, treatment and disposal of faecal sludge in the individual residential sector in the Russian Federation. At the federal, regional and municipal levels, there is no adequate regulatory system and management of such waste type. In most cases authorities try to solve the problem at the local level, involving private companies for collecting and transporting of faecal sludge only if

* Corresponding author: alena.basamykina@gmail.com

https://doi.org/10.1051/e3sconf/202016902008

owner pays for this service. Moreover, the collected faecal sludge is discharged to sewage treatment plants or worse to a water body or on the ground. Thus, enormous damage is caused to the environment that, at the moment, cannot be estimated due to the lack of a monitoring system [1,5,6].

To solve this problem, an integrated approach is required. It should include not only the implementation of planning and management systems, but also the applicability of the resource potential of faecal sludge using the best available technologies. From an economic feasibility point of view, the most important problem is the reducing of transportation costs. With this objective in view, the use of small semi-centralized plants for the primary processing of faecal sludge generated in the individual residential sector allows to minimize the FS volume and transportation costs [1,5]. Faecal sludge co-treatment with wastewater is possible to optimize the technological process of FS treatment [4,7,8].

The most efficient method for minimizing the faecal sludge volume is dewatering. This process leads to a decrease of the moisture content from 98% to 75% (that is tantamount to an increase of the solids content from 2% to 25%). Moreover, the volume of dewatered faecal sludge that must be transported for another treatment step will be 12 times less than the unprocessed volume. That can significantly reduce transportation costs. [1,6]

This article analyzes the most common methods of faecal sludge dewatering, and also makes a criteria comparative analysis to determine the most multi-operated and effective processing method.

2 Materials and methods

2.1 The analytic hierarchy process (AHP)

In this article we used the analytic hierarchy process to conduct criteria evaluation to determine the most multi-operated and effective dewatering method of faecal sludge.

The analytic hierarchy process (AHP) is a mathematical structured technique for complex decisions. The method does not imply any "correct" solution, but allows interactively finding the best decision consistent with the given problem essence and the requirements for its solution. The analytic hierarchy process is used to compare objects, to solve management problems, forecasting, etc [9].

The AHP generates a weight for each evaluation criterion (dewatering efficiency, operation time, process automation, required space, climate dependent, capital expenditures, operating costs, energy efficiency, emissions and process waste). The higher the weight, the more important the corresponding criterion. Next, for a fixed criterion, the AHP assigns a score to each dewatering methods (unplanted drying beds, planted drying beds, solar drying, mechanical dewatering, geotubes) The higher the score, the better the performance of the option with respect to the considered criterion. Finally, the AHP combines the criteria weights and the dewatering methods scores, thus determining a global score for each method, and a consequent ranking. The global score for a given dewatering method is a weighted sum of the scores it obtained with respect to all the criteria [9].

Each criterion is expressed by an attribute. The larger the value of the attribute, the better the performance of the dewatering method with respect to the corresponding criterion.

2.2 Dewatering methods

2.1.1 Unplanted drying beds

Unplanted drying bed is a permeable layer where faecal sludge is dried by natural evaporation. The site is filled with filter media, most commonly gravel is a bottom layer and sand is a upper layer (Fig. 1). The bottom has a slope where the drainage system is located [4,6,10].

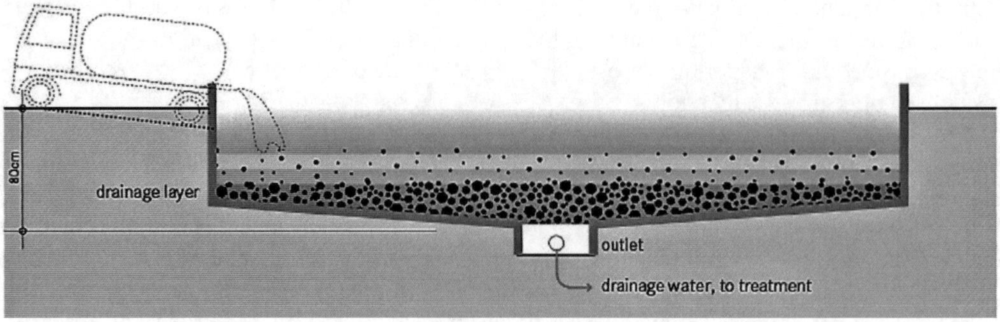

Figure 1. Schematic of an unplanted drying bed [11]

Unplanted dying beds are operated in periodic mode. Faecal sludge is located on the surface, liquid infiltrates through sand and gravel for several days. The bulk of solid fraction remains on the surface. Approximately 50-80% of faecal sludge (volume) infiltrates as liquid through the filter media or evaporates. Hence, dehydrated faecal sludge is removed from the surface manually (by shovels) or mechanically (by scraper system or equipment), once every few weeks or months depending on operating conditions [6, 8,10,12].

There is a disadvantage of this technology –regularly the filter media treatment before a new process cycle. Moreover, faecal sludge has no effective stabilization or disinfection [6,12].

Unplanted drying bed should have tent covers or roofs to protect from rain, otherwise the dewatering process will be ineffective [6,8].

2.1.2 Planted drying beds

The ongoing process is similar to the process occurring on a standard drying beds. But difference is that there is plants cultivation on planted drying beds (Fig. 2). Plants selected for a specific climate. Moreover, there is additional evaporation and more efficient faecal sludge dewatering instead of standard drying beds. The filter material does not need to be cleaned after each feeding/ drying cycle [6,7,8].

Planted drying bed must be acclimatized oneself for at least three months before faecal sludge loading. Raw material can be laid directly on the previous layer. Plants and their root systems maintain the filter porosity. They must be collected from the site in accordance with the growth cycle [4,6,8,14].

Studies [8,14] demonstrate that plant growth accelerates with increasing loading frequency of planted drying bed. Plant growth can have a significant impact on both site performance and the potential use of end product.

Extended period of dewatering time can lead to increased inactivation of helminth eggs. Before using of faecal sludge after treated on planted drying beds in agriculture, it is necessary to realize a composting process [14].

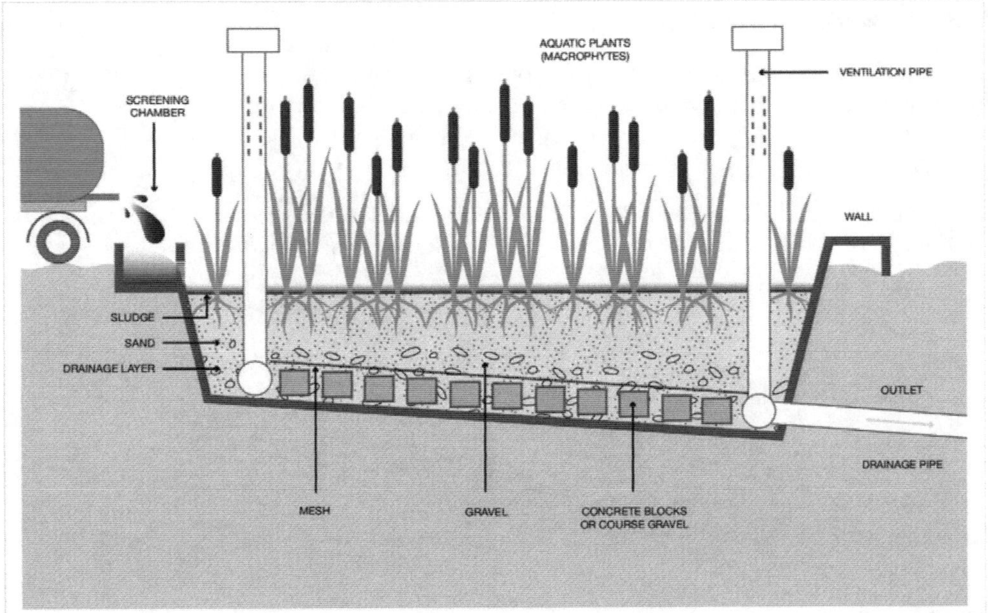

Figure 2. Schematic of a planted drying bed [13]

2.1.3 Solar drying

Solar drying is the process of moisture removing by means of sunlight. There is a banch process. A special textiles (geotextiles) and a perforated layer are added for dewatering between the layer of faecal sludge and sand (Fig. 3).

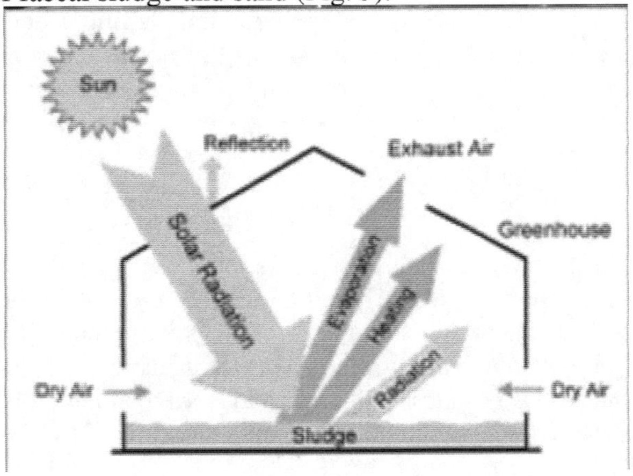

Figure 3. Solar drying process [16]

This will ensure preventing of layers mixing and improving the process efficiency of dewatering and filtration [3,4,15].

The intake of rainwater can significantly reduce the rate of faecal sludge dewatering. Therefore, a transparent roof should be provided above the site for optimal removal of evaporated moisture. The roof surface must be made of durable plastic for optimizing the sunlight penetration [3,15].

2.1.4 Mechanical dewatering

Mechanical dewatering is usually carried out on a screw press (Fig. 4). Dewatered faecal sludge must proceed to additional stage (for example, drying or composting) before usage as organic fertilizer. Centrifuges, belt presses or filter presses may be used instead of a screw press, but this machines is less efficient [4,7,8,17].

Figure 4. Screw press for sludge dewatering [17]

The main disadvantage of this method is a coagulant (polyelectrolyte) demand to stabilize the dewatering process as well as to increase its effectiveness. The results show that conditioning does not increase dewatering rate, but increases moisture reduction rates [8,17,18,19].

Not only polymeric materials can be used as conditioners for faecal sludge, but also waste from local plant resources (for example, seeds, oilcake or chitosan). They are as effective as polymers for dewatering sewage sludge [18].

The *M. Oleifera* seeds, oilcake and chitosan are natural organic materials that can increase the useful properties of the end processed products, such as nutrient content (C, N, P) and / or calorific value while inorganic polymers reduce the calorific value and fuel potential [18,20].

2.1.5 Geotubes

Geotubes are highly effective because of they capture up to 98% solids. Geotextile is a permeable textile made of polyethylene fiber. It has capability to infiltrate moisture and, thus, dewatering process of faecal sludge is more effective that proceeding on drying beds. The filtrate is collected and purified. The process proceeds inadvertently, under the gravity, has no mechanical parts, and it is noiseless (Fig. 5). Therefore method is potentially suitable for decentralized FS processing. Additional external heat accelerates the faecal sludge dewatering [6,7,8,22,23].

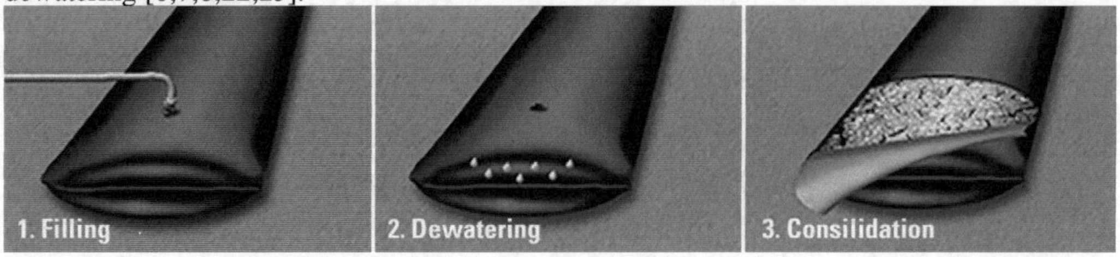

Figure 5. Dewatering with geotube technology [21].

A polymer is added to faecal sludge to precipitation increase and avoid clogging of the geotextile layer. For this purpose, study [22] demonstrates usage of chitosan obtained from shrimp waste. It turned out to be quite effective in the dewatering of faecal sludge from septic tanks and sewage sludge, but ineffective for faecal sludge from pit latrines.

The geotubes can increase the processing intensity and reduce the required area for dewatering at treatment plants [22]. There is a disadvantage: geotubes become a waste themselves after filling the entire volume with solids [6,7,8,22].

3 Results and discussions

The analytic hierarchy process is associated with a consistency ratio. Assuming that there are ten criteria (dewatering efficiency, operation time, process automation, required space, climate dependent, capital expenditures, operating costs, energy efficiency, emissions and process waste) (Table 1) and five dewatering methods, the weights of the criteria are first computed through pairwise comparisons. (Table 2).

Table 1. Criteria definition

Dewatering efficiency (DE)	The maximum moisture removal from fecal sludge due to the dewatering process
Operation time (OT)	Process time of dewatering
Process automation (PA)	Applicability of automated units for fecal sludge feeding, for removing of dewatered sludge from the unit, for liquid discharge
Required space(RS)	The area required for main unit and supporting facilities
Climate dependent(CD)	Climate and weather effects (rain, drought, cold temperatures, etc.) on the dewatering process effectiveness
Capital expenditures (CE)	Costs of the site construction and the purchase of equipment, including supporting facilities
Operating costs(OC)	Costs of chemicals, purchase of plant seeds, geobags, air supply, etc.
Energy efficiency (EE)	Energy consumption of main unit and supporting facilities
Emissions (E) and Process waste (PW)	Environmental impact includes two criteria: air emissions and waste generation (plant waste, used geobags, etc.). This work does not consider the environmental impact of generated wastewater, since most commonly they are treated at the nearest wastewater treatment plant

The criteria must be evaluated in pairs so as to determine the relative importance between them and their relative weight to the global goal. Table 2 shows the relative weight data between the criteria that have been determined. In order to interpret and give relative weights to each criterion, it is necessary to normalize the previous comparison matrix. The normalization is made by dividing each value by the total column value (Sum). The contribution of each criterion to the organizational goal is determined by calculations made using eigenvectors. Eigenvector indicates the relative weights between each criterion. it is obtained in an approximate manner by calculating the mathematical average of all criteria.

The RI (Retention Index) is fixed and is based on the number of evaluated criteria (10), the value is 1.49. The maximum eigenvalue (λmax) is calculated by summing the product of each element in the Eigenvector by the respective column total (sum) of the original comparison matrix. The Consistency Index (CI) is defined as minimum number of changes divided by the number of changes required on the parsimony score. In order to verify whether the consistency index (CI) is adequate, it was suggested [9] the consistency rate (CR), which is determined by the ratio between the consistency index (CI) and the random consistency index (RI). The matrix will be considered consistent if the resulting ratio is less than 0.15 (15%).

Table 2. Evaluation of criteria eigenvectors

	DE	E	OT	CD	CE	OC	EE	RS	PW	PA	*Geometric means*	*Eigenvectors*
DE	1	3	5	5	4	4	7	8	8	8	4.606	**0.287**
E	0.33	1	3	5	6	7	7	8	8	9	3.911	**0.244**
OT	0.2	0.33	1	2	4	5	5	7	7	8	2.352	**0.147**
CD	0.2	0.2	0.5	1	2	3	3	4	5	7	1.480	**0.092**
CE	0.25	0.167	0.25	0.5	1	2	4	6	7	7	1.285	**0.080**
OC	0.25	0.143	0.2	0.33	0.5	1	3	5	6	5	0.939	**0.059**
EE	0.143	0.143	0.2	0.33	0.25	0.33	1	4	5	3	0.606	**0.038**
RS	0.125	0.125	0.143	0.25	0.167	0.2	0.25	1	3	5	0.384	**0.024**
PW	0.125	0.125	0.143	0.2	0.143	0.167	0.2	0.33	1	4	0.279	**0.017**
PA	0.125	0.111	0.111	0.143	0.143	0.2	0.33	0.2	0.25	1	0.200	**0.012**
Total (Sum)	2.748	5.344	10.547	14.753	18.203	22.897	30.78	43.53	50.25	57	16.042	
RI	1.49										λ_{max}	**11.587**
											Consistency index (CI)	**0.176**
											Consistency rate (CR)	**0.118**

Then, all of dewatering methods are pairwisely compared with respect to each criterion and set out in separate tables using the scale. The sum of each row is computed, normalised, then placed in the last column and labelled local weights. The column is used to build a new table with the criteria set out along the top row and dewatering methods build the left-hand column (Table 3). The value in each cell of each column is multiplied by the weight of the criteria associated with the columns and each of dewatering methods, and the sum of each row is computed. The computed numbers are set out in the last column of the final table, which represents the level of attention that should be paid to the alternatives or global weights. As a result, the final ranking for each dewatering technology is based on the global weights.

The Consistency Index (CI) is defined as minimum number of changes divided by the number of changes required on the parsimony score. The Consistency Rate (CR) (Table 3) is based on priority eigenvalues and the consistency index of each method. The Generalized Consistency Rate (GCR) is the ratio CR to RI. Since the Generalized Consistency Rate value (0.066) is less than 0.15 (or 15%), the comparisons are confirmed.

Thus, using the analytic hierarchy process, we determined that mechanical dewatering is the most multi-operated and effective dewatering method of faecal sludge. Despite the low energy efficiency, significant capital expenditures and great ttechnological infrastructure, this method is most acceptable due to the best dewatering efficiency, low emissions, process automation and does not depend on climate conditions.

Table 3. Final priority evaluation of faecal sludge dewatering methods

	DE	E	OT	CD	CE	OC	EE	RS	PW	*PA*	*Final Priority Results*
Priorities eigenvalues	0.287	0.244	0.147	0.092	0.080	0.059	0.038	0.024	0.017	0.012	
Mechanical dewatering	4.580	3.677	4.474	5.156	0.195	0.214	0.220	0.402	0.729	5.665	**3.469**
Geotubes	2.339	0.537	2.277	1.904	1.420	1.182	1.585	0.564	0.241	1.000	**1.585**
Planted drying beds	0.844	0.436	0.807	0.536	2.087	2.352	2.237	2.012	1.947	0.740	**0.998**
Unplanted drying beds	0.411	0.295	0.446	0.536	3.936	3.545	2.237	3.882	2.631	0.561	**1.059**
Solar drying	0.268	0.833	0.271	0.354	0.491	0.474	0.574	0.564	1.105	0.425	**0.479**
Consistency index (CI)	0.124	0.056	0.130	0.072	0.163	0.097	0.052	0.052	0.047	0.041	
Consistency rate (CR)	**0.098**										
Generalized consistency rate (CR)	**0.066**										

4 Conclusions

Based on the the analytic hierarchy process to determine the most multi-operated and effective dewatering method of faecal sludge (unplanted drying beds, planted drying beds, solar drying, mechanical dewatering, geotubes), the highest priority result (3.469) was determined for the mechanical dewatering process. Principally, it is link with high dewatering efficiency, the rapid operation time, as well as the possibility of process automation and low environmental impact. Moreover, this technology requires the smallest operating areas and does not depend on climatic conditions. At the same time, mechanical dewatering has some disadvantages relative to other processes: the need for great technological infrastructure and the reagents usage, as well as high capital and operating costs.

The lowest total points (0.479) was determined for solar drying. It is link with high depending on climate conditions, as well as the lack of process automation.

In conclusion, one should be guided by the end product proceeded from faecal sludge when making a selection of technologies. It is also worth considering local climatic conditions, on-site installation features, capital and operating costs. The accumulation place features of faecal sludge also important: FS from the septic tank has a higher humidity, while FS from the pit latrine has high solids content. In any case taking into account all the priority factors, it is possible to select the optimal technology for a specific area taking into account environmental, economic and technological features.

References

1. M. Cacanoski, *Small-scale wastewater and faecal sludges treatment and management in low-income rural and peri-urban communities in developing countries* (IRC International Water and Sanitation Centre, Hague, 2003)
2. J.T. Radford, S. Sugden, Water SA J. **40**, 183-188 (2014)
3. H. Harada, L. Strande, S. Fujii, *Challenges and Opportunities of Faecal Sludge Management for Global Sanitation* (Towards Future Earth: Challenges and Progress of Global Environmental Studies. Kaisei Publishing, 2016)
4. I. Blackett, P. Hawkins, *FSM Innovation Case Studies - Case Studies on the Business, Policy and Technology of Faecal Sludge Management (second edition)* (Bill & Melinda Gates Foundation, USA, 2017)
5. K. Ross, K. Abeysuriya, N. Mikhailovich, C. Mitchell, *Governance for decentralized sanitation: Global Practice Scan* (Institute for Sustainable Futures, ADRAS, 2014)
6. *Faecal sludge and septage management – An orientation module (Part A: learning notes)* (NIUA, India, 2017)
7. K. Tayler, *Faecal Sludge and Septage Treatment - A guide for Low and Middle Income Countries* (Practical Action Publishing Ltd, UK, 2018)
8. *Standard Operating Procedure for Fecal Sludge Management for Municipalities in Gujarat* (CEPT University, 2015)
9. T.L. Saaty, *The Analytic Hierarchy Process* (McGraw-Hill, New York, 1980)
10. A. Gueye, M. Mbéguéré, S. Niang, C. Diop, L. Strande, Ecol. Eng. J., **94**, 617-621 (2016)
11. E. Tilley, L. Ulrich, C. Lüthi, Ph. Reymond, C. Zurbrügg, *Compendium of Sanitation Systems and Technologies. 2nd Revised Edition* (Eawag, Switzerland, 2014)
12. *Implementer's guide to lime stabilisation for septage management in the Philippines* (USAID, Philippines, 2015)
13. J. Nikiema, O. Cofie, R. Impraim, *Technological options for safe resource recovery from fecal sludge* (IWMI, CGIAR, WLE, 2014)
14. E.H.M. Sonko, M. Mbéguéré, C. Diop, S. Niang and L. Strande, J. of W. San. and Hyg. for Dev., **4**, 633-641 (2014)
15. A.M. Muspratt, T. Nakato, C. Niwagaba, H. Dione, J. Kang, L. Stupin, J. Regulinski, M. Mbéguéré, L. Strande, J. of W., San. and Hyg. for Dev., **4**, 223-230 (2014)
16. http://www.wendewolf.com/klssystem.php?lang=en
17. https://www.huber.de
18. M. Gold, P. Dayer, M.C.A. Sene Faye, G. Clair, A. Seck, S. Niang, E. Morgenroth, L. Strande, Env. Tech. J., **37**, 2802–2814 (2016)
19. A. Muspratt, A. Miller, *Mechanical Dewatering for Fecal Sludge Treatment Applications Process & Technology Review* (30degrees, 2018)
20. A. Arias-Hoyos, J.L. Hernández-Medina, A.F. Castro-Valencia, N.E. Sánchez-Peña, Biot. en el Sect. Agr. Agr. Ed. Esp., **1**, 29-39 (2017)
21. https://www.tencategeotube.com/en/technical-info/how-it-works
22. F. Ziebell, M. Gold, J. Matovu, J. Maiteki, C. Niwagaba, L. Strande, *Dewatering of faecal sludge with geotextiles: Results from laboratory and bench-scale experiments in Kampala, Uganda* (Eawag, Switzerland, 2016)
23. A.A. Larionova, V.V. Filatov, V.N. Zhenzhebir, I.P. Palastina, N.A. Zaitseva, A.I. Kurbatova, L.E. Glagoleva, Ekoloji J., **106**, 571-578 (2018)

Production of high protein supplementary feed from landfill gas

Elena Klevanova*, and *Alexander* Lukanin

Peoples' Friendship University of Russia (RUDN University), Faculty of Ecology, 6 Miklukho-Maklaya St, Moscow, 117198, Russian Federation

Abstract. The article describes the problem of feed protein deficiency. The main sources of the methane for the biomass production are considered. The chemical composition, the haprin main characteristics and its production technological scheme are presented. The solid municipal waste volume disposed of annually at the landfills is estimated. The quantitative characteristics of atmospheric pollutant emissions from the JSC "Polygon" landfill (Vladimir region) are presented. It is proposed to use landfill gas as the main raw material in the haprin production. The haprin amount obtained during the operation of the JSC "Polygon" landfill was calculated. The economic benefits of this production were estimated

1 Introduction

The intensification of animal agriculture requires a sharp increase in the production of feed protein compensating for its lack in the traditional vegetable feed. 30 million tons per year is the deficit of this product in the world, while in Russia the shortage of feed protein currently exceeds 2.5 million tons per year [1].

The main way to reduce and eliminate feed protein deficiency is to produce biomass using microbial synthesis: strains of methane-oxidizing (methanotrophs for example Methylomonas meyhanica, Methylobacter tundripaludum) and methane-producing (methanogen, for example, Methanococcus capsulatus bacteria). These bacteria have a record biomass accumulation rate thereby creating a large amount of protein. The biosynthesis process is less time-consuming and proceeds continuously regardless of the season and weather conditions while obtaining a protein that is stable in quality [2].

Currently experts from the different countries are actively studying the vital activity of the methanotrophic bacteria. For example, scientists from Calysta Inc. in California and String Bio in Bangalore (India) are developing new methods for converting methane to protein [2].

In Russia the largest producer of feed bacterial biomass from natural gas was the Svetloyarsk plant "Volgogradbiosytez" (Volgograd region). About 40.000 tons of haprin (short for "gas protein") were produced and sold during the period of its functioning from 1985 to 1994. From 1996 to 2006 the Norwegian company Norferm produced a complete analog of the haprin called "bioprotein" for feeding fish and farm animals using Soviet

* Corresponding author: eklevanova@mail.ru

technology. The Danish company UniBio for the pilot installation development of modified Soviet technology the production of the bio proteins under the name "UniPrin" ("uniprotein") for use in feed and food products [1].

2 Materials and methods

The haprin compositionally is a high-protein product containing in terms of absolutely dry substance at least 70% of crude protein and also having almost all essential amino acids (total not less than 55%) (Table 1). The haprin also has lysine 4.0 - 5.3%, sulfur-containing amino acids (serine, methionine) - 1.7%. The haprin contains a large amount of vitamin B complex: thiamine (B1) - 14.1 mg/kg, cyancobalamin (B12) - 5.6 mg/kg and micro and macro elements. Also it is non-toxic, does not have carcinogenic and cumulative effects [3].

Table 1. The amino acids included in the haprin.

Amino acids	Content, %	Amino acids	Content, %
lysine	4.0 – 5.3	histidine	1.7 – 2.5
tryptophan	1.4 – 1.6	aspartic acid	5.3 – 5.8
arginine	2.3 – 3.5	threonine	2.4 – 3.0
serine	1.2 – 2.3	glutamic acid	6.8 – 7.8
proline	2.3 – 3.1	valine	4.1 – 4.2
alanine	4.2 – 4.8	leucine	4.5 – 4.9
cystine	0.3 – 0.5	phenylalanine	2.3 – 2.9
isoleucine	2.6 – 3.0	methionine	1.3 – 1.7
tyrosine	1.6 – 2.1	glycine	3.0 – 3.8

The haprin is a full-fledged substitute for an animal protein in contrast to an animal feed (including cereals and legumes). It provides balanced amino acid nutrition of animals, birds and fish. For example it contains 5 times more protein than barley or oats. The haprin is superior to fishmeal and soybean meal in its nutritional value and balance in amino acid composition.

The haprin production is based on the implementation of cultivation processes in bioreactors of various designs using non-pathogenic strains of methane-assimilating cultures. The bacteria of the genus Methylomonas, Methylococcus, Methylocystts, Methulosinus, Methylobacter can be used as methane-assimilating cultures by continuously supplying oxygen-containing gas (the air, the oxygen-air mixture, the oxygen) and methane-containing gas (the natural gas, the methane) to the bioreactor. Further stages of processing the resulting biomass are traditional and include processes of concentrating cells from biosuspension, thermal sterilization and drying to obtain the finished product in the form of powder or granules [4, 5].

The technological process of obtaining haprin is continuous and consists of the following main stages:

1. The preparation of a nutrient substrate and the cultivation of seed pure culture.
2. The biomass fermentation.
3. The culture fluid separation.
4. The microbial suspension thickening.
5. The biomass inactivation.
6. Drying, packaging, packaging of the finished product.
7. Recycled process water purification.

To ensure the main technological process stages, auxiliary production stages are provided as part of the installation: reception, preparation of solutions and supply to the fermenters of ammonia water, phosphoric acid, salts and other chemicals in the required concentrations; preparation and supply of oxygen; cultivation of sowing "pure culture"; the exhaust gases conclusion of the fermentation process for disposal; purification of gas-air emissions at the stages of thickening, drying and packaging of the finished product; washing and disinfection of equipment.

Figure 1 shows a structural block diagram of one of the options for the biotechnological production of the haprin.

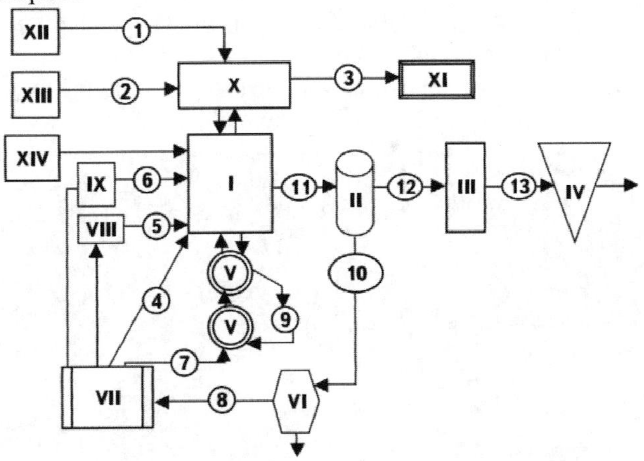

Fig. 1. Block diagram of the haprin production: streams: 1 - natural gas for cultivation; 2 - oxygen source for cultivation; 3 - waste gas for energy utilization; 4 - process water for cultivation; 5 - culture medium solution for cultivation; 6 - ammonia-water solution for cultivation; 7 - circulating water to make up for losses; 8 - biologically purified water in the water treatment system; 9 - circulating water in the heat transfer system; 10 - waste cultured medium after concentration; 11 - biomass suspension from fermenters; 12 - concentrated biomass suspension for drying; 13 - commercial product. Stage: I - production fermentation; II - concentration (microfiltration and separation); III - drying; IV - packaging; V - heat removal system; VI - effluents biological treatment; VII - water treatment system; VIII - the system for the preparation and supply of nutrient salts solution; IX - the system for preparing the supply of ammonia water; X - gas-phase recirculation system; XI - exhaust gas utilization system; XII - the system for the preparation and supply of natural gas; XIII - the system of preparation and supply of an oxygen source; XIV - sowing system.

The standard biotechnology for the methanotroph biomass production with a difference in some technological nuances can be based on various sources of methane, for example, from the gas pipeline system, from processing agricultural waste, from associated petroleum gas and others.

This article proposes the use of the methane of landfill gas (LG) landfills resulting from the biological conversion of organic substances with the participation of a methane-forming consortium of microorganisms in the landfill body, which causes odor rejection and harms people living in a radius of 500-1000 meters from the landfill [6]. Moreover, the process of organic waste fractions decomposition has been going on for many decades [7], which makes the landfill a stable LG source [8-10].

Nowadays in Russia there are more than 14 thousand large landfills with an area of more than 4 million hectares. These data are comparable with the sizes of some countries, for example, the area of Switzerland is 4.12 million ha, the Netherlands is 4.15 million ha, and the size of Slovenia (2 million ha) or Israel (2.2 million ha) is 2 times smaller.

The landfill gas is a mixture of gases, the main ones are CH_4 (volume fraction - 45 ÷ 75%), CO_2 (volume fraction - 25 ÷ 55%), N_2 (up to 5%, often over 10%), H_2S, O_2, H_2 and

CO [11, 12]. Moreover since the LG composition includes carbon dioxide before it is fed to the fermenter it is necessary to purify carbon dioxide.

As the study object for calculating the annual LG emissions from the landfill we took the JSC "Polygon" landfill (55.930125, 39.709838) located in the Vladimir Region on the border of the Petushinsky and Sobinsky districts about 3 kilometers from the village of Boldino and Babanino and 1.300 meters northwest of the M7 highway The Volga (Fig. 2).

The JSC "Polygon" landfill is a recess relatively isometric in shape a width of about 90 meters and elongated from west to east by about 150 meters.

This landfill began to operate on January 1, 2015. The amount of waste disposed per day is 1000 tons. The landfill area in the modern kennel is 17 hectares.

Fig. 2. The JSC "Polygon" landfill location.

The pollutant emissions calculation from the landfill was made based on the "Methodology for calculating the quantitative characteristics of atmospheric pollutant emissions from the solid municipal and industrial waste landfills" [13].

To carry out the calculation we used the data obtained including from the operating organization. The data are shown in table 2.

Table 2. The data used in the calculation

Data	Value
Average importance of MSW	47 %
Organic content in waste, total	55 %
lipoid	2 %
carbohydrate-like	83 %
protein	15 %
Average monthly temperature for the warm season	15 °C
Number of warm days	228 days
Duration of the warm period (more than 8°C)	5 months
Duration of the cold period (from 0°C to 8°C)	7 months
Amount of waste disposed per year	365 000 tons
Start of landfill operation	2015 year
End of landfill operation	2022 year

3 Results and discussion

As a result of the calculation made on the Methodology basis, the specific yield of the landfill gas was 0.17 kg/kg of waste. Moreover the complete fermentation period of the waste organic part is 20 years.

Main characteristics:

- the quantitative LG output - 8.6 kg / t of waste per year;
- the methane percentage in the total LG volume - 52.9%.

Also table 3 shows the values of the maximum one-time and the gross emissions of LG components.

Table 3. Maximum single and gross emissions of landfill gas components.

Component	Maximum single emission, g/s	Gross emission, t/year
Methane	420.61	15 585.63
Toluene	5.75	212.95
Ammonia	4.24	156.99
Xylene	3.52	130.48
Carbon Monoxide	2.00	74.22
Nitrogen dioxide	0.88	32.69
Formaldehyde	0.76	28.28
Ethylbenzene	0.76	27.98
Sulfur anhydride	0.56	20.62
Hydrogen sulfide	0.20	7.66

During the 5 years of the landfill's operation the gross methane emission amounted to 77 928 tons. Given that from 1.6 m^3 of methane 1 kg of haprin is obtained, from the landfill gas of the JSC "Polygon" landfill would be possible to produce 72 889.9 tons of haprin (taking into account methane losses of 30% with gases) from 2015 to 2019.

The cost of production of 1 kg haprin is 41 070 rubles taking into account the raw materials cost, wages, amortization and other additional costs. At the same time the market price of 1 kg haprin is 90 000 rubles. In this regard the economic benefit in the high-protein product production from the methane of the landfill is about 713 million rubles per year or 3.567 billion rubles over five years.

4 Conclusion

The methane using from the landfill as the main product for the haprin production is due to the significant annual volume of LG formation, minimizing the landfill's negative impact on the environment, the high quality of the resulting feed product as well as economic benefits.

The research was supported financially by Russian Foundation for Basic Research (Project No. 17-01-00693a). The publication has been prepared with the support of the "RUDN University" Program 5-100.

References

1. U. Lee, J. Han, M. Wang, J. of Cleaner Prod., **166**, 335-342 (2017)
2. S.P.Bulgakov, N.N. Gavrilov, *Innovation: prospects, problems, achievements:,* 192-197 (Plekhanov RUE, Moscow, 2013) (in Russian)

3. A.Y. Vinarov, Current trends in science and technology, **2-2** (2017) (in Russian)
4. M. Aresta Carbon Dioxide as Chemical Feedstock, 417 (2010)
5. E.S. Klevanova, M.D. Kharlamova, International Research Journal, 135-140 (PFUR, Moscow, 2016) (in Russian)
6. Ya.I Vaysman, S.V. Maksimova, Management of methanogenesis at landfills, 235 (Permian, 2003) (in Russian)
7. Ya.I.Vaysman, Waste management. Sewage and biogas of landfills, 259 (Perm national polytechnic university, Permian, 2012) (in Russian)
8. A.A. Larionova, V.V. Filatov, Ecology, **108**, 571-578 (2018) (in Russian)
9. H. Omar, S. Rohan, International Journal of Greenhouse Gas Control, **58**, 159-168 (2017)
10. P.A. Trubayev, O.V. Verevkin, B.M. Grishko, *Energy Systems,* 436-443 (Belgorod, 2017) (in Russian)

https://doi.org/10.1051/e3sconf/202016902010

Municipal Waste Management in Towns of Switzerland: a Foreign Ecologist's View

Roman Mamadzhanov[1], *Yulia* Zakirova[1], and *Mykhadi* Umarov[2]

[1] Peoples' Friendship University of Russia (RUDN University), Faculty of Ecology, 6 Miklukho-Maklaya St, Moscow, 117198, Russian Federation
[2] Complex Institute Named After Kh.I. Ibragimov of the Russian Academy of Sciences, Staropromyslovskoe schosse, 21a, Grozny, the Chechen Republic

Abstract. The article presents the results of the research carried out by the foreign ecologists, particularly the study of the system of municipal waste management in the municipality of Maienfeld, Switzerland. The total amount of municipal waste has been calculated. The systems of municipal waste collection, sorting, transportation and recycling have been studied. Besides, four possible scenarios of waste recycling in Maienfeld have been developed. Moreover, the environmental assessment of the waste recycling system by GHG/SLCP emissions has been conducted, and the impact of recycling on the climate has been determined. Finally, taking into account the results of the research, the best scenario of recycling has been proposed.

1 Introduction

Currently, Switzerland is one of the countries in Western Europe with the highest per capita annual municipal waste generation [1, 2]. On average, in Switzerland one person produces more than 714 kg of waste per year, in Germany – 638 kg, in Austria – 610 kg, Sweden – 471 kg [1, 2].

Nevertheless, in spite of economic and population growth from 2001 to 2010 the total amount of municipal waste in Switzerland increased only by 16%, from 4,79 million tons in 2001 to 5,56 million tons in 2010 [3]. It is a good example of the stability of the municipal waste collection and recycling system over time.

According to the Federal Office for the Environment (FOEN), municipal waste covers household waste and generated by businesses waste similar in composition to household waste.

Marton Herczeg from Copenhagen Resource Institute notes that Switzerland has a long tradition in diverting waste from landfill to waste recycling and incinerating plants [3].

The above can be confirmed by the fact that the average percentage of recycled waste in 2001-2010 was approximately 50%, recycled secondary materials (glass, metal, plastic, paper and cardboard) – more than 30%, recycled organic waste – more than 15% (Fig. 1) [4]. For comparison, in 2011 the total percentage of recycled waste in Germany was 45%, in Belgium – 36%, in Denmark – 31%. [4].

Before recycling rubbish should be separated and collected in containers.

According to the Federal Office for the Environment (FOEN), there are eight types of waste which should be collected (Table 1) [5].

Table 1. Types of collected waste in Switzerland

Type of Waste	Recycling Management	Profit	Collection site
Paper and cardboard	Village/City	10-50 CHF/1 ton	Public collection
Glasspack	VetroSwiss	2-6 Rappen/1 pack	Public collection
WEEE[1]	SWICO[2]/S.E.N.S.[3]	Depending on the type	Points of purchase
Fluorescent lamps	SLRS[4]/S.E.N.S.	Depending on the type	Points of purchase
PET bottles	PRS[5]	1,8 Rappen/Bottle	Public collection/ points of purchase
Tin cans	FERRO recycling	1 Rappen/Can	Public collection
Aluminum cans	IGORA[6]	2 Rappen/Can	Public collection/ points of purchase
Batteries	INOBAT[7]	3,20 CHF/1 kg	Public collection/ points of purchase

[1]Waste electrical and electronic equipment
[2]Recycling of waste electrical and electronic equipment used in the IT, consumer electronics, etc.
[3]Recycling of household appliances
[4]Light Recycling Foundation in Switzerland
[5]PET recycling in Switzerland
[6]Aluminum recycling company
[7]Batteries recycling company

It is apparent from Table 1 that biodegradable (organic) waste is not included in this system. Besides, end-of-life vehicles should also be recycled. Since 2006, the average cost of recycling a car is more than 30 CHF.

It is important that each year FOEN spends more than 7,2 million CHF on waste recycling management in Switzerland [6].

According to the FOEN report, the total amount of municipal waste in Switzerland is rapidly increasing [6]. For example, annually there are more than 840 million items of waste made of plastic, metal, cardboard, paper or cigarettes in Swiss streets or on highways. On average, every year there are more than 100 items of rubbish per capita.

Despite the high rates of municipal waste recycling, in towns with a population of less than five thousand people and a rapid increase in the total amount of waste, municipal waste is collected only once a week. Actually, refuse is collected in large rubbish bags accumulating in the streets before transportation in a few days. This can lead to accidental spread of insects, rodents or pathogenic compounds in the streets.

The aim of our research is to reveal advantages and disadvantages of municipal waste management in towns of Switzerland and consider applying the best ideas in Russia.

The main objectives are:

- to study the system of collection and segregation of municipal waste,
- to study the system of transportation of municipal waste and to develop proposals for its modernization;
- to describe the system of recycling of municipal waste;
- to do calculations related to the environmental assessment of the waste recycling system by GHG/SLCP (greenhouse gas / short-lived climate pollutant) emissions;

to determine the impact of GHG emissions from the waste recycling system on the climate and propose the best recycling scenario for Maienfeld.

2 Methods

To study the system of collection and separation of municipal waste the open data set and our research data were used [7].

For studied the municipal waste transportation and recycling systems we used the knowledge and experience of employees of the A&M AG recycling center in Landquart Switzerland.

The study of the system of transportation of municipal waste was based on knowledge and experience of employees of the A&M AG recycling center in Landquart, Switzerland.

For the environmental assessment of the system of waste recycling and the impact of GHG emissions (CO_2, CH_4, BC and N_2O) on the climate, we used the Emission Quantification Tool (EQT) for estimation of GHGs / SLCPs from solid waste sector developed by the Institute for Global Environmental Strategies (IGES).

We compared BAU (business as unusual) practice of waste recycling with four possible scenarios (scenario 1, 2, 3, 4). The key data set for this research is presented in Table 2. [8].

Table 2. The key data on recycling of municipal waste in Maienfeld.

Parameter	Tons/day				
	BAU	**Scenario 1**	**Scenario 2**	**Scenario 3**	**Scenario 4**
Collected/Uncollected					
Total amount of generated waste	5,50	5,51	5,51	5,51	5,51
Total amount of waste collected and treated by the municipality	3,05	3,38	3,47	4,31	4,56
Total amount of waste collected and treated by commercial companies	2,01	1,73	1,47	1,02	0,86
Total amount of uncollected waste	0,44	0,4	0,57	0,18	0,09
Treatment					
Recycling	1,00	2,00	2,00	3,00	4,00
Incineration	1,00	1,00	1,00	1,00	1,00
Landfill disposal	1,00	0,00	0,00	0,00	0,00

It is important that BAU was calculated using the IGES computer model [8]. Scenarios were as close as possible to real-life conditions of waste management in Maienfeld.

As can be seen from Table 2, all scenarios have remained unchanged taking into account the data on the total amount of waste in 2019. Meanwhile, other parameters have significantly changed.

The GHG emissions from waste recycling were measured per one ton of generated waste (kg/ton), the impact on the climate – kg of CO_2-eq/ton.

It should be mentioned that the same technological treatment method (including extraction of nonferrous and ferrous metals, glasspacks, treatment of paper and cardboard) is used in each of the above scenarios.

3 Results

The object of our research is Maienfeld, a municipality in the Landquart Region in the eastern part of Switzerland, in the canton of Graubünden. The total area of Maienfeld is 32,33 km^2, the population is 2587 people [9].

Maienfeld is bordered on the west by Bad Ragaz, Canton of St. Gallen and on the west and north by the community of Fläsch, Canton Graubünden. To the north it is bordered by Balzers, Triesen, Schaan, the country of Liechtenstein, Nenzing and Vorarlberg, Austria. To the east is the community of Seewis im Prättigau, Canton Graubünden. To the southwest Jenins und Malans, Canton Graubünden. In the south it is bordered by Igis and Zizers, Canton Graubünden, as well as in the southwest the community of Mastrils, Canton Graubünden.

From the environmental point of view, the agrolandscape occupies 44,3% of the area, forests - 31,9 %, social infrastructure facilities – 5,4 % [9]. In addition, the Rhine river flows through Maienfeld. Consequently, the environmental issues (particularly municipal waste management) are among the major ones for Maienfeld.

The total amount of municipal waste in Maienfeld averages 1841,9 tons per year.

All categories of municipal waste should be separated (Table 1). Nevertheless, people can buy special blue bags (for 14 CHF per 10 bags) to collect all types of municipal waste without separation. It should be noted that the price of a kitchen rubbish bag (only for organic waste) is 2 CHF per 20 bags. Thus, Maienfeld has a special tax on unsorted waste.

All containers for municipal waste are divided into:

- containers for unsorted (sorted) municipal waste;
- containers for dog excrement;
- containers for glass, cans, paper and PET;
- containers for wooden items, street dust.

As we can see, the bulk of municipal metal waste, WEEE, including batteries and fluorescent lamps is excluded from this waste separation system. These categories of waste should be separated in Landquart city in the A&M AG recycling center situated 5,5 km away from Maienfeld. People need to go there on their own. Due to that, the transportation cost of recycling of batteries and fluorescent lamps is higher.

The system of municipal waste transportation in Maienfeld includes:

- transportation of waste from waste containers to the recycling center (A&M AG recycling center);
- waste treatment including reuse of secondary materials, incineration of unrecyclable materials at incineration plants;
- landfill disposal.

It is important that collected waste is transported to the recycling center only once a week in summer, and once a fortnight in winter. All rubbish collected into bags and special containers from households is left in the streets by residents to be transported in 3-5 days. According to our study, more than 40% of these bags are disturbed by rodents, cats and dogs. They look for organic remains to feed on them. This can lead to accidental spread of pathogenic bacteria in the environment.

The strategy for waste management in Maienfeld aims at reducing the total amount of unrecyclable waste after treatment. The hierarchy of waste management in Maienfeld is the same as all over Switzerland and consists of reduction, reuse, recycling, incineration and landfill disposal technologies.

The main idea of waste recycling system in Maienfeld is to reduce the total amount of waste and to make the most of reuse of secondary materials. Besides, «recycling» takes third place after collection and separation. In addition, incineration and landfill disposal methods are used only to reduce unrecyclable materials. Overall, this approach prevents the ingress of toxic compounds (which are dioxins, chlorine and benzene) into the atmosphere and the topsoil.

The results of environmental assessment of the waste recycling system by GHG/SLCP emissions are shown in Fig. 1.

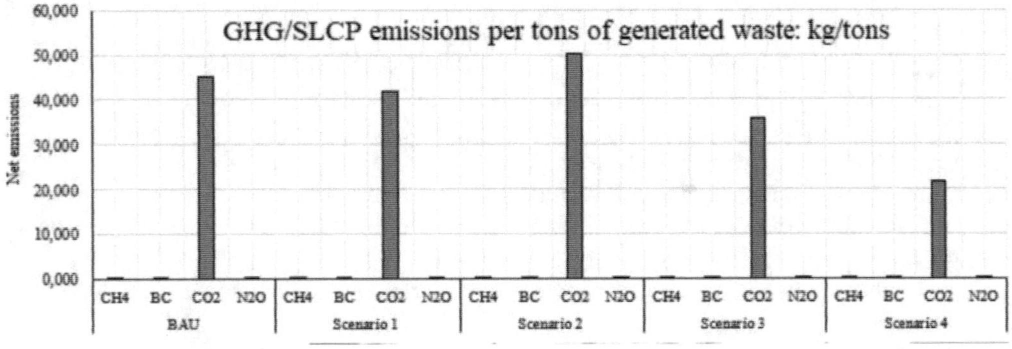

Fig. 1. GHG emissions from waste recycling.

Fig. 1 shows that the highest GHG/SLCP emission level is in Scenario 2 (50,98 kg/ton) and it is higher than in BAU (45,82 kg/ton), while the GHG emission in Scenario 4 is lower (22,38 kg/ton). Scenarios 1 and 3 also show the higher GHG/SLCP emission level with 42,49 and 36,46 kg/ton correspondingly, but it is lower than in BAU.

As we have noted, upon recycling toxic compounds (which are GHG/SLCP emissions) enter the environment. One of the main GHGs is CO_2 (<20,00 kg/ton of waste in each scenario). Meanwhile, the total amount of N_2O and CH_4 is lower (>1,00 kg/ton of waste in each scenario). This can be explained by the significant amount of CO_2-eq due to fuel (gasoline or diesel) use for transportation of waste to the recycling facility and recycling itself.

The impact of waste recycling on the climate taking into account CO_2-eq/ton shows, that the minimal impact of GHGs, particularly CO_2-eq emissions upon recycling is in Scenario 4 (21,70 kg of CO_2-eq/ton), which is less than in BAU. The maximum impact on the climate is in Scenario 2 (50,09 kg of CO_2-eq/ton).

It could depend on the decrease in the total amount of waste collected and treated by the municipality (3,47 tons/day) and the increase in the total amount of uncollected waste (0,57 tons/day) in Scenario 2. In Scenario 4 the total amount of waste collected and treated by the municipality is 4,56 tons/day and the total amount of uncollected waste is 0,09 tons/day.

It should be noted that Scenario 4 has got one of the highest recycling rates (more than 70,00% of the total amount of waste/day, or 4,00 tons/day (Table 2). Meanwhile, the recycling rate in Scenario 2 is only 36,30% of the total amount of waste/day, or 2,00 tons/day (Table 2).

4 Conclusion

In conclusion, we should say that the system of collection and separation of municipal waste in Maienfeld is modern and eco-friendly. It is confirmed by the high rates of collection of municipal waste (more than 50,00% in each scenario), and by waste sorting into four containers. In view of all that has been mentioned so far, it would be interesting to consider the possibility of implementation of the waste sorting system in towns of Russia.

The transportation system needs to be improved. Collected household waste is transported to the recycling center only once a week in summer, and once a fortnight in winter. All the waste collected into bags and special containers (Fig. 1) from households is left in the streets by residents to be transported in 3-5 days. According to our study, more than 40% of these bags are disturbed by rodents, cats and dogs. They look for organic remains to feed on them. This can lead to accidental spread of pathogenic bacteria in the environment.

The results of our study of the waste recycling system in Maienfeld allowed us to determine four scenarios of recycling. We consider Scenario 4 as the best recycling scenario for Maienfeld with the lowest level of GHG/SLCP emissions (21,70 kg of CO_2-eq/ton of waste) and the minimal impact on the climate taking into account GHG emissions (21,70 kg of CO_2-eq/ton). As a result, technological options used in this Scenario seem to be the most appropriate choice for Maienfeld.

Our recommendations are the following:

- to compare systems of waste management in Maienfeld and in other towns of Switzerland:
- accordingly, to identify the best system of waste management for towns of Switzerland;
- besides, the environmental assessment of waste incineration and landfill dumping technologies should be carried out;
- cost-effectiveness of waste management including collection, separation and recycling should be studied;
- based on this experience to consider the possibility of implementation of the best ideas of waste management in towns of Russia.

References

1. Swiss are champions in waste management URL.: https://www.swissinfo.ch
2. W. Joos, V. Carabias, H. Winistoerfer, A. Stuecheli, Waste management, **19**, 417-425 (1999)
3. *Municipal waste management in Switzerland* URL.: https://www.bafu.admin.ch/bafu/en/home
4. *Separate waste system* URL.: http://www.separate-wastesystems.eu
5. *Ordinance on the Avoidance and the Disposal of Waste* URL.: https://www.admin.ch/
6. *Zero littering* URL.: https://www.bafu.admin.ch
7. *A&M AG Recycling center* URL.: https://recycling-center.ch
8. *Emission Quantification Tool (EQT) for estimation of GHGs/SLCPs from solid waste sector* URL.: https://www.iges.or.jp
9. *Stadt Maienfeld* URL.: http://www.maienfeld.ch/de/

Development of the Technology for Processing Plant Breeding By-Products to Obtain Biosorbent

Tatiana Bauer [1,*], *Svetlana* Sushkova[2] , *Aleksei* Fedorenko[1, 2] , *Ilya* Lobzenko[2] , *Anatoly* Barakhov[1] , *Vera* Butova[1], and *Ekrem* Tuşat[3]

[1]Federal Research Center the Southern Scientific Center of the Russian Academy of Sciences, 344006 Rostov-on-Don, Russian Federation
[2]Southern Federal University, Academy of Biology and Biotechnology, 344090 Rostov-on-Don, Russian Federation
[3]Konya Technical University, Department of Geomatics Engineering, 42075 Konya, Turkey

Abstract. This study focuses on the potential transforming sunflower husk wastes into innovative green products to be used for the recultivation of contaminated soils. This transformation makes it possible to utilize the large quantity plant breeding by-products to obtain a highly porous biosorbent. Optimal conditions for the preparation of bisorbents can be determined based on measurements of their specific surface and porosity. It is shown that structural characteristics of sorbent depend on conditions of the pyrolysis of primary raw material. Optimal conditions for obtaining biosorbent with a specified proportion of meso- and micropores and high specific surface are performed under three-stage pyrolysis as given by following conditions: sample holding time of 10-30 min, final temperature 700°C, and heating rate 15°C/min. A 3-month incubation study was conducted to investigate its performance in stabilization of copper contamination in Fluvisol. The sequential extraction procedure showed that the 1% sunflower husks biosorbent treatment reduced the mobile pool (exchange and carbon-bonded fractions) of Cu from 8 to 4%, respectively, compared to the contaminated soil. Cu is found to be tightly bonded owing to the increase of the residual and organic matter-bonded metal fractions. Owing to its adsorptive properties, biosorbent amendment of copper-contaminated soils appears as a promising approach to reduce the pollutant mobility and bioavailability.

1 Introduction

At present, huge amounts of renewable plant-related wastes and reworking products of carbon-containing raw material are produced every year in the world [1]. Therefore, a rational solution to their utilization with the simultaneous recovery of upgraded products is an urgent issue [2, 3].

* Corresponding author: bauertatyana@mail.ru

It is interesting to note that huge amounts of wastes are accumulated currently in each region based on the cultivation and production of organic feedstock. In particular, vast (tonnage-scale) source of wastes leftover after the reworking of sunflower husk are available now in the Rostov region (southern European part of Russia). On the average, their accumulation rate is as much as 300 thou ton/year, and this value is growing each year due to intensification of the production of oilseeds. This huge potential, however, is not fully exploited (not more than 25–30%) because due to a lack of economic and efficient waste utilization technologies. Some part of the waste is used for the production of fuel briquettes, but a major part is dumped, resulting in economic inefficiency and ecological hazards [4].

Promising methods are considered shifting towards technologies that facilitate the reworking of plant biomass into biochar (biosorbent). Currently, biochar is widely applied for the amelioration and remediation of soils subjected to technogenic pollution [5, 6]. Biochar commonly is obtained by the pyrolysis of plant waste in an inert atmosphere. In this process, the composition of primary raw material and conditions of its pyrolysis are of the crucial factors that control parameters of the pore structure of sorbent [7]. Among these parameters, the maximum temperature of biomass treatment in the pyrolysis reactor exerts the greatest influence on the properties of the produced biochar [8]. The lower the pyrolysis temperature, the lesser the percentage content of carbon and the degree of its aromatization in the final product. The pyrolysis duration shows an inverse correlation with the amount of water and oxygen-containing compounds in the ultimate product. Moreover, the end product is characterized by high acidity and low thermal stability [7].

Studies of the remediation of Cu-contaminated soil are particularly important. Copper is a priority pollutant of soils in southern Russia [9, 10]. High technical applicability of copper and its extensive use in different industrial fields are responsible for the soil pollution with this metal. The release of Cu compounds to riverine systems can result in extensive accumulations of metal in surrounding riverbank sediment and floodplain soils. The remediation of floodplain soils polluted with heavy metals such as Cu is necessary for mitigating their environmental and human health risk. The complexity of the soil formation process, its high dynamism, the specifics of water supply, the significant influence of intrazonal factors are the main reasons for the poor knowledge of the floodplain soils detoxification methods.

The aim of the present work is to develop a technology for obtaining highly porous biosorbent from the sunflower husk waste to recultivate the Cu contaminated soils.

2 Materials and Methods

Experimental biochar samples were obtained by the thermal decomposition of sunflower husk based on the variation of temperature (300°, 500°, 700°, 900°C), heating rate (from 5 to 30 °C/min) and hold period (from 10 to 75 min) of biomass. For this purpose, the primary raw material was subjected to a steel retort that was hermetically sealed with a welded flange for gas removal and placed in a muffle furnace, supplying the vessel with nitrogen to create an inert atmosphere. The vessel was heated in several stapes with a specified temperature increase rate and incubated at the desired heating temperature for 10-75 min. After termination of the calcination, the vessel was cooled down to room temperature and the obtained biochar was unloaded. Then, the chemical composition, physicochemical properties and structural characteristics were determined.

The yield of biosorbent per dry mass of primary raw material was calculated according to the following formula [11]:

$$M_b / [M_r\ (1\text{-}W_r / 100)]\ 100,\ \% \tag{1}$$

where M_b is the yield of biosorbent, g; M_r is the weighed portion of the primary raw material, g; and W_r is the moisture content in the primary raw material, %.

The specific surface and porosity of the carbon-bearing samples were measured with an ASAP 2020 volumetric analyzer, and micrometric properties were determined by the low-temperature nitrogen adsorption method. The specific surface and porosity parameters were determined using the N_2-based BET specification method in the equilibrium range of P/P0 = 0.05-0.97. Volumes of micro- and mesopores in the sorbents were determined with the relative t method based on the Hurkins-Jura equation for computing thickness of the statistic adsorbate layer [12]. The computation of the pore size distribution was based on the nonlocal density functional theory (NLDFT) [13].

Specific features of the topography (surface) and morphology (microgeometry) of the obtained sorbent fragments were analyzed with a Carl Zeiss EVO-40 XVP scanning electron microscope under the standard conditions (for nonconducting low-contrast samples). The microscope analysis was done in a low vacuum, 15 kV, and high emission mode.

We accomplished model experiments to study the efficiency of the recultivation of the contaminated soil based on biosorbent. For this purpose, we sampled the upper humus horizon (0–20 cm) of the clay loamy Fluvisol in the Kamenka district (Rostov region). This soil is characterized by the following physicochemical properties: C_{org} 4.3%, pH 7.5, exchange cations (Ca^{2+}+Mg^{2+}) 38.1 cM(+)/kg, $CaCO_3$ 0.6%, physical clay fraction (particles < 0.01 mm) 55.8%, clay fraction (particles < 0.001 mm) – 32.0%, and Cu 43.7 mg/kg. The pollutant was represented by Cu (300 mg/kg per pure element). Before starting the experiment, soil samples were dried and ground to <1-mm-size particles; plant remains and large inclusions were removed; weighed soil portions (100 g each) were placed into plastic containers; and an aqueous solution of copper acetate was added and thoroughly stirred. After a 2-week-long metal incubation in the soil, the vessels were mixed with 1% of biochar corresponding approximately to 200 kg/100 m^2 or 20 t/ha if recalculated to the 20-cm-thick topsoil. The experiment was carried out with three-replicate measurements over 3 months. Then, the dry soil was again ground and sieved (pore size 1 mm) for further analyses. Analogous procedures but without addition of metal and biosorbent were performed with the control sample. In the course of the entire experiment, optimal moisture content (60% of the total moisture capacity) was maintained in the vessels. The contaminated soil was incubated at room temperature and natural lighting.

The composition of Cuu compounds in soil was determined using the sequential fractionation method proposed by Tessier et al. [14]. This method allows extracting five fractions of metal compounds: exchange (1M $MgCl_2$) related with carbonates (1M CH_3COONa), Fe-Mn (hydr)oxides (0.04M $NH_2OH \cdot HCl$ in 25% CH_3COOH), organic matter (0.02M HNO_3+30% H_2O_2, pH 2, then 3.2 M CH_3COONH_4 in 20% HNO_3) and residual (HF+$HClO_4$, then HNO_3conc.) fractions. The Cu content in leachates was determined by the atomic absorption spectrometry with a Kvant-2AT AAS.

3 Results and Discussion

Processing of the sunflower husk samples at the different temperature, heating rate and hold period gave the following results: yield of biochar 13-29%, moisture content 1-6%, yield of volatile matter 7-23%, carbon content 60–89%, ash content 0.5-12.7%, and pH 7.1-9.6. In this process, pyrolysis exerted a significant influence on the structure and properties of the obtained biosorbents. Parameters of the pore structure of biochars obtained from the sunflower husk are presented in Table 1. In all studied range of heating rate and hold period (10-45 min) at the final heating temperature (300°C), the specific surface of biochar samples computed from the BET method is characterized by small values: 5-40 m^2/g. The increase of the final temperature to 700°C fosters an appreciable growth of the specific surface to 660 m^2/g depending on the heating rate and hold period. The total volume of pores in sorbents varies from 0.47 to 0.89 cm^3/g (Table 1).

Table 1. Parameters of the pore structure of biosorbent samples recorded at different conditions of pyrolysis.

SSA, m^2/g	**V_Σ, cm^3/g**	**V_{micro} (>2 nm), cm^3/g**	**V_{meso} (2-500 nm), cm^3/g**	**V_{macro} (>500 nm), cm^3/g**
5-660	0.47-0.89	0.12-0.63	0.04-0.19	0.09-0.31

SSA: specific surface area (m^2/g); V_Σ: total pore volume (cm^3/g); V_{micro}: micropore volume (cm^3/g); V_{meso}: mesopore volume (cm^3/g); V_{macro}: macropore volume (cm^3/g).

Based on electron microscopy, the studied biochar samples are marked by heterogeneity in terms of the disperse and morphological composition and pore size, depending on the conditions of their production. At the pyrolysis temperature of 300-700°C, the samples show abundant isometric and elongated pores from 3 to 12 µm across. At high final temperature (900°C) of pyrolysis, the pore structure is distorted: the surface is broken into separate microblocks, with pores between them reaching 38 µm. They are marked by significant surface alterations that deteriorate the sorption properties of the biochar.

Analysis of structural characteristics of the obtained biochar samples revealed the following point: optimal conditions for their preparation with the specified proportion of micro- and mesopores and the high specific surface can be provided by the three-stage pyrolysis with the sample hold period ranging from 10 to 30 min at the final temperature of 700°C and heating rate of 15°C/min. Pyrolysis at 900°C seems to be less economic than heating to 700°C. Moreover, the increase in the sorbent processing temperature provokes not only the pore structure distortion in the sorbent but also the increase in the cancerogenic compounds therein.

For the model experiment a biosorbent with the greatest specific surface was chosen at the optimal conditions of pyrolysis with further properties as follows obtained: moisture content 3.5%, ash content 10.1%, yield of volatile matter 9.6%, carbon content 87%, and pH 9.0. The studied samples are marked by the following parameters: specific surface area of 660 m^2/g and total pore volume of 0.81 cm^3/g. The major part of pores (0.63 cm^3/g) is composed of micropores (>2 nm).

The results of scanning electron microscopy demonstrated that the studied biosorbent sample is heterogeneous in terms of morphological composition and pore size compared to the characteristics of the primary raw material. The biosorbent particles show the anisotropic structure of pores that represent long cylindrical cavities extending along the longitudinal axis of sunflower husk particles. The sagittal section (Fig. 1a) shows a surface with numerous fractures and slotted pores 1-2 µm wide. The segmentary chip (Fig. 1b) contains numerous rounded pores (10-30 µm) and smaller pores (0.5-1 µm).

The sequential fractionation of copper compounds in the noncontaminated Fluvisol (Control sample) revealed explicit domination of metal in the residual silicate-related fraction (Table 2). A significant contribution to the robust bonding of Cu is also made by the soil-hosted organic matter that accounts for 16% of the total sum of fractions. In the first two fractions marked by the weakest bond with soil, the relative metal content is just 3%. Low mobility and domination of Cu in the residual fraction suggest the absence of a notable amount of technogenic metal in the background soil.

 https://doi.org/10.1051/e3sconf/202016902011

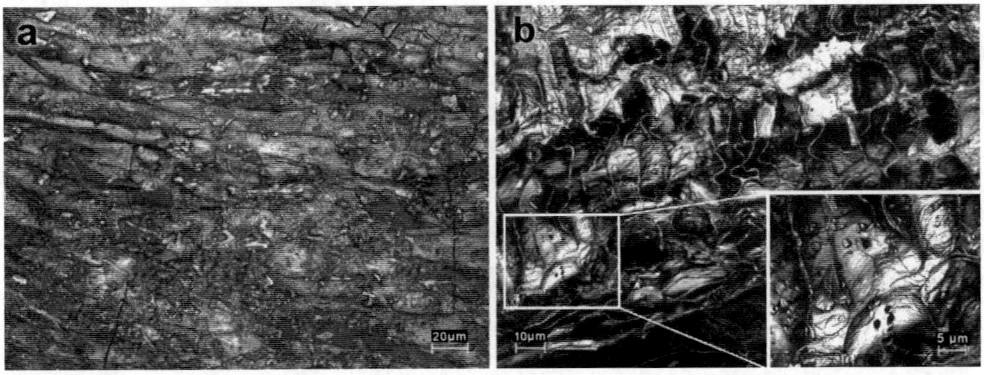

Fig. 1. SEM photomicrographs of the biosorbent: (a) Fragment of sample in the sagittal plane, (b) transverse cleavage.

The increase of the technogenic load (300 mg/kg) upon soil provokes alteration of the fractional composition of Cu. The share of the residual fraction decreases due to the increase in the relative content of all other metal fractions (Table 2). Analogous reverse correlation between the most conservative residual fraction of Zn versus the technogenic pollution of soils is also reported in [15, 16].

The attenuation of metal mobility was recorded in the experiment performed with the addition of biosorbent. Exchange forms and carbonate-bonded copper species decrease by 2% (Table 2). In addition to the high specific surface of the carbon-bearing sorbents [17], the alkaline reaction of medium is also responsible for the weakening of metal mobility. For example, the increase of pH related to the addition of biochar in the Pb-contaminated soils promoted a decrease in the bioavailability of heavy metals and positive influence on microbiological processes [18].

Table 2. Fractional composition of copper compounds in a Fluvisol during the addition of biosorbent.

Variant	Fractions					
	Exchangeable	Carbonate-bound	Fe/Mn (hydr)oxide-bound	Organic matter-bound	Residual	Sum
mg/kg						
Control	0.2±0.04	1.2±0.1	3.0±0.2	6.8±0.7	32.5±2.7	43.7
Cu	8.7±0.9	15.9±1.6	32.4±4.1	80.3±6.9	209.7±23.1	347
Cu + sorbent	4.6±0.7	11.2±1.0	23.9±4.2	89.7±7.5	215.6±18.4	345
% of the sum fractions						
Control	0	3	7	16	74	100
Cu	3	5	9	23	60	100
Cu + sorbent	1	3	7	26	63	100

Similar data are reported elsewhere [19], demonstrating that the addition of biochar obtained from rice straw and bamboo stems for the sandy loam of paddy soil also appreciably decreases concentrations of the mobile Cu, Cd, Pb and Zn species. The addition of the carbonaceous sorbent to the contaminated soil leads to a 3% growth of the Cu content in the organic-related fraction. This trend can be caused by the addition of organic matter to soil that actively absorbs Cu ions. More than two-thirds of coal is composed of the amorphous carbon, a part of which is burned during pyrolysis with the formation of differently sized pores. It results not only in the development of a large specific absorbing surface, but also in the increment of organic matter in the soil [20, 21]. The increase of soil organic matter due to biosorbent addition can convert labile Cu into the less mobile fractions, e.g., organically bound fractions. The functional groups (e.g., –OH, –COOH,

–C=O-, and C=N) in biochar provide also binding sites for heavy metals to form complexes, which increase the specific adsorption of metals [22]. The residual fraction content also tends to increase in the experiment with sorbent addition (Table 2). These results suggest a more robust bonding of Cu in the soil at the presence of a highly porous biosorbent. Treatments of carbonaceous sorbent increased the transformation of the unstable Cu (exchangeable and carbonate-bound fractions) into stable metal (residual fraction), which had resulted from the elevated pH.

4 Conclusions

The technology was developed for the sunflower husk transforming into biosorbent to promote the sustainability and remediate the technogenic contaminated soils. Optimal conditions for the three-stage pyrolysis of the primary raw material were chosen for the preparation of biosorbent with specified proportions of micro- and mesopores and high specific surface. The model experiment demonstrated that the addition of green produced biochar to the contaminated Fluvisol exerts a positive influence on the immobilisation of the mobility Cu compounds. This is manifested in decrease of the first two most mobile fractions of metal and the increase of the share of Cu tightly bonded with organic matter and clay minerals. The results demonstrated that the produced biosorbent can effectively immobilize Cu, thereby reducing its mobility in contaminated floodplain soils. This indicated that these amendments can directly transform the highly availability metal speciation to the stable speciation in soils.

The reported study was funded by RFBR, project number 19-34-60041 and Grant of President of Russian Federation, no. MK-2244.2020.5.

References

1. Y. Dai, Q. Sun, W. Wang, L. Lu, M. Liu, J. Li, S. Yang, Y. Sun, K. Zhang, J. Xu, W. Zheng, Z. Hu, Y. Yang, Y. Gao, Y. Chen, X. Zhang, F. Gao, Y. Zhang, Chemosphere, **211**, 235-253 (2018)
2. T. Väisänen, A. Haapala, R. Lappalainen, L. Tomppo, Waste Management, **54**, 62-73 (2016)
3. Q. Zhang, D. Zhang, H. Xu, W. Lu, X. Ren, H. Cai, H. Lei, E. Huo, Y. Zhao, M. Qian, X. Lin, E.M. Villota, W. Mateo, Ind Crops Prod., **146**, 112185 (2020)
4. K. He, J. Zhang, Y. Zeng, Sci. Total Environ., **670**, 236-244 (2019)
5. S. Bashir, M. Shaaban, S. Mehmood, J. Zhu, Q. Fu, H. Hu, Bull. Environ. Contam. Toxicol., **100**, 834-838 (2018)
6. M. Zhu, L. Zhang, L. Zheng, Y. Zhuo, J. Xu, Y. He. Front. Microbiol., **9**, 579 (2018)
7. M. Ahmad, A.U. Rajapaksha, Ju.E. Lim, M. Zhang, N. Bolan, D. Mohan, M. Vithanage, S.S. Lee, Y.S. Ok, Chemosphere, **99**, 19-33 (2014)
8. M. Uchimiya, D.I. Bannon, L.H. Wartelle, I.M. Lima, K.T. Klasson, J. Agric. Food Chem., **60**, 5035-5044 (2012)
9. T. Bauer, T. Minkina, D. Pinskii, S. Mandzhieva, S. Sushkova, J. Geochem. Explor., **176**, 108-113 (2017)
10. Minkina T, Pinskii D, Bauer T, Nevidomskaya D, Mandzhieva S, Sushkova S, J. Geochem. Explor., **174**, 107-112 (2017)
11. J. Velebil, J. Malaťák, J. Bradna, Res. Agr. Eng., **62 (4)**, 179-184 (2016)

12. C. Scherdel, G. Reichenauer, M. Wiener, Microporous Mesoporous Mater., **132 (3)**, 572-575 (2010)
13. J. Landers, G.Y. Gor, A.V. Neimark, Coll. Surf., A: Physicochem., Eng. Aspects, **437**, 3-32 (2013)
14. A. Tessier, P. Campbell, M. Bisson, Anal. Chem., **51**, 844-850 (1979)
15. M.V. Burachevskaya, T.M. Minkina, S.S. Mandzhieva, T.V. Bauer, V.A. Chaplygin, S.N. Sushkova, P. Orlović-Leko, L.Yu. Mashtykova, V.D. Rajput, J. Soils Sediments, **18 (6)**, 2379-2386 (2018)
16. M.T. Nielsen, J.J. Scott-Fordsmand, M.W. Murphy, S.M. Kristiansen, J. Soils Sediments, **15**, 1558-1570 (2015).
17. M. Pukalchik, F. Mercl, V. Terekhova, P. Tlustoš, Chemosphere, **203**, 228-238 (2018)
18. A.D. Igalavithana, S.E. Lee, Y.H. Lee, D.C. Tsang, J. Rinklebe, E.E. Kwon, Y.S. Ok, Chemosphere, **174**, 593-603 (2017)
19. K. Lu, X. Yang, G. Gielen, N. Bolan, Y.S. Ok, N.K. Niazi, S. Xu, G. Yuan, X. Chen, X. Zhang, D. Liu, Z. Song, X. Liu, H. Wang, J. Environ. Manage., **186 (2)**, 285-292 (2017)
20. D. O'Connor, T. Peng, J. Zhang, D.S.W. Tsang, D.S. Alessi, Z. Shen, N.S. Bolan, D. Hou, Sci. Total Environ., **619**, 815-826 (2018)
21. C.S. Lwin, B.H. Seo, H.U. Kim, G. Owens, K.R. Kim, Soil Sci. Plant Nutr., **64 (2)**, 156-167 (2018)
22. Z. Tan, Y. Wang, L. Zhang, Q. Huang, Environ. Sci. Pollut. Res., 24, 24844-24855 (2017)

Technology of landscaping in arid zones by using biohumus from sheep wool

Alibek Ydyrys[1*], *Birlikbay* Yeszhanov [2], *Nurlan* Baymurzaev[2], *Sayat* Sharakhmetov[2], *Askar* Mautenbaev[2], *Bekzat* Tynybekov[2] and *Tilek* Baidaulet[2,3]

[1]Al-Farabi Kazakh National University, Biomedical Research Centre, 71 al-Farabi Ave. 050040 Almaty, Kazakhstan
[2]Al-Farabi Kazakh National University, 71 al-Farabi Ave. 050040 Almaty, Kazakhstan
[3]Asfendiyarov Kazakh National Medical University, 94 Tole bi 2 050000 Almaty, Kazakstan

Abstract. The problem of greening the arid zones in Kazakhstan and in the world, as well as addressing water shortages in agriculture in these areas, requires new ideas or innovative technologies. We used sheep's wool to create biohumus in combination with desert soils. In Kazakhstan sheep's wool is considered useless, although it is rich in bioresources. Our research shows that biohumus obtained from sheep's wool is highly fertile in laboratory experiments, and for it the need to use water is 3 times less than for other soils. Under laboratory conditions, we have proven that biohumus can be used to grow plants in different ecological zones. Its use in the wild/field can solve several problems of greening dry areas and growing food crops in low humidity conditions. In addition, it can increase the value of sheep's wool as a bio resource. It is thus an economically promising technology that meets the environmental standards of a green economy.

1 Introduction

The development of safe food production and increasing agricultural productivity for human are considered in parallel with the current biodiversity conservation problem [1, 2]. This is particularly important for countries with a harsh climate located in arid zones such as Kazakhstan. Currently, the problem of the impact of cold as a limiting adverse environmental factor is one of the important problems in Kazakhstan for the development of agricultural areas, and there are also acute problems of desertification, declining fertility and soil degradation, as well as lack of irrigation water. These problems are caused by irrational water use, outdated irrigation systems and are aggravated by growing rates of deforestation, excessive grazing and high vulnerability of the region to climate change. There are several works in this direction, which are related to the impact of low positive temperatures, severe environmental and climatic conditions on plant growth processes [3]. New ideas or innovative technologies are needed to address water shortages in agriculture, greening and agricultural development in the arid zones of Kazakhstan and the world. Some researchers have attempted to quantify the impact of the "Grain for Green" program on water ecosystem

* Corresponding author: ydyrys.alibek@gmail.com

services for arid and semi-arid China [4] and recommended an alternative system called "Ecological Agriculture" to address the problems of modern, highly productive agriculture [5]. Nano-hydrophobic sands allow achieving a rich crop yield per unit area and reducing the amount of irrigation water [6].

Humus is a "medicine" for soil, "bread" for plants. Biohumus is the basis of ecologically safe agriculture and the way to sustainable agriculture in the 21st century. There are different scientific methods of its creation, using different raw materials. We have used sheep's wool, which is an easily accessible and cheap resource in Kazakhstan, to create a new biogumus. Sheep wool is considered as useless but rich bio-resources in Kazakhstan. The results of laboratory testing of the biogumus obtained in combination of desert soils with purified and synthesized sheep's wool are presented below.

2 Materials and Methods

For the production of biohumus we used waste in the form of contaminated sheep wool as raw material. First, the sheep's wool was cleaned in the LLUFHB006-100 washing and drying machine. Then we mixed sheep's wool powder and dry soil in a ratio of 1:10.

For methodical study, we collected four types of soil on which we cultivated seedlings of legumes. The first control group of soils was taken from the arid zones and was well moistened, the second group was represented by black soil, the third group of soils consisted of our biohumus, and the fourth group had arid soils.

The research is conducted by direct biotesting. By the growth of root system and seedlings in relation to the control we can judge about the possibility of using biohumus mixed with soil for growing grains.

First of all, we took bean grains for germination as a test object. 100 g of biohumus was mixed with 50 ml of distilled water and placed in a Petri dish, covered with filter paper, then 20 grains of bio-test were placed. Control was carried out by germination of bio-test grains in distilled water. By the growth of root system and seedlings in relation to the control we can determine the degree of impact of pure biohumus on the test object.

Then we grew the seedlings for four weeks. Each week we measured the height of the plants, marked the phases of plant development (flowering and fruiting) taking into account the total weight of the crop per 1 m2 of each experimental group, the area of each experimental group and the nitrogen, phosphorus and potassium content.

Testing of protein, phosphorus and potassium content was conducted by biochemical analysis in laboratories.

3 Results and discussion

The experiments were conducted by laboratory of Faculty of biology and biotechnology of Al-Farabi Kazakh National University. The plot area is 10 m^2.

The table 1 shows control group growing well in all four weeks. Since dry soils evaporate water quickly, the first control group we had to swim three times as often as the other groups. The second group grew well in the 1st week, but from the 2nd week the condition of the plants becomes average. The third group's performance is excellent all week long. As it turned out, our biohumus has the ability to preserve water for a long time. Plants in group 4 slowly die within four weeks due to lack of water. Research conducted by direct biotesting.

Table 1. Growth dynamics of seedlings of beans on different soils

	1st week		2nd week		3rd week		4th week	
	seedling length mm	root system length, mm	seedling length mm	root system length, mm	seedling length mm	root system length, mm	seedling length, mm	root system length, mm
Control	3.2± 0.3	4.1± 0.2	10.1± 0.52	12.5± 0.81	21.11± 0.63	26.1 ± 0.22	39.13 ± 0.24	40.5 ± 0.71
Black soil	3.5± 0.6	4.8± 0.51	8.6± 0.31	10.7± 0.26	17.4± 0.51	19.3± 0.46	31.81 ±0.46	38± 0.37
Biohumus	3.11± 0.4	4.4± 0.7	9.7± 0.33	11.3± 0.31	24.61± 0.28	32.14 ± 0.35	56.5 ± 0.76	67.3 ± 0.27
Desert soil	2.4± 0.6	3.1± 0.53	5.7± 0.41	8.4± 0.33	7.6± 0.4	10.7± 0.23	0	0

After 4 weeks of cultivation, we analyzed the chemical composition of the bean sprouts in a biochemical laboratory (percentage by dry matter). We measured the amount of potassium, phosphorus and protein fraction per 100 mg. As can be seen from Table 2The chemical composition of plants grown in the new biohumus is slightly inferior to that of the control and chernozem groups.. At the same time, as can be seen from Table 1, the length of seedlings grown in biohumus (56.5 ± 0.76) and the length of their root system (67.3 ± 0.27) is significantly higher than in other groups, probably due to water retention properties of biohumus. This indicates that when using sheep's wool biohumus, the minerals required for plant growth remain available.

Table 2. Macronutrients and protein content in the bean sprouts

Groups	Macronutrients (100 mg ,%)		
	Potassium	Phosphorus	Protein
Control	1.8	6.1	0.57
Black soil	2	6.8	0.6
Bio-Humus	1.7	5.6	0.54
Desert soil	0	0	0

The first signs of nitrogen, potassium and phosphorus deficiency are weak and stunted shoots, leaves with yellowness or pale green. Potassium as an element is needed to increase productivity, resistance to drought, low temperatures, fungal diseases. The first signs of potassium starvation are the barely noticeable withering of leaves and a decrease in their elasticity, the appearance of white borders along the leaf edge, which subsequently turns brown. Phosphorus-containing fertilizers strengthen the root system of plants, increase their resistance to adverse weather factors, provide full ripening of fruits. They are applied in autumn when digging the soil or in nearbore circles with organic matter. Nitrogen is one of the most important elements for plant growth. Fertilisers are applied directly to the soil during spring digging (urea) and in dissolved form (ammonium nitrate). Based on the results of the experiment, we believe that the application of fertilizers will be appropriate with biohumus.

Using this technology in a wildlife/field can solve several problems. First, it is greening the drylands. Secondly, when used in agriculture, it will reduce the need for irrigation. This technology can be of great marketing value due to its low cost. This technology meets environmental standards. Physiological and biochemical activation of seeds and seedlings of beans by treating them with nanoconcentration of new active biohumus increases their salt resistance and increases their competitiveness in terms of available oxygen and photosynthetic properties. It has also been experimentally established that when sowing beans from seeds treated with biohumus, due to the high synthetic ability of biomass above and below ground, much more root mass and plant residues are left behind than when sowing from conventional seeds. At the same time, the bean sprouts themselves act as a phytomeliorant, helping to enrich the soil with organic matter and thus improving its biological activity, chemical, physical and water-physical properties. In other words, this method is a new approach to old problems of land reclamation and is proposed to be introduced into production.

4 Conclusion

According to this study, a group of legume seedlings grown using biohumus during a month showed the highest rates of sprout length and root system length compared to groups grown on other soil types. In particular, they are approximately one and a half times higher than seedlings grown without using biohumus but with abundant moisture, which required three times more frequent watering. Without active moistening, plants in desert soil groups begin to dry up from the first week of seedling growth. Biochemical analysis of potassium, phosphorus and protein content in the seedlings of the group of beans grown with the use of biohumus is not significantly different from the control and black soil.

Thus, our experiments have shown that the new bihumus, containing sheep's wool, has a great influence on the growth of seedlings of beans with a decrease in chafing in irrigation. New fertilizers are important for irrigation and landscaping of agrocenoses and are an effective way of cultivating such soils. It should be noted that studies on the application of a mixture of fertilizers and biohumus on different soils will continue in the field.

References

1. A. Akhmetova, N. Mukhitdinov, A. Ydyrys, Pakistan J. Of Botany. **47(2)**, 511 – 515 (2015).
2. A.B. Akhmetova, N.M. Mukhitdinov, A. Ydyrys, A.A. Ametov, Z.A. Inelova, M. Öztürk, J. of Animal and Plant Sc. **28(5)**, 1400-1404 (2018).
3. E.A. Bukenova, Z.M. Bassygarayev, A.B. Akhmetova, Z.K. Zhunusbayeva, A. Ydyrys, Research on Crops, **20(1)**, 210-214 (2019).
4. Yu chunWang, JunZhao, JiewenFu, WeiWei, Ecol. Indicators, **104**, 659-668 (2019).
5. M. Kiley-Worthington, Agriculture and Environment, **6(4)**, 349-381 (1981).
6. M. Myrzabaeva, Z. Insepov, K.K. Boguspaev, D.G. Faleev, M. Nazhipkyzy, B.T. Lesbayev, Z.A. Mansurov, Eurasian Chemico-Technol. J. **19**, 91-98. (2017).

Evaluation of the biological activity of meadow chernozem soils after the application of biochars with different pyrolysis temperatures in a model experiment

Vladislav Zinchenko[1*], *Andrey* Gorovtsov [1,2], *Svetlana* Sushkova [1], *Svetlana* Antonenko [1], *Elena* Fedorenko [1], *Pavel* Pogonyshev[1], and *Anatoly* Barakhov [1]

[1] *Southern Federal University, Rostov-on-Don, Russia*
[2]*Federal Rostov Agricultural Scientific Center, settl. Rassvet, Russia*

Abstract. The paper aims to study the effect of fresh and aged biochar with different pyrolysis temperatures and metal resistant bacteria on the numbers of microorganisms and the dehydrogenase activity of meadow chernozem soils. The use of fresh biochar leads to an increase in the number of soil bacteria and suppression of dehydrogenase activity; after ageing biochar by incubation in the soil, dehydrogenase activity increases significantly.

1 Introduction

Biochar is a carbon sorbent that is made by pyrolysis of organic residues. A key aspect of creating biochar is the raw materials and the pyrolysis temperature. For example, the temperature of pyrolysis directly affects the physical properties of the biochar: specific surface area, pore sizes, etc.

Biochar creates new structures for microorganisms that ensure soil aeration and the use of mineral nutrition elements. On the other hand, in the process of pyrolysis, a large number of substances can be produced that can exert an inhibitory effect on the soil and microorganisms living in it. One of the most studied and common groups of pyrolysis products are volatile organic compounds (VOCs). It is known that the content of these substances in biochar does not depend primarily on raw materials, but rather on the conditions for biochar production [1]. VOCs in the biochar are often represented by methanol, acetone, benzene, sorbed gases. Various aliphatic and aromatic acids, phenols are often also present [2, 3]. The negative influence of VOCs on microbial communities and plants is well established. That is why soil microbial communities undergo significant changes in the first months after the biochar application.

A variety of approaches are used for the remediation of contaminated soils, including the use of microorganisms as one of the agents for the remediation of contaminated soils [4]. The activity of introduced microbial cultures leads to the conversion of pollutants into forms inaccessible to the plants and accelerates the decay of organic pollutants, reducing soil toxicity and increasing its suitability for agricultural use.

* Corresponding author: zinj007@gmail.com

The aim of this study was to study the effect of biochars obtained at different pyrolysis temperatures and strains of metal-resistant bacteria on the biological parameters of soils.

2 Object and research methods

The soils of the Seversky Donets River floodplain have been studied. The model experiment was laid to study the effects of biochar and microorganism strains application on the biological parameters of these soils. In the experimental variants, biochar produced at a final pyrolysis temperature of 500 C° (biochar-500) and 700 C° (biochar-700) was uniformly mixed with soil at a dosage of 2.5% by weight. The biochar was produced from oilseed waste - sunflower husks. For the preparation of biochar, pyrolysis was carried out in a muffle furnace in a special thick-walled metal vessel (retort) under limited oxygen conditions. The heating was carried out stepwise; when the final pyrolysis temperature was reached, the oven was turned off and left for complete cooling. Spores of metal-resistant bacteria belonging to the *Bacillus* genus were introduced into the soil in an amount of 10^{10} CFU / kg of soil. Inoculation of the strains was implemented as follows: the biomass of bacteria was grown on a solid nutrient medium for 3 days, then the biomass was washed into the test tubes with water. The number of cells for inoculation was determined by densitometry. Calibration curves were prepared for each strain prior to the experiment using densitometry in combination with the plate counts on a solid nutrient medium. Suspensions of three strains were mixed proportionally to ensure an equal share of each strain in the final inoculum. The amount of water in the inoculum was calculated to adjust the soil moisture to 60% of the maximum moisture capacity. The sieved air-dried soil was introduced into the pot layer by layer, and each layer was wetted with a suspension of bacterial strains. This allowed us to achieve the most uniform distribution of bacterial cells throughout the soil. Experimental variants with the biochars only and the control variants were placed in the same 2-liter pots and brought up to 60% of the total moisture capacity with water instead of inoculum. Spring barley (*Hordeum vulgare* L.) was grown in the pots. After 30 days of vegetation, the plants were harvested and soil samples were taken for analysis. This was followed by a period of maintaining optimal soil moisture without plants for 90 days, then the sampling procedure was repeated.

The numbers of copiotrophic bacteria were determined on nutrient agar, prototrophic bacteria capable to assimilate inorganic nitrogen source were enumerated on starch-ammonia agar. To determine the soil dehydrogenase activity, the amount of red triphenylformazane formed by dehydrogenation of colorless triphenyltetrazolium chloride was measured at wavelength 490 nm after extraction with ethanol.

3 Results and discussion

According to the data obtained by plate counts, it becomes evident that in the sample with the application of fresh biochar-500, the number of microorganisms was 1.54 times higher than in control. The increase in the number of microorganisms can be explained by the colonization of new ecological niches created by the pore space of the biochar. The application of fresh biochar-700 didn't cause any significant differences from the control. Such trends were observed for both copiotrophic and prototrophic bacteria. The use of microorganism strains increased the numbers of copiotrophic and the protorophic bacteria by 29.3% and 36.8% respectively. (Table 1).

Table 1. The effect of biochar with different pyrolysis temperatures on the number of microorganisms in the soil.

Experimental variant	The number of bacteria, million CFU / g of absolutely dry soil	
	Copiotrophic	Prototrophic
Control	37.86±1.48	47.43±4.16
Bacteria	48.99±2.24	64.93±1.4
Biochar 500	58.56±1.21	62.84±6.17
Biochar 700	39.56±2.45	39.56±6.41

The study also evaluated the activity of microorganisms and the level of cellular respiration by measuring soil dehydrogenase activity.

According to the results of the analysis of dehydrogenase activity immediately after plant harvesting, it becomes clear that, despite the increase in the number of microorganisms, the activity of dehydrogenase in the experimental variants with the addition of microorganisms, fresh biochar 500 and 700 was significantly lower than in the control, by 41.8%, 18.3%, and 77% respectively. (Fig. 1).

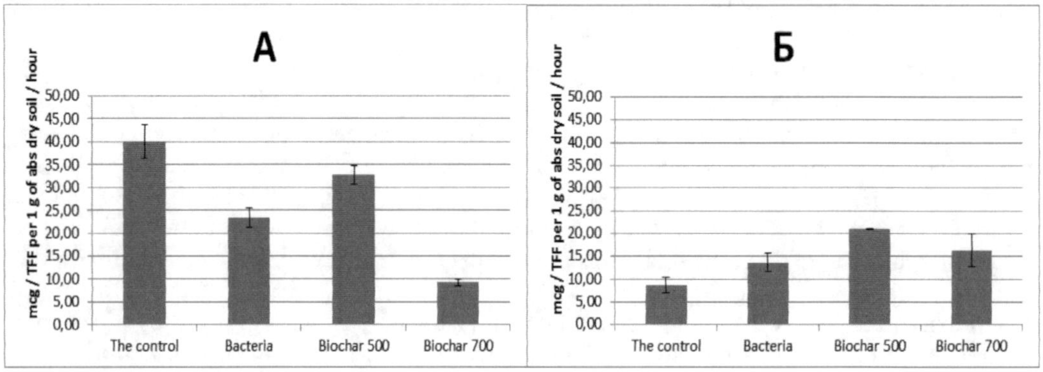

Fig. 1 Dehydrogenase activity of the soil immediately after harvesting the plants (A) and after a 3-month incubation without plants (B)

A significant decrease in dehydrogenase activity indicates the inhibition of oxidative metabolism. The cause of the inhibition was probably various volatile organic substances, as well as polycyclic aromatic hydrocarbons (PAHs) accumulated during pyrolysis. The reason for the inhibition of dehydrogenase activity in the variant with the application of bacteria was probably due to antagonistic biotic interactions between soil bacteria and newly introduced strains. After a 3-month period of incubation of biochar in soil, while maintaining optimal moisture without plants opposite trends were observed. The analysis showed a decrease in dehydrogenase activity in the control sample and in the sample with the introduction of microorganism strains, which is explained by the lack of stimulation from the root system of the plants. However, after the period of incubation without plants, all remediation variants exceeded the control in the level of dehydrogenase activity. In the samples with the application of biochar 500, the dehydrogenase activity also slightly decreased in comparison with the results immediately after plant harvesting. However, it remained at a level 2.4 times higher than the control, and for the sample with the application of biochar 700, an increase in dehydrogenase activity can be noted. In both cases, the destruction of toxic volatile organic substances that had a negative effect on the microbial communities of soils, as well as the colonization of the pore space of the biochar can explain the observed results. Similar data are also available in the literature: the application of biochar increases the metabolic activity of soil microbial communities in drought conditions, and the "aged" biochar turned out to be more effective than fresh one [5]. In addition, biochar can retain nutrients through adsorption and makes them available to the microbiota that colonizes it. The initial substrates from which biochar is made, such as wood, straw, composts, sewage sludge, etc. always contain a significant amount of ash mineral elements that remain in the biochar after pyrolysis. Carbon begins to transform into a gaseous state at 100 °C, nitrogen at 200 °C, sulfur at 375 °C, phosphorus and potassium only at 700-800 °C [6]. Nutrient elements can then slowly diffuse into the soil and are used by plants and microorganisms. It has been shown that a significant proportion of potassium and phosphorus contained in biochar granules ends up in soil solution [7]. The stability of the bonds between biochar granules and nutrients is directly related to the pH of the soil solution, and with its change (i.e. due to root exudation), the process of sorption and desorption of nutrients can occur. The introduction of mineral elements with wood ash has a pronounced positive effect on soil microbial communities [8]. A similar effect can be expected in the case of the introduction of biochar rich in ash elements. An indirect confirmation of this is the data [9], which showed an increase in the activity of phosphate and sulfate mobilizing bacteria when biochar is introduced into the soil.

4 Conclusion

After analyzing all the data obtained during the study, we can conclude that the biochar creates additional niches for microorganisms and stimulates the biological activity of soils; however, it is recommended to use aged biochar that has undergone preliminary incubation to prevent temporary negative effects from pyrolysis products. Biochar produced at a final pyrolysis temperature of 500 °C is more effective and less toxic than biochar with a final pyrolysis temperature of 700 °C. This is probably due to the physicochemical properties of biochars – pore size, specific absorption surface, and the amount of toxic pyrolysis products accumulated during the manufacturing process.

The study was supported by a grant from the Russian Science Foundation 19-74-10046.

References

1. K.A. Spokas et al., Chemosphere, **85**, №. 5, P. 869-882, (2011)
2. W. Buss, O. Mašek, Env.Sci.Pollut.Res., **23,** №. 19, P. 19580-19589, (2016)
3. M. Ghidotti, et al., ACS Sust. Chem. Engin., **5,** №. 1, P. 510-517, (2017)
4. X. Chen, & V. Achal, Biostimulation of carbonate precipitation process in soil for copper immobilization. *J. Hazard. Mater.*, *368*, 705-713. (2019).
5. L. Paetsch et al., Scientific reports, **8**, №. 1, P. 1-11, (2018)
6. T.H. DeLuca et al., Biochar for environmental management: science, technology and implementation, **2**, P. 421-454. (2015)
7. T. Limwikran, et al., Geoderma, **312**, P.139-150. (2018).
8. S. Mahmood et al., FEMS Micr. Ecol., **43(1)**, P.121-131. (2003)
9. A. Fox et al., FEMS microbiology ecology, **90(1)**, P. 78-91.(2014)

The effect of No-till technology on the mineral nitrogen content in the Lower Don Chernozem

Anna Medvedeva[1,*], *Olga* Buryukova[1], *Yaroslav* Ilchenko[1], *Tatyana* Minkina[1], *Roman* Kamenev[2], and *Coşkun* Gülser[3]

[1] Southern Federal University, Stachki Ave, 194/1, Rostov-on-Don, 344090, Russia
[2]Don State Agrarian University, Krivoshlykova St., 24, Rostov Region, Oktyabrsky District, Persianovsky Village, 346493, Russia
[3]Ondokuz Mayis University, 55139, Samsun, Turkey

Abstract. The paper presents the results of a five-year study of the impact of various agricultural technologies (No-till, minimum and traditional using moldboard ploughing) on the content of mineral nitrogen in Haplic Chernozem in southern zone of Rostov Region. It has been revealed that the content of ammonium and nitrate nitrogen in the winter wheat areas cultivated by various agricultural technologies does not significantly change in samples collected both in spring and in summer. However, the content of nitrate nitrogen under resource-saving technologies (both minimum and No-till) has been higher than under ploughing throughout the whole period of study. The trend identified has not been mathematically confirmed. Nevertheless, the impact of No-till technology on the intensity of ammonification and nitrification should not be unequivocally denied, since a significant amount of mineral nitrogen is extracted by crops, and their crop yields under minimum and zero tillage was higher, than when ploughing was applied.

1 Introduction

Soil is one of the most important natural resources. It performs vital biogeocenotic functions. High soil fertility is the basis for the existence of the civilization on our planet. Providing the population with high-quality and environmentally friendly food requires an active nutrient balance in ecosystems, optimization of the biological cycle of substances of the pedosphere [1]. Agriculture uses the soil as a substrate to produce crops through the application of mineral fertilizers and significantly limits its ecological function in the biosphere, disrupting the vegetation cover, reducing the humus content, and significantly polluting with pesticides, heavy metals, oil and other substances.

The most important element of the technology that affects soil fertility and crop yields is the primary soil cultivation, which has a wide range of methods from traditional ploughing to No-till technique with many options for boardless cultivation and various minimization conditions. Traditional cultivation methods, as a rule, have an adverse effect on the long-term soil productivity due to erosion, loss of organic matter and increase soil degradation risk [2,

* Corresponding author: medvedeva.estelior@yandex.ru

3]. Nowadays no tillage-seeding technique - No-till has been widely recognized to be the highest level of minimizing tillage and seeding.

Many studies find that the use of No-till can improve the agrophysical, physicochemical and biological properties of soils by optimizing the structure of the soil and reducing the risk of erosion processes [4, 5].

Available research on the dynamics of mobile nitrogen accumulation depending on the primary cultivation method shows that replacing ploughing with other methods gives ambiguous results. Thus, Kiryushin and Zabolotsky [6, 7] pointed out a steady decreasing tendency in the nitrate nitrogen content when flat cutting and minimum tillage were applied. It was associated with the presence of crop residues in the upper soil layer that slow down soil heating and promote the immobilization of available nitrogen by microorganisms. A number of researchers point out that no significant differences in nitrate nitrogen content in the meter soil layer under deep soil loosening and No-till techniques have been established [8, 9].

As reported by Novikova [10] flat-cutting and No-till methods contributed to an increase in nitrate nitrogen content, compared with ploughing. Apparently, such disagreement is associated primarily with differences in soil properties, environmental conditions, and features of agricultural practices.

The purpose of the study was to investigate the content of mineral nitrogen in Haplic Chernozem when No-till technique was used under insufficient moisture conditions.

2 Materials and methods

The present studies were conducted in agrocenoses of the steppe zone in the European part of southern Russia. This zone is characterized by a continental climate, unstable and insufficient moisture. Annual precipitation is 410-460 mm, HTC = 0.7-0.76, the average annual temperature is 8.7-9.5 degrees, the sum of temperatures for the active growing season is higher than 3400 degrees. Groundwater is 7-10 m below the ground surface. The formation of high yields of field crops in these landscapes depends primarily on the state of the water conditions, as well as on the conditions of mineral nutrition. The object of study was Haplic Chernozem calcareous heavy loamy formed on loess loam.

A comparative analysis of the intensity of ammonification and nitrification processes in the soil under various agricultural technologies (traditional, minimum and No-till) has been carried out for 5 years (2013 - 2017) in Peschanokopsky district of Rostov Region. The traditional technology has been used in the study area for more than 50 years, minimal - since 2000, No-till has been applied since 2008. The reference plot was the virgin area.

The ammonium nitrogen content was determined according to GOST 26489 in the edition of the Central Scientific Research Institute of Agricultural Chemistry. The method is based on preparing a complex compound by the interaction of ammonium chloride with Nessler's reagent in an alkaline medium and its subsequent photocolorimetry (wavelength 400-440 nm, soil:solution ratio 1:30). Nitrate nitrogen was determined according to the Grandval-Lajoux method. The method involves the interaction of nitrates with disulfophenolic acid with subsequent formation of trinitrophenol that causes solution colour change to yellow when placed in alkali due to the formation of potassium (KOH) or sodium (NaOH) salts, their amount being equivalent to the nitrate content, followed by photocolorimetry (wavelength 400-440 nm, soil:solution ratio 1:5).

Mathematical processing of the results was carried out by correlation analysis and ANOVA methods using the STATISTICA 10 software package.

https://doi.org/10.1051/e3sconf/202016902014

3 Results and discission

On average the content of ammonium nitrogen in the 0-20 cm layer has been ranging from 16.8 - 19.8 mg/kg throughout the period of study (Figure 1). Down the profile there has been a uniform decrease in the amount of ammonium. Over the years of research the trend identified has continued.

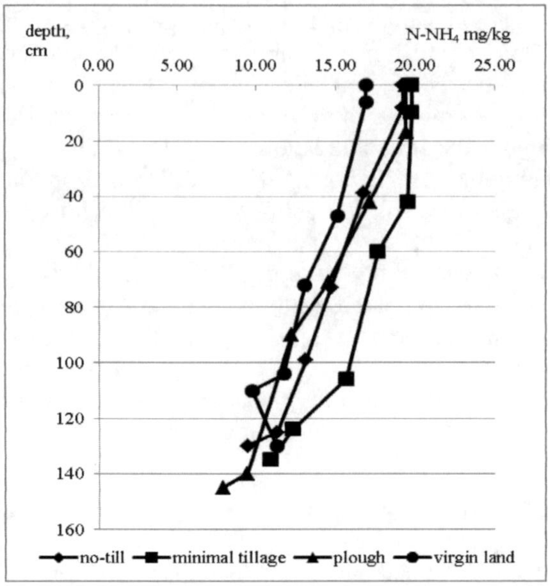

Fig. 1. The average depth-wise distribution of ammonia nitrogen in Haplic Chernozem over the years of research, mg/kg of soil.

The results obtained indicate that the intensity of ammonification when using various cultivation methods is almost the same (Figure 2).

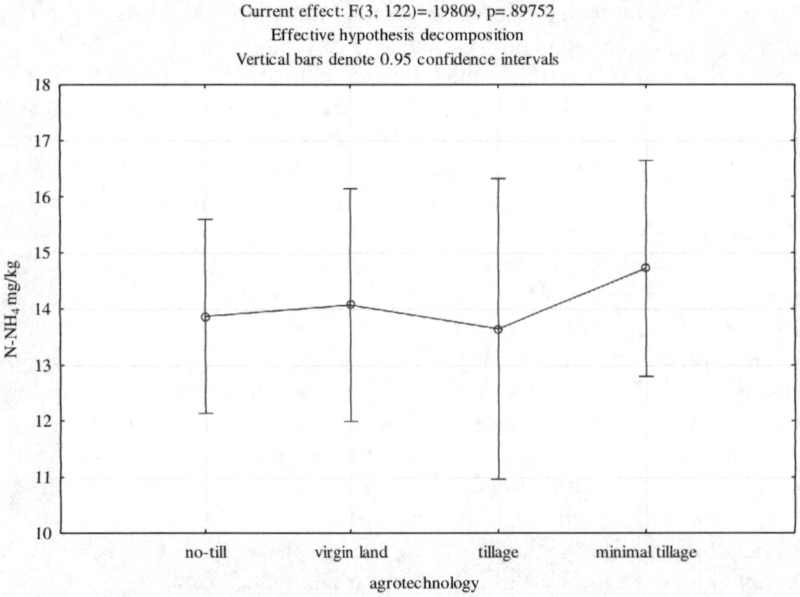

Fig. 2. The effect of the primary cultivation method on the content of ammonium nitrogen in Haplic Chernozem (average profile).

The ammonia form of nitrogen, having a number of advantages, cannot fully provide plants with nitrogen. There is much evidence that in many cases it is necessary to combine plant nutrition with ammonium and nitrate nitrogen. Soils in Rostov region are characterized with high biological activity, and, therefore, are able to produce nitrates in large quantities. It has been found that in most cases the nitrate form prevails over the ammonium ($N\text{-}NO_3/N\text{-}NH_4>1.0$), which is typical for the studied agrocenoses.

5-year average content of nitrate nitrogen in calcareous Haplic Chernozem of a virgin land has been 20.1 mg/kg (Figure 3 a).

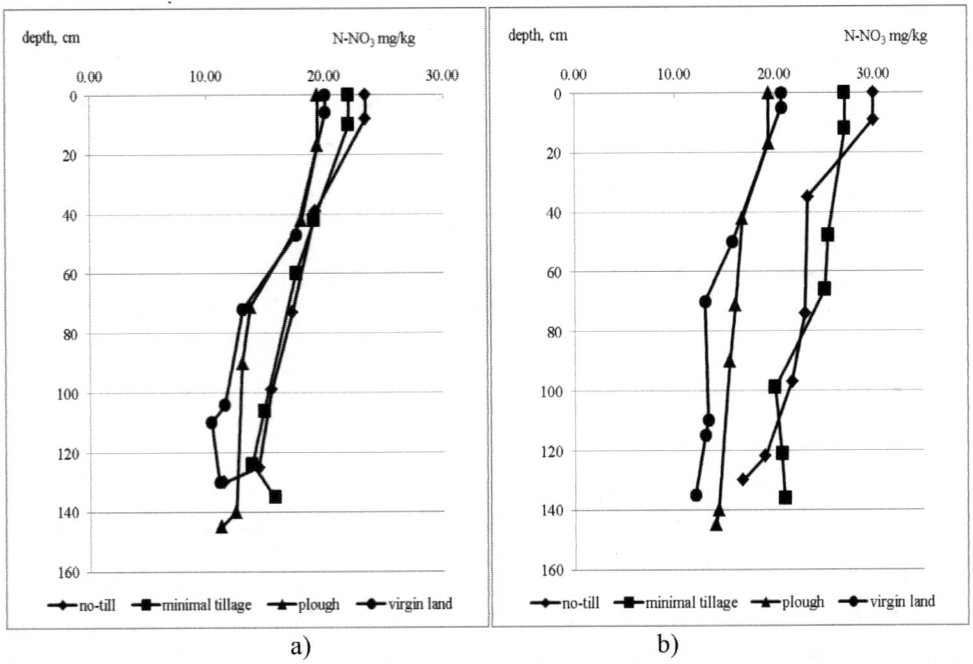

Fig. 3. Depth-wise distribution of nitrate nitrogen in Haplic Chernozem, mg/kg of soil (a) on average over the years of research; b) 2014 sampling).

Under No-till technique the level of nitrate nitrogen in the upper horizon increases to 23.5 mg/kg, whereas under minimal cultivation - up to 22.3 mg/kg. In case of ploughing the content of nitrate nitrogen (19.4 mg / kg) slightly decreased. The nature of nitrate nitrogen distribution along the chernozem profile was similar for all cultivation methods.

The content of nitrate nitrogen under resource-saving cultivation was slightly higher than in the virgin land and under ploughing. The maximum differences were recorded in 2014 during the summer period (August 5), with a significant lack of precipitation (Figure 3 b). The use of resource-saving cultivation in 2014 allowed to increase nitrate nitrogen content to 27.0-30.0 mg/kg (20.7 mg/kg in virgin land, and 19.4 mg/kg when ploughing was used).

Although, when comparing the results, the advantage of resource-saving cultivation over traditional ploughing seemed certain, the mathematical processing of the entire data obtained showed no significant effect of the cultivation method on nitrate nitrogen (Figure 4).

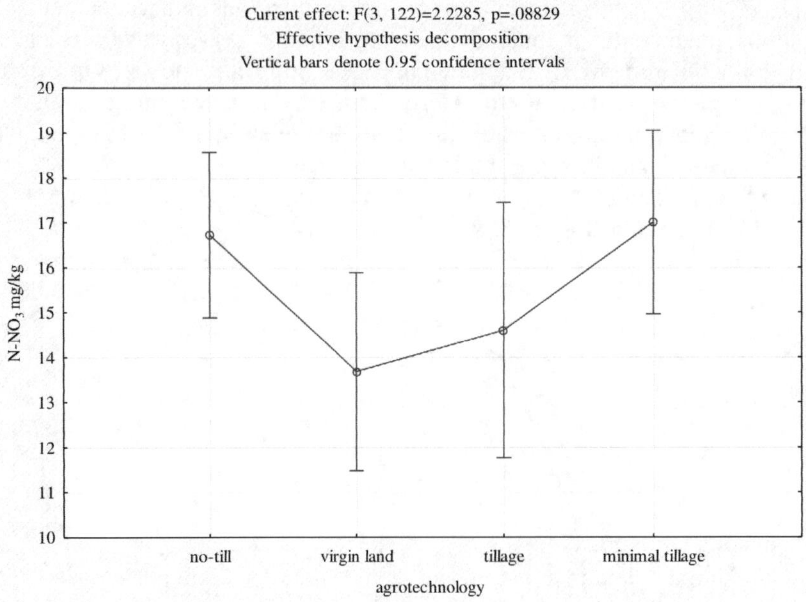

Fig. 4. The influence of the primary cultivation method on nitrate nitrogen content in Haplic Chernozem (average profile).

However, it is impossible to affirm categorically that there is no No-till effect on nitrate nitrogen accumulation, since its content also depends on plant absorption intensity during yield formation, and the crop yield with resource-saving cultivation was higher than when ploughing method was applied.

Malhi et al. [11] demonstrated that soil retained the smaller amount of mineral nitrogen at a depth of 0-10 and 20-30 cm when ploughing was used compared No-till method. It was concluded that reducing or even eliminating ploughing would improve N-reserve of the soil.

The results obtained are quite understandable, since nitrification intensity is mainly determined by soil hydrothermal conditions (moisture and aeration). It is well known that key conditions for nitrification to proceed intensively are favorable soil moisture of 60-70% of capillary moisture capacity, good aeration, and an optimal temperature of 25-32 ° C.

The depth and cultivation method significantly affect the water, temperature and microbial activity of the soil. Tereshchenko [12] found out that resource-saving technologies including No-till were responsible for a greater moisture accumulation in the soil, which led to a high intensity of biological processes.

Despite the fact that the rate of nitrification activity is very dynamic and depends on the agricultural technology used, fertilizers, meteorological conditions, Nieminen [13] pointed out a closer relationship between the content of organic matter and the nitrification in soils when No-till technique was used compared to conventional cultivation.

The major source of ammonia and nitrate nitrogen is the pool of organic matter and organic N-compounds. Therefore, the supply of plants with nitrogen depends on the rate of decomposition of soil organic matter to compounds assimilated by the plant. Our previous studies have shown that the humus content in Haplic Chernozem depends on the primary cultivation method [14]. Correlation analysis has revealed a close relationship between the content of ammonium and nitrates and the amount of humus.

4 Conclusion

The supply of plants with nitrogen depends on the rate of decomposition of the soil organic matter to compounds assimilated by plants. Close relationships has been found between the intensity of ammonification and nitrification processes and humus accumulation (r=0.58; r=0.57, respectively, at p=0.05). The content of ammonia nitrogen does not significantly depend on the agricultural technology used, with all processing methods it was practically at the same level. Cultivation methods have showed no significant effect on nitrate nitrogen. However, the impact of resource-saving technologies on nitrogen mobilization in the soil should not be unequivocally denied. Mineral nitrogen is the major source of nitrogen for plants and has great effect on yield formation, and the yield with minimum and zero tillage was higher than when moldboard ploughing is used.

It should also be noted that under a favorable combination of agrometeorological conditions, differences in the intensity of ammonification and nitrification between the cultivation methods were smoothed out. Under a significant shortage of precipitation or plentiful precipitation and low temperatures, the positive effect of resource-saving cultivation manifested itself more clearly and contributed to a certain increase in the content of mineral nitrogen.

Thus, agrocenoses, characterized by a significant alienation of biomass, in the absence of special agricultural techniques for regulating soil fertility, are not able to maintain a sufficient level of nutrients. Therefore, the efficient use of agricultural land requires continuous improvement of technologies for increasing soil fertility and optimizing the mineral nutrition of plants. The use of No-till technology in conditions of insufficient moisture can improve nitrogen nutrition of plants.

The research was supported by the project of the Ministry of Science and Higher Education of the Russian Federation within the framework of the state task of the Southern Federal University for 2020-2022 (№ FENW-2020-0028)

References

1. V.G. Mineev, *Selected writings*, 601 (Moscow, 2005) (in Russian)
2. R.P. Mathew, Y. Feng, L. Githinji, R. Ankumah, K.S. Balkcom, Appl. Environ. Soil Sci., **2012,** 10 (2012) doi:10.1155/2012/548620
3. B.D. Soane, B.C. Ball, J. Arvidsson, G. Basch, F. Moreno, J. Roger-Estrade, Soil and Tillage Research, **118,** 66-87 (2012)
4. Y.P. Dang, H. Liu, L. Carvalhais, V. Rincon-Florez , M. Crawford, P.G. Dennis, M. Peer, P.M. Schenk, Soil and Tillage Research, **158,** 91–99 (2016) doi:10.1016/j.still.2015.12.007
5. J. Bareche, D. Plaza-Bonilla, C. Cantero-Martínez, J. Lampurlanés, J. Álvaro-Fuentes, Field Crops Research, **203,** 74–85 (2017) doi:10.1016/j.fcr.2016.12.008
6. V.I. Kiryushin, Agronomic soil science. 668 (2010) [In Russian]
7. V.V. Zabolotskikh, Siberian Bulletin of Agricultural Science, **1,** 95-98 (2013) (in Russian)
8. V.T. Maltsev, E.N. Dyachenko, Achievements of Science and Technology of AIC, **12,** 8-11 (2011) (in Russian)]
9. A.N. Vlasenko, N.G. Vlasenko, Agriculture, **3**, 17-19 (2016) (in Russian)
10. A.S. Novikova, V.T. Lobkov, *The influence of modern energy-saving technologies for soil treatment on agrochemical indicators of the soil. Innovative fundamental and applied research in the field of chemistry in agricultural production*, 8-12 (Orel, 2011) (in Russian)

11. S.S. Malhi, A. Légère, A. Vanasse, G. Parent, The J. of Agricultural Sc., **156**, 472–480 (2018) doi: 10.1017/S0021859618000448
12. V.V. Tereshchenko, O.A. Biryukova, D.V. Bozhkov, *Agrophysical properties of chernozem under various cultivation methods* (LAP, 2015) (In Russian)
13. J.K. Nieminen, P. Pohjola, Boreal Environ. Res, **19**, 257–266 (2014)
14. A.M. Medvedeva, O.A. Biryukova, Y.I. Ilchenko, A.V. Kucherenko, E.V. Kuchmenko, Successes in modern natural sciences, **1**, 29-34 (2018) (In Russian).

Identifying priority management of Ecuadorian forests based on the environmental integrated assessment

Silvia Llerena[1*], Priscila Arias[1], Jhonn Cueva[2], Georgina Almeida[3], and Cristian Salazar[1]

[1]Peoples' Friendship University of Russia (RUDN University), Faculty of Ecology, 6 Miklukho-Maklaya St, Moscow, 117198, Russian Federation
[2]Peoples' Friendship University of Russia (RUDN University), Agrarian and Technological Institute, 6 Miklukho-Maklaya St, Moscow, 117198, Russian Federation.
[3]Peoples' Friendship University of Russia (RUDN University), Engineering academy, 6 Miklukho-Maklaya St, Moscow, 117198, Russian Federation.

Abstract. Anthropogenic impacts, such as deforestation, soil erosion, and carbon dioxide emissions, have a negative influence over global warming due to the increase of CO_2 levels in the atmosphere. The sustainable forest management is a way to mitigate climate change owing to the carbon storage capacity of forests. This study highlights the priority of forest management according to the integrated assessment of carbon storage under anthropogenic impacts in the administrative units of Ecuador. In the obtained map, the provinces Guayas, Esmeraldas, and Manabí showed the highest values of 25, 22.85 and 19.9, respectively, followed by two Amazon provinces, Morona Santiago and Sucumbíos. Therefore, we concluded that deforestation, soil erosion, and carbon dioxide emissions were more pronounced on the coast mainly due to agriculture and livestock activities and the forests in these provinces must have priority management. This analysis is useful for planning environmental practices in order to increase carbon storage as a strategy of mitigation for global warming.

1 Introduction

The world is suffering several global changes due to an unbalance between the human population increase and the capacity of the resources to sustain the increasing demand in the ecosystem services. Magrin and colleagues explain how the anthropogenic impacts such as deforestation, soil degradation, and human emissions have contributed to climate change [1], being deforestation the second-largest source of anthropogenic greenhouse gas (GHG) emissions [2]. The direct effect of deforestation is the decrease of carbon reservoirs and sinks. Furthermore, the loss of trees, which retains the soil with its roots, causes soil erosion that leads to additional carbon losses [3]. The sustainable management of forests, which includes forest conservation and restoration, reducing hazards, and maintaining forest

[*] Corresponding author : alellerenag@gmail.com

health, is one mechanism to reduce the impacts of anthropogenic pressures. It is an effective way of climate change mitigation because forests have the capacity for carbon uptake and storage it in biomass and soils.

In Ecuador like in other Latin American countries, the anthropogenic impacts have released GHG to the atmosphere. Ecuador emits approximately 80,6 million tons per year of greenhouse gases because of the energy sector, land-use change and forestry, agriculture, industry, bunker fuels, and wastes. The second place of GHG emissions is occupied by land degradation and deforestation with 25,35% [4,5]. Even though Ecuador is responsible for only 0.15% of greenhouse gases, it is one of the most affected by the fragility of its ecosystems. In response, Ecuador pledged to limit 2019-2025 GHG emissions to 9% in energy, industrial processes, wastes, and agriculture sectors; and additionally, a 4% reduction in the land use and land-use change sector, which includes deforestation and land degradation. In addition, a total reduction of 20.9% and 20% in the above-mentioned sectors could be achieved with the support of alliances and international cooperation management [4].

To reduce emissions in the sector of land-use change and forestry, governmental and non-governmental institutions have initiated various measures. In 2016, Ecuador presented its REDD+ Action Plan: Forests for the Good Living 2016-2025 to fight deforestation and protect forests. The aim of the REDD + Action Plan is to contribute to national efforts to reduce deforestation and forest degradation through conservation, sustainable forest management, and optimization of other land uses to reduce pressure on forests [6]. The Socio Bosque Program (SBP) is another forest conservation strategy, which gives monetary incentives to private forest owners and indigenous community. Additionally, Ecuador has 54 protected natural areas in forest ecosystems [7].

The identification, assessment, forecast, and protection of forests with high carbon sequestration can optimize resources and ensure the effectiveness of forest management strategies [8, 9, 10, 11]. The priority conservation areas for carbon sequestration, which are territory defined by environmental characteristics threatened by human activities or natural causes [12], have already been used to promote the reduction of high rates of deforestation and restore degraded areas in other countries [13,14]. In Ecuador, there are deforestation studies related to change of land use, demography, and agricultural production [15]. Nevertheless, an integrated assessment of environmental indicators such as deforestation, carbon dioxide emissions and soil erosion over forest carbon storage is non-existent, which is the objective in this study.

The integrated assessment allows evaluating multiple parameters of different regions in order to identify situations with more urgent attention [16, 17]. This study includes the integrated assessment of the environmental indicators of anthropogenic impacts over forest carbon storage in administrative units of Ecuador. The result is a map of forest management for identifying conservation priority areas for natural resource protection.

2 Materials and Methods

Our research consisted of three stages: (1) collecting data of deforestation, anthropogenic soil erosion, carbon dioxide emissions, biomass, and organic carbon; (2) an integrated assessment with environmental indicators based on the collected data; and (3) the development of a forest management priority map based on integral factor values.

n the initial stage, the data were acquired from different sources. The deforestation data between 2008 and 2014 were obtained from the Ministry of the Environment of Ecuador [18], the erosion index from a scientific study about erosion [19] and the CO_2 emissions from the National Institute of Statistics and Censuses of Ecuador [20]. The data of biomass and carbon storage in tons per hectare belong to the National Forestall Evaluation Project [21]. Table 1 summarizes the data used in this work.

Table 1. Deforestation, erosion, CO_2 emissions, biomass and organic carbon data

Provinces	Deforestation (ha/year)	Erosion index (%)	CO_2 emissions (t)	Biomass (t/ha)	Organic Carbon (t/ha)
Azuay	1039	75	4303.45	246.2	123.1
Bolívar	2939	75	839.32	246.2	123.1
Cañar	365	25	938.19	246.2	123.1
Carchi	74	75	1231.61	246.2	123.1
Chimborazo	42	755	1914.28	246.2	123.1
Cotopaxi	1619	75	1882.1	246.2	123.1
El Oro	4274	75	10240.93	74.08	37.04
Esmeraldas	5476	75	2206.53	166.68	83.34
Guayas	2570	50	44114.44	173.26	85.63
Imbabura	246	75	4167.13	246.2	123.1
Loja	1815	75	1457.26	95.82	47.91
Los Ríos	395	25	7625.12	245.54	122.77
Manabí	1411	100	12291.16	74.08	37.04
Morona Santiago	7925	75	2440.2	320.8	160.4
Pichincha	408	75	22912.15	246.2	123.1
Santo Domingo	679	75	3228.23	246.2	123.1
Sucumbíos	5416	75	2440.2	151.6	75.8
Tungurahua	487	75	2994.34	210.2	105.1

Once all data had been collected, interval scales for each Environmental Indicator (E.I.) were developed by using SPSS software (Table 2):

Table 2. Interval scales for indicators

Environmental indicator 1. Deforestation between 2008 and 2014 (ha/year).

1.- < 500	2.- 500 -1000	3.- 1000-2500	4.- 2500-5000	5.- >5000

Environmental indicator 2. CO_2 emissions (t) 2011.

1.- < 1000	2.- 1000-2000	3.- 2000-4000	4.- 4000-10500	5.- >10500

Environmental indicator 3. Erosion index (%) 2011.

1.- 0 None	2.- 25 Potential	3.- 50 Act.	4.- 75 Pot. a.	5.- 100 Very a.

Environmental indicator 4. Biomass in forests (t/ha) 2013.

1.- < 95	2.- 95-150	3.- 150-200	4.-200-250	5.- >250

Environmental indicator 5. Carbon Capture by forest (t/ha) 2013.

1.- < 75	2.- 75-100	3.- 100-125	4.-125-150	5.- >150

According to the interval scales, a numerical value between 1 and 5 was assigned to each E.I. of each province. Then, using these assigned values, weight coefficients regarding the urgency of management were proposed for each E.I. based on the Delphi technique and the qualimetry method in [22]. A weight coefficient whose numerical value is '2' corresponds to the most urgent state and the less immediate state is quantified with '0,5'. Finally, the interval scale number (1 to 5) was multiplied by the weight coefficient for each environmental indicator. The results of these multiplications and the integral factors, which are calculated as the sum of the mentioned products by province, are reported in Table 3.

The map of forest management priority according to the environmental indicators of anthropogenic impacts over forest carbon storage in Ecuador (Fig. 1) was developed based on the integral factor and using the graphics editor InkScape.

Table 3. Integral factor for forest management priority

Provinces	E.I. 1	E.I. 2	E.I. 3	E.I. 4	E.I. 5	Integral factor	Priority Management
Azuay	2.4	2.8	6	2	3	16.2	Low
Bolivar	6	0.5	6	2	3	17.5	Low
Cañar	0.6	0.5	1	2	3	7.1	Very low
Carchi	0.5	1.2	6	2	3	12.7	Low
Chimborazo	0.5	1.2	6	2	3	12.7	Low
Cotopaxi	3	1.2	6	2	3	15.2	Low
El Oro	7.5	4	6	0.5	1	19	High
Esmeraldas	8.75	2.1	6	3	3	22.85	High
Guayas	6	10	3	3	3	25	High
Imbabura	0.5	2.8	6	2	3	14.3	Low
Loja	3	1.2	6	2	2	14.2	Low
Los Ríos	0.6	2.8	1	2	3	9.4	Very low
Manabí	2.4	5	10	0.5	2	19.9	High
Morona Santiago	10	2.1	6	1.3	1.3	20.7	High
Pichincha	0.6	7.5	6	2	3	19.1	Moderate
Santo Domingo	1.4	2.1	6	2	3	14.5	Low
Sucumbíos	8.75	2.1	6	3	3	22.85	High
Tungurahua	1.4	2.1	6	2	3	14.5	Low

3 Results and discussion

Our results revealed that the forest management priority is concentrated on Guayas region (Fig. 2), the principal environmental indicator that encourages this behavior is the carbon dioxide emission (44114,44 ton), followed by the deforestation rate (Table 1 and Figure 1). Agricultural and aquaculture expansion has contributed to the deforestation and soil erosion, which release large amounts of carbon retained in the soil. In this area, mangrove zones predominate, which at a global scale can retain almost 1000 t C /ha, approximately three to four times higher than other tropical forest types [23]. Thus, forest management strategies are needed to conserve these remaining forests, an initiative is the implemented plan in 2009 by the National Forest Assessment of Ecuador (ENF, for its initials in Spanish language: Evaluación Nacional Forestal del Ecuador). The main aims of the ENF were to provide data and information for decision-making and development of public policies for Sustainable Forest Management and to meet the requirements for accessing to international carbon market for periodical forest monitoring over the long term [24].

Esmeraldas' forest was ubicated in second place of forest management priority (Fig. 2). In this case, the anthropogenic impact that has affected the most was the deforestation (Table 1 and Fig. 1). A high forest management priority was observed for Manabí province, too (Fig. 2). The results of our study can be related to those exposed in the Homogeneous Deforestation Process Zones project [15], where the deforestation is projected for 2020 based on two variables: stability of the productive landscape (regeneration) during the period 2000-2008 and changes in the population distribution. The resulting projected annual deforestation between 2008 and 2020 for Guayas, Esmeraldas, and Manabí was high (20.35, 13.21 and 8.24 km^2, respectively) due to pasture and permanent crops of cocoa, sugar cane, banana, rice, and corn. In comparison to our findings, these three provinces have a high level of management priority mainly due to the high perpetrated deforestation.

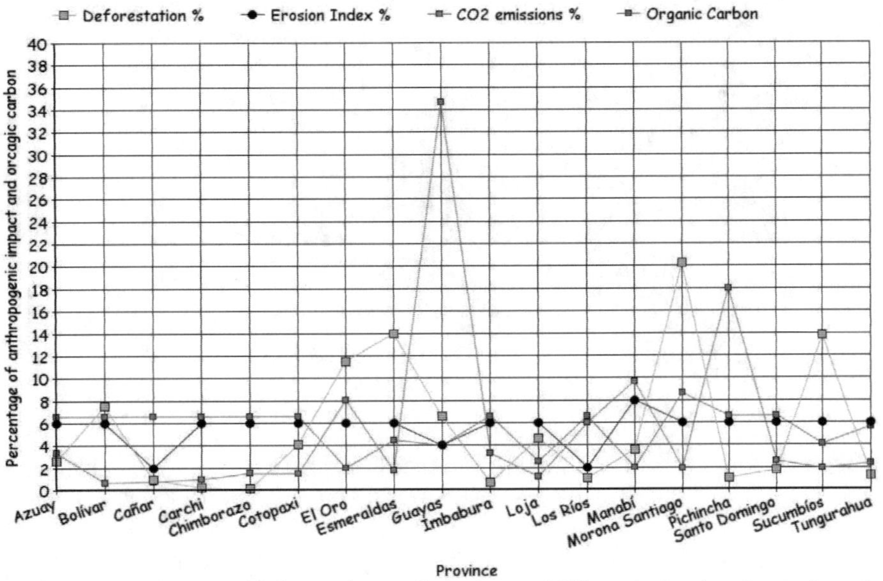

Fig. 1. Anthropogenic impacts (deforestation, soil erosion and CO_2 emissions) and organic carbon absorbed by provinces in Ecuador.

In the Amazon region, Sucumbíos and Morona Santiago show a high management priority (Fig. 2). It can be explained by the deforestation as a result of the illegal logging, agricultural activities, and the oil industries that build roads and exploit petroleum. The importance of forest conservation lies in the fact that this region has the Amazon Lowland Evergreen forest with the highest amount of organic carbon (106.4 ton/ha, Figure and Table 1). In the Homogeneous Deforestation Process Zones project [15] the estimated annual deforestation was high not only for Sucumbíos (123.19 Km^2), but for Morona Santiago (62.27 Km^2).

A moderate management priority belongs to El Oro and Pichincha provinces. The anthropogenic impacts that mostly affect them are deforestation and CO_2 emissions, respectively (Fig. 1). In general terms, the map of forest management priority (Fig. 2) is in line with a spatial model that projects deforested fractions for 2020 [25], where the Coast region (including Guayas, Esmeraldas, and Manabí) has the highest deforested fraction in Ecuador.

Fig. 2. Map of forest management priority according to environmental indicators of anthropogenic impacts over forest carbon storage in administrative units of Ecuador.

4 Conclusions

In accordance with the integrated assessment between the carbon storage capacity of forests and the anthropogenic impacts, we have concluded that the coastal provinces Guayas, Esmeraldas, and Manabí present the highest values 25, 22.85 and 19.9, respectively, which means that the three provinces have a high management priority. A high management priority is also observed in the Amazon region, in the provinces Morona Santiago and Sucumbíos. The Andes or mountains region is mainly characterized by a low management priority. Deforestation, soil erosion, and carbon dioxide emissions were more pronounced on the coast. At the mountains and on the coast, the increase in deforestation was principally due to agriculture and livestock activities. These areas have the potential to retain carbon owing to the high registered carbon storage of 160,4 t/ha in Morona Santiago and 75,8 t/ha in Sucumbíos forests [21]. Furthermore, these forests offer other ecosystem services such as conservation of biodiversity, protection of hydrographic basins and the regulation of regional climates. Thus, it is necessary to conserve them because the stability of forest ecosystem contributes not only to the ecosystem equilibrium of the country, but of the world.

For this reason, we suggest that the government must promote and enhance public policies that encourage access to technical assistance, supplies, improved seeds, and credits. Besides, the intensification of the efficient use of agricultural space to avoid the extensive and unnecessary expansion of the agricultural frontier through deforestation must be promoted. The forest monitoring systems can be enhanced in order to protect the areas with high priority of conservation. In addition, we recommend determining the carbon budget of the forests (absorption - emission) that will allow to understand the past behavior, when deforestation had the highest rate, and in this way, to avoid repeating this behavior in the future.

References

1. G. Magrin, J. Marengo, J. Boulanger, M. Buckeridge, E. Castellanos, G. Poveda, F. Scarano, S. Vicuña, *Climate Change 2014: Impacts, Adaptation, and Vulnerability*. Cambridge University Press, 1499-1566 (2014)
2. F. Pendrill, U. Persson, J. Godar, T. Kastner, D. Moran, S. Schmidt, R. Glob, Env. C., **56,** 1-10 (2019)
3. E. Lugato, P. Smith, P. Borrelli, P. Panagos, C. Ballabio, A. Orgiazzi, A. Jones, Sci. Adv., **4**, 11 (2018)
4. *UNFCCC. Primera contribución determinada a nivel nacional para el acuerdo de París bajo la convención marco de naciones unidas sobre cambio climático. [First nationally determined contribution to the Paris agreement under the United Nations framework convention on climate change].* URL: https://www4.unfccc.int/sites/ndcstaging/PublishedDocuments/Ecuador%20First/Primera%20NDC%20Ecuador.pdf. (2019) Web site (in Spanish).
5. *World Resources Institute (WRI). "Climate Watch".* URL: https://www.climatewatchdata.org/. (2014)
6. L. Tapia, A.Valverde, C. Terán, C. Rosero, V. Placencia, J. Armijos, M. García, P. Sangoluisa, J. Alvear, M. Chinchero, D. Jiménez, *Submission of the Republic of Ecuador to the United Nations Framework Convention on Climate Change. Ecuador's Forest Reference Emission Level for Deforestation.* (Quito, 2015). URL: https://redd.unfccc.int/files/2014_december_frel_submission_ecuador.pdf
7. M. Bertzky, C. Ravilious, A. Araujo, V. Kapos, D. Carrión, M. Chíu, B. Dickson. *Carbono, biodiversidad y servicios ecosistémicos: Explorando los beneficios múltiples.* [Carbon, biodiversity and ecosystem services : Exploring multiple benefits] (UNEP-WCMC, Cambridge, Reino Unido, 2011). (in Spanish).
8. S. Llerena, A. Tarko, A. Kurbatova, P. Kozhevnikova. *E3S Web of Conferences*, **96**, 02002 (EDP Sciences, Les Ulis, 2019).
9. A. Tarko, A. Kurbatova, S. Llerena. *E3S Web of Conferences*, **116**, 00090. (EDP Sciences, Les Ulis, 2019).
10. A.I. Kurbatova, A.M. Tarko. *IOP Conference Series: Earth and Environmental Science*, **272(2)**, 022003 (IOP Publishing Ltd, Bristol, 2019)
11. A.I. Kurbatova, A.M. Tarko, H.A. Qdais, E.A. Grigorets, P.V. Kozhevnikova. *International Multidisciplinary Scientific GeoConference Surveying Geology and Mining Ecology Management, SGEM,* **19(3.2)**, 633-640 (SGEM, Sofia, 2019).
12. L. Morgan, S. Maxwell, F. Tsao, A. Wilkinson, P. Etnoyer, *Marine priority conservation areas: Baja California to the Bering Sea*, 125 (Commission for Environmental Cooperation of North America and the Marine Conservation Biology Institute, Montreal, 2005).
13. S. Alvarez, A. Rubio, Rev. Chap., **19**, 125-137 (2013).
14. C. Cruz-Huerta, M. González-Guillén, T. Martínez-Trinidad, M. Escalona-Maurice, iFor.-Biogeo. and For. **10,** 6, 923 (2017).
15. M. Castro, R. Sierra, O. Calva, J. Caamacho, F. López, *Zonas de Procesos Homogéneos de Deforestación del Ecuador: Factores promotores y tendencias al*

2020. [Homogeneous Deforestation Process Zones of Ecuador: Promoting Factors and Trends to 2020] (Quito, 2013) (in Spanish).

16. V. Dmitriev, N. Kaledin. Balt. Reg., **8**, 2: 87-98 (2016).
17. V. Dmitriev, V. Kulesh, Y. Sergeev. *IOP Conference Series: Earth and Environmental Science* **302(1)** (IOP Publishing Ltd, Bristol, 2019).
18. *MAE. Estadísticas de Patrimonio Natural, datos de bosques, ecosistemas, especies, carbono y deforestación del Ecuador continental. [Natural Heritage statistics, data on forests, ecosystems, species, carbon and deforestation of continental Ecuador].* URL: http://suia.ambiente.gob.ec/documents/10179/346525/ESTADISTICAS +DE+PATRIMONIO+FINAL.pdf/b36fa0a7-0a63-4484-ab3e-e5c3732c284b. (2015) (in Spanish).
19. S. Jiménez. *La erosión del suelo en el Ecuador. Estudio científico de los indices de erosión en el Ecuador .Observatorio de política ambiental.* (2011) (in Spanish).
20. *Instituto Nacional de Estadísticas y Censos (INEC). Información Ambiental de los Hogares. [Household Environmental Information].* URL: https://www.ecuadorencifras.gob.ec/documentos/web-inec/Encuestas_Ambientales/ Hogares/Hogares_2015/DOCUMENTO_TECNICO_ENEMDU_MODULO_AMBIENTAL_2015.pdf. (2011) (in Spanish).
21. *FAO. Evaluación de los recursos forestales mundiales 2015. [Global forest resources assessment 2015].* (Roma, 2015). URL: http://www.fao.org/3/a-i4808s.pdf (2015) Web site. (in Spanish).
22. G.G. Azgaldov, A.V. Kostin, V.V. Sadovov, *Kvalimetriya dlya vsekh [Qualimetry for all]*, 165 (2012) (in Russian).
23. S. Hamilton, J. Lovette., PLoS ONE, **10**, 3 (2015).
24. D. Segura, D. Jiménez, M. Chinchele, J. Iglesias, A. Sola. *XIV Congreso Forestal Mundial* (Durban, 2015) (in Spanish).
25. R. Sierra., *Patrones y factores de deforestación en el Ecuador continental, 1990-2010. Y un acercamiento a los próximos 10 años. [Deforestation patterns and factors in continental Ecuador, 1990-2010. And an approach to the next 10 years]. Conservación Internacional Ecuador y Forest Trends* (Quito, 2013). URL: https://www.forest-trends.org/wp-content/uploads/2013/03/rsierra_deforestacionecuador1950-2020_180313-pdf.pdf (in Spanish)

Economic efficiency of the water monitoring system in Nigeria

Nnabueze Kalu Nwankwo, and *Yulia* Zakirova*

Peoples' Friendship University of Russia (RUDN University), Faculty of Ecology, 6 Miklukho-Maklaya St, Moscow, 117198, Russian Federation

Abstract. Currently, many areas in Nigeria face strong water stress, and to reduce the pressure exerted on water, making water a reusable ecosystem service is essential. Economic efficiency of a water monitoring system can involve having a program to influence and manage water resources and the environment based on the utilization of modern information technologies and techniques. This article aims at demonstrating the importance of Economic efficiency of a water monitoring system, cases and deaths from water-related diseases by decade. A significant factor of managing water safety is understanding water availability, demand, utilization, how it acts in response to change, climate and pressures. Comprehensive understanding of a water system correctly and deeply depends on professionalism, monitoring, documentation and reporting of water system parameters. Human health risk assessment model was applied to estimate cases and deaths from water-related diseases.

1 Introduction

Natural systems including human economic and social activities are influenced by water in the perspective of a river basin. Water systems are designed by physical basins, climate systems, human use, and environmental changes [1]. Economic efficiency takes place when the cost of producing a given output is as low as possible [2]. Economic efficiency of monitoring system allows for a proactive response, data security, information gathering and the overall good health of people. It prevents unpleasant incidents and when these incidents occur, they are identified faster, which saves lives, time and money. Ecologists, Environmentalists and Economists supporting values on Eco-system service such as water have a different opinion to whether price and fares should be introduced for water protection and who is to pay for the cost of the protection. These various opinions are a result of consideration of the triple bottom line, which differs from traditional reporting frameworks and includes ecological, economic and social aspects that are often difficult to measure.

* Corresponding author: yulia_nigma@mail.ru

2 Methods

The U.S. Environmental Protection Agency (EPA) health risk assessment methodology was followed for this study. This method focuses on assessing the risk of certain behaviours and exposures of a specific population by categorizing uncertainties, monitoring exposure to a specific infection, and evaluating the characteristics of infection. The evaluation consists of three steps: hazard identification, exposure assessment, and risk characterization, in this case, all based on pre-existing literature. We then carried out the calculation on Death-to-Case Ratios for a target population, estimating years of potential life lost and the number, percentage, and ranking of deaths from water-related causes of death.

2.1 Data Collection

Most of the data used in the assessment were collected by document review. Data related to water infection identification and documents from authorities such as the World Health Organization.

2.2 Data Analysis

Two main data analysis methods were used in this study: risk characterization and risk comparison. Risk comparison refers to the evaluation of the risk of exposure to each identified infection as well as collectively estimating and comparing the health risk. Risk characterization refers to the analysis and summarizing of data collected in each step of the risk assessment.

3. Results

3.1 Hazard Identification
In this step, toxicokinetic and toxicodynamic data were collected for water-related infections identified.

3.1.1 Toxicokinetics

Most of the identified infections are typically absorbed during water metabolism from the digestive tract. All identified infections can travel to different parts of the body once absorbed. There isn't limited information about how the infection is distributed in the human body. Some water infections are life-threatening while some do not pose a significant health-threatening risk. Most infection usually accumulates instead of going through any specific metabolic processes in the human body. Most identified infections can't be excreted by the renal system.

3.1.2 Toxicodynamics

All identified water infections may contribute to different types of adverse excretory outcomes. Diarrhoea has been identified as having a significant impact on the digestive and excretory system.

3.2 Exposure Assessment

In this study, the number of deaths from estimated water-related causes of death, all ages in Nigeria, 2017 was 226,484. The data considered in the assessment are based on studies that were reviewed by the World Health Organization.

Based on studies, out of the infections identified cholera, typhoid and dysentery were found to have more than 87% of deaths from estimated water-related causes of death. The infection levels observed were considered to pose a significant risk to populations within the ages of 5 to 14 and 25 to 34. On the other hand, the government's incompetency makes the infection rate more complicated as little or nothing is done to provide clean water to communities deprived of access to clean water.

3.3 Risk Characterization

Based on all findings above, all infection identified calls for concerns. The infection levels are higher than the yearly projections of deaths from water-related death. Water-related infection is linked to major types of adverse health concerns such as cancer, digestive diseases and excretory diseases. Assuming a daily consumption of contaminated water, all identified chemicals' maximum exposure levels (arsenic, asbestos, radon, agricultural chemicals, and hazardous waste) would exceed the guideline levels.

4 Economic Value and limitations in Water Protection Nigeria

4.1 Current state and scenarios

In general, the economic value of water ranges from socio-economic value to environmental value such as providing habitable space for biodiversity. Limitations regarding a fully developed water monitoring system include government intervention, price instability, and steady changes in the climate. In many regions, the price of water varies from year to year due to the cost of production, which is mainly based on the price of fuel. Water degradation has cost effects in Nigeria and on Nigerians, this goes a long way affecting the gross national product (GNP). In general, the degradation of water and water services contributes to a reduction in GNP, regional crisis and increases the infant mortality rate. Priority factors estimated in the evaluation of water as ecosystem service includes health risk, environmental water risk and economic impact assessment (Table 1).

Table 1. Estimated number of cases and deaths from water-related diseases by decade — Nigeria, 1940–2019.

Decade	Number of new cases	Number of deaths	Death-to-case ratio (× 100)
1940-1949	1,956,000	1,034,659	52.90
1950-1959	2,120,001	1,182,794	55.79
1960-1969	3,509,535	1,942,134	55.34
1970-1979	2,744,213	1,320,600	48.12
1980-1989	2,265,120	1,131,999	49.98
1990-1999	2,031,117	985,910	48.54
2000-2009	1,919,742	910,711	47.43
2010-2019	1,801,924	800,875	44.45

In Nigeria in 2018, a total estimate of 598,543 deaths occurred. The estimated population was 196,752,840 [8]. The crude mortality rate in 2018 was, therefore, (598,543÷196,752,840) × 100,000 with a total of 304.21 deaths per 100,000 population.

In 2018 a total estimate of 198,329 deaths was attributed to water-related diseases, yielding a cause-specific mortality rate of 100.80 per 100,000 population. In addition, an estimated total of 130,000 deaths happened among persons aged 4–7 years, and the age-specific mortality rate was 419.35 per 100,000 4–7-year old (Table 2).

Table 2. Estimated deaths attributed to cholera or typhoid by age Group in Nigeria, 2017.

Age group (years)	Population (× 1,000)	Number of cholera deaths	Number of typhoid deaths
0-4	11,861	1510	1821
5-14	31,523	1001	2353
15-24	29,251	723	1332
25-34	33,203	822	1221
35-44	30,117	731	1012
45-54	15,248	812	942
55-64	20,022	903	750
65+	20,611	810	782
Total	191,836	7312	10213

Table 3. Estimated deaths and years of potential life lost attributed to typhoid by age group in Nigeria, 2017.

Age group (years)	Deaths	Age midpoint	Years to 65	YPLL
0-4	1821	2.5	62.5	113,813
5-14	2353	10	55	129,415
15-24	1332	20	45	59,940
25-34	1221	30	35	42,735
35-44	1012	40	25	25,300
45-54	942	50	15	14,130
55-64	750	60	5	3,750
65+	782	-------	--------	--------
Total	10213			389,083

The calculation of years of potential life lost (YPLL) and YPLL rates.

The data in Tables 2 and 3 were used to calculate the Typhoid-related mortality rate for all ages, the mortality rate for persons under age 65 years, YPLL, and YPLL rate.

1. ***Typhoid related mortality rate, all ages*** = (10213÷191,836,000) × 100,000 = 5.3 typhoid deaths per 100,000 population
2. ***Typhoid related mortality rate for persons under age 65 years***

$$\left[\frac{1821+2353+1332+1221+1012+942}{(11861+31523+29251+33203+30117+15248+20022)} \times 10^5\right]$$

$$= \left[\frac{8681}{171225}\right] \times 10^5$$

= 5,069.9 Typhoid deaths per 100,000 persons under age 65 years. (1)

3. Typhoid related YPLL

a) Calculating the midpoint of each age interval. The midpoint of the age group 0–4 years is (0 + 4 + 1) ÷2, or 5 ÷2, or 2.5 years. The same is repeated for all age groups.

b) Subtracting the midpoint from the endpoint will determine the years of potential life lost for a particular age group. For the age group 0–4 years, each death represents 65–2.5= 62.5 years of potential life lost. The same is repeated for all age groups.

c) To calculate age-specific years of potential life lost, the number of deaths in a given age group is multiplied by its years of potential life lost. For the age group 0–4 years: 1821 deaths x 62.5 =113,813

d) The total years of potential life lost attributed to typhoid in Nigeria in 2017 were 389,083.

4. Typhoid related years of potential life lost rate:

= years of potential life lost at age 65 rate in Nigeria;

= years of potential life lost divided by population to age 65.

$$= \left[\frac{389083}{171225} \right] \times 10^3$$

$$= 2{,}272.3 \text{ YPLL per 1,000 population under age 65} \qquad (2)$$

Annual freshwater withdrawals in Nigeria total 12.470 m³ bn in Dec 2010. This records an increase from the previous number of 11.570 m³ bn for Dec 2005. Nigeria's annual freshwater withdrawals data is updated yearly, averaging 10.940 m³ bn from Dec 1987 to 2010, with 4 observations (Fig.1, 2)

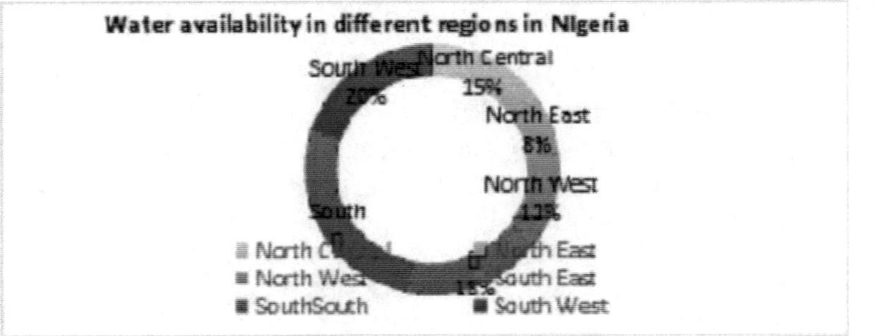

Fig. 1. Water availability in different regions in Nigeria.

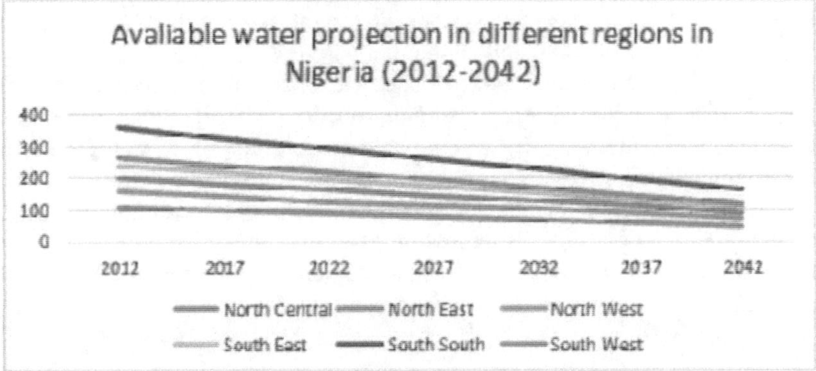

Fig. 2. Available water projection in different regions in Nigeria (2012-2042).

Several regions in Nigeria are facing a severe water crisis. Data reveals that the crisis is getting worse by the day. The per capita water availability in the country has fallen by 73% in 55 years from 1962 to 2017 (Fig.3)

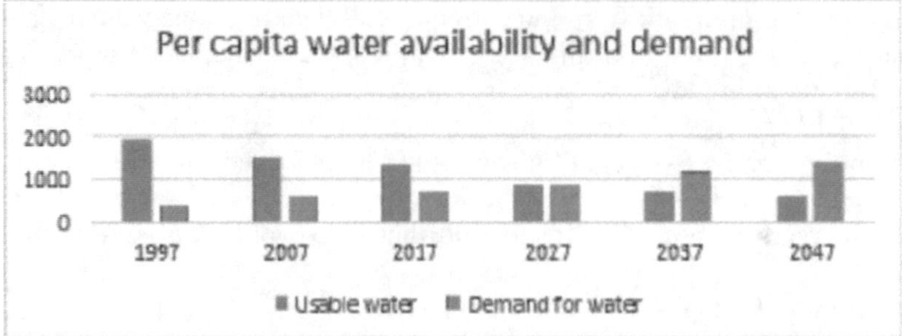

Fig. 3. Per capita water availability and demand.

The demand for water for various uses in Nigeria is estimated to be about 700 BCM, 1200 BCM, and 1400 BCM by the year 2017, 2037 and 2047 correspondingly. According to an ongoing temperature analysis conducted by scientists at NASA's Goddard Institute for Space Studies (GISS), the average global temperature on Earth has increased by about 0.8° Celsius (1.4° Fahrenheit) since 1880. Two-thirds of the warming has occurred since 1975, at a rate of roughly 0.15-0.20°C per decade [3].

4.2 Economic Limitations in Water Protection Nigeria

Limitations regarding a fully developed water monitoring system include government intervention, price instability, and steady changes in the climate. In general, the fundamentals of water supply are based on expectations, and if the outcome of the expectations is wrong, the excessive supply or scarcity cannot be revised. In many regions, the price of water varies from year to year due to the cost of production, which is mainly based on the price of fuel. Providing prices cover operating costs and cost of water distribution, it benefits water supply companies to carry on with production and supply if prices for water drop to a very low level.

The instability of portable water prices is accompanied by the instability of public income. Most times local and state government workers are not paid for several months making it impossible for them to buy clean water. Systems and infrastructures constructed to support and improve availability and predictability of water demand and supply help decrease vulnerability of water in times of droughts and shortages. Most rural communities incline in improvement and development of water infrastructure as alternatives for the sustenance of economic, social and environmental activities.

Within the perspective of climate change, the frequent occurrences of floods and droughts are increasingly changing, causing insecurity in realization and application of more modern water resources infrastructure. Some results of a changing climate are expected to include more frequent spring flooding, shifts in the growing season, and changes in the hydrological cycle. These changes in climate conditions influence social, ecological, economic and physical structures in a way that makes a more active approach to be needed for water management to ensure that demands for water in different regions in Nigeria are reached.

5 Conclusion

Economic efficiency of water monitoring system is greatly influenced by the water management system. Despairingly, the Nigerian government fails to live up to its policies and promises. Civil servants, poor people and forgotten rural communities majorly don't have access to good quality water. Professionals, policymakers, stakeholders and planners need to create a platform for decisions about activities concerning water usage and distribution, waste disposal, use of fertilizers, pesticides and insecticides in the farm as well as to decide on the allocation of ecosystem service such as water resources to different regions. Decisions and balance on water usage have to be made considering the three-bottom line that is social, environmental and economic to recognize and categorize tolerable tradeoffs.

References

1. D.R. Carter Borden, *Water Quality Monitoring System Design* (International Institute for Sustainable Development, 2015). URL: https://www.iisd.org/sites/default/files/publications/water-quality-monitoring-system-design.pdf
2. *Earth observatory* (Nasa earth observatory 2010). URL: https://earthobservatory.nasa.gov/world-of-change/decadaltemp.php
3. *Cholera – Nigeria* (World Health Organization, 2017). URL: https://www.who.int/csr/don/12-july-2017-cholera-nigeria/en/
4. *The situation of women and children in Nigeria* (Unicef, 2018) URL: https://www.unicef.org/nigeria/situation-women-and-children-nigeria
5. Wise, *Methodologic Alternatives for Measuring Premature Mortality* (Centers for Disease Control and Prevention 2004). URL: https://www.cdc.gov/csels/dsepd/ss1978/Lesson3/Section3.html#_ref8
6. O.A. Kabiru, O.O. Akeeb, B.M. Wasiu, A.I. Bamidele, S.A. Edward & O.A. Samuel, *Typhoid Fever: Tracking the Trend in Nigeria* (PubMed, 2018). URL: https://www.ncbi.nlm.nih.gov/pmc/articles/PMC6128359/

E3S Web of Conferences **169**, 02017 (2020) https://doi.org/10.1051/e3sconf/202016902017
APEEM 2020

Determining the environmental situation as a factor in the differentiation of housing prices in Nur-Sultan by using the Analytic Hierarchy Process

Kamila Akhmedinova*

Peoples' Friendship University of Russia (RUDN University), Faculty of Ecology, 6 Miklukho-Maklaya St, Moscow, 117198, Russian Federation

Abstract. Modern cities, being centers of intellectual activity, trade, culture, science and development, are becoming places of attraction for people. It is in urban conditions that the economic and social evolution of man takes place. At the same time, as urbanization grows, a number of problems arise, including environmental ones. Therefore, it is necessary to take environmental factors into account when planning urban development, housing construction and creating infrastructure. This study is devoted to the application of the method of hierarchy of stochastic systems using the technology of geographic information systems, allowing to assess the adequacy of the cost of housing in relation to the environmental situation in the city of Nur-Sultan.

1 Introduction

Environmental problems in the modern world are becoming widespread, thereby emphasizing relevance now. The ecological situation, especially in large urban agglomerations, is dangerous for the townspeople due to non-compliance with environmental standards. Consequently, the priority policy direction is the efficient use of land resources and the corresponding consideration of environmental factors in determining the prices of real estate [1]. The environmental factors in the context of this work are natural and natural-anthropogenic factors that are not means of labor, commodities or sources of energy and raw materials, but that directly affect the efficiency and usefulness of using the property.

Thus, the environmental factor becomes the most important factor in the pricing of real estate and takes on a new meaning in a new post-industrial society. Modern trends, in this context, serve as an additional indicator for determining the development of society, since environmental education is an integral part of a new stage of development [2].

* Corresponding author: akhmedinovakamila@gmail.com

2 Methods

The aim of the work is to study the relationship between environmental load and property value. For this, the Pearson Correlation Criterion will be used.

The main tasks set to achieve the goal:

- Differentiation of habitats obtained using the Analytic Hierarchy Process
- Identification of the main types of pollution of environmental components.

The AHP method allows to structure and understand in a clear and rational way a complex decision-making problem in the form of a hierarchy, to compare and quantify alternative solutions. In our particular case, compare the areas of the city by environmental characteristics. AHP application procedure:

- Construction of a qualitative model of the problem in the form of a hierarchy that includes a goal, alternative options for achieving the goal and criteria for assessing the quality of alternatives.
- Prioritization of all elements of the hierarchy using the method of pairwise comparisons.
- Synthesis of global priorities of alternatives by linearly folding the priorities of the elements in the hierarchy.
- Verification of judgments for consistency.
- Making decisions based on the results obtained.

Table 1. Initial value of factors.

	Stationary sources of pollution, t / year	Mobile sources of pollution, t / year	Hydrosphere Pollution / MPC	Soil pollution, Kc	Radiation pollution,Sv	Noise pollution, dB	Mechanical pollution, S
Area 1	x	x	x	x	x	x	x
....	...	...	...	...	...	...	...
Area N	N	N	N	N	N	N	N

Having determined the main goal, we proceed to a pairwise comparison of environmental factors. It is also important to accept the following ratios of dominant factors [3, 4].

The radiation factor has (RP):

- absolute preference over soil contamination factor;
- obvious preference over mechanical contamination;
- a preference, a compromise between the essential and the obvious, over the noise pollution factor;
- significant preference over the hydrosphere pollution factor;
- preference, a compromise between insignificant and substantial, over the stationary pollution factor;
- a preference, a compromise between insignificant and substantial, over the factor of mobile pollution.

The stationary source factor has (SSP):

- absolute preference over soil contamination factor;
- obvious preference over mechanical contamination;
- preference, a compromise between the insignificant and the obvious, over the factor of mobile sources of pollution;
- a preference, a compromise between the essential and the obvious, over the pollution factor of the hydrosphere;

- obvious preference over noise pollution;

The factor of mobile sources of pollution has (MSP):

- a preference, a compromise between the essential and the obvious, over the pollution factor of the hydrosphere;
- absolute preference over soil contamination factor;
- obvious preference over noise pollution;
- obvious preference over mechanical contamination.

The hydrosphere pollution factor has (HP):

- obvious preference over noise pollution;
- slight preference over soil contamination;
- preference, a compromise between the essential and the obvious, over the factor of mechanical pollution.

The noise pollution factor has (NP):

- obvious preference over soil contamination;
- preference, a compromise between the essential and the obvious, over the factor of mechanical pollution.

The mechanical pollution factor has (MP):

- obvious preference over soil contamination (SP)

Based on the above, we can compare the weight of factors:

Table 2. The weight of factors.

Factors	SSP	MSP	HP	SP	RP	MP	NP	Op	Exp (1/9)	Value
SSP	1	2	3	7	0.5	0.5	6	63	1.8069	0.2082
MSP	0.5	1	3	6	0.5	2	6	54	1.7676	0.2037
HP	0.3333	0.3333	1	2	0.2	2	3	0.2663	0.8278	0.0954
SP	0.1428	0.1665	0.5	1	0.1428	0.3333	0.5	0.0002	0.2963	0.0342
RP	2	2	5	7	1	5	5	1400	2.8136	0.3243
MP	2	0.5	0.5	3	0.2	1	2	0.6	0.7573	0.0874
NP	0.1666	0.166	0.333	2	0.2	0.5	1	0.0018	0.4055	0.0467
	6.1427	6.1665	13.3333	28	2.7428	11.3333	23.5		8.675	1

3 Results and discussions

After a pairwise evaluation of factors, the product of the values of their evaluation is calculated in rows, and then the root of the degree to which the number of factors is estimated is extracted from the obtained value. The weight of the factor is determined by dividing the value obtained by extracting the root by the sum of these valuesfor all factors. As a result, the sum of the weight of all factors must be equal to unity [5].

To control the accuracy of factor comparisons, the Consistency Indicator (CI) is calculated as the sum of the products of the sum of the values of the factor assessment in the row and column for each factor. Then, the consistency index (CIndex) is calculated as the quotient of dividing the difference between the CI and the number of factors by the difference in the number of factors and units.

Next, the consistency ratio (CR) is calculated as the quotient of dividing the Cndex by a random index (RI). CR should be less than 0.1; otherwise, it is necessary to check the correctness of the assessment of factors. RI depends on the number of factors and is selected from table 3.

Table 3. Random Index Values.

Number of factors	1	2	3	4	5	6	7	8	9	10
Random index	0	0	0.58	0.9	1.12	1.24	1.32	1.41	1.45	1.49

Since the CR obtained is <0.1, the estimates do not have significant contradictions and can be taken for further calculations [6].

In order to make a final decision on the choice, it is necessary to transfer the value of priority vectors from all tables for each factor to the final table and calculate the global priority. It is defined as the sum of the products of the factor weighting by the criterion priority factor.

Based on the above methodology and taking into account all factors, the following map was compiled for all factors, both intermediate and informative (Fig. 1).

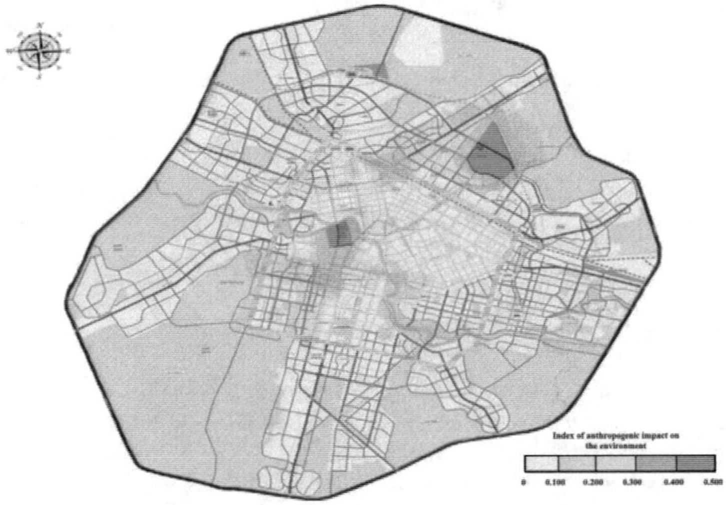

Fig. 1. Index of ecological load on natural environmental components in Nur-Sultan (according [7, 8, 9, 10, 11]).

As a result, taking into account all factors, we obtain the full environmental load in the city of Nur – Sultan. The map was overlaid with a map of the cost per square meter, which we received using open sources (Krysha.kz).

For representativeness of the analysis, we used the BCG diagram of B. Henderson [12]. The data are presented in the fig. 2 [13].

Results were highlighted in 5 categories:

- 1 category - groups C-I, D-1, D-II. This category includes a high pollution index and low property value. Only 2 districts of the city fall into category 1, which is 4.2%.

- 2 category - groups D-III, D-IV, C-IV. This category includes a high pollution index and the highest value of real estate. In the 2nd category, only 1 district

- 3 category - groups B-IV, A-IV, A-III. This category is characterized by the minimum values of the pollution index and the highest cost per square meter. 3 microdistricts of the city belong to the 3 category, 6.25% - Med.gorod,Astana Arena, and EXPO.

- 4th category - groups A-I, A-II, B-I. This category is characterized by the lowest values of the pollution index and the lowest cost per square meter. This category includes 16 microdistricts of the city, which is 33.3%.

-5 category - groups B-II, C-II, C-III, C-IV. This category is characterized by average pollution index and cost per square meter. This category includes 26 microdistricts or 54.2%. This is the main category [14].

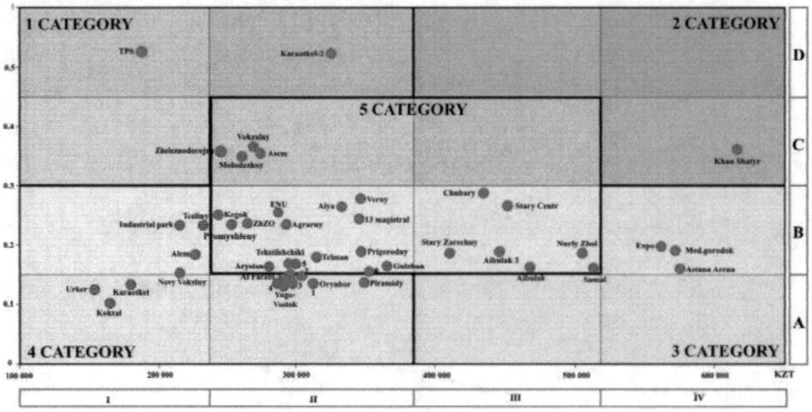

Fig. 2. A diagram of the relationship between the pollution index and the average cost per square meter in different microdistricts of Nur-Sultan.

The use of the Analytic Hierarchy Process allowed us to conclude the following. Only 10.4 % of cases the state of the environment was adequate to the environmental situation, and high real estate prices coincided with the optimal ecological characteristics of the area [15].

4 Conclusions

It can be concluded that at present the environmental situation plays a weak role in the pricing of Nur-Sultan real estate market, while prestige is an important factor. It should also be noted that the impact of the environmental factor is often limited to "visible" environmental problems rather than "invisible" ones. Therefore, for example, the contribution of radiation exposure is levelled out and, on the contrary, the perception of industrial enterprises, thermal power plants, etc. is aggravated.

The city of Nur-Sultan is positioning as a city of the future, but a necessary condition for achieving this result is the creation of a safe and secure environmental situation. For this, some recommendations for solving the main environmental problems in the city can be offered.

- Tightening the legal framework in the field of environmental protection is necessary, as well as the stable development of public authorities in the field of environmental safety.
- The condition of "invisible" factors, such as the radiation environment, should be monitored and the results made available to the public.
- The transition to more environmentally friendly vehicles can help to resolve a problem of emission. However, the low standard of living of the population will not allow this in the near future.
- It is necessary to increase in density of green spaces, primary along the roads.

The influence of the environmental factor is often limited by the action of “visible” environmental problems than “invisible” ones. Therefore, the contribution of radiation exposure is leveled and, on the contrary, the perception of industrial enterprises, thermal power plants, etc. is sharpened.

References

1. D. L. Volkov, *Real Estate Economics and Finance*, 32 (SPbGU, St. Petersburg, 1999) (in Russian)
2. A.V. Sevost'yanov, *Economic valuation of real estate and investments* (Akademiya, Moscow, 2008) (in Russian)
3. M. Kac and J. Logan, *In Fluctuation Phenomena*, eds. E.W. Montroll & J.L. Lebowitz (North-Holland Publ.Co, Amsterdam, 1979)
4. E. Nelson, *Quantum Fluctuations* (Princeton University Press, Princeton, 1985)
5. Wei G.W, Planck equation. J. Phys. A: Mathematical and General, **33**, 4935– 4953 (2000)
6. M.E. Shaikin. J. Automation and Remote Control, **63**:5, 766-776 (2002)
7. The newsletter on the state of the environment of the Republic of Kazakhstan for 2018 Available at: https://kazhydromet.kz/ecology/informacionnye-byulleteni-o-sostoyanii-okruzhayuschey-sredy-respubliki-kazahstan
8. The newsletter on the state of the environment of the Republic of Kazakhstan for 2019 Available at: https://kazhydromet.kz/ecology/informacionnye-byulleteni-o-sostoyanii-okruzhayuschey-sredy-respubliki-kazahstan
9. J.N. Zaturanov, T.N. Antipova, Scient.J. NRU ITMO, **1**, 15 (2013)
10. Ko Wang, Marvin L. Wolverton *Real Estate Valuation Theory* (Springer US. Kluwer Academic Publishers, 2002)
11. N.I Ivanov, M.V. Butorina, N.N. Minina, Vestnik MGSU **3-**1 (2011)
12. B. Bultena, B. Grünbaum and F. Ruskey, J., Comp.sciense, Math. CCCG (1999)
13. N. J. A. Sloane and S. Plouffe, *The Encyclopedia of Integer Sequences* (Academic Press, New York, 1995)
14. D. Hubbard *How to Measure Anything: Finding the Value of Intangibles in Business* pg. 46 (John Wiley & Sons, 2007)
15. P. Silaeva, K. Akhmedinova, M. Redina, A. Khaustov. *18th International Multidisciplinary Scientific GeoConference SGEM2018*, **18,** 727-732 (2018)

Palaeogeographical heritage of the Permian Period in the network of protected areas and museums of the Perm Region

Yuliya Glazyrina[1,2 *], *Larisa* Zhuzhgova[3], *Sergey* Buzmakov[1], and *Pavel* Sannikov[1]

[1]Perm State University, Biogeocenology and Nature Protection Department, 614000 Perm, Russia
[2]Perm Regional Museum, Natural History Department, 614000 Perm, Russia
[3]Perm Regional Museum, Natural History Collections, 614000 Perm, Russia

Abstract. Article presents an overview of palaeogeographical heritage of the Permian Period in the network of protected areas and museums of the Perm Region. Permian Period, received its name after the city of Perm in Russia, includes the so-called "Main Permian Field". It contains two types of palaeogeorgaphical heritage: marine and terrestrial. The representation of palaeogeographical objects of different types is analysed for the geoheritage of the protected areas *in situ*. Geoheritage *ex situ*, in the network of the regional museum collections, is estimated from the representative aspects. Recommendations on optimization of the protected areas network and museum acquisition is given from the point of view of paleo-geographical heritage objects representation.

1 Introduction

Objects of palaeogeographical heritage of the Permian Period in the territory of the Perm Region has the international significance as they refer to the so-called "Main Permian Field", sediments that have become a basis for the Permian Period discovery, now included into the International Stratigraphical and Geochronological Timescales [1, 2, 3]. Permian System and Period was named after the city of Perm in Urals, Russia, by the Scottish geologist Sir Roderick Murchison in 1841 [4, 5, 6]. The Permian (299-251 million years ago) is the time of the of terrestrial and marine sedimentation process, which is described all over the world. The most significant of the Permian deposits and outcrops, including boundaries of stratigraphic subdivisions and Global Stratotype Section and Point (GSSP) are protected areas.

Permian outcrops in the territory of the modern-day Perm Region are included into the regional network of the protected areas. It refers to the palaeogeoraphical heritage *in situ*, while geological and palaeontological natural history museum collections across the Perm Region represents geoheritage *ex situ*. To provide the most efficient model of geoheritage research, preservation and presentation there should be a balance rather than a substitution of both.

* Corresponding author: glazyrina_yuliya@mail.ru

1.1 Geographical distribution of Permian deposits in the world

The Permian Period was named after the city of Perm, Western Urals, Russia. This name is a geographical toponym, given by the Scottish geologist Sir Roderick Impey Murchison in 1841, after a 20,000-km-long expedition across the European part of Russia and the Ural Mountains [2, 5, 6, 7]. Since late 1800s, the Permian had been widely used in stratigraphical nomenclature.

For quite some time, the Permian was divided into two parts: Lower and Upper Permian. Currently, a tripartite subdivision is used – the Lower Permian, the stratotype of which is situated in the western Urals, Middle Permian with the stratotype in the southwestern United States and the Upper Permian with the stratotype in southern China.

Permian terrestrial deposits are generally correlated with the former Pangaea supercontinent, while evaporites and marine deposits connected to lagoons and shallow seas tend to be found along its margins. Permian deposits are observed on all continents, with the "Main Permian Field" in the Pre-Urals in Russia (mostly Perm Region, as well as Bashkiria, Chelyabinsk Region, Orenburg Region and others).

The GSSP of the Lower Permian, Asselian Stage, is in the valley of the River Aidaralash, near Aqtube, the Ural Mountains [3, 8]. The Subcommission on Permian Stratigraphy of the International Commission on Stratigraphy co-ordinated working groups on the approval of the Permian GSSP. The latest one was the ratification of the Sakmarian-base GSSP (Lower Permian) in the southern Urals, Russia, in 2018 [3, 9].

The Middle Permian GSSP of the Guadalupian was ratified in the early 2000s. All GSSPs, Roadian, Wordian and Capitanian sections, are located in the Guadalupe Mountains National Park, Texas, USA [3, 10].

The Upper Permian, the Lopingian, is based on the GSSPs of the Changhsingian and Wuchiapingian stages were ratified in 2004 and 2005, respectively [1, 3]

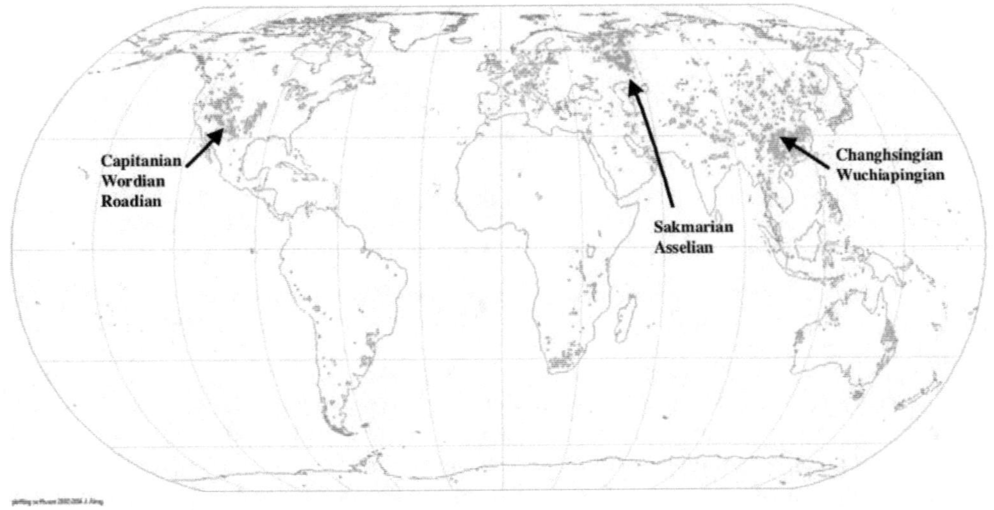

Fig. 1. Map of Permian palaeographical fossil sites around the world, with the GSSP (based on the Fossil Works data base, 2002-2014; accessed February 20, 2020).

1.2 Palaeogeographical and geological heritage in the Perm Region

The deposits of the Permian System are united into the "Main Perm Field", a historically significant area of distribution of sedimentary rocks and fossils of Permian Period, which are cropped out to the surface. Palaeogeographical and geological heritage of the Permian in the Perm Region could be divided into two types of objects. The first type contains objects *in situ*, which are located at the (1) protected areas, (2) territories suggested for the protected areas, (3) territories that contain geoheritage objects but not recommended for the protection. The second type contains objects from the museum collections. They could be divided into two main groups: (1) geological collections, (2) paleontological collections, the last one consists of different types such as palaeoentomological, palaeobotanical, etc.

2 Overview of protected areas for the Permian Period palaeogeographical heritage *in situ*

The network of protected areas in the Perm Region consists of two strict nature reserves of federal level, 259 protected areas of regional significance and 102 protected areas of local significance [10-14].

Protected areas of geological importance in the Perm Region can be divided into several types: tectonic, stratigraphical, palaeontological, cosmological, geomorphological, petrographic, mineralogical, hydrogeological, mining-geological and karstological [14].

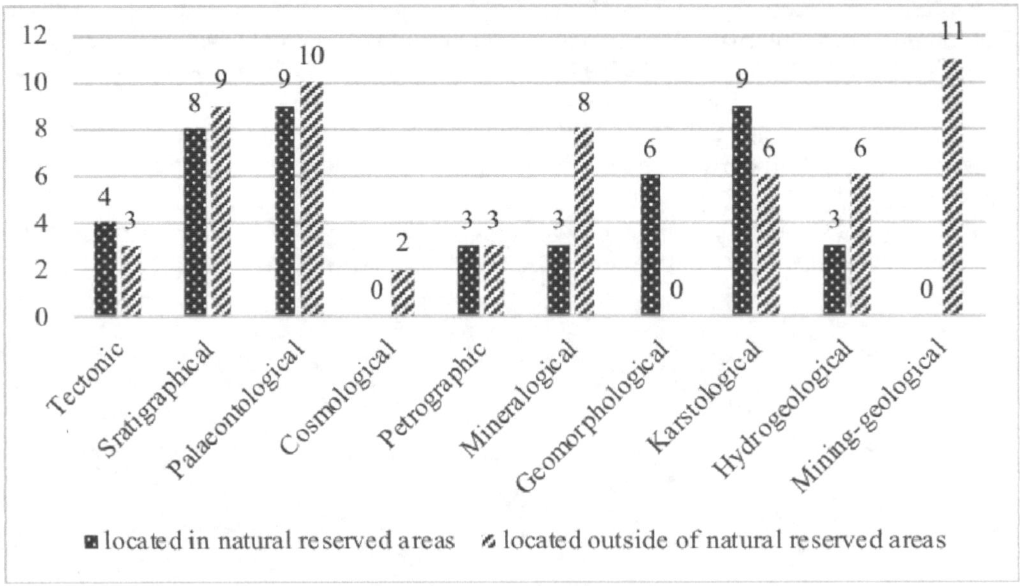

Fig. 2. Palaeogeographical and geological monuments of various types in the network of protected areas [15].

At least 45 of protected areas are devoted to the palaeogeographical and geological heritage, 3/5 of palaeogeographical heritage objects of the Permian Period are included into the paleontological, stratigraphic, mineralogical and karstological monuments. More than half of protected areas, which include objects of the Permian Period, have a specialized geological profile (geomorphological, karstological, etc.), the other part of palaeogeographical heritage objects of the Permian belong to the complex protected areas.

Objects of palaeogeographical and geological heritage of the Permian are classified as two main types of sediments: marine and terrestrial.

2.1 Marine Permian outcrops on the protected areas

Marine sediments are concentrated to the Pre-Ural foreland basin, which stretches sub-meridionally along the Urals. The most significant sections of the Lower Permian are marine sediments: the brine section (*Goniatida, Nautiloidea, Anthozoa*), the Polazna section (*Stromatolites*), and the Chikali section (*Bryozoa*). The lower border of the Asselian is represented in the "Plakun" section. The section "Kyn" includes the Gzhelian stage of the Upper Carboniferous and the Asselian stage of the Lower Permian. The "Filippovsky" quarry represents the Filippovsky horizon of the Kungur stage of the Lower Perm. The section "Holodny Log" is a stratotype of the cold horizon of the Asselian stage of the Lower Permian. The "Belaya Gora" is a hypostratotype of the Sakmarian stage in the Lower Permian. Irene horizon is presented in the section "Elkino" [12, 14, 16].

2.2 Terrestrial Permian outcrops on the protected areas

Continental deposits are allocated towards the western and southern parts of the region. Among the most significant there are fossil sites of the Middle Permian "Chekarda" (palaeoentomology, palaeobotany) and "Ezhovo" (vertebrates) [17-19] (Fig. 3).

3 Museum collections: the Permian Period palaeogeographical heritage *ex situ*

Network of the regional and municipal museums of the Perm Region includes 38 museums, which hold in storages 1,690,330 objects belonging to 15 types of collections. Natural history collections total amount is 73,513 objects, which is 4.3% of the total number.

Collections of the palaeographical significance are geological and palaeontological ones, with the total amount of 11,043 objects, which is 1% from the total amount. Mineralogical collections are represented in 1/2 museums of the region. More than 1/3 of the region's museums complete and present items of the Permian Period (Fig. 4).

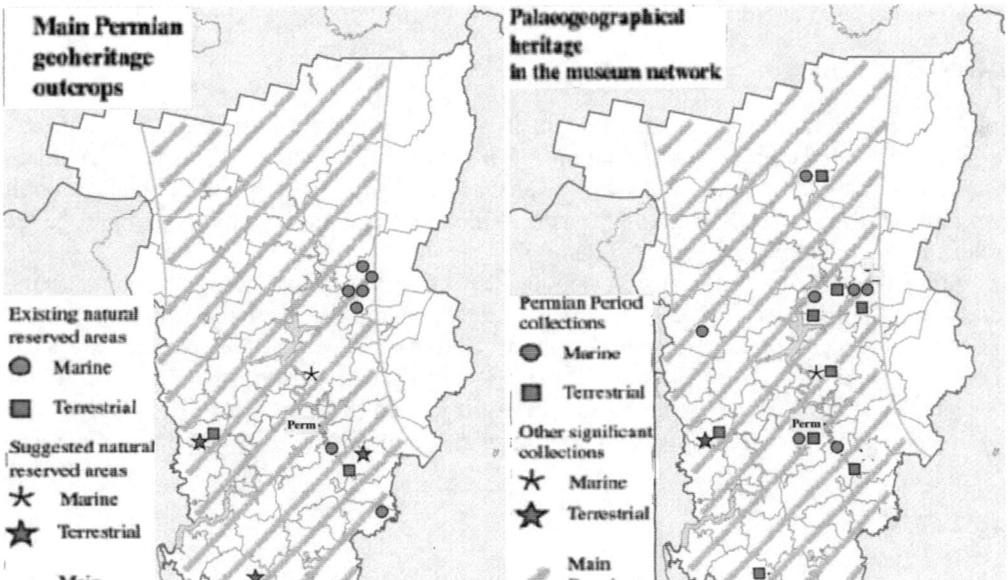

Fig. 3. Permian Period palaeographical heritage in the Perm Region: protected areas network *in situ* and museum network *ex situ*

4 Gaps in the Permian palaeogeographical heritage at the network of protected areas and museums

For more than 40 years, protected areas optimization research is leading by the scientists of the Perm State University (Biogeocenology and Nature Protection Department of the Geographical Faculty and Regional, Oil and Gas Geology Department of the Geological Faculty) and by the Mining Institute of the Ural Branch of the Russian Academy of Sciences. Their joint efforts revise the existing network of the protected areas to analyse the representativeness of the geoheritage preservation efforts and suggest new geosites to be reserved under the regional and national law [12, 15, 17].

Among them are the Middle Permian locality of fossil vertebrates, paleo-flora and paleotopes "Klyuchiki" (Kuedinsky district), Middle Permian locality of tetrapods and fossil flora "Kokuy" (Ochor district), Lower Permian locality "Polazna" (Dobryansky district), continental flora transects along the Barda river (Lysvensky district) [12, 17]. Mining geological monuments, related to the history of the Permian mineral resource development (including salts, evaporites, gypsum, copper mining, etc.), are still not represented in the protected areas network. It is expedient to create a complex protected area, specialized on palaeogeographical heritage of the Permian, based on the UNESCO Global Geopark model of sustainable development [12, 20].

Table 1. Representation of the Permian Period palaeographical heritage in the museums and protected areas of Perm Region

Representation of the Permian in:	Total amount	Including Permian	Percentage
Regional museums	38	14	37 %
Protected areas of regional significance	259	27	10 %

5 Conclusion

Palaeogeographical identity of the Permian Period at the place of its discovery suppose a complex system dynamic representation of the territory's natural heritage *in situ*, on protected areas, and *ex situ*, in museums. Museums contain retrospective documentation of geographical diversity of the territory (natural history collections, holotypes, photo, video archives, materials scientific expeditions, etc.).

Recommendations on optimization of the protected areas network and museum acquisition is given from the point of view of palaeogeographical heritage representation:
(1) To include recommended geoheritage objects into the network of the protected areas.
(2) To strength the representation of palaeogeographical aspect in collections in the museums located within the geographical area of the Main Permian Field deposits.
(3) To complete typological objects of geological collections, including common minerals.

References

1. S.G. Lucas, S. Shen, *The Permian chronostratigraphic scale: history, status and prospectus* (Geological Society, London 2016)
2. V.P. Ozhgibesov, R.J. Kaffi, *Tectonics and palaeogeography of the Main Permian Field and the Perm Basin of the Colorado-Mexican Paraplatform* in *Modeling of geological systems and processes* (Perm State Univ, Perm, 1996) (in Russian)
3. *International Chronostratigraphic Chart*. Version 01/2020.

4. M.J. Benton, A.G. Sennikov, A.J. Newell, Proceedings of the Geologists' Ass., **121**, 313–318 (2010)
5. M. Collie, J. Diemer (eds), *Murchison's wanderings in Russia. His geological exploration of Russia in Europe and the Ural Mountains, 1840 and 1841* (British Geological Survey, 2004)
6. R.I. Murchison, E. de Verneuil, A. von Keyserling. *The Geology of Russia in Europe and the Ural Mountains* (V.1 London, V.2 Paris, 1845).
7. C.R. Scotese, R.P. Langford, *The Permian of Northern Pangea, volume 1, Paleogeography, Paleoclimates, and Stratigraphy*, 3-19 (Springer-Verlag, Berlin, 1995)
8. V.I. Davydov, B.F. Glenister, C. Spinosa, S.M. Ritter, V.V. Chernykh, B.R. Wardlaw, W.S. Snyder, Episodes **21(1)**, 11–18. (1998)
9. *The Sakmarian-base GSSP has been formally ratified* (International Commission on Permian stratigraphy newsletter, 2018-08-25, 2018)
10. *Three GSSP markers settled down in the Guadalupe National Park* (International Commission on Permian stratigraphy newsletter, 2013-07-21, 2013)
11. G.A. Voronov, Zh.A. Chistyakova, S.P. Stenno (eds.). *List of the specially protected and recommended natural territories of the Perm Region as of July 1, 1988* (Perm State University, Perm, 1989). (in Russian)
12. S.A. Buzmakov (ed.) *Atlas of protected areas of the Perm Region* (Aster, Perm, 2017) [in Russian]
13. S.A. Buzmakov, A.A. Zaitzev, P.Yu. Sannikov, Voprosy stepevedenia, **15.** 55-58 (2019).
14. I.I. Tchaikovsky (ed.) *Geological Monuments of the Perm Region.* (Mining Institute of the Ural Branch of the RAS, Perm, 2009) [in Russian]
15. P.Yu. Sannikov, S.A. Buzmakov. *Prospects for the development of protected areas network of Perm Region*. (Perm State University, Perm, 2015)
16. G.Yu. Ponomareva, A.S. Alekseev, T.N. Isakova, O.L Kossovaya, *Middle Urals. Carboniferous and Permian marine and continental successions. Field Trip Guidebook of XVIII International Congress on the Carboniferous and Permian*, 36-47 (Aster, Perm, 2015)
17. S.V. Naugolnych, *Geological Monuments of the Perm Region. Encyclopedia*, 146-196 (Mining Institute of the Ural Branch of the RAS, Perm, 2009) (in Russian)
18. L.V. Zhuzhgova, G.Yu. Ponomareva, D.S. Aristov, S.V. Naugolnykh. *Chekarda is a location of fossil insects and plants from the Permian period. Monograph on the geology; paleobotany and paleoentomology of Chekarda* (Perm State University Press, Perm, 2015) (in Russian)
19. P.K. Tchudinov. *Early Terapsids* (Nauka, Moscow, 1983) (in Russian)
20. B.I. Chuvashov, V.A. Chermnykh, V.I. Kopnin, V.P. Ozhgibesov, P.A. Sofronitsky. *Permian System: guides to geological excursion in the Uralian type localites.* (University of South Carolina, Columbia, 1993)

The role of specially protected natural areas in solving environmental problems of the Russian arctic (using the example of Vaygach Island)

Anna Aleinikova[1,*], *Maksim* Byashkin[2] , *Tatiana* Gaivoron[1], *Galina* Mainasheva[2] and *Natalia* Marsheva[1]

[1]Peoples' Friendship University of Russia (RUDN University), Faculty of Ecology, 6 Miklukho-Maklaya St, Moscow, 117198, Russian Federation
[2]Moscow City University, Institute of natural science and sport technology, Chechyulina St 1, 105568, Moscow, Russia

Abstract. The Russian Arctic is a unified physical and geographical system of land and sea areas of Northern Eurasia, within which unique natural and historical and cultural objects are concentrated. The system of specially protected natural areas of the Russian Arctic preserves vulnerable arctic, tundra landscapes under conditions of various anthropogenic activities. In the face of modern climate change, monitoring of the transformation of the Arctic landscapes in specially protected natural areas is important. The landscape map compiled as a result of our own field and office studies shows 26 natural complexes. The most interesting for ecological tourism is the north-western part of the island. .By the example of Vaygach Island, the natural and historical-cultural possibilities of organizing a protected area, including the solution of environmental problems, are considered.

1 Introduction

Environmental problems in the Russian Arctic are caused by both natural and anthropogenic factors.

In the Arctic region, there are signs of global climate change: an increase in the temperature of surface air layers, a decrease in the area of land glaciers and sea ice, and permafrost.

Global climate change is leading to a widespread reduction in the permafrost area in the western Arctic. Cryogenic geomorphological processes – thermocarst, thermoabrasion, thermoerosion, solifluction – are activated [1]. Changes in abiotic components and factors in the development of geosystems cause changes in soils, vegetation, wildlife, and, as a result, natural complexes in general.

Main materials for our research were obtained during field studies in scientific expedition of the World Wildlife Fund (WWF) in 2013. The interpretation of satellite images and literary analysis were also used.

---

* Corresponding author: anshur@mail.ru

 https://doi.org/10.1051/e3sconf/202016902019

2 Results and discussion

Warming changes plant communities in the Arctic, e.g. dwarf shrubs replace mosses, lichens, pollster plants, which reduces the resistance of tundra and Arctic soils to external influences. The natural habitats of Arctic animals are changing as they are forced to look for new sources of nutrition [2].

At the same time, under the conditions of a significant reduction in the anthropogenic impact on Vaygach Island (since the 90s of the 20th century) due to climate warming, the restoration of affected natural complexes has begun. Boreal and hypoartic flora elements are spreading across the territory, high Arctic and Arctic species are being moved to the tops of the ridges.

Anthropogenic changes in the geosystems of the Arctic are caused not so much by traditional methods of nature management of the indigenous population as by modern intensive exploitation of natural resources. The development of mineral deposits on the Kola Peninsula, in the Timan-Pechora, Vorkuta, Pur-Nadym, Norilsk regions (production of oil, gas, coal, ferrous, non-ferrous, and precious metal ores) leads to the pollution of air, surface water, and soil. The area of natural geosystems is reduced due to the construction of cities and towns, laying of roads and pipelines.

In the 1990s, industrial activity in the Arctic was significantly curtailed, but there were destroyed and dilapidated buildings, up to 5 million tons of construction waste and 13 million rusty metal barrels, as well as rusty ships, cars, warehouses with tanks of fuel and lubricants. All this causes irreparable damage to Arctic geosystems. UNEP/GEF experts have identified more than 100 "hot spots" (impact areas) of accumulated environmental damage in the Russian Arctic, associated with industrial facilities where the environmental situation has reached a crisis or pre-crisis state [3]. The most relevant ones are, for example, Kovdor, the Pechora Bay, and the Taz Bay, which are located directly in the area of specially protected natural areas.

Conservation of Arctic landscapes is largely ensured by specially protected natural areas.

In the Russian Arctic at the beginning of 2012 there were 450 protected areas with a total area (excluding marine areas) of over 94.6 million hectares, which is about 16.2% of the total area of the Russian Arctic region. [4]. Among protected areas there are federal, regional, local protected areas, as well as reserves, national and natural parks, wildlife preserves and reserves (Fig. 1).

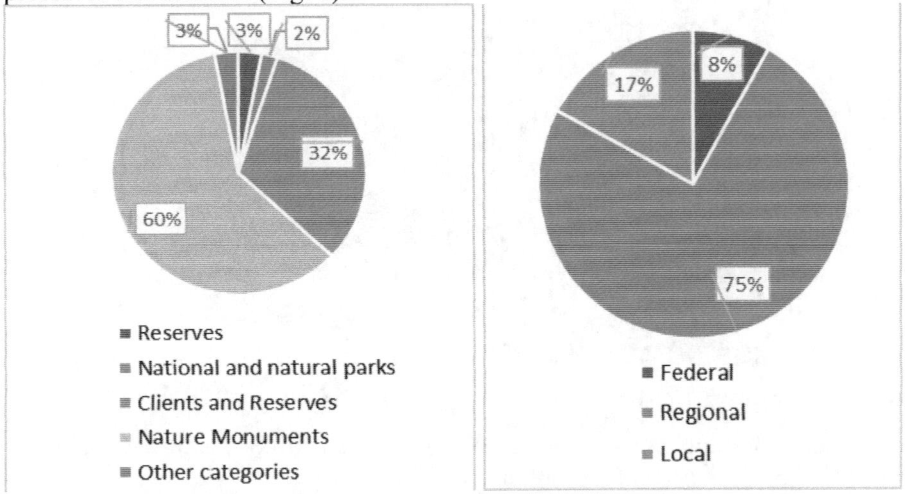

Fig. 1 The ratio of the number of specially protected natural areas of various categories of the Arctic region of Russia.

The creation of new protected areas in the Arctic region will contribute to both preservation of Arctic landscape diversity and development of educational environmental tourism.

A variety of unique landscapes are located on the island of Vaygach, which were studied by a scientific expedition of the World Wildlife Fund (WWF) in 2013 to justify the establishment of a protected area.

Space images show clear north-western linearity of relief morphostructural elements [5].

The main features of the landscapes of Vaygach Island are:

• spread of permafrost rock mass;

• diversity of the island relief – from 60-metre-high ridges separated by canyon-like river valleys to flat frost mound bogs (Fig. 2);

Fig. 2. a) Ranges and limestone massifs of Vaygach Island; b) Canyon-like valley with rich forb groups

• young and undeveloped river valleys;

• availability of periodically flooded saline coastal meadows – marshes or laidas;

• abrasion and abrasion-accumulative shores (Fig. 3); landscapes of floodplains of deep-cut valleys of small rivers and streams rich in biodiversity and endemics

Fig. 3. Abrasive shores of Vaygach Island

A wide variety of arctic tundra landscapes are prevalent on the island, and the establishment of a protected area will ensure their conservation.

The landscape map compiled as a result of our own field and office studies shows 26 natural complexes (Fig. 4). The island is characterized by a variety of landscapes, the youth of river valleys, the presence of permafrost soils, flooded coastal meadows. The most interesting for ecological tourism is the northwestern part of the island.

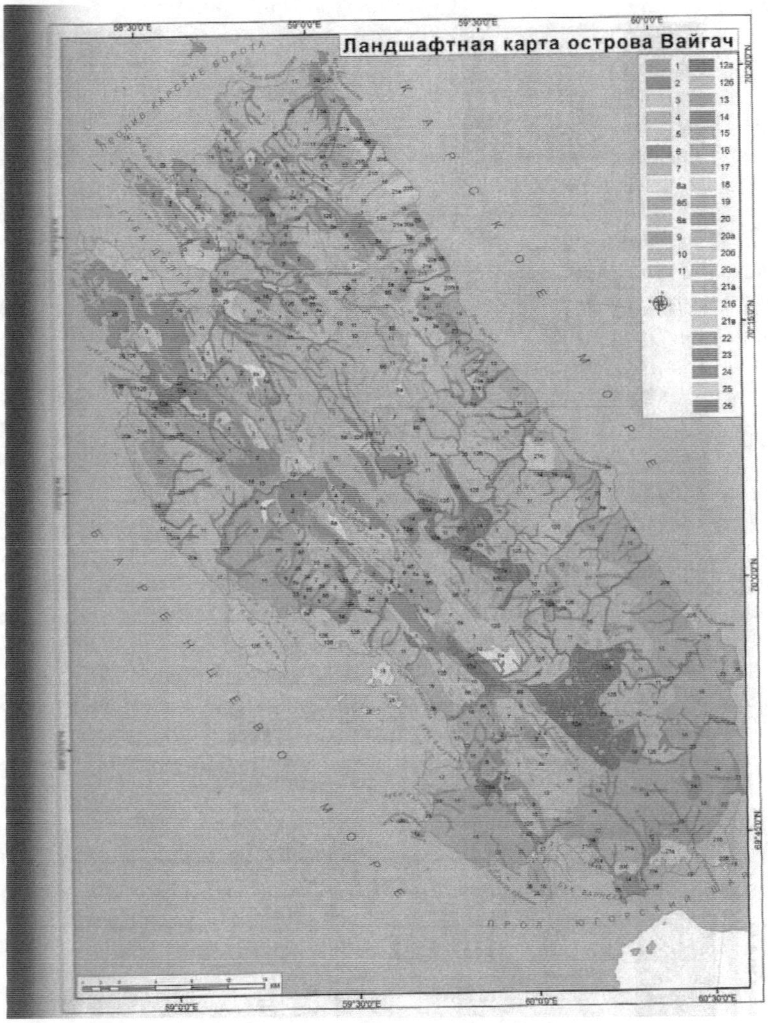

Fig.4. The landscape map of Vaygach Island

Historical and cultural sites on Vaygach Island have thus been preserved in slightly modified natural landscapes. The combination of the cultural component and the landscape of almost untouched polar tundra led to the formation of a unique phenomenon called the Arctic cultural landscape [6].

The Marine Arctic Complex Expedition (MACE) conducted field research on the Vaygach Island between 1986 and 2000 in order to search for and record historical and cultural sites [7].

Archaeological sites are:

• Litosalya settlement with flint tools of the Stone Age;

• Nenets sanctuaries with stone idols (houris), finds of wooden and iron objects, animal bone accumulations (Zayatsky Kamen, Great Bolvanskaya Mountain).

Historical monuments of Vaygach Island development were also discovered:

• Pomor settlements of the 19th-20th centuries with finds of iron forged axes, wooden arrows for hunting fur-bearing animals, fragments of wooden oars, fragments of porcelain and glassware;

• Polar station (1912), camp of hydrographic expedition, trading station (20th century).

3 Conclusion

Modern climate change will lead to the transformation of landscapes of the island such as changes in the coastline, increased abrasion, swamping of coastal areas, lowland and flat landscapes of the island.

The landscape map compiled as a result of our own field and office studies shows 26 natural complexes. The island is characterized by a variety of landscapes, the youth of river valleys, the presence of permafrost soils, flooded coastal meadows. The most interesting for ecological tourism is the north-western part of the island.

In combination with natural resources of educational ecological tourism, i.e. unique polar landscapes, historical and cultural monuments provide an opportunity to study natural features in connection with the history of the development of Arctic regions. In many specially protected natural areas of Russia, protected objects also include historical and cultural sites (for example, protected areas of the Moscow region, Central Russia). Thus, the existing and newly organized protected natural areas of the Russian Arctic contribute to solving environmental problems, monitoring landscape changes following climate changes, as well as preserving not only Russia's natural, but also historical and cultural heritage.

References

1. M.M. Byashkin., T.D. Gavoron, *Ecological Problems of Specially Protected Natural Areas of the Russian Arctic. Geography of Arctic Regions 2017*, 66-69 (Stary Gorod, St. Petersburg, 2017) (in Russian)
2. G.M. Mainasheva, V.A. Tyurina, *Trend of Changes in Some Natural Conditions in the Arctic. Geography of Arctic Regions 2017,* 37-39 (Stary Gorod, St. Petersburg, 2017) (in Russian)
3. *The Russian Arctic in the 21st century: environmental conditions and development risks: a reference manual publication* (Feoria, Moscow, 2013) (in Russian)
4. M.S. Stishov, *Specially Protected Natural Areas of the Russian Arctic: Current State and Development Prospects* (WWF, Moscow, 2013) (in Russian)
5. A.A. Aleinikov, A.M. Aleinikova, M.V. Bocharnikov, P.M. Glazov, P.P. Golovlev, V.O. Golovleva, etc., *Vaygach Island: Nature, Climate and People* (WWF, Moscow, 2014) (in Russian)
6. M. Gavrilo, E. Ermolov, Protected by the State, **3**, 16-34 (2015) (in Russian)
7. *Vaygach Island. Cultural and Natural Heritage. Monuments of History of Arctic Exploration.* Book 1. (Russian Research Institute of Cultural and Natural Heritage named after D.S. Likhachev, Polar Research Foundation, Moscow, 2000) (in Russian)

Cognitive analysis of the socio-ecological and economic system of the region for choosing a sustainable development strategy (on the example of oil-producing regions of Russia)

Tatiana Ledashcheva[1*], and *Vladimir* Pinaev[1,2]

[1]Peoples' Friendship University of Russia (RUDN University), Faculty of Ecology, 6 Miklukho-Maklaya St, Moscow, 117198, Russian Federation
[2]Environmental Consulting Company FRECOM Ltd, Moscow, Russia

Abstract. To correspond with Strategy of Russian regional development and sustainable development aims, in the article we propose to base strategy development on the results of modelling and analysis of socio-environmental-economic region systems in the form of oriented graphs. Authors developed model for oil exploring regions Urals Federal district, based on open source data and lacking information is identified, as well as types and ways of it's gathering for completion of the modelling process.

1 Introduction

In line with the law «On strategic planning in Russian Federation» administrations of all regions of Russia elaborated Development strategies, containing priority development tasks and each task has complex of measures aimed at solution. However, the law itself does not define sustainable socio-economic development or the criteria for its achievement, which makes it difficult to assess the appropriateness of the approved development strategy for a particular region. The criteria defined in the strategy and the planned values of individual parameters characterize mainly the development of the economy and, to a certain extent, the development of society in the region. However, it is not even certain whether the set of social criteria is sufficient, and the lack of sufficient and justified parameters characterizing the environmental aspects of regional development does not allow us to assert that the planned regional development is sustainable.

It is interesting, however, that social aspects are in some way connected with economic aspects. With the absence of a sufficiently developed economy, there are no means to implement social support. On the other hand, the presence of a developed economy does not guarantee social welfare.

So we believe that in order to develop a strategy for regional development, it is necessary, first of all, to have a system of criteria that characterize the region as a socio-environmental and economic system as a whole. It is clear that for different regions such systems will be different, depending on the structure of the gross regional product, social

* Corresponding author: ledashcheva-tn@rudn.ru

structure, geographical location and many others. In this article as an example we will consider the oil-producing regions.

As it is well-known, the concept of sustainable development appeared in 1987 in report «Our common future» by World Commission on Environment and Development [1]. Since that time scientific community, including Russian, spent a lot of time improving this concept and identifying criterions for it's assessment. Common trend of different "sustainable development" definitions is the unity of economic, environmental and social aspects. At the same time, as noted above, the criteria listed in the development strategies of the Russian regions, which mainly include indicators of economic and social development, can not be considered as criteria for sustainable development. Ignoring or not paying enough attention to environmental aspects in regional management can lead to environmental problems. This can endanger environmental security of Russia - component of national security.

The most complete system of sustainable criteria was developed in 1996 by the UN Commission on sustainable development. It includes 132 indicators (environmental, social, economic, institutional). In practice, the use of this system is not rational, primarily because of its cumbersome and difficult to collect information on all criteria. In 2001, at the conference on European statistics in Canada, a shortened system of 58 criteria was proposed. However, the application of this system for a particular region is not rational – some criteria may not be applicable, some criteria may, on the contrary, be crucial for the region, and statistical information is not available for the evaluation of many criteria. Thus, each region should have its own system of criteria. Many scientific works are devoted to the development of criteria for different regions of Russia. for example [2,3], in which some approaches have a sufficiently deep methodological development [3].

However, we believe that even a fairly complete set of indicators that solves the problem of the description of condition or sustainable development of the region, cannot be a sufficient basis for the regional development strategy, because it does not take into account the relationship between the indicators. It should be borne in mind that each region is a unique system where each criterion is somehow related to the others, which means that measures aimed at improving one target criterion will affect all the others. Ignoring this fact can negatively affect the expediency of the development strategy of the region.

Therefore, it is not sufficient to collect a list of indicators to assess the state of the region in order to develop a regional development strategy. It is necessary to understand how important this or that factor is for the state of the region as a whole, based not on the views of the decision maker, but on the mechanism of functioning of the region as a whole. It is necessary to be able to predict, to what extent the change of this or that indicator will affect all other parameters as a whole. Actually, it is necessary to model development of the situation in the region at change of this or that parameter, i.e. to build a model of the region taking into account all parameters.

It should be noted that attempts to build a system of sustainable development indicators based on their mutual impact are also being made by different researchers, while statistical models of the impact of some indicators on others are being considered [4]. In our view, however, this approach lacks consistency, which implies that each indicator influences each one, directly or indirectly, and in some cases we can change the power of influence through some economic, social or legislative measures.

2 Methodology

In our work we are rely on methodology of socio-environmental-economic systems modelling in the form of oriented graphs (methodology of cognitive modelling), developed

under supervision of V.I. Gorelov and described in detail [5]. Modelling process includes following stages:
1. Problem (goal) statement.
2. Identification of factors (subsystems) covering all aspects of the system operation within the framework of the task; factors are presented as vertices of the directed graph.
3. Each factor is associated with a numerical characteristic (indicator), the value of which can be measured
4. Figuring out the direct connections between the factors, they are represented by the arcs of the directed graph.
5. Investigation of strong connectivity of the directed graph and sufficiency of a set of factors.
6. Assignment to arcs of signs " + " or " - " depending on the nature of mutual influence - obtaining a sign oriented graph.
7. Estimation of weight coefficients of arcs of the directed graph by direct calculation, statistics or expert evaluation. Obtaining a weighted directed graph and its vertex adjacency matrix.
8. Calculation of the pulse process and the system response to a single impact on each vertex. This includes an assessment of the stability of the simulated system, which is a necessary condition for the correctness of the model.
The reaction of the system is calculated by the formula (1)

$$X = \lim_{k \to \infty} p(E + \sum_{i=1}^{k} A^i) \quad (1)$$

Here X is a vector of values of indicators in fractions of initial values, A is a matrix of adjacencies of vertices of the directed graph, p is a vector of initial pulse.
The reactions X(i) (i=1,…,n) of the system is calculated for the initial single effects on each factor separately.
9. Calculation of system weight coefficients characterizing the influence of each factor on the state of the system as a whole, using the formulas (2) developed by the authors of the method:

$$W_i = \sum_{j=1}^{n} \frac{X(i)_j \cdot sgn(i)}{\max(X(i)_j) - \min(X(i)_j)} \quad (2)$$

Here, sgn(i) is 1 if factor i is considered positive to assess the state of the region and -1 otherwise.
10. Analysis of the obtained weight coefficients, evaluation of the system. It can be concluded about the stable development of the system, stagnation or pre-crisis / crisis state of the system and the optimal directions of system management. Thus, the state of the system is a crisis, if for any factor i signs Wi and sgn(i) are different.
11. Proposals for the management of the system, the calculation of the response of the model to the proposed management activities, which may consist in both the impact on the factors and the change in the weights of connections between the factors.

Finally, if apply proposed methodology to the task of region modelling, we receive proved strategy for region development, based on current state of region.

3 Results and Discussion

The basic model of the region was developed to support decisions on oil production project management and presented in [6]. Based on it, and having studied the available strategic and statistical data available for the regions of KHMAO and YANAO [7], we have chosen to modelling the following factors: the number of population, quality of life (estimated as the average level of income, including the consumer price index); the level of satisfaction of citizens with the quality of life; recreational resources (the area suitable for recreation of citizens); forest resources; biodiversity; cattle breeding; traditional land use; oil production; real estate (number of dwellings built per year); morbidity; regional budget; infrastructure (length of paved roads); volume of pollutants emitted into the air per year; average level of air pollution (percentage of maximum permissible concentration (MPC); volume of wastewater discharge; Quality of water resources; soil contamination with oil products; municipal and industrial waste; unemployment rate; energy production. Fishing is also important for YANAO.

Analysis of open source data on selected regions allows to make following conclusions on quantitative content of the model.

There are data on oil and gas exploration, housing construction, morbidity, regional budget, average income, the length of paved roads. Environmental factors - emissions and discharges - are worse. Information on environmental investments does not provide data on how these investments will be used or information on the effectiveness of such investments. There is also no information about oil spills and oil pollution.

At the same time, reports on the socio-economic situation of the Yamal-Nenets Autonomous Okrug contain information on various budget items, the amount of social payments, profitability of enterprises, etc., which allows to directly calculate some of the weight coefficients of the economic sphere. There is insufficient data to assess social connections. For example, data on the average length of a sick leave are needed to directly assess the link "morbidity-standard of living". In order to assess the links with the "standard of living" factor, in addition to data on average income by industry, it is necessary to obtain information on the approximate number of employees in a particular industry. There is no information about the presence or absence of oil and gas industry allocations for the improvement and construction of roads.

For the statistical evaluation of certain economic relations (e.g., the impact of the budget on infrastructure), the financial data on the growth of consumer prices were recalculated.

As the result of open source data analysis we were able to calculate 14 weight coefficients of the bonds:

«unemployment-living standard» = -0,03;

«morbidity-living standard» = -0,04348;

«living standard – regional budget» = 0,286;

«living standard – real estate» = 0,186;

«regional budget – living standard» = 0,032282;

«regional budget – real estate» = 0,824;

«infrastructure – living standard» = 0,454375;

«infrastructure – regional budget» = 0,003184;

«real estate – living standard» = 0,409799;

«real estate – regional budget» = 0,05;

«cattle breeding – living standard» = 0,2906;

«cattle breeding – regional budget» = 0,000549;

«oil exploration – living standard» = 0,836033;

«oil exploration – regional budget» = 0,115585.

In addition, as a result of the calculation of statistical coefficients from open source data, we estimated 8 weight coefficients:

« living standard – morbidity» = -0,017;

« living standard –cattle breeding» = -0,5;

«reginal budget – infrastructure» = 2,1;

«infrastructure – unemployment» = -0,3;

«infrastructure– cattle breeding» = -0,26;

«cattle breeding – morbidity» = -0,022;

«oil exploration – unemployment» = -0,1;

«oil exploration – cattle breeding» = -0,48.

As already mentioned, some data on the above parameters were missing, we removed them from the model, replacing the connections associated with direct impact on disease, livestock, recreational resources. So, the cumulative impact of oil exploration on morbidity, which is increasing due to emissions, pollution of water bodies and various pollution from oil spills, is estimated at 0.0936.

When using coefficients calculated by the statistical method, it should be borne in mind that they include both direct and indirect relationships. For example, we reduce the regional budget-infrastructure ratio for the model by the product of the regional budget-real estate and real estate - infrastructure ratios.

Processing of the expert survey results allowed to estimate the remaining arc weights. However, it should be noted that the composition of the expert group was not included neither the administration nor the people of the region. To correctly assess the coefficients, it is necessary to obtain information on the above aspects and conduct sociological research and surveys in the region. This work can be finished only with region administration support.

The calculation of the model give us next system weights:

W(oil production)=2

W(cattle breeding)=1,2

W(traditional land use)=1,7

W(recreational resources)=1,1

W(unemployment rate)=-0,7

W(morbidity)=-1

W(regional budget)=4

W(infrastructure)=1,1

W(real estate)=2,5

W(level of satisfaction with the quality)=0,9

W(life quality)=2

W (number of population)=2,5

W(energy production)=1.

The analysis of the results allow us to draw some preliminary conclusions, for example: the general state of the region can be characterized as stable development; the volume of oil exploration has a positive impact on the overall state of the region, but to a lesser extent than the volume of the regional budget. This is probably due to the fact that oil production is not a project of the region, but is carried out by state-level companies, that is, first of all, most of payments are sent to the state budget (not to the local budget), and secondly, most of the employees of these companies are not inhabitants of the region. Accordingly, it is necessary to maintain the exploration of hydrocarbons at a stable level (since the increase in exploration is currently impossible for objective reasons) and to develop various sources of income. The same analysis of region model with included suppositive projects of another activities can help to choose the best ones for the moment. Attention should be paid to the preservation of traditional land use, which is even more important than livestock farming. Morbidity in the region plays a more negative role than unemployment. Видимо это связано с недостаточной населенность региона (показатель численности населения оказывает положительное влияние даже в большей степени чем нефтедобыча).

In the course of the calculations, we saw that migration in the region is unsustainable, probably because of complex climatic conditions and simply because of the quality of life (its increase is as important for the region as oil production). But the increase in the population can be achieved in another way - a temporary increase due to tourism development. The obvious solution is to develop ecological tourism taking into account the maintenance of traditional land use. It should be noted that the introduction of the ecotourism mechanism will lead to a change in the structure of the region's connections, it can be assumed that as a result the number of recreational zones will increase.

However, it should be noted that these conclusions can not be absolutely reliable, because there is no reliable information about more than half of the model's weight coefficients and an expert method of their evaluation is used. At any rate, even if such scarce information is available, it is interesting to look at the dynamics of the coefficients, recalculating them in accordance with the new annual statistics, which is the subject of further research.

4 Conclusions

To summarize, we can make some conclusions.

We proposed a model of oil-producing regions of Russia as a socio-ecological and economic system, sufficient in terms of integrity.

The available official statistical information allows to directly calculate part of the coefficients related to the social and economic sphere without relying on the estimated regression coefficients.

However, the available information is categorically insufficient to estimate, even statistically, the coefficients of the environmental-social and environmental-economic parts of the model.

At the same time, the availability of the model makes it possible to quickly determine what kind of statistics or other information is necessary for its accumulation and use in strategic planning of sustainable development of a region and to develop forms for collecting this information.

Based on the available information, we have estimated weight coefficients of indicators of the region's condition, which could be taken into account in the strategic planning. Thus, the high weight of the budget indicator of the region with half the weight of the oil production indicator indicates the need to fill the budget from other sources, i.e. to develop other economic activities. Traditional land use, the weight of which is not much lower than the weight of oil production, may be underestimated in this regard. In addition, we should not forget about the possibility to influence the values of the connections coefficients of the model. Thus, in the current situation, investments in recreational zones (a simple increase in their number) will not pay off practically. But if we take into account the priority of traditional land use, the logical solution would be to develop ecotourism, which can change the model's relationships and bring both indicators to a higher value level for the region.

However, in order to be sure to draw similar conclusions and to integrate them into the sustainable development strategy, it is necessary to supplement the model with the currently missing information.

References

1. *Our common future. Report of WCED.* URL: http://xn--80adbkckdfac8cd1ahpld0f.xn--p1ai/files/monographs/OurCommonFuture-introduction.pdf
2. Manyushis A.Y., Barsukov I.E., *Regulation of sustainable development of a large region, city: problems, ways of reforming, innovations.* (Mezhdunarodnyj Soyuz ekonomistov, Moscow, 2016) (in Russian).
3. Zhukov I.V., Bajbakov E.I., Ekologicheskij konsalting, **4**, 2-10 (2008) (in Russian)
4. O.V. Bakanach, N.V. Proskurina and Y.A. Tokarev, *SHS Web of Conferences*, **62**, 05001 (EDP Science, Les Ulis, 2019)
5. Gorelov V.I., Karelova O.L., Ledashchaeva T.N. *System modeling in social and economic sphere*, (Logos, Moscow, 2012) (in Russian)
6. T. Ledashcheva, V. Pinaev, *International Multidisciplinary Scientific GeoConference Surveying Geology and Mining Ecology Management, SGEM,* **18(1.5)**, 11-18 (STEF92, Sofia, 2018)
7. Official web-page Department of State statistic service for Tyumen oblast, Khanty-Mansyisky autonomous region – Yugra and Yamalo-Nenets autonomous district http://tumstat.gks.ru

E3S Web of Conferences **169**, 03001 (2020) https://doi.org/10.1051/e3sconf/202016903001
APEEM 2020

Sustainability as a basic property of the phenomenon of life organized in supraorganismal systems

Alexander Nikol'skii,* *Konstantin* Belovezhets, and *Elena* Vanisova

Peoples' Friendship University of Russia (RUDN University), Faculty of Ecology, 6 Miklukho-Maklaya St, Moscow, 117198, Russian Federation

Abstract. The phenomenon of life is organized in supraorganismal systems, populations and ecosystems. The main mechanisms of sustainability of these systems are: 1) the influence of limiting factors on population growth and ecosystem development; 2) the allocation of plant and animal species into ecological niches; 3) co-adaptation of plants and animals; 4) species genetic integration; 5) biocommunications. Limiting factors inhibit populations' growth and distribution of organisms, so that populations obey the logistic growth, reaching a relatively stable climax stage. The space of limiting factors restricted by the tolerance limits of organisms forms an ecological niche of species. During long-joint evolution (co-evolution), species form co-adaptations (mutual adaptations). As a result of species genetic integrity, each new generation inherits not only adaptations for its specific ecological niche, but also a determined physical space including all the environmental factors to which the species is adapted. Consolidation of individuals with a related genotype is supported by intraspecific communication processes. Means of communication are genetically determined. Humankind is the only species on Earth that has subjugated limiting factors. As a result the population of humans shows an unlimited exponential growth, population explosion. It is a basic factor in the instability of the Nature - Society system.

1 Introduction

Resolution of the United Nations [1] "Transforming our world: the 2030 Agenda for Sustainable Development" adopted by the General Assembly in 2015 calls the world community to the sustainable development without explaining what is to be developed sustainably. And only from the context of this document it is possible to understand that the sustainability of the system "Nature – Society" is discussed.

Sustainability is a natural state of the phenomenon of life organized in *supraorganismal systems* (populations and ecosystems) of different levels. The purpose of our report is to discuss some mechanisms of the sustainability of supraorganismal systems. Due to metabolism and interactions in ecosystems, plants and animals are sustainably using the energy of the Sun that comes to Earth over millions of years.

*Corresponding author: bobak@list.ru

German agricultural chemist Justus von Liebig was the first to justify the stability of supraorganismal systems, 25 years before another great German, Ernst Haeckel, uttered the word "ecology" [2]. Considering the atmosphere as the habitat of organisms, Liebig emphasized that this habitat *was created by the organisms themselves* and maintained by them in a stable condition [3, p. 21]: "Plants not only afford the means of nutrition for the growth and continuance of animal organization, but they likewise furnish that which is essential for the support of the important vital process of respiration; for besides separating all noxious matters from the atmosphere, they are an inexhaustible source of pure oxygen, which supplies the loss which the air is constantly sustaining. Animals on the other hand expire carbon, which plants inspire; and thus the composition of *the medium in which both exit* (italics are ours), namely, the atmosphere, is maintained constantly unchanged".

Ecology, *the science of the structure and functioning of supraorganismal systems* [4], clearly indicates the basic mechanisms of stability of the phenomenon of life, such as, in particular: 1) limiting factors, 2) distribution of species and populations into ecological niches, 3) coadaptations, 4) genetic integration of species; 5) intraspecific and interspecific communicative processes. And it also clearly indicates the role of humankind in the destruction of mechanisms of sustainability.

Further, on the results of our own studies published earlier, and on the results of studies of other authors, we discuss the main mechanisms of sustainability of the phenomenon of life organized in populations and ecosystems.

2 Main

2.1 Limiting factors

The key idea of ecology is the concept of *limiting factors*, a mechanism that restrains population growth and expansion of plants and animals. The concept of limiting factors explains the fundamental difference between the dynamics of the population of humans and the dynamics of the populations of other organisms' species. Fig. 1 illustrates this idea, showing that populations of plants and animals have a pattern of logistic growth. Having reached the limit of growth in certain environmental conditions, the population remains in this relatively (wavy line) stable state of abundance and productivity for an unlimited time. And, on the contrary, humankind, controlling the limiting factors and having subordinated resources to itself, thereby created the conditions for unlimited, exponential population growth, deprived of a relatively stable stage.

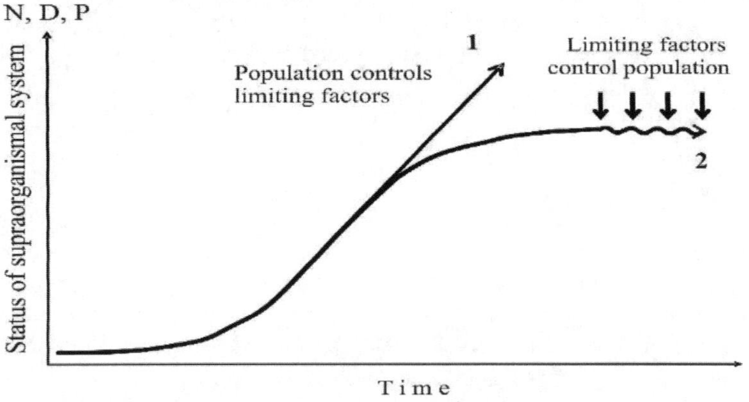

Fig. 1. Development patterns of supraorganismal systems. Growth of human population (1) and other species of organisms (2). N – population size, P – productivity of population/ecosystem. D – species diversity of ecosystem.

The idea of limiting factors can easily be extrapolated from populations to ecosystems, which in addition to abundance and productivity are also characterized by species diversity, and the limit of growth is associated with a relatively stable stage of climax. Eugene Odum [5] paid attention to it, emphasizing that in general the status of ecosystems is determined by the flow of energy coming from the Sun, which is clearly illustrated on the example of zonal ecosystems. Indeed, for many millions of years, rainforests near the equator remain rainforests, with their huge biological productivity and almost unlimited species diversity and, for example, tundra at high latitudes remains tundra, with its poor productivity and a limited number of plant and animal species.

The main contribution to the development of the idea of limiting factors was made at the beginning of the last century by J. Johnstone, V. Shelford, A. Tinemann, I.K. Pachosky, and G.F. Gause [6-10].

An outstanding role in comprehension of the mechanisms of population growth was played by the equation of the Belgian demographer Pierre Verhulst [11], rediscovered at the beginning of the 20th century by American scientist Raymond Pearl [12]:

$$\frac{dN}{dt} = \frac{bN(K-N)}{K},$$

where K is the maximum possible population size, N is the population size at time t, b is a reproduction coefficient (proportionality coefficient).

When $N = K$ the growth stops, and the populations enter an equilibrium state. In this relatively stable condition they can be for an unlimited time.

Logistical, or "Verhulst", curve depicted in Fig. 1:2 shows the future of a population or an ecosystem restricted by limiting factors. For the ecosystem, the main limiting factor is the energy of the Sun. After reaching the climax stage, the ecosystem remains at this level of productivity and species diversity until the flow of energy coming from the Sun changes [5]. Due to the fact that humankind controls the limiting factors, its population grows exponentially to infinity (Fig. 1).

2.2 Distribution of species and populations into ecological niches

A population of any species of plants and animals is relatively stable exactly because it is restricted by limiting factors that together form an *ecological niche* — hypervolume, or multidimensional space of factors [13, 14]. The tolerance range for each of the factors and, what is not less important, for their combined effect ("the effect of constellation" [15]) "locks" the population within the limits of the ecological niche. Eurybiontism of organisms is not absolute and has its limitations. The prefix "eury", being understood literally, is sometimes misleading. Only humans could spread across the entire surface of our planet by controlling of the limiting factors.

2.3 Coadaptations

The stability of supraorganismal systems is also indicated by such a ubiquitous phenomenon as *coadaptations* or mutual adaptations. A mutual adaptation of insects and pollinated plants is a good example. Coadaptations are the result of co*evolution*. Evolution is a long process. Therefore, a *stable* condition of the species composition of the subjects of the coevolutionary process is necessary in order for coadaptation to take place. That is why on the evolutionary time scales the species structure of ecosystems cannot change too quickly.

2.4 Genetic integration of species

A huge influence on the stability of supraorganismal systems is exerted by the *genetic integration* of species, regulated by the flow of genes [16]. Each new generation inherits not only a complex of adaptations that allows it to assimilate the ecological niche peculiar to the species, but also, at the place of birth, a certain physical space with all the limiting factors to which the species is adapted.

2.5 Intraspecific and interspecific communicative processes

Consolidation of individuals with an affined genotype is facilitated by intraspecific communicative processes. The species specificity of the communicative means, for example sound signals, corresponds to the genotype. So the phonotype, which is the complex of features of the sound signal, specific for the species, corresponds to the genotype. In early bioacoustic studies the species specificity of sound signals was already shown in various groups of vertebrate and invertebrate animals [17].

Fig. 2 shows, as an example of the correspondence between the phonotype and the genotype, cluster diagrams of the characteristics of alarm call, as well as nuclear and mitochondrial DNA of 7 species of marmots.

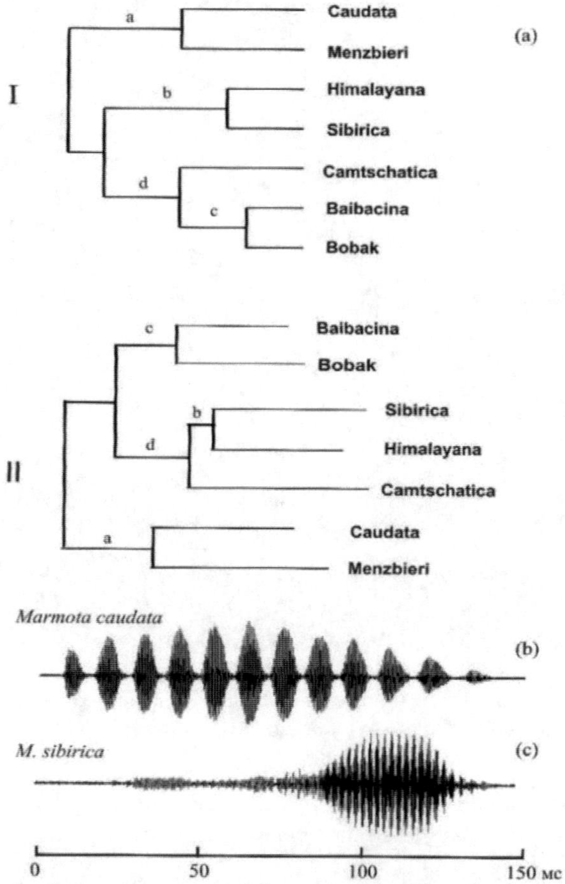

Fig. 2. Correspondence of the phonotype to the genotype. (a) – dendrogram of the sound signal characteristic of marmots (I, according to [18]) and mitochondrial and nuclear DNA (II, according to [19]); a, d, c, d – clusters; oscillograms of the alarm call of Red (b) and Mongolian (c) marmots are an example of the species-specific characteristics of the signal.

As it can be seen from the Fig. 2, the distribution by species on both dendrograms coincides. The number of clusters coincides too. The difference is only in the position of the black-caped marmot (*Marmota camtschatica*). It falls into different clusters at the dendrograms, which is quite explainable by its huge and strongly torn range [20], within which significant intraspecific differences are possible. It is possible that these differences reach an interspecific level.

A special case of biocommunications is a biological signal field. It also supports the stability of supraorganismal systems. The biological signal field represents the traces of vital activities left by each generation of animals. These traces of vital activities create an odor-visual image of the territory, forming a matrix of stable elements, so that each new generation reads information from it [21-24]. As a result of such ecological inheritance [25], information about the territory with the resources located on it, is sustainably transmitted from generation to generation.

Fig. 3 shows an example of a matrix of stable elements in the biological signal field. On the family site of the steppe marmots (*M. bobak*), which is shown there, the matrix of stable elements that organizes the trajectory of animals moving around the territory is formed by paths and burrows. Both of them are updated and steadily used from generation to generation for a very long time, up to many thousands of years [26].

Fig. 3. Example of a matrix of stable elements of the biological signal field is the territory of a marmot family. Traces of vital activities (burrows and paths), transmitted from generation to generation, create a visual (and image of space, organizing the trajectory of the movement of animals through the territory with resources on it.

The biological signal field formed by a multi-species community supports the sustainability of ecosystems, facilitated by *interspecific communicative processes.*

3 Conclusion

At the beginning of our report, we drew attention to the fact that humanity controls the limiting factors, including resources in the traditional sense of the word, for example, food resources. The earliest forms of farming and cattle breeding appeared about 12 to 8 thousand years ago [27, 28]. Many thousands of years ago [29] the first power engineers created the first power plant, a fire, subjugating the most important of the limiting factors, the temperature. In fact, the fire was the first alternative energy source, the alternative to the Sun. After mastering the fire, clothes and dwellings, people started to control temperature, the limiting factor that restricts ranges of absolutely all species of plants and animals.

Today, there are about 7.5 billion of us on the Earth, and nothing holds back the further increase in the number of people. But in the UN Resolution on Sustainable Development [1] nothing is said about the demographic explosion that destroys the biosphere: in this heterogeneous international organization, it is not easy to discuss sensitive issues of demography.

We find an in-depth analysis of the demographic explosion problem, a politically sustained and delicately formulated solution to it, from academician V.M. Kotlyakov in his remarkable work "Geography in a Changing World" [30]. Kotlyakov offers the only possible way for the sustainable development of the "Nature – Society" system [30, p. 345-346]: "... Comprehensive approaches show the need for a revision of the value system that guides mankind ... The traditional fear of depopulation and political incentives for the prestige of large families are associated with the inevitable in the past loss of competitiveness of the nation reducing the population. This danger disappears during the transition to a proportional reduction in the number of all nations (while maintaining the populations of small nations). Recently, the prestige of large families has been rapidly declining throughout the world, and humanity is getting closer to the moral realization of the possibility of transition to one-child birth. Thus, the opportunity under consideration is real from an economic, demographic, environmental, and moral and ethical point of view."

In conclusion, we would like to draw attention to the importance of systematizing the knowledge accumulated in archeology over the past 200 years on the history of the exemption of *Homo sapiens* from limiting factors, as an independent section of human ecology, which will contribute to a better understanding of the role of human in the future of our planet.

References

1. *Resolution adopted by the General Assembly on 25 September 2015" Transforming our world: the 2030 Agenda for Sustainable Development".* URL: https://www.un.org/ga/search/view_doc.asp?symbol=A/RES/70/1&Lang=E
2. E. Haeckel, *Generelle Morphologie der Organismen*. Bd. 1, 2 (Berlin: Verlag von Georg Reimer, 1866)
3. J. Liebig, *Organic chemistry in its applications to agriculture and physiology* (L.: Taylor and Walton, 1840)
4. N.P. Naumov, Zh. Obshch. Biol., **32(6)**, 651-666. (1971)
5. E. Odum, *Fundamentals of Ecology* (WB Saunders, Philadelphia, 1971)
6. J. Johnstone, *Conditions of life in the sea. A short account of quantitative marine biological research* (Cambridge: The University Press, 1908)
7. V.E. Shelford, Biol. bulletin, **25(2)**, 79-120 (1913)
8. A. Thienemann, *Das Leben im Süßwasser* (Breslau: Ferdinand Hirt, 1926)
9. I.K. Pachosky, *Fundamentals of Phytosociology*. The course, taught at the Agronomy Faculty of the Polytechnic Institute (Kherson: Studencheskii komitet Sel'skkhoztekhnikuma, 1921) (in Russian)
10. G.F. Gause, *The Struggle for Existence* (Baltimore: The Williams and Wilkins Company, 1934)

11. P.F. Verhulst, Mémoires de l'Académie Royole des Belgique, **18(1), 1-38** (1845) (Citation: Hutchinson, 1978).
12. R. Pearl, L.J. Reed, Proceedings of the National Academy of Sciences, **6(6)**, 275-288 (1920):.
13. G.E. Hutchinson, Cold Spring Harbor Symposia on Quantitative Biology, **22(2)**, 415-427 (1957)
14. G.E. Hutchinson, *An Introduction to Population Ecology* (New Haven, L.: Yale Univ. Press, 1978)
15. S.S. Schwartz, *Current problems of ecology: Fifth All-Union Ecological Conf.,* 21-31 (Moscow: Mosk. Gos. Univ., 1973) (in Russian)
16. E. Mayr, *Animal species and evolution* (Oxford Univ. Press, London,, 1963)
17. *Acoustic behaviour of animals* (R.-G. Busnel, ed., Elsevier, Amsterdam, 1963)
18. A.A. Nikol'skii, Russian Journal of Zoology, **93(8)**, 1026-1043 (2014) (in Russian)
19. S.J. Steppan, G.J. Kenagy, Ch. Zawadzki, R. Robles, E.A. Lyapunova, R.S. Hoffmann, J. of Mammalogy, **92(5)**, 1028-1037 (2011)
20. I.M. Gromov, D.I. Bibikov, I.I. Kalabukhov, M.N. Meier, *Fauna of the USSR. Mammals.* T. III. Vyp.2. *Terrestrial Sciurids (Marmotinae)* (Moskva, Leningrad: Nauka, 1965) (in Russian)
21. N.P. Naumov, Development of the concept of structural levels in biology, 322-331 (Moscow: Nauka, 1972):. (in Russian)
22. N.P. Naumov 1973. **Current problems of ecology: Fifth All-Union Ecological Conf**., 3-20, (Moscow: Mosk. Gos. Univ., 1973) (in Russian)
23. N.P. Naumov, M.E. Gol'tsman, E.P. Kruchenkova, N.G. Ovsyanikov, S.V. Popov, V.M. Smirin, *Theriology issues. Ecology, Population Structure and Intraspecific Communications between Mammals*, 31-75 (Moscow: Nauka, 1981) (in Russian)
24. E.A. Vanisova, A.A. Nikol'skii, Biology Bulletin Reviews, **3(5)**, 335-346, (2013)
25. A.A. Nikol'skii, Russian J. of Ecology, **45(1)**, 76-79, (2014)
26. Yu.A. Dubrovskii, *The study of the geography of natural resources of the animal and plant world (For 60th Anniversary of A.N. Formozov)*, 80-94 (Moscow: USSR Academy of Sciences, 1962). (in Russian)
27. *Agriculture*: Manual G.I. Bazdyrev ed. (Moscow: INFRA, 2013), 608 p. [in Russian]
28. N.V. Grikhina, I.A. Sorokina, *History of zootechnical science*: *manual* (Michurinsk: Michurinskii GAU, 2019) (in Russian)
29. R. Shahack-Gross, F. Berna, P. Karkanas, C. Lemorini, A. Gopher, R. Barkai, J. of Archaeological Science, **44**, 1 (2014): 12-21.
30. V.M. Kotlyakov, *Selected Works in Six Books*. Book 3. *Geography in a changing world* (Moscow: Nauka, 2001) (in Russian)

Ecosystems Transformation and Assessment of Bird Resources in the Republic of Tatarstan (Russia)

Ilgizar Rakhimov[1], *Kadrya* Ibragimova[2]

[1]Kazan (Volga region) Federal University, 420008, Kazan, Russia
[2] Associate professor, Kazan (Volga region) Federal University, 420008, Kazan, Russia

Abstract: The intensification of economic activities, wide scales of the destruction of the woods, the growth of the areas of the worked lands, creation of water basins, urbanization and other forms of transformation of natural landscapes for last centuries have essentially changed the conditions of animals' habitation. 305 species of birds can be found within the territory of Tatarstan. We performed an analysis of the state of the ornithofauna over the past 120 years, which allowed us to get estimates of species density and distribution in the republic. The research material is a gain to the knowledge about ornithofauna of Tatarstan, the state of each species populations and may be used to develop a regional monitoring system.

1 Introduction

Among the problems facing to ornithology, the major is preservation of biological diversity. It has found reflexion in the program and declarations of United Nations Conference on Environment and Development (Rio de Janeiro, 1992) about preservation of biological diversity, Governmental decree of the Russian Federation «On Procedure of governmental accounting, the state cadastre and the state monitoring of fauna objects».

Nowadays the opinion of native scientists about strengthening of the regional faunistic researches has got wide spread, which is directed on preservation of a biodiversity and creation of faunistic cadasters. During epoch of intensive anthropogenous influence on fauna, including birds, death of many species, serious reorganizations in communities have forced to realize that biodiversity preservation was one of aspects of a sustainable development of a modern society and the major principle of nature protection activity.

The pattern of anthropogenic factors influence on ornithological complexes in Tatarstan has increased over the past century due to certain reasons. The key factor of changes in the avifauna of the Middle Volga region is an intensive human economic activity. The processes related to the active development of the region's territory cause disruptions of avifauna. As a result of deforestation and tilling of wild land, creation of reservoirs, growth and expansion of urban settlements, the prevalence of monocultures in agriculture, significant expansion of windbreaker areas, as well as creation of even-aged monoculture forest plantations, open pits, the landscape of Tatarstan has changed significantly. Now the most of its territory is represented by "cultivated forest plantations" [1], with a complex mosaic of scattered spots of dark coniferous, broad-leaved and pine forests, alternated with secondary birch grove and aspen wood. The remainders of former zonal vegetation types, allopatric due to human activity. Tessellation of landscapes, diversity of habitats contributes to fluctuation of natural habitat of many species. Non-periodic fluctuations of faunal boundaries of certain bird species in the context of anthropogenic transformation of natural systems are scrutinized by ornithologists. Small populations of birds in forest outlier, once numerous in natural wildlife habital areas, now leave them, then repopulate once again.

Revealing of the laws of avifauna complexes formation, reaction of some species to the anthropogenous influences, in this connection forecasting of the close and remote consequences of these processes, the development of the regulation measures of mutual relations of the mankind and birds in anthropogenous landscapes are actual problems of ornithological researches.

2 Methods

The Republic of Tatarstan is located in the middle Volga basin and forms part of the Russian plain. There are no high mountains there, it has rather smooth relief: broad flat watersheds alternate with shallow, well-developed river valleys. The main natural green areas of the Republic of Tatarstan are forest and forest-steppe. Forests cover about 17 % of the entire territory. The main types of planted forest are represented by conifer and leaf forest. The Middle Volga region is located in the transition zone from taiga forests to forest-steppe, which determines bird species diversity in Tatarstan.

The authors have been conducting targeted research and collecting material for more than 40 years. During the expedition trips to different regions of the Middle Volga region, observations were made on the avifauna, its species and quantitative structure, spatial distribution, and ecology of individual species and groups of birds. Fauna and the population of birds were studied by repeated routing accounts in the territories with various degree and character of anthropogenous transformation: in nature reserves, in water basins, in agrolandscapes and settlements. Reliability of nesting was defined according to the criteria, which were recommended by Committee of the European Ornithological Atlas (The EBCC Atlas of European breeding birds, 1997). Nesting considered to be proved when there were found nests, lying of eggs, baby birds, and meetings of flights, i.e. an actual material. Nesting with demonstration by birds of the elements of nested behavior was considered as the probable: singing, gathering of a nested material, forage transportation, etc. Birds' summer staying in suitable conditions for nesting has allowed to consider that birds were probably nesting.

The general extent of constant routes has made 285 km, there were passed about 10 thousand km. There were held supervision over flight and daily movings of birds. There were defined the basic laws of spatial distribution of nesting populations.

3 Results

Resettlement of some of bird species across anthropogenic landscapes is a relatively recent phenomenon that can be observed over large areas even within the life-span of one generation, which cannot be compared with any of the known natural processes in terms of speed and outcomes of the said changes. For the Middle Volga region, a collared turtledove illustrates this model, having settled quickly enough in vast areas [2]. A similar process is currently observed with a black redstart.

On January 15, 2009, the State Council of the Republic of Tatarstan adopted the framework nature conservation law "Environmental code of the Republic of Tatarstan", which defined the prospects of a regional strategy for rational nature management and nature protection. The Chapter 6, Article 122 Wildlife Conservation defines the main task of State control in the field of wildlife protection, including state registration and maintenance of the state cadastre and state monitoring of wildlife objects [3]. This monitoring of wildlife objects is impossible without specific materials on quantitative assessment of animal resources. This task determined the relevance and significance of the research.

For the first time, the estimates of the number of 305 birds' species inhabiting Tatarstan were obtained, the status of their habitation was determined and clarified, while the last reports on the avifauna of Tatarstan were dated 2001 [4]. The status of nesting habitats was determined based on biotope analysis of the territory of Tatarstan, the structure and distribution of its land fund and the conducted research on model territories and routes. The state of study of certain species and ecological groups of birds in the Republic of Tatarstan is assessed. The directions of avifauna formation in Tatarstan and the prospects for preserving birds' diversity in the region are presented.

4 Discussion

Multi-year research in the territory of the republic allowed us to get data on birds' population and to identify trends in changes in some of avifauna forms. Following the analysis of available data for the last 120 years, several groups of species have been identified in the territory of the Republic of Tatarstan, having the following population indicators.

1. Increasing population or they are more seen within the territory of Tatarstan. For such species, like swift *Apus apus*, sky lark *Alauda arvensis*, greenish warbler *Phylloscopus trochiloides,* great-crested grebe *Podiceps cristatus*, this is due to the expansion of nestable biotopes areas. The population of common partridge *Pedrix pedrix*, listed in the first edition of the Red data book of the Republic of Tatarstan, has increased and this species was removed from the list of rare and protected avifauna forms [5]. This group also includes species that populated Tatarstan in recent decades, for example, Black redstart *Phoenicurus ochruros*, Eurasian penduline tit *Remiz pendulinus.* This group includes 36 species of birds, including *Egretta alba, Cygnus olor, Anas platyrhynchos, Milvus migrans, Accipiter nisus, Buteo buteo, Crex crex, Fulica atra, Larus ridibundus, Asio otus, Garrulus glandarius, Acrocephalus arundinaceus, Sylvia atricapilla, Erithacus rubecula, Chloris chloris, Carpodacus erythrinus* etc.

2. The next group is bird species with stable population over the past decades (may be either low or high). Low population is observed in most bird of prey, for example, owls *(Accipiter gentilis, Pernis apivorus, Aquila rapax, Falco tinnunculus, Glaucidium passerinum, Strix uralensis*). High population is observed in most titmouses, finches. The group includes 117 species, such as: *Anser albifrons, Anser fabalis, Anas acuta, Aythya ferina, Lyrurus tetrix, Porzana porzana, Tringa nebularia, Actitis hypoleucos, Xenus cinereus, Limosa limosa, Larus argentatus, Chlidonias nigera, Columba livia, Cuculus canorus, Dendrocopos major, Picoides tridactylus, Riparia riparia, Anthus trivialis, Motacilla flava, Corvus cornix, Corvus corax, Acrocephalus dumetorum, Phylloscopus trochilus, Regulus regulus, Muscicapa striata, Luscinia luscinia, Parus major, Fringilla coelebs, Spinus spinus, Acanthis cannabina, Emberiza citrinella* etc.

3. A group with a declining population or they are less seen within Tatarstan. These species are the matter of particular concern and require additional protection measures. The main reason for such changes is intensive economic activity. The processes related to the active development of the region's territory cause disruptions of avifauna [6]. As a result of deforestation and tilling of wild land, creation of reservoirs, growth and expansion of urban settlements, the prevalence of monocultures in agriculture, significant expansion of windbreaker areas, as well as creation of even-aged monoculture forest plantations, open pits, the landscape of Tatarstan has changed significantly. We can observe a mosaic of scattered spots of dark coniferous, broad-leaved and pine forests, alternated with secondary birch grove and aspen wood. Studies conducted in

various forests of Tatarstan showed a significant reduction in cavity nesting birds number in the young forest stands. The group includes 63 species. They are Yellow-breasted blinding *Emberiza aureola*, Greylag goose *Anser anser*, Hen-harrier *Circus cyaneus*, House-sparrow *Passer domesticus*. For this reason, Greylag goose is listed in Red data book of the Republic of Tatarstan [7]. This group includes: *Botaurus stellaris, Anas crecca, Cygnus cygnus, Circaetus gallicus, Circus aeruginosus, Aquila chrysaetos, Falco subbuteo, Falco vespertinus, Otis tarda, Numenius arquata, Chlidonias hybrida, Columba oenas, Bubo bubo, Asio flammeus, Strix aluco, Alcedo atthis, Dryocopus martius, Hirundo rustica, Delichon urbica, Sturnus vulgaris, Corvus frugilegus, Cinclus cinclus, Locustella naevia, Turdus viscivorus, Parus cianus* etc.

4. A small group of birds with fluctuating population. It was observed in 20 species, including: *Buteo lagopus, Lagopus lagopus, Streptopelia decaocto, Nyctea scandiaca, Loxia curvirostra, Pyrrula pyrrula, Bombycilla garrulus* etc. Natural factors are the reason for population fluctuations, mainly related to feeding and weather conditions.

5. The last group includes 70 species with an unclear trend due to poor study of species and lack of information about the population within Tatarstan. This group is represented by small rare species that can be included in the list of protected birds of Tatarstan: *Gavia stellate, Pelecanus crispus, Rufibrenta ruficollis, Clangula hyemalis, Melanitta nigra, Mergus merganser, Circus macrourus, Accipiter badius, Porzana parva, Porzana pusilla, Glareola nordmanni, Syrrhaptes paradoxus, Anthus hodgsoni, Prunella montanella, Phyllo Scopus inornatus, Pinicola enucleator, Emberiza leucocephala* etc. Many of them are listed in the last edition of the Red data book of the Republic of Tatarstan, for example, red-necked grabe *Podiceps grisegena*, gerfalcon *Falco rusticolus*.

Thus, the main reasons for changes in quantitative figures are anthropogenic factors that affect the animals' habitat and their feeding. The most severe impact of anthropogenic factors on natural ecosystems of Tatarstan was in the last 100-120 years. The following changes have occurred over the past period of record:

- As far anthropogenic landscapes grew and developed, their fauna grew and developed, too.
- Settling of birds in anthropogenic-transformed territories was due to ecological advantages of the anthropogenic landscape, besides, natural areas were included in the anthropogenic landscape with their fauna.
- A number of bird species actively settled in urbanized territories and formed urban populations: Great tit, Gray crow, Fieldfare, Linnet, White wagtail, Common wheatear, Lesser whitethroat.
- Both in the past and now, synanthropic species that traditionally inhabit anthropogenic landscapes laid the basis of bird population.
- The creation of reservoirs, deforestation, creation of agrocenoses, urbanization of the region and other forms of economic impact on natural ecosystems have significantly affected the state of avifauna in the Middle Volga region.
- The process of anthropogenic territories population with species formely extrinsic to the region will continue, perhaps, in the near future such species as Common serin, Syrian woodpecker and others will appear in the Middle Volga region.

5 Conclusion

The obtained estimate data on the number of birds in Tatarstan allow us to identify the main research and practical directions for formation of most preferable human and bird relationships in terms of intensive anthropogenic impact on natural complexes. The solution of this issue lies in scientific management of both the number and behavior of birds. Primarily it considers dominant species of economic importance, mainly for agriculture. The results obtained shall be used in a wide range of environmental measures, primarily for long-term monitoring, as a database for formation of designated conservation areas of Tatarstan.

References

1. V.D. Ilyichev *Practical use and protection of birds in the South Ural region* (1983)
2. K.N. Blagosklonov *Geography and ecology of terrestrial vertebrates*, **3** (1978)
3. *The Environmental Code of the Republic of Tatarstan*, Kazan (2015)
4. I.I. Rakhimov *Avifauna of the Middle Volga in the conditions of anthropogenic transformation of natural landscapes*, Kazan (2002)
5. *The Red Book of the Republic of Tatarstan*. Kazan (2006)
6. V.M.Konstantinov, W.Nowicki, A.G. Pichurin, *Acta ornithological*, **31** (1996)
7. *The Red Book of the Republic of Tatarstan.* Kazan (2016)

E3S Web of Conferences **169**, 03003 (2020) https://doi.org/10.1051/e3sconf/202016903003
APEEM 2020

About the reasons for the rare plant species disappearance in the Upper Volga region

Elena Borisova[1], *Nikita* Konotop[1], *Julia* Vinogradova[1], *Dmitriy* Markov[1], and *Alexey* Sivukhin[1]
[1]Ivanovo State University, Ivanovo, 153025, Russia

Abstract. The problems of the extinction of aquatic plant species in Ivanovo region glacial lakes are discussed. As a result of research a sharp decrease in the number of rare hydrophytes (*Elatine hydropiper, Isoetes lacustris, I. echinospora, Sparganium gramineum*) has been established. The reasons are complex, they are associated with the eutrophication of ponds, heavy metals soil pollution, uncontrolled recreational loads and *Elodea canadensis* invaded into lake ecosystems. The ecological niche of *Sparganium gramineum* in the lakes is occupied by *S.* × *longifolium*, the hybridogenic origin of which is proved by modern molecular genetic analysis.

1 Introduction

In the Upper Volga region the ways to preserve natural biodiversity and the natural functioning of natural systems are very relevant. Due to increasing anthropogenic pressures and pollution of natural communities rare plants species are the most vulnerable. Every year there is decrease of their populations, areas of distribution, eviction of indigenous species by invasive ones. The Red Book of the Ivanovo Region includes 149 species of vascular plants, including 12 aquatic species – hydrophytes [1]. In recent decades some aquatic species associated with glacial lakes (*Elatine hydropiper*, *Isoetes lacustris*, *I. echinospora*, *Potamogeton praelongus*, *Sparganium gramineum*, etc.) have been endangered. Similar problems are observed in other regions. The rare species *Sparganium gramineum* is replaced by a stable hybrid of *S.* × *longifolium* by the lakes eutrophication [2–3]. Everywhere there is a displacement of invasive species by invasive ones [1, 4].

In the Ivanovo region, all glacial lakes are protected and have a status of specially protected natural areas. As part of a comprehensive environmental survey of the Valdayskoe and Vysokovskoe lakes, the reasons for the reduction in the number of rare hydrophytes were studied. This problem especially concern the rare glacial species (*S. gramineum*), which is decreasing in number from year to year and disappears completely [1–3, 5].

The purpose of this study was to conduct comprehensive monitoring of the glacial lakes Valdayskoe and Vysokovskoe with the habitat of rare hydrophytes and determine the causes of their disappearance.

2 Materials and methods

Studies were conducted in the Ivanovo region, Ivanovo district, in the vicinity of a large city - Ivanovo. Glacial lakes Valdayskoe and Vysokovskoe have been researched. Water analysis was carried out and soil samples were taken in the surroundings of these lakes.

Analysis of soil samples was carried нby standard method [6]. Laboratory preparation of extracts from soil samples was carried out on the biological and chemical faculty of Ivanovo State University basis, the Hardware analysis – with 210 VGP atomic absorption spectrophotometer.

Water samples were taken in the lakes Valdayskoe and Vysokovskoe in 2018. Organoleptic and chemical analyses of water were performed in a specialized laboratory of the branch of the Federal State Health Institution in the city of Shuya, analysis of the content of heavy metals in water was performed in the laboratory of geographic information technologies using a PE-5300VI spectrophotometer [7].

The hybrids were detected at the Institute for Biology of Inland Waters RAS Laboratory of Geography and Systematics of Aquatic Plants; morphological and molecular genetic studies of *S. gramineum* from Valdayskoe and Vysokovskoe lakes were carried out. Standard molecular techniques were performed: DNA isolation, PCR, and sequencing [8]. Polymerase chain reaction – ISSR method (Inter simple sequence repeats) [9]. The material for analysis was samples of leaf tissue. DNA was isolated by the sorbent method with some modifications [10].

3 Results and discussion

Lake Valdayskoe is located in the north-east 3 km of Ivanovo near the Sergievskoe and Byakovo villages. The lake is small (the area – 27.1 ha, the maximum depth – 13 m) and closed. In the 1920s Valdayskoe Lake and its coastal zone been studied in detail [11]. Large populations of *Isoëtes echinospora* were noted, the plants were found almost along the entire perimeter of the lake and in some places formed large groups. *Sparganium gramineum* formed a continuous ring 3–16 m wide at a depth of 140–240 cm, in association with *Ceratophyllum demersum*.

Lake Vysokovskoe is located in 12 km north of the city of Ivanovo near the village of Vysokovo (the area – 36.4 ha, the maximum depth – 10 m). In the 1920-1980s *Isoëtes lacustris* was noted in the lake. *Sparganium gramineum* formed small loose groups within an area of 20–80 m.

In 2012–2019 the flora of the lakes sharp decrease in the number of *Isoetes echinospora* in Valdayskoe Lake was noted. It is found in solitary specimens, superseded by *Elodea canadensis*. *Isoëtes lacustris* in Lake Vysokovskoe was not found, it might have disappeared.

Sparganium gramineum specimens were observed annually in the lakes during the studies, but morphologically they differ from the typical *Sparganium gramineum* with keeled leaves and with rising inflorescences [5]. Therefore, molecular genetic studies were performed to obtain amino acid sequences using the ITS marker to identify polymorphic positions that distinguish hybrids well [8]. The result of molecular genetic analysis (polymorphic positions) is presented in Table 1. An analysis of the results on the ITS nuclear marker revealed that *S. gramineum* from Lake Valdayskoe is a hybrid of *S.* × *longifolium*. About 12 polymorphic positions were found. Molecular genetic studies of plants from Lake Vysokovskoe have not been conducted.

Table 1. Polymorphic positions of *S.* × *longifolium* from Lake Valdayskoe by ITS marker.

Sample number	The location of polymorphic positions in the gene											
	20	33	39	45	85	87	91	16	383	388	405	451
NK9	Y	K	Y	Y	Y	Y	R	N	Y	Y	Y	Y
NK10	Y	K	Y	Y	Y	Y	R	N	Y	Y	Y	Y
NK11	C	T	T	C	T	Y	A	T	Y	Y	Y	Y
NK12	C	T	T	C	T	Y	A	Y	Y	Y	Y	T

**Note*: *G+T=K, T+C=Y, G+A=R, A+T=W, N=3, 4 ANY*

Being in close proximity to the regional center, the lakes are places of mass summer recreation for people. On the shores of the lake daily in the summer, there are more than 50-70 cars. Most cars stop near the water's edge, spontaneous picnic spots, tent camps are organized, household rubbish is left. This leads to soil pollution in these lakes. When analyzing soils in the vicinity of the investigated lakes, excesses of the maximum permissible concentration) of some heavy metals were found (Table 2.) There are almost 5 times more mobile forms of cobalt in the soils near Lake Valdayskoe than in the soils of the lake near Vysokovskoe. The content of nickel in the soil is exceeded in the soils by Lake Valdayskoe by 20%, by almost 50% by Lake Vysokovskoe.

Table 2. The content of heavy metals in soil samples.

Lake name	The metal content in mg / kg							
	Cu	**Zn**	**Ni**	**Co**	**Pb**	**Fe**	**Mn**	**Cd**
Valdayskoye	4,23*/ 6,60**	18,60/ 29,90	7,51/ 5,11	8,32/ 11,30	2,43/ 2,89	5321/ 78,90	130,1/ 73,70	0,704
Vysokovskoye	5,21/ 3,53	55,38/ 30,38	13,70/ 11,22	24,04/ 26,91	5,75/ 2,86	3777,2/ 69,30	218,34/ 88,90	0,676

* *Note*: *gross forms, **movable forms

Heavy metals salts from soils fall into lakes and plants actively absorb some heavy metals (lead and cadmium) from the soil and from water and accumulate them.

When studying the composition of water in lakes in 1920. Extreme poverty of the salt regime was observed, as well as high transparency up to 7–9 m [12]. The results of analyzes of water samples in lakes made in 2018 are presented in Table 3.

Analyzing the data of chemical indicators of water in the lakes, it was found that the composition of the water changed and salts of nitric and nitrous acids appeared in it, because the lake is used for swimming. Constant presence of nitrites and phosphates in the water is not observed. Transparency has decreased and amounts to 1.5–1.8 m in the summer months.

Despite the fact that the lakes are specially protected natural areas, protection regimes are constantly violated and not controlled. Therefore, ecosystems of lakes are disturbed, water and soil are polluted. As a result of anthropogenic changes and degradation of coastal vegetation in the studied lakes, the number of rare vulnerable plant species has sharply decreased. *Sparganium gramineum* is superseded by a more adapted hybrid of *S.* × *longifolium*. Glacial relics of *Isoëtes lacustris*, *I. echinospora* almost disappeared in these lakes due to the spread of the invasive species – *Elodea canadensis*. *Elatine hydropiper* is sporadically noted into the Valdayskoe Lake.

Table 3. The amount of dissolved substances in the water of the lakes.

Indicators(Units)	Value		Hygienic standard
	Valdayskoe Lake	Vysokovskoe Lake	
Organoleptic indicators:			
Smell at 20°C (point)	1	2	≤ 2
Smell at 60°C (point)	2	2	≤ 2
pH	7,2	6,8	6,5÷8,5
Turbidity	0,96	0,98	≤ 20
Color (grad)	18,1	20,6	≤ 35
Dry residue (mg/dm^3)	175,8	194,1	≤ 1000
Total hardness	3,1	3,4	≤ 7,0
Permanganate oxidation (mg/dm^3)	4,0	3,9	≤ 7,0
Petroleum products in total (mg/dm^3)	<0,1	<0,1	≤ 0,3
Surfactant (mg/dm^3)	< 0,01	<0,01	≤ 0,5
Phenolic index (mg/dm^3)	-	-	≤ 0,25
Dissolved oxygen (mg/dm^3)	5,4	5,7	≥ 4,0
Chemical composition of water:			
NH3 (mg/dm^3)	0,05	0,1	≤ 1,5
Fe (mg/dm^3)	0,32	0,47	≤ 1,0
Cu (mg/dm^3)	0,0014	0,005	≤ 1,0
Mn (mg/dm^3)	0,05	0,05	≤ 0,1
Nitrates (mg/dm^3)	0,17	0,68	≤ 45,0
Nitrite (mg/dm^3)	0,07	0,08	≤ 3,3
Pb (mg/dm^3)	0,0004	0,0006	≤ 0,01
Sulphates (mg/dm^3)	33,4	24,6	≤ 500
Chlorides (mg/dm^3)	6,4	6,9	≤ 350
Fluorides (mg/dm^3)	0,17	0,11	≤ 1,5
Zn (mg/dm^3)	0,013	0,018	≤ 1,0

To preserve the data of the unique lakes of the Upper Volga region and the rare species in them, strict control of compliance with protection regimes, organizational regulation of recreational loads, environmental monitoring of water, soils, and populations of rare species is necessary. It is advisable to allocate special areas – the reserves with a special protection regime will be established prohibiting any anthropogenic interference.

4 Conclusion

As a result of research in the glacial lakes of Valdayskoe and Vysokovskoe a sharp decrease in the number of rare hydrophytes (*Elatine hydropiper, Isoëtes lacustris, I. echinospora, Sparganium gramineum*) has been found. The reasons for these processes are complex.

Based on a comparison of the main indicators of water composition in the lakes, the presence of nitrates and phosphates, turbidity of the water and a sharp decrease in transparency are shown. Soils in the vicinity of the lakes are contaminated with heavy metals, lead, manganese and copper are also present in the water of the studied lakes. This is due to the eutrophication of water bodies, soil pollution with heavy metals, uncontrolled recreational loads. The negative effect on rare species distribution is caused by large groups of *Elodea canadensis*, displacing *Isoëtes lacustris, I. echinospora, Elatine hydropiper.*

The ecological niche of *Sparganium gramineum* in the lakes is occupied by *S.* × *longifolium*, whose hybridogenic origin has been confirmed by molecular-genetic analyses. Ecological monitoring of the water composition, the state of rare species populations in the lakes and control of the protection regimes of these unique lakes are thought necessary.

References

1. *Krasnaya Kniga Ivanovskoy oblasti*. Ivanovo: IPK «PresSto», 2 (2010)
2. A. A. Bobrov, O. A. Mochalova, E. V. Chemeris. Botanicheskiy jurnal, **99(9)**, 1025–1043 (2014)
3. N. K. Konotop, J. S. Vinogradova. *Problemy botaniki: istoriya i sovremennost*. Pod red. V.A. Agafonova. Voronezh: Tsifrovaya poligrafiya, **391**, (2020)
4. E. A. Borisova. Rus. Jur. of Biolog. Invasions, **2(1)**, 1–5 (2011)
5. C. D. K. Cook, M. S. Nicholls. Botanica Helvetica, **96(2)**, 213–267 (1986)
6. S. A. Barry and S. C. Clark. New Phyto, **81**, 773–783 (1978)
7. N. V. Yakovenko, D. S. Markov, E. P. Turkina. Ecology, Environment and Conservation Paper, **21**, 111-119 (2015)
8. J. D. Sulman, B.T. Drew, C. Drummond, E. Hayasaka, K. J. Sytsma. American Journal of Botany, **100(10)**, 2023–2039 (2013)
9. E. Zietkiewicz, A. Rafalski, D. Labuda. *Genome fingerprinting by simple sequence repeat (SSR)-anchored polymerase chain reaction amplification*. Genomics, **20**, 176-183 (1994)
10. M. Höss, S. Pääbo, Nucleric Acid Research, **21**, 3913–3914 (1993)
11. N. V. Corde, D. A.Lastochkin Д.А. *Valdayskoye ozero. Ocherk naseleniya ozera i yego gidrologicheskaya kharakteristika Ivanovo-Voznesensk*, 129 (1930)
12. A. N. Sivukhin, D.S. Markov, E. A. Borisova. *Problems of regional ecology*, **3**, 81–87 (2019)

Segetal complex – pioneer stages of syntrophic communities in the Meshchovskiy Opole of the non-Chernozem Zone of Russia

Valeriy A. Burlutskiy[1*], *Vladimir N.* Mazurov[1], *Alexander F.* Peliy[2], *Polina S.* Semeshkina[1], *Ekaterina S.* Borodina[3], *Ameth* Diop[2]

[1]Kaluga Research Institute of Agriculture, 249142, Kaluga region, village of Kaluga Experimental Agricultural Station, Russian Federation
[2]Peoples' Friendship University of Russia, 117198, Moscow, Russian Federation
[3]RSAU – MTAA named after K.A. Timiryazev, 127550, Moscow, Russian Federation

Abstract. In the conditions placor soil Meshchovskiy Opole the Central economic region of the Nonchernozem zone of the Russian Federation on agro-gray forest middle loamy soils defined by the main species composition of segetal complex field agrophytocenoses in the structure of five fields grain and grain-grass crop rotation in the limit of syntaxon *Stellarietea mediae* R. Tx. et al. ex von Rochow 1951. The combinations of weed and field species specific to field crops are established and the influence of minimization of basic tillage is shown, including No-till, and applying calculated norms of mineral fertilizers (NPK) on the dynamics of the species composition and number of the main components of the segetal complex is shown. The original synanthropic plant communities on the fallow lands of Meshchovsky Opole develop from the segetal complex of previous agrophytocenoses. The core of the cenoflora is represented by 104 species of higher herbage plants from 29 families. Synanthropic communities have a low species saturation – 8 species per 100 m^2 and a projection coverage of 57.6%. In acesite territories the advantage is given to the types of anemochory edificatory. Phytomass of secondary phytocenoses without comprehensive measures to improve their floristic composition cannot be used for bulk feed production

1 Introduction

Field community segetal species are formed in specific agro-climatic conditions and depending on between the cenofloras floristic regions, agricultural crops, structure of crop rotation, intensification of agricultural technology and level of environmental and economic development entity, developing floristically impoverished local segetal complexes, but with a relatively constant set of species, one of the regular causes of the declining productivity of arable land.

Modern adaptive landscape systems of agriculture are aimed at regulating the number of segmental price elements in agrophytocenoses at a level below their economic threshold of

* burlutskyvalery@gmail.com

harmfulness [1-5]. The floral composition of pioneer communities developing on fallow lands is diverse and varies from the previous segetal complex, a set of crops on arable land, and the ecological and soil conditions of the area. The analysis of successional variability of naturally formed vegetation cover on a temporarily abandoned arable land removed from active agricultural turnover is of scientific and practical significance for assessing the economic potential and developing technologies for their targeted accelerated development [6-8].

2 Materials and methods

Research on the floristic composition of weed-field communities was conducted in 1999-2019 at the Kaluga research Institute of agricultural in biologized five-field crop rotation (total area - 4950 m^2, area of 1 plot - 110 m^2, repeatability - 3 times, location-tiered) with varying degrees of saturation of the structure with leguminous and leguminous perennial grasses: crop rotations without complete fallow – 30% (*Pisum sativum* L. – *Solanum tuberosum* L. – *Fagopyrum esculentum* Moench – *Vicia sativa* L. + *Avena sativa* L. – *Triticum aestivum* L. / *X Triticosecale* Witm. ex A. Gamus) and 40% (*Vicia faba* L. – *S. tuberosum* L. – *Hordeum vulgare* L. + *Trifolium pratense* L.- *T. pratense* – *T. aestivum*), grain-grass – 60% (*L. angustifolius* – *T. aestivum* + *T. pratense* - *T. pratense* - *T. pratense* – *Triticosecale*) on agro-gray forest middle loamy soils on placor.

Research on the floristic composition of the pioneer stages of synanthropic plant communities was conducted in 2006-2019 at the Kaluga research Institute in a field experiment to study the evolution of secondary phytocenoses and the development of alternative technologies for accelerated development of abandoned land on a plot of 110.0 hectares located on a smooth hillslope of the over-floodplain terrace of the South-Eastern exposure in the Oka – Vyssa rivers interfluve. Analysis (cenoflora) of abandoned lands by a set of plant species forming communities of any rank and any type of vegetation in 2009-2019 in the Peremyshlskiy, Sukhinicheskiy, Babyninskiy and Meshchovskiy districts of the Kaluga region.

3 Results and discussion

In agrophytocenoses, synanthropic field communities classified by J Braun-Blanquet [9, 10], derivatives were formed in the syntaxon limit *Stellarietea mediae* R. Tx. et al. ex von Rochow 1951. The core of the complex was determined by the types presented by us in descending order of their projective coverage:

- moderately (*numerosae*) - *Achillea millefolium* L., *Capsella bursa-pastoris* (L.) Medik, *Centaurea cyanus* L., *Chenopodium album* L., *Cirsium arvense* L., *Convolvulus arvensis* L., *Echinochloa crusgalli* (L.) Beauv., *Elytrigia repens* (L.) Nevski, *Equisetum arvense* L., *Galeopsis tetrahit* L., *Galium aparine* L., *Matricaria chamomilla* L., *Polygonum aviculare* L. s. l., *P. bistorta* L., *P. convolvulus* L., *Rumex acetosella* L., *Stelaria media* L., *Taraxacum officinalis* Wigg., *Thlaspi arvense* L., *Trifolium arvense* L., *Stachys arvensis* L.;
- few (*pauces*) - *Agrostis tenuis* Sibth., *Atriplex patula* L., *Cirsium setosum* (Willd.) Bess., *Consolida regalis* S.P. Gray, *Galeopsis speciosa* Mill., *Linaria vulgaris* Mill., *Matricaria perforate* Merat., *Menta arvensis* L., *Polygonum lapathifolium* L., *P. persicaria* L., *Raphanus raphanistrum* L., *Sonchus arvensis* L., *Spergula arvensis* L.;
- rarely (*rarae*) - *Barbarea vulgaris* R. Br., *Bunias orientalis* L., *Euphorbia helioscopia* L., *Fumaria officinalis* L., *Stachys palustris* L.

Revealed increased phytocoenotic role rhizomatous segetal components in the period 2009 - 2019, relative to previous decades (figure 1).

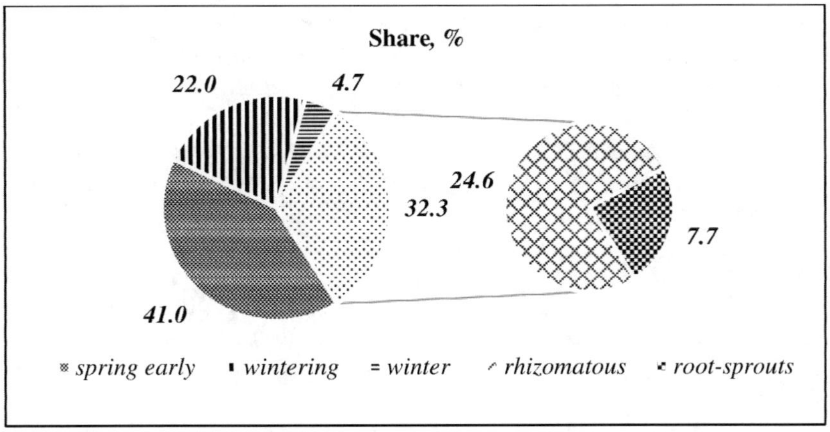

Fig. 1. Structure of a segmental complex

In agrophytocenoses, stable culturally specific combinations of segetal complex species were formed after the third rotation in the crop rotation (table 1).

Table 1. Main intra-field combinations of segetal species

Agrophytocenos	Combination of species
H. vulgare, T. aestivum, Triticosecale, V. sativa + A. sativa	I – *A. millefolium, E. repens, Eq. arvense, C. cyanus, Ch. album, C. arvensis, M. arvensis, P. aviculare, P. convolvulus, S. arvensis, S. media,*
	II – *A. tenuis, B. vulgaris, Eu. helioscopia, P. persicaria, T. officinalis, Th. arvense*
F. esculentum	I – *C. cyanus, Ch. album, P. aviculare, P. convolvulus, S. arvensis, S. media, T. arvense*
	II – *A. millefolium, E. repens, Eq. arvense, M. arvensis, R. acetosella*
L. angustifolius, P. sativum, V. faba	I – *C. setosum, R. acetosella, S. arvensis*
	II – *A. tenuis, B. vulgaris, Eq. arvense, E. repens, Eu. helioscopia, M. arvensis, P. persicaria*
T. pratense	I – *A. millefolium, E. repens, Eq. arvense*
	II – *C. arvensis, P. persicaria, R. acetosella, S. arvensis, S. media, T. officinalis*

Note: I – characteristic, II – accompanying

In the link of crop rotation legumes - winter cereals, the use of basic shallow tillage (disking-BDU-2.5) contributed to an increase in weed infestation of crops on average in one field by 52%, and the technology (direct seeding) No-till (RAPID A600) – by 65%, relative to the annual plow up (PLN-3-35) – 28 copies/m^2. The application of mineral fertilizers (NPK) in the norms calculated using the full balance method increased the participation of components of the segetal complex by 6 ... 15% in the structure of agrophytocenoses.

Communities of fallow period were assigned to syntaxons to the class Stellarietea mediae Tx. et al. ex von Rochow 1951, later ones belong to the class *Artemisietea vulgaris* Lohmeyer et al. ex von Rochow 1951. The syngenetic series was limited by pyrogenic influence. Derivatives were prone to frequent fluctuations. Cenoses had a low saturation – 8 species per 100 m^2, the projection coverage-57.6% on average. The spectrum consisted of 104 species of higher plants of 29 families, the most frequent of which were: *Achillea millefolium* L., *Agrostis capillaris* L., *Alchemilla vulgaris* L., *Apera spica-venti* (L.) Beauv., *Artemisia absinthium* L., *A. abrotanum* L., *A. vulgaris* L., *Barbarea vulgaria* R. Br., *Berteroa incata*

(L.) DC., *Capsella bursa-pastoris* (L.) Medik, *Carex* spp., *Carum carvi* L., *Centaurea jacea* L., *Chenopodium album* L., *Convolvulus arvensis* L., *Cichorium intybus* L., *Cirsium setosum* (Willd.) Bess., *Dactylis glomerata* L., *Deschampsia caespitosa* (L.) Beauv., *Echinochloa crus-galli* (L.) Beauv., *Equisetum arvense* L., *Elytrigia repens* (L.) Nevski, *Erodium cicutarium* (L.) L` Her., *Erigeron canadensis* L., *Euphorbia virgata* Waldst. et Kit., *Galingosa parviflora* Cav., *Galeopsis tetrahit* L., *Glechoma hederacea* L., *Festuca* spp., *Hypericum perforatum* L., *Juncus* spp., *Leontodon autumnalis* L., *Leucanthemum vulgare* Lam., *Matricaria discoidea* DC., *Phleum pratense* L., *Plantago media* L., *P. major* L., *Poa annua* L., *P. pratensis* L., *P. trivialis* L., *Polygonum aviculare* L. s. l., *Potentilla anserina* L., *Ranunculus acer* L., *R. repens* L., *Rumex acetosella* L., *R. confertus* Willd., *Sonchus arvensis* L., *Spergula arvensis* L., *Stellaria media* (L.) Vill., *S. graminea* L., *Tanacetum vulgare* L., *Taraxacum officinale* Wigg., *Thlaspi arvense* L., *Trifolium* spp., *Tussilago farfara* L., *Vicia* spp., *Viola arvensis* Murr.

The community was presented agrobiological groups of grasses (55%), cereals (25%), legumes (15%), spore (10%) and sedge-juncaceae (5%). Cenoses with adventive participation were more variable in productivity and quality of biomass, the yield was 1.5 times higher than that of native plants [11, 12]. The number of species in communities was due to the diversity of the segetal complex, and its increase was due to the possibility and availability of immigration of diasporas. Species abundance was determined by the prevailing groups. Mutually complementary relationships appeared between cereals and legumes, herbs, and spores, sedge-juncaceae in other combinations - antagonistic (Table 2).

Table 2. Correlation coefficient ($r \pm r_s$) relationships of agrobiological groups

Agrobiological group	Legumes	Motley grass	Sedge-juncaceae	Spore
Cereal	0.95±0.04	-0.98±0.01	-0.60±0.21	-0.88±0.10
Legumes	-	-0.99±0.01	-0.33±0.26	-0.95±0.04
Motley grass		-	0.40±0.24	0.92±0.06
Sedge-juncaceae	-	-	-	0.32±0.29

In the motley grass group *Compositae* the proportion species was 20% and 5 advents (*Erigeron annus* (L.) Pers., *E. canadensis* L., *Galingosa parviflora* Cav., *Matricaria discoidea* DC., *Solidago gigantea* Ait.). Share of species *Amaranthaceae, Chenopodiaceae Convolvulaceae, Fumariaceae, Hypericaceae, Onagraceae, Papaveraceae, Violaceae* was low – 6,4% (Fig. 2).

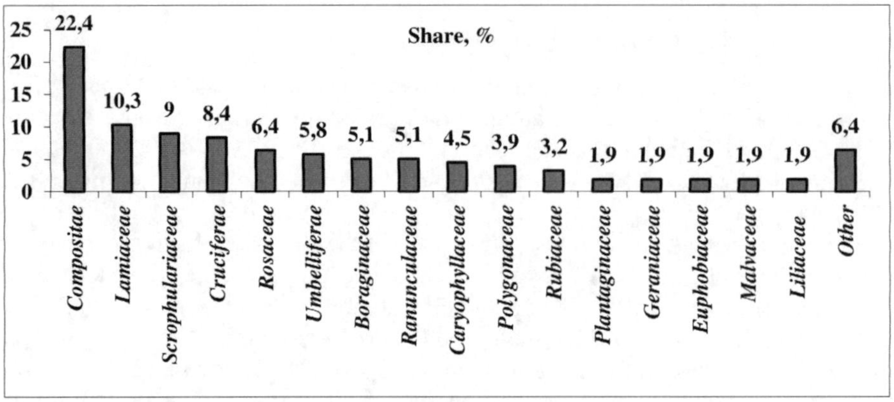

Fig. 2. Distribution of families in the motley grass group

Communities *in statu nascendi* were subjected to anemochoric invasion of diasporas. Aggregations developed that correlated with the structure of bordering communities. The degree of abundance on the Drode scale for *Salix* spp. was – Sol_{gr}/Sp-Cop_1, *Betula* spp. – Sol_{gr}-Sp/Cop_1, *Populus tremula* L. – Un-Sol/Sp, *Pinus sylvestris* L. – Un_{gr}/Sol_{gr}, and (neophyte) *Acer nedugo* L. – Un/Un-Sol. Advantages in ecesis were anemochora species - edificators *Salicaceae*, *Betulaceae*, *Pinaceae*.

4 Conclusion

The main floral composition of the segetal complex of field agrophytocenoses in the syntaxon limit in the conditions of Meshchovskiy Opole was determined *Stellarietea mediae* R. Tx. et al. ex von Rochow 1951. Cultural-specific combinations of weed-field vegetation species have been established. Minimization of the main tillage and introduction of calculated norms of mineral fertilizers contributes to an increase in the degree of contamination of crops by 52 ... 65% in biologized farming systems. Characteristics of secondary phytocenoses indicate that without a complex of agrotechnological techniques to improve their floral composition, phytomass cannot be important in the production of bulky feed. Communities are formed by 104 species of higher herbage plants from 29 families have low species saturation and low projective coverage and develop from the segmental complex of previous agrophytocenoses. In ecessis anemochores edificators have an advantage.

References

1. A. Kashtanov, O. Sizov, E. agr. in Rus., **11,** 174 (2008)
2. N. Semenov, N. Muromtsev, A. Snitko, For. prod., **3**, 3 (2016)
3. J. Alix-Garcia, Land Use Pol., **59**, 284 (2016)
4. A. Prishchepov, D. Müller, M. Dubinin, Land Use Pol., **30**, 4, 873 (2013)
5. Y. Helin, D. Juan, J. RUDN: Ecol. & l. s. **3**, 17 (2014)
6. V. Burlutskiy, V. Mazurov, P. Semeshkina et all, Bul. Rus. agr. scien., **3**, 18 (2019)
7. M. Makarova, M. Redina, A. Popkova, J. RUDN, Ecol. & l. s., **25**, 1, 7 (2017)
8. G. Sigora, L. Nichkova, T. Khomenko, J. RUDN, Ecol. & l. s., **25**, 1, 88 (2017)
9. J. Braun-Blanquet, *Pflanzensociologie* (Wien, 1964)
10. N. Ermakov, B. Mirkin, L. Naumova, *The current state of the vegetation* (Ufa, Gilem, 2012)
11. V. Burlutskiy, V. Mazurov, I. Osokin et all, J. RUDN, Agr. & anim. husb.,**14**, 2, 114 (2019)
12. E. Garnier, M.-L. Navas, K. Grigulis, *Plant Functional Diversity: Organism traits, community structure, and ecosystem properties* (Oxford, Oxf. Un. Pr., 2016)

E3S Web of Conferences **169**, 03005 (2020) https://doi.org/10.1051/e3sconf/202016903005
APEEM 2020

Do Diversity and Bio-morphological Spectrum of Psammophytes Matter in Restoring Vegetation of Sand Pits?

Alaa Dabbagh*

Peoples' Friendship University of Russia (RUDN University), Faculty of Ecology, 6 Miklukho-Maklaya St, Moscow, 117198, Russian Federation

Abstract. Open mining of sand is executed in the sand pits. After completion of the work sand pits are usually recultivated: firstly they are graded and then planted with trees and shrubs, but very often sand pits are abandoned to regenerate themselves. The sand pit becomes a source of dust and a hardly healed wound on the earth's body. In order to provide an adequate revegetation on such substrates, it is necessary to study bio-morphological peculiarities of growing plants in a very specific environment characterized by high stiffness of slopes and high mobility of the substrate. The material for our study collected from Moscow Region sand pits in the Russian federation. Plants were collected from both northern and southern exposure slopes. Our work is based on the geobotanical descriptions made according to standard method on experimental sites of 5x5 m^2. The results of the undertaken shows that *Compositae, Gramineae, Fabaceae* are the most representative of the surveyed one hundred twenty seven species of higher vascular plants from 29 families. Diversity on the southern slopes was more than on the northern ones. Species diversity is determined by the successions occurring spontaneously after a rough anthropogenic intervention, under the influence of such environmental factors as lighting, moisture, texture and richness of soil. The spectrum of life forms showed that the overwhelming majority according to K. Raunkiru are hemicryptophytes. Perennial herbaceous polycarpics are dominant in the spectrum of life forms, according to I.G. Serebryakov.

1 Introduction

A huge impact on the environment is resulted from the unfair exploitation of sand pits during which changes on the cycle of substances and energies, the dynamics of natural processes, the structure and productivity of bio-systems occur, and the species diversity of plants, also, gets lost. This impact is, clearly, embodied in land scenery, and changes in sand pits hydrogeological and climatic regime [1, 2, 3]. Under a good fitting set of circumstances, sand pits vegetation self-recovery will take not less than one hundred years [4, 5]. Among the stages of reclamation, the biological stage, in the restorations process of sand pits vegetation

* Corresponding author: alaa_dabbagh@hotmail.com

is considered to be the most important one [6]. In order to identify sustainable species that are promising for the restoration of disturbed lands, floristic studies are to be done.

Sand pits can be defined as unusual habitats with high substrate mobility and sharp fluctuations in temperature and water conditions of the slopes, a high degree of steepness, and with peculiar vegetation. It is similar to that characteristic of well-aerated, fine-grained substrates with a good washing regime [7]. Sand pits plants are adapted to life on moving sands: they are either covered with sand, or their underground organs are exposed [8]. To provide water and mineral nutrition a deep and highly branched root system were developed as a form of adaptation (for example, *Artemisia vulgaris* L.). Farther more it fixes it on a mobile substrate [9].

The aim of this research is to study sand pits plants bio-morphological characteristics in order to be used later on as the scientific basis for any project aiming at the restoration of vegetation cover of any disturbed land.

2 Materials and methods

The materials of our research have been collected from five sand pits located in different direction of Moscow region (Zvenigorodsky, Lyzlovo located in the west of Moscow and Dzerzhinsky, Ramensky, Lytkarinsky located in the southeast of Moscow). The Fixed Route method and the Geobotanical descriptions method were used; a list of vascular plant species was compiled in all above mentioned cites. Geobotanical descriptions on an area of 5x5 m^2, performed according to standard methods, located along the northern and southern slopes were the basis of the work. The total cover percentages of the found plants species determination, the dominant and subdominant species allocation and a list of species indicating the cover percentage, height and phenophase compilation were obtained through the implementation of the geobotanical description. K. Raunkier and I.G. Serebryakova life forms were adopted as the literature bases to perform the biomorphological analysis of the studied flora [10, 11].

3 Results and Discussion

3.1. Species diversity of sand pits vegetation

Insignificant species diversity of sand pits vegetation has been noticed. During 4 years, 127 species belonging to 29 families were found, which, in general, is common of sandy landscapes. Diversity ratio is consistent with Demina O.N. et al. [8]. Ten of the 127 species are adventitious plants for example *Robinia pseudoacacia* L. *Acer negundo* L. Beside the smooth progress of the landscape development (due to the rough technological intervention that has occurred), the number of species is determined also by the succession shifts under the influence of environmental factors, such as light, moisture and soil richness, and topological characteristics of the sand pits, of which steepness of the slopes can be mentioned. It should be noted that species diversity of southern exposure slopes, is higher compared to the northern exposure slopes. It is explained by habitat more contrasting conditions. A wide range of environmental factors (light exposure, soil moisture, etc.) allows more species of plants to settle on the southern slope.

Vascular plants grouped by families, growing in the Moscow region sand pits, are distributed unevenly. *Compositae*, *Gramineae*, *Fabaceae* are the largest number of species which fall within families (Table 1).

Table 1. Taxonomic spectrum of native species vascular plants of sand pits in the Moscow region.

Nr.	Family name	Number of species	% of the total number of species	Number of genera	% of the total number of genera
1	*Compositae*	21	17.9	15	17.4
2	*Gramineae*	19	16.2	14	16.3
3	*Fabaceae*	14	12.0	7	8.1
4	*Caryophyllaceae*	6	5.1	5	5.8
5	*Polygonaceae*	6	5.1	2	2.2
6	*Rosaceae*	6	5.1	4	4.6
7	*Labiatae*	5	4.2	5	5.8
8	*Scrophulariaceae*	5	4.2	4	4.6
9	*Umbelliferae*	5	4.2	5	5.8
10	*Cruciferae*	4	3.3	4	4.6
11	*Salicaceae*	4	3.3	2	2.2
12	*Betulaceae*	2	1.7	1	1.2
13	*Hypericaceae*	2	1.7	1	1.2
14	*Liliaceae*	2	1.7	2	2.2
15	*Rubiaceae*	2	1.7	1	1.2
16	*Boraginaceae*	1	0.9	1	1.2
17	*Chenopodiáceae*	1	0.9	1	1.2
18	*Cyperaceae*	1	0.9	1	1.2
19	*Dipsacaceae*	1	0.9	1	1.2
20	*Equisetaceae*	1	0.9	1	1.2
21	*Oleaceae*	1	0.9	1	1.2
22	*Onagraceae*	1	0.9	1	1.2
23	*Pinaceae*	1	0.9	1	1.2
24	*Plantaginaceae*	1	0.9	1	1.2
25	*Primulaceae*	1	0.9	1	1.2
26	*Ranunculaceae*	1	0.9	1	1.2
27	*Ulmaceae*	1	0.9	1	1.2
28	*Urticaceae*	1	0.9	1	1.2
29	*Papaveraceae*	1	0.9	1	1.2
	Total	**117**	**100**	**86**	**100**

3.2. K. Raunkiaer life forms spectrum of plants in sand pits

A typical life form that dominates the phytocenoses of Central Russia is hemicryptophyte. Hemicryptophytes, according to the data obtained by the analysis of K. Raunkier life forms for plants of sand pits in the Moscow Region, dominate on the northern and southern slopes of the pits (63%). They are well adapted even to the peculiar and often extreme growing conditions on the sandy substrate. (Fig.1). Hemicryptophytes rosette shoot serves successful wintering and also contributes to the economical use of energy. Such plants can begin to grow earlier. They make use of spring moisture efficiently and at the same time they avoid competition with stronger species [12].

 https://doi.org/10.1051/e3sconf/202016903005

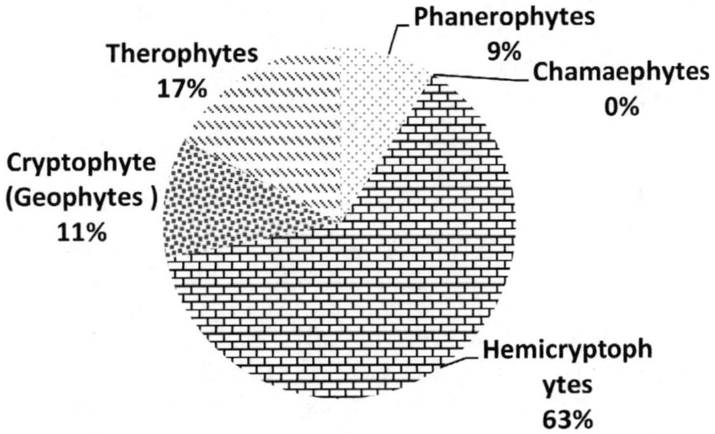

Fig. 1. Spectrum of life forms according to the system of K. Raunkiaer on different habitats.

Therophytes have a significant proportion (19 species, 17%). In general and despite of the difficult environmental conditions of sand pits (strong lighting, sand movement, harsh winter conditions in the Moscow region), therophytes are actively undergoing unfavorable conditions in the form of seeds. They also actively take control over anthropogenic violated territories. Curiously, chamaephytes are completely absent in most of the sand pits. Such differences, most likely, are not associated with the exposure of the slopes but with vegetation species diversity, that spontaneously develops in each sand pit [13].

3.3. I.G. Serebryakov life forms spectrum of plants in sand pits

According to I.G. Serebryakov life forms spectrum, all sand pits were diferentiated by the predominance of perennial herbaceous polycarpics, among which large share of tap rooted plants *Taraxacum officinale* L., followed by long-rhizome *Tussilago farfara* L., then short-rhizome *Artemisia vulgaris* L. (Fig.2).

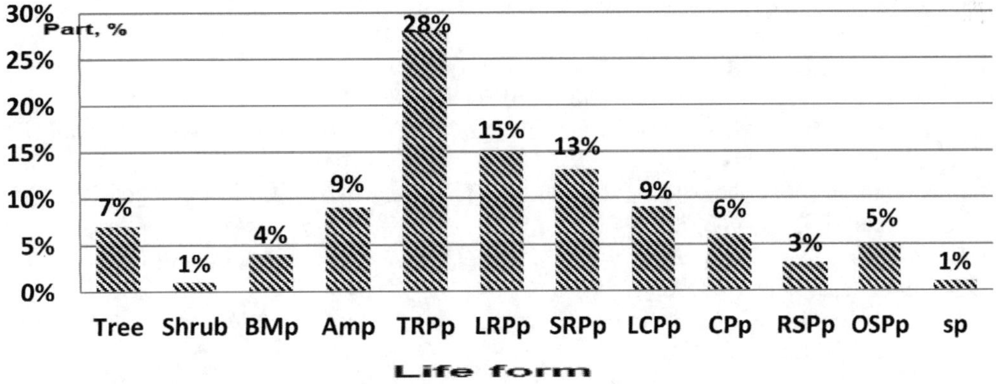

Fig.2. the spectrum of life forms by I. G. Serebryakov on different habitats
Notation keys: BMp – biennial monocarpic; Amp – annual monocarpic; TRPp – tap root polycarpic; LRPp – long rhizome polycarpic; SRPp – short rhizome polycarpic; LCPp – loose cespitose polycarpic; CPp – creeping polycarpic; RSPp – root sucker polycarpic; OSPp – overground stoloniferous polycarpic ; Sp – semishrub

The dominance of tap rooted plants is, mainly, due to their ability to adapt to strong light conditions and resistance to drought, and the capability to get water from deep layers of the soil. Long and short rhizome plants tend to loose substrates and soft soils, although the latter are more adaptive to the growing conditions [14].

4 Conclusions

To sum up, it is worth mentioning that sand pits plants, perform a pioneering function; they bind sand grains, prevent or reduce substrate movement by forming subordinate roots and vegetative shoots, they quickly start growing after falling asleep with sand, have a branched or deeply penetrating root system, are resistant to drought and mechanical damage, can improve soil properties by holding it or enriching leaf litter, change wind intensity. The aforesaid conclusions can clearly answer our question "Do Diversity and Bio-morphological Spectrum of Psammophytes Matter in Restoring Vegetation of Sand Pits?" Theoretically, yes they matter, where implementation is left for further empirical research to answer the question.

Relying on the results of our research we recommend the use of herbal perennial plants to obtain vegetation on sand pits. They are able to form turf and prevent the movement of sand slopes quickly. Cereals are considered of primary importance, for example: *Festuca rubra* L. *Bromopsis inermis* (Leyss.) Holub. *Festuca pratensis* Huds. *Agropyrum tenerum* Vaseg. *Dactylis glomerata* L. *Elymus fibrosus* Schrenk. *Elytrigia repens* (L.) Nevski. *Calamagrostis epigeios* (L.) Roth. *Phleum pratense* L. In terms of appropriacy and importance, legumes occupy the second place. For example, *Trifolium pratense* L. *Melilotus officinalis* (L.) Pall. *Trifolium hybridum* L. Finally, oligotrophic species can be mentioned from the tree layer.

References

1. M.V. Makannikova, Z.A. Ionova, Recl. Of. Dist. Lands. E **2**, 131 (2018)
2. F. Boscutti, A.Vianello, F. Bozzato, Rest. Eco. 25. E **4**, 595 (2017)
3. O. Sumina, E. Koptseva, Cur. Tren. In. Land. Resear. E **3**, 559 (2019)
4. R. Gentili, S. Sgorbati, C. Baroni, Rest. Eco. 101. E **19**, 32 (2011)
5. A. Bradshaw, Land. And. Urb. Plan. **51**, 89 (2000)
6. E. Diana, Ear. And. Env. Scien. **194** (092007) (2018)
7. L.A. Maslennikova, *The psammophilic flora of the central part of the Volga Upland* (PhD thesis, Moscow, 1999)
8. O.N. Demina, P.A. Dmitriev, L.L. Bul. Of. The. Sam. Scien. Cent. Of. The. Rus. Acad. Of. Scien. E **4**, 1004 (2012)
9. B.A. Bykov, *The main ecobiomorphs of desert plants of the Turan lowland (brief review)* (Alma-Ata- Science, Moscow, 1987)
10. A.M. Aleinikova, E.A. Vanisova, E.Yu. Vasilieva, *Study of natural ecosystems* (House of RUDN Moscow, 2015)
11. Yu.E. Alekseev, P.Yu. Zhmylyov, E.A. Karpukhina, *Plants of the city of Dubna* (International University of Nature, Society and Man "Dubna", 2014)
12. N.A. Berezina, N.B. Afanas'eva, *Plant ecology* (Academa, Moscow, 2009)
13. O.B. Калашникова, Bul. Of. The. Sam. Scien. Cent. Of. The. Rus. Acad. E **1**, 839 (2011)
14. H.T. Gajrabekov, S.B. Macaev, M.V. Geroeva. Vest. KrasGAU. E**4**, 128 (2012)

The study of the zoobenthos of the Tsraudon river basin (the Terek river basin)

Ia E. Dzhioeva[*], *Susanna K.* Cherchesova , *Oleg A.* Navatorov, and *Sofia F.* Lamarton

North Ossetian state University named after K.L. Khetagurov, Vladikavkaz, Russia

Abstract. The paper presents data on the species composition and distribution of zoobenthos in the Tsraudon river basin, obtained during the 2017-2019 research. In total, 4 classes of invertebrates (Gastropoda, Crustacea, Hydracarina, Insecta) are found in the benthic structure. The class Insecta has the greatest species diversity. All types of insects in our collections are represented by lithophilic, oligosaprobic fauna. Significant differences in the composition of the fauna of the Tsraudon river creeks and tributary streams have been identified. 7 families of the order Trichoptera are registered in streams, and 4 families in the river. It is established that the streamlets of the family Hydroptilidae do not occur in streams, the distribution boundary of the streamlets of Hydropsyche angustipennis (Hydropsychidae) is concentrated in the mountain-forest zone. The hydrological features of the studied watercourses are also revealed.

1 Introduction

The biocenoses of flowing reservoirs of the North Caucasus, and especially small rivers, remain insufficiently explored today; particularly, there is no information about the systematic composition, biology and ecology of amphibiotic insects (mayflies, stoneflies, caddisflies and dipterous) of the studied basin. Amphibiotic insects are an essential link in the food chain of our reservoirs and at the same time can be attributed to reliable indicators of water quality. In this regard, a comprehensive study of mountain streams and, in particular, small rivers is a very important tool for biomonitoring, on the basis of which we can judge both the state of a particular watercourse and the river network as a whole. Small rivers are a source of drinking water, and they are unique refugium's of amphibious fauna. The diversity of amphibiotic communities is a guarantee of the stability of aquatic ecosystems. With such a sensitive tool for long-term monitoring, we can judge about the well-being of the watercourse based on the numerical changes in zoobenthos groups and their species composition. One interesting waterways in the study of dynamics of communities of zoobenthos, is a small river Tsraudon (the Terek river basin).

[*] Corresponding author: cherchesova@yandex.ru

2 Materials and methods

The paper presents the results of research carried out by us in the basin of the river Tsraudon in the 2017-2019. Data on the hydrology and composition of amphibiotic insects of the Tsraudon river were given earlier [1, 2]. In this paper, we consider the species composition of amphibiotic insects of the Tsraudon river basin (the Tsraudon river and its tributary streams). When comparing the fauna of streams and the river Tsraudon, we rely on the data of the above-mentioned work. During the research, 5 stations were installed in various altitude zones (from 550 m above sea level to 1300 m above sea level). In total, more than 2000 specimens of amphibiotic and aquatic insects, as well as representatives of related fauna (Crustacea, Mollusca, Turbellaria, Hydracarina) were collected.

3 Results

The Tsraudon river is a typical small river of the Terek basin [1, 4], which flows in the mountain-forest and foothill zones of the Republic. The river is fed by underground water and rain. In the mountainous part of the basin-the river is transparent, the rocky bottom is composed of large boulders and pebbles of light rocks. The speed of water flow in this section, which passes in the mountain forest, where valuable tree species are concentrated (beech, hornbeam, birch) is 1.5 m/s, the depth of the stream reaches 0.3-0.5 m. spring and summer floods are typical for the watercourse. On the foothill section, the flow depth is 0.1-0.3 m, the water transparency is reduced, and the water flow rate is reduced to 0.5-1.0 m/sec. During the period of snowmelt (may-june) and heavy rains (late may – mid – july), the river becomes mudslide-the flow speed during this period is 5-7 m/sec. Tributary streams also flow in mountain-forest and foothill zones, they are characterized by the following indicators: flow rate – 0.4 – 1.2 m/s, depth – 0.1-0.3 m, water consumption 0.1 – 0.9 m^3/s. In hot summer and winter, the water in the streams becomes shallow and often dries up, and during heavy rains the water level in the streams reaches critical values. The water temperature in summer ranges from 8^0C to 12^0C. The bottom of the streams is stony with sandy and muddy scythes, along the banks there are thickets of blackberries, raspberries, elderberries, hazel, there are walnut plantings.

As a result of cameral processing of the material, we have established 5 classes for the Tsraudon river basin: Gastropoda, Turbellaria, Crustacea, Arachnida and Insecta. The leading role in the structure of zoobenthos is played by insects (Insecta) – 96%, represented by 4 orders, 19 families and 42 species. All other representatives of zoobenthos (Mollusca, Turbellaria, Amphipoda, Hydracarina) make up 4% of all gathers.

4 Discussion

As mentioned above, in the structure of the zoobenthos of the Tsraudon river basin, the most numerous is the class Insecta, which unites 4 orders (Ephemeroptera, Plecoptera, Trichoptera and Diptera). The order Ephemeroptera is represented by 5 families and 11 species: Ephemerellidae (1), Baetidae (3), Heptageniidae (5), Oligoneuriidae (1), Caenidae (1). The order Trichoptera is represented by 6 families and 12 species: Rhyacophilidae (3), Hydropsychidae (4), Glossosomatidae (1), Apataniidae (1), Lepidostomatidae (1), Limnephilidae (2), and Hydroptilidae (1). The order Plecoptera includes 5 families and 13 species: Perlidae (2), Perlodidae (1), Nemouridae (5), Taeniopterygidae (2), Capniidae (1), and Leuctridae (2). The order Diptera is mainly represented by 3 families and 6 species: Chironomidae (2), Simuliidae (3), Blepharoceridae (1).

In addition to the representatives of the insect class, the structure of zoobenthos includes gastropods, amphipods, planarians, and aquatic mites, which are significantly inferior both in numbers and in species diversity to the insect class (Fig. 1, 2).

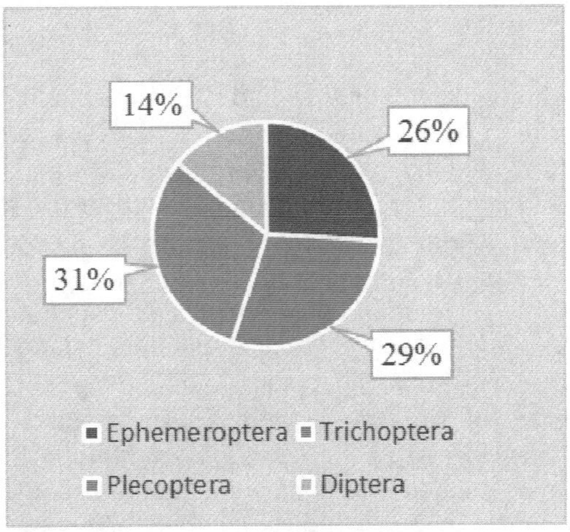

Fig. 1. Taxonomic ratio of orders of the class insects

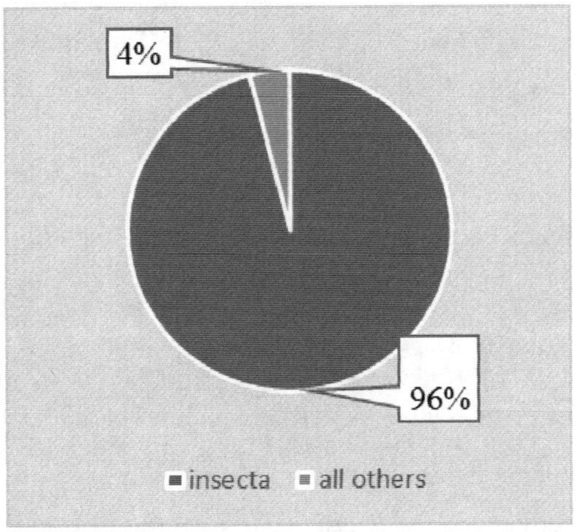

Fig. 2. Percentage of insects and other representatives of zoobenthos

The fauna of tributary streams differs from that of rivers by the predominance of representatives of the order Trichoptera, for the development of which the stability of the biotope and temperature regime are favorable. There are 7 families registered here: Rhyacophilidae, Hydropsychidae, Glossosomatidae, Philopotamidae, Limnephilidae, Lepidostomatidae, Apataniidae, which make up 74% of the total mass of benthic streams. Only 4 families (Rhyacophilidae, Hydropsychidae, Glossosomatidae and Hydroptilidae) have been recorded for the Tsraudon river. The density of caddisflies in streams is quite high (more than 1000 copies/m^2). The dominant species in the streams of the basin are: *Glossosoma capitatum* Mart., *Pothamophylax latipennis* Curt., rare ones are: *Dinarthrum longiplicatum* Mart.

The difference in species composition of the river fauna is the dominance of benthos species of caddisflies of the family Hydropsychidae (it should be noted that the species *H. angustipennis* Curt. is distributed only in the mountain-forest zone), in spring there are a lot of Hydroptilidae caddisflies in the Tsraudon river (in the streams-tributaries of the pool family Hydroptilidae are not registered).

5 Conclusion

Comparing the features of the faunas of streams and rivers, it should be noted that typical stream species are: *Wormaldia subnigra* McL. (rare in our collections); *Pothamophylax latipennis* Curt., *Limnephilus extricates* VcL., *Dinarthrum longiplicatum Mart.*; species from the families Rhyacophilidae, Glossosomatidae, Hydropsychidae are found in the river and in the streams; the family Hydroptilidae in our collections – numerous, spring typical river view.

Zoogeographic composition of trichopteran fauna: *Agapetus oblongatus* Mart. - endemic to the Caucasus; *G. capitatum* Mart., *D. longiplicatum* Mart. - subendemic; *H. sciligra* Mal. - Central Asian; *R. fasciata* Hag., *W. subnigra* McL., *H. angustipennis* Curt. – European; *R. nubile* Zett., *H. contubernalis* McL., *P. latipennis* Curt. – Palearctic.

Ecology of species: issues of ecology of amphibiotic insects living in small rivers were discussed earlier [3, 5, 6]. the analysis of the fauna of amphibiotic insects of the river Tsraudon and streams-tributaries of the studied basin showed that the larvae of mayflies, caddisflies, stoneflies and dipteran live in the surveyed reservoirs all year round. Biotopes of collected larvae are typical mountain streams with a high flow rate, and clear water, summer temperature is 12-14^{0}C.

Thus, the main faunal core of hydro biocenoses of the Tsraudon river basin is represented by oligosaprobic, stenothermic and litophilous fauna.

References

1. Kornoukhova I.I., Cherchesova S.K. *Fauna and ecology of animals of the Caucasus*, 41-45 (Ordzhonikidze, 1987)
2. Cherchesova S.K. *Amphibiotic insects (Ephemeroptera, Plecoptera, Trichoptera) of the rivers of North Ossetia* (Moscow, 2004)
3. Tsibirova L.L., Khablieva A.A., Kompantsev A.A., Cherchesova S.K. Bulletin of Tambov University. Series: Natural and technical Sciences, **22(5-1)**, 1031-1034 (2017)
4. Phallagova Dz.M. *Proceedings of the Institute of the History, Economics and Literature* (Ordzhonikidze, 1973)
5. Cherchesova S.K. Proceedings of the Russian State Agrarian University - Moscow Timiryazev Agricultural Academy, **3**,182-189 (RSAU, Moscow, 2002)
6. Cherchesova S.K., Tuaeva S.R., Kornoukhova I.I. Proceedings of Gorsky State Agrarian University, **52(4)**, 385-390 (2015)

Productivity dynamics of oak forests of the Crimean Peninsula

Roman Gorbunov*

IBSS, 299011 Sevastopol, Russia

Abstract. The work proposes a methodology for studying the productivity dynamics of forest ecosystems based on remote sensing data. The possibility of using the NDVI vegetation index to study the interannual variability in productivity of regional ecosystems is shown. The long-term productivity dynamics of oak forests of the Crimean Peninsula is analyzed. On the basis of the data series obtained, the periods characterized by peculiarities of the productivity dynamics of forest ecosystems are identified.

1 Introduction

The history of GIS-modeling of ecosystem functioning and dynamics is extremely extensive. NDVI (Normalized Difference Vegetation Index) is often used as an indicator of ecosystem productivity, which is one of the many indexes developed at present to determine the biological ecosystems productivity based on the calculation of multispectral space images channels ratio [1-6]. NDVI index among the whole variety of vegetation indexes is one of the basic ones and has a great practice in determination, well-established methodology of interpretation data obtained and is already built as automatic scripts into modern geoinformation software packages.

Consideration of the index methodology within this work is not interesting, since it is formed and tested. It should only be noted that in criticizing the practice of using this index the question is raised concerning the necessity of comparing data with test sites to determine the actual biomass value, since the index is relative. In this study we are more interested in the productivity dynamics in relative values when climatic parameters change rather than actual biomass. It should also be noted that for the territory of Crimea, including natural plant communities, there is wide field experience in determining the communities' productivity, and the index interpretation is also possible, but in our studies is not necessary.

Many works are devoted to the productivity studying of forest ecosystems in Crimea. The most systematized data on the dynamics of forest communities and forests' productivity on the Crimean Peninsula are presented in the works of Yu.V. Plugatar [7-10]. At the same time, the presented works are focused on studying the productivity of specific trees' species, rather than ecosystems as a whole. In this regard, the aim of this work is to

* Corresponding author: karadag_station@mail.ru

develop approaches and assess the interannual dynamics of the oak forests productivity on the Crimean Peninsula using remote sensing data.

2 Materials and methods

Landsat space images for the period from 1979 to 2017 were used as the research material. The images were obtained by downloading from the USGS open source (https://www.usgs.gov/). Images with the maximum set of color channels of Level 1 GeoTIFF Data Product type were selected from the entire satellite data catalogue. A total of 277 space images were received for different periods with different territorial coverage of the Crimea. The sensor resolution of each image is 30x30 m. Specifics of Landsat satellite orbit and camera operation peculiarities divide the territory of the Crimea into 4 overlapping segments, which, when overlaid, allow getting quite a detailed display of the territory in multichannel spectrum. The images were processed in ArcGis 10.4.2 program.

Features and history of the Landsat project do not provide an opportunity to get a complete set of all four satellite image segments for each of the interesting periods. In some cases 2, 3 or even 1 image segment may be presented for one date, depending on the image orbit passage. In addition, there is a certain shift of some images when switching from satellite to satellite (the period from 1979 to 2017 covers the operation of Landsat3, Landsat4, Landsat5, Landsat7 and Landsat8 satellites).

In addition, the presence of clouds in the images is very important when selecting images. The cloud minimization rule in the image was used.

It is important to select the date range, within which the image segments can be identical to the connection. Taking into account that the main purpose of image analysis is to estimate the dynamics of biological productivity, the target period for our studies is the vegetation period.

Besides, analysis of the images presented in the project allows concluding that in winter and even in autumn and spring almost all space images have cloud cover. In some cases it reaches 100% and does not allow using the image.

Within the vegetation period, we chose August as the month. In the period of this month for the period from 1979 to 2017 the images are available on the following dates: 01.08, 05.08, 06.08, 12.08, 16.06, 20.08, 26.08, 29.08, 31.08. The whole range of images was processed, images with cloudiness was rejected.

The choice of sequences overlay images is conditioned by the rule as the cloudiness is removed and the date approaches the middle of the vegetation period and the middle of August.

The space images database analysis shows that it is not appropriate to use the June month images, as in this case the maximum value of biological diversity for this period has not been reached. In July the biomass value is considerably higher, however in practice almost every space image contains cloud cover, especially within the Mountainous Crimea, which is associated with high precipitation value in this period in Crimea. Taking this into account it is rather difficult and not appropriate to use July images.

Thus, both from a technical and logical point of view, the using of images for August month is the most optimal, because on the one hand, during this period a higher biomass value has been achieved, and on the other hand, images for this period are "the cleanest" from the clouds, and for almost every year of the selected period it is possible to collect at least one "strip" of image over the territory of the Crimea.

To determine the NDVI index, the ArcGIS 10.4.2 image analysis module is used. Automatic calculation of the index value in this module is classified on a scale from 0 to 200 units (this range is designed for an 8-bit structure and can be displayed using a color scale or color grade, which makes its use more convenient), which can later be converted to the value of the index values in the range from 0 to 1 in the classical version of the index interpretation. Given that the task of our work is to compare the value of biological productivity, rather than interpretation the vegetation cover, the use of standard ArcGis 10.4.2 scale is more convenient for work.

A database of ecosystem productivity values has been compiled in order to analyse the multiannual spatio-temporal dynamics of ecosystem productivity as a reflection of changes in multi-year ecosystem states. The landscape and typological map of the Crimean Peninsula compiled by G.E. Grishankov [11] was used as an operational and territorial unit of the analysis.

3 Results

As a result of the research, a database on the productivity value of Crimean ecosystems was compiled based on multispectral space images' data for the period 1979-2017.

The results of productivity studies of forest ecosystems in Crimea, obtained on the basis of their multiannual field study, are presented in the works of Yu. V. Plugatar. These results were compared with data obtained from remote sensing over a multiannual period.

In the most general form, all oak forests on the Crimean peninsula should be divided into two large predominant subgroups - forests of *Quercus petraea* and forests of *Quercus pubescens*.

3.1 *Quercus petraea* forests

Yu.V. Plugatar, considering the value of productivity of this forests group notes [9] the exactingness of this species to soil humidity, as well as the fact that Quercus petraea tree stands are not sufficiently productive and are characterized by a spatial prevalence of forests with low bonitet. It is noted that among the tree stands the most productive are forests in fresh sudubrava with an average stockpile of 153 m^3/ha.

Within the landscape differentiation, Quercus petraea forests are represented within contours № 60-71 in the mid-mountain landscape belt and contours № 48-55 in the low-mountain landscape belt [11]. As an example of multiannual dynamics of changes in ecosystem states, expressed in the form of the "effect" of productivity, let's choose contour No. 68 (mid-mountain slope, dissected beams with oak and mixed broad-leaved forests). These contours by landscape and ecotopic conditions most fully correspond to fresh sudubrava.

In this case it is necessary to make a certain correction - Yu.V. Plugatar conducts an analysis of forest productivity in this case and subsequent variants for multiannual timber stockpile with corresponding calculations of its stockpiles. In our case, with the help of the NDVI index it is clearly possible to fix and consider in the multiannual dynamics the amount of biomass contained in the plant leaf apparatus, which as a result, in the multiannual section provides growth and accumulation of biomass in wood.

Fig. 1 shows the values of forest productivity changes in this contour as of August month.

The average value of the NDVI index is in the range of 140-150 units, while the established multiannual norm is possible to consider the value of 146-148 units, which on average may correspond to the value of actual productivity of 153 m^3/ha, determined on the basis of multiannual observations of Yu V. Plugatar [9]. The minimum values for this landscape contour are in the range of 70-120 units, and the maximum values are in the range of 165-170 units, which indicates a decrease in actual productivity to 70 m^3/ha in certain landscape parts and its possible increase to 165-170 m^3/ha in some areas depending on changes in ecotopic conditions.

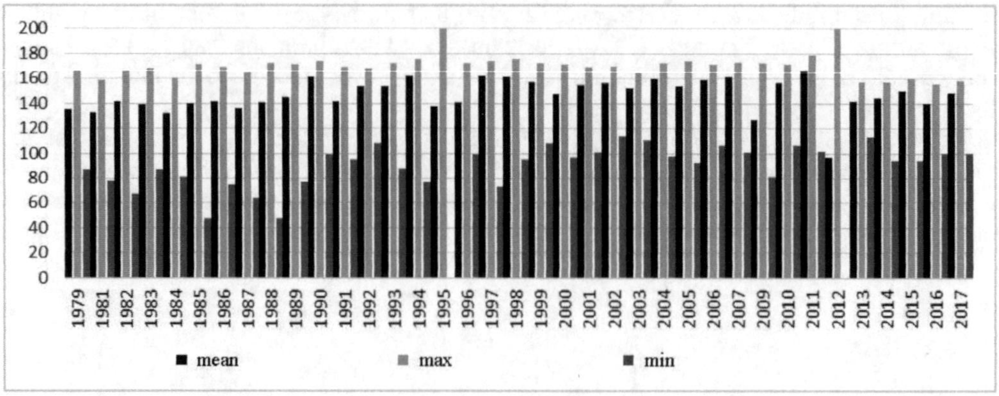

Fig. 1. Multiannual changes in productivity values for mid-mountain slope, dissected beams with oak and mixed broad-leaved forests landscape.

Thus, for landscape types of the Crimean territory, it is possible to identify types of ecosystem conditions depending on the of productivity value at the level of normal productivity, decreased productivity and increased productivity.

Consideration of the multiannual indicator dynamics reveals certain periods in the form of peculiar waves. The period from 1979 to 1988 is characterized by a normal level of productivity ecosystems under consideration with insignificant parameter fluctuations; the period from 1989 to 1998 tends to increase parameter fluctuations and the appearance of single states of reduced and increased productivity. Further the period from 1999 to 2007 is characterized by sustainable parameter values in the range close to the increased productivity, and in the period from 2008 to present days there is a sharp fluctuation of parameters towards chaotic changes in the increased and decreased direction, which can testify to certain tendencies of parameters' change of heat and moisture supply towards fluctuation of temperature norms and values of precipitation.

3.2 *Quercus pubescens* forests

Within the landscape differentiation of the Crimean Peninsula Quercus pubescens forests are represented within the contours № 92-98 in the mid-mountain landscape belt and contours № 48-55 in the low-mountain landscape belt [11]. As an example of the multiannual dynamics of changes in ecosystem states, expressed as an "effect" of productivity, we choose contour № 95 (mid-mountain stepped-slope with oak and mixed broad-leaved forests).

Yu.V. Plugatar notes [9] that these forests are represented in 8 types of ecotopes with spatial predominance of one of them - dry sudubrava, and the average actual productivity is 83-110 m^3/ha depending on age.

By its habitat conditions landscape contour № 95 is close to the ecotope of dry sudubrava. Fig. 2 shows the multiannual dynamics of the NDVI index for this ecosystems type in Crimea.

Consideration of the presented diagram shows more significant average values' fluctuations of the considered value in the given ecosystems type at the level of 120-150 units, with average normal values in the range of 135-138 units. The maximum values are more sustainable in time and fluctuate in the range of 150-160 units with local maximums up to 200 units. The minimum index values for Quercus pubescens forests are lower than those for Quercus petraea forests and range from 30 to 90 units, with a minimum normal value of about 50-55 units.

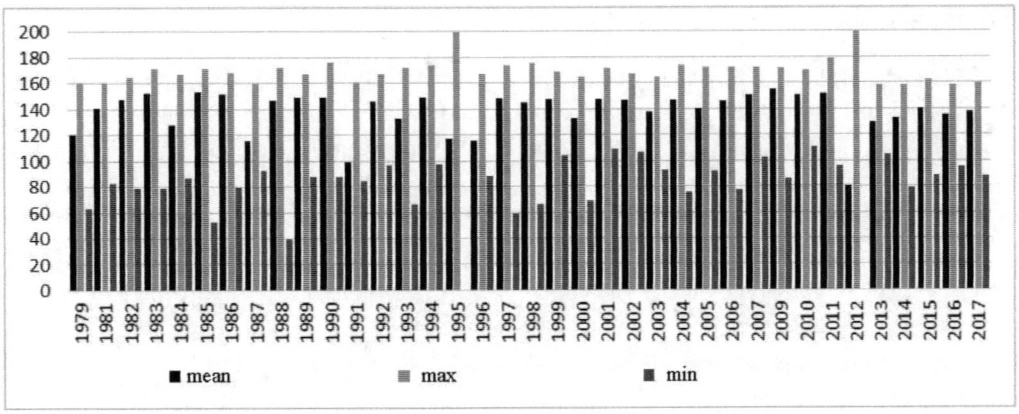

Fig. 2. Multiannual changes in productivity values for mid-mountain steppe-slope landscape ecosystems with oak and mixed broad-leaved forests.

Empirical generalization allows to conclude that depending on ecotopic conditions the actual productivity value at normal range of states is within 90-95 m^3/ha, minimum 85-90 m^3/ha, maximum 95-100 m^3/ha, with the necessary correction for the age of planting.

Consideration of the temporal dynamics of this parameter for Quercus pubescens forests shows great differences in comparison with Quercus petraea forests, which confirms a more significant productivity dependence not only on moisture conditions, but also on temperature states.

From 1979 to 1986, there was a gradual increase in the productivity value with a local decline in 1984 (this year is distinguished by excellent circulation conditions and local minimum precipitation) from normal values to higher values.

The period from 1987 to 1996 is characterized by significant fluctuations in productivity, which as a result, in interrelation with hydrothermal conditions leads to the formation of an unstable picture of the ecosystems state in productivity value with the presence of sharp skips from year to year.

The period from 1997 to 2010 is characterized by sustainable increased productivity values in the range of 138-152 units of the index, with minor decreases in 2000, 2002, 2005.

The period from 2010 to present is characterized by some decline in productivity of Quercus pubescens forests after a sharp local decline in 2012 (obviously unfavorable dry conditions prevailed in this year during almost all seasonal ecosystem states) and the approaching of productivity to the normal ecosystems state.

4 Conclusion

Thus, on the basis of remote sensing data the series of data on productivity values spatial dynamics of oak forests growing on the territory of the Crimean peninsula were obtained, the periods characterizing the main regularities of temporal productivity values variability of the considered forest ecosystems were revealed.

The results obtained will allow to reveal further connection between spatio-temporal dynamics of oak formations productivity and regional climate change manifestations. The proposed research methodology is universal for any type of regional ecosystems.

The study was made as part of the IBSS research (registration number: AAAA-A19-119061-190081-9).

References

1. N. Pettorellia, J. Vik, A. Mysteruda, J.-M. Gaillardb, C.J. Tuckerc, N.C. Stensetha, Trends in Ecology and Evolution, **20**, 503-510 (2005)
2. G. Hmiminaa, E. Dufrênea, J.Y. Pontaillera et al., Remote Sensing of Environment, **132**, 145-158 (2013)
3. G. MeeraGandhia, S. Parthibanb, Nagaraj Thummalu, A. Christyd, Procedia Computer Science, **57**, 1199-120 (2015)
4. D.P. Roya, V. Kovalskyya, H.K. Zhanga, E.F. Vermoteb, L. Yana, S.S. Kumara, A. Egorova, Remote Sensing of Environment, **185** (2016)
5. N. Mishraa, D. Heldera, J. Barsib, B. Markhamc, Remote Sensing of Environment, **185** (2016)
6. G.V. Lobanov, A.I. Bashkirskii, A.Yu. Zvereva, B.V. Trishkin, A.P. Protasova, Scientific notes of the Bryansk State University, **2**, 57-61(2016) (in Russian)
7. Yu.V. Plugatar, *From Crimean forests* (Novoe Slovo, Kharkiv, 2008) (in Ukrainian)
8. A.F. Polyakov, Yu.V. Plugatar, *Forest formations of Crimea and it ecological role* (Novoe Slovo, Kharkiv, 2009) (in Russian)
9. Yu.V. Plugatar, *Forests of Crimea* (IT ARIAL, Simferopol, 2015) (in Russian)
10. Yu.V. Plugatar, S.P. Korsakova, O.A. Ilnitsky, *Ecological monitoring of Soutern coast of Crimea* (IT ARIAL, Simferopol, 2015) (in Russian)
11. E.A. Pozachenyuk (Ed.), *Modern landscapes of the Crimea and adjacent water areas* (Business-Inform, Simferopol, 2009) (in Russian)

Assessment of ecological status of street plantings in Lipetsk city

Elena Parakhina[1*], *Zhanna Silaeva*[2], *Ludmila Kiseleva*[3], and *Elena Zhidkova*[4]

[1]Peoples' Friendship University of Russia (RUDN University), Faculty of Ecology, 6 Miklukho-Maklaya St, Moscow, 117198, Russian Federation
[2]Orel State Agrarian University named after N.V. Parakhina, Department of landscape architecture, Orel, Russia
[3]Orel State University named after I.S. Turgenev, Department of Plant Botany, Physiology and Biochemistry, Orel, Russia
[4]Lipetsk State Pedagogical University named after P.P. Semenov - Tyan-Shansky, Department of Geography, Biology and Chemistry, Lipetsk, Russia

Abstract. The purpose of the work is to evaluate the ecological state of the prevailing species of woody plants in the city of Lipetsk and to elaborate recommendations for improving green spaces. There were examined thirty street landscaping facilities, including 8537 trees. In the course of the study, there were identified five classes of trees vitality. There were also calculated the average scores of the prevailing tree species and the overall ratio of the city street plantings. As a result of the study, there were identified 31 species of trees. 14.5% of the surveyed trees were classified as "healthy" by vital condition. The "weakened trees" group got 38.8 %. "Severely weakened" group counted 24.7%. The category" shrinking" trees had 19.2%. Completely dry woody plants scored 2.8%. Based on the obtained data, there were made recommendations how to improve the sustainability of green spaces in an urbanized environment.

1 Introduction

Creating a comfortable urban environment is a priority in many cities of our country, since human health depends directly on the conditions in which they live. Increasing the comfort of a person's stay in the city conditions is possible through the preservation and development of the system of planting. An important component of the system of planting of any city is played by woody plants. In addition to aesthetic and sanitary and hygienic functions, wood plantations also play a recreational role.

However, at present the pressure on urbanized areas has increased due to the increasing influence of human economic activity, which reduces the comfort of the urban environment. In addition, the modern system of planting in most cities of the European part of Russia, including Lipetsk, was formed in the 60-70s years of the last century.

In the last decade, interest to the system of planting has been growing: the range of plants used is expanding, interesting landscape compositions (topiaries and arboplastics) are appearing, and modern technologies of care work are being developed.

* Corresponding author: parakhina-ea@rudn.ru

In this regard, it becomes relevant to study the current state of urban tree vegetation in order to determine the degree of harmful anthropogenic impact, predict the fate of plantations and theoretical basis for conservation measures.

Many researchers are engaged in studying of a condition of woody plants in the conditions of the urbanized environment, increasing of its comfort, development of an ecological framework of cities with participation of woody plants, studying of anthropogenic influence on green plantations [1-4]. At the same time, only some authors are engaged in identifying the current state of plants and their degree of disturbance [5, 6]. The state of plantations was assessed in small towns of the Lipetsk region [7, 8]. No such work was carried out in Lipetsk. Although the city is a major industrial center. There is a work " Otsenka..." [9], which addresses the impact of the urban environment on plants, but does not address the issues of plant vitality assessment in urbanization.

The purpose of our work is to assess the ecological condition of the prevailing species of woody plants in Lipetsk and to recommend improvements to green plantations for a comfortable environment.

2 Research methods

The vitality assessment of wood plantations was carried out in 2018-2019 using the method of the state of trees and forest stands diagnostic [10]. A total of 8,537 trees belonging to 31 species were examined. We calculated the average state score for each tree species (CI) and the total coefficient of the green plantings condition (Ctot.) in Lipetsk. It was established borders of wood plantations, determined the species and were determined scores status of individual trees of each species using the scale visual assessment of trees by their appearance (overall condition of the plants, density of leafy, the degree of stem and crown damage, etc.).

The trees examined were part of thirty public plantations: planting strips along the main streets, including areas in the citywide shopping and administrative centers of Lipetsk. A total of 30 streets were investigated.

3 Results and discussion

Lipetsk is one of the largest cities in the Central Chernozemic region of Russia, the agglomeration center, where the production of full-cycle iron industry is concentrated. It is the city-resort. It was founded in 1703, although the first settlements on this territory date back to the II Millennium BC. Initially, Lipetsk was founded around iron factories during the reign of Peter I and only in 1779, by decree of Empress Catherine II, it received the status of a district city of the Tambov province. Lipetsk became the center of the Lipetsk region in 1954 [9].

During the Great Patriotic War (1941-1945), the city's wood plantations did not suffer much. Many trees and shrubs have been preserved. For example, there is an oak grove in the Verhniy Park with trees more than 200 years old.

In the following years, about 30 species of tree and shrub species were used in city landscape construction, mostly wild, for example, *Quercus robur* L., *Tilia cordata* Mill., *Acer platanoides* L. and others, as well as the most widely distributed introducers - *Acer negundo* L., *Caragana arborescens* Lam., *Larix sibirica* Ledeb. etc. This is typical of many cities [5, 6].

When assessing the of woody plants condition in Lipetsk, it was found that trees predominate in green plantations, which create a favorable environment for human health. This is why we first assessed the sustainability of trees. Most often, 31 species of trees are found in Lipetsk (table 1).

Table 1. Average points (Ki) of state of tree species and total state coefficient (Kt) of studied green plantations in Lipetsk

№	Species	Ki	Kt
1	*Acer negundo* L.	2.78	2.01
2	*Acer platanoides* L.	1.91	
3	*Acer saccharum* Marshall	2.31	
4	*Aesculus hippocastanum* L.	2.45	
5	*Betula pendula* Roth.	1.76	
6	*Caragana arborescens* Lam.	2.86	
7	*Crataegus submollis* Sarg.	1.32	
8	*Fraxinus excelsior* L.	2.89	
9	*Fraxinus pennsylvanica* March.	2.33	
10	*Larix sibirica* Ledeb.	2.44	
11	*Malus domestica* Borkh.	2.86	
12	*Picea abies* (L.) Karst.	2.49	
13	*Picea pungen* Engelm.	1.47	
14	*Populus balsamifera* L.	2.14	
15	*Populus × berolinensis* (C. Koch) Dipp.	2.56	
16	*Populus nigra* L.	1.89	
17	*Populus tremula* L.	1.47	
18	*Pseudotsuga menziesii* (Mirb.) Franco	2.14	
19	*Quercus robur* L.	1.66	
20	*Robínia pseudoacacia* L.	2.35	
21	*Salix alba* L.	1.81	
22	*Salix babylonica* L.	1.72	
23	*Salix caprea* L.	1.03	
24	*Salix fragílis* L.	1.58	
25	*Sorbus aucuparia* L.	2.56	
26	*Thuja occidentalis* L.	3.04	
27	*Tilia cordata* Mill.	1.78	
28	*Tilia platyphyllos* Scop.	1.63	
29	*Ulmus glabra* Huds.	2.93	
30	*Ulmus laevis* Pall.	2.74	
31	*Ulmus pumila* L.	2.15	

Currently, the average trees age in the city is 50-70 years.

During the work, a visual condition assessment of 8537 trees growing on 30 streets of the city was carried out and 5 vitality classes were identified (Fig.1).

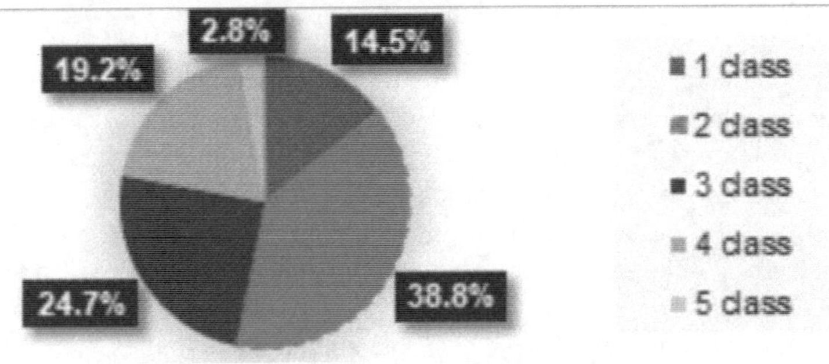

Fig. 1. Percentage distribution of the studied trees in Lipetsk by vitality classes (1 class – healthy, 2 class – weakened, 3 class – severely weakened, 4 class – shrinking, 5 class – dry).

Only 1238 (14.5%) of the examined trees are classified as "healthy". 3312 (38.8%) trees refer to the "weak trees" class. They have minor damage: shrinking branches, thinning crown, possibly the presence of ectoparasites.

"Severely weakened" trees account for 2109 trees (24.7%). This category is characterized by significant shrinking of branches, dryness, boundary and central chlorosis, and in some places there is the bark dying off.

The category" shrinking " includes – 1639 trees (19.2%). There is a shrinking of branches throughout the crown, there is no growth, and there are numerous bodies of marsupials and basidial fungi on the trunks.

Woody plants that are completely dry (without leaves, the bark is peeling or completely absent) are 239 trees (2.8 %). It should be noted that this class includes mainly representatives of the genus *Populus* L. (*Populus nigra* L., represented in the Lipetsk plantations mainly pyramidal form, *P.* × *berolinensis* (C. Koch) Dipp. etc.).

Based on the results of calculating the average state score for each species of tree and the overall state coefficient of green plantations, it was found that the tree plantations of Lipetsk streets are weakened (C tot. = 62.35 / 31 = 2.01) requiring preventive measures, pruning and sanitation (table.1).

The research also noted that low vitality is observed in species that grow along main streets with heavy traffic. Plants of the following species were most affected: *Betula pendula* Roth, *Fraxinus excelsior* L., *Acer platanoides* L., *Populus nigra* L., *Acer negundo* L., *Caragana arborescens* Lam.

The main reasons for mass drying, thinning, and dying of trees in the city's plantations are: compaction and salinization of the soil, lack of proper agricultural equipment, and mechanical damage. As a result, vulnerability to pests and diseases increases. The following diseases are most common on trees and shrubs in Lipetsk: necrotic-cancerous on branches and trunks, vascular diseases, leaf spots, rust, powdery mildew, rotten diseases, etc. A similar situation is noted in another Russian cities [1, 2, 3, 4, 6].

The system of increasing the sustainability of woody plants and protecting plantationss in the city should include a set of measures. The main ones are: organization of general supervision over the appearance and spread of pests and diseases; quarantine measures and certification of planting material; preventive activities aimed at increasing the plants resistance to negative factors of the urban environment; measures for active plants protection from pests and diseases [6].

It should be noted that in recent years, various measures have been carried out in Lipetsk to improve the urban environment and replace plantatons. In the future, this will have a positive impact on the urban environment state, increase its comfort, and, consequently, it will have a positive impact on human health.

4 Conclusions

Thus, according to the results of the study, the following conclusions can be drawn:

1. The species composition of tree plants in Lipetsk, used in landscaping street plantings in Lipetsk, is represented by 31 species of trees, with the predominant native species: *Quercus robur* L., *Tilia cordata* Mill., *Acer platanoides* L.

2. The result of the visual evaluation 8537 trees growing on 30 city streets found that: 14,5% of the surveyed trees are classified as "healthy"; the share of damaged trees accounted for 38.8%; the category of "severely weakened" trees is 24.7% of the trees, and "drying" is 19.2%; woody plants completely dry 2.8% of the studied breeds.

3. Based on the results of calculating the average state score for each type of tree and the overall coefficient of the state of green spaces, it was found that the tree stands of the streets of Lipetsk are weakened, requiring preventive measures, pruning and sanitation.

4. Based on the obtained data, the following recommendations are proposed: organization of General supervision over the appearance and spread of pests and diseases; quarantine measures and certification of planting material; preventive activities aimed at increasing the resistance of plants to negative factors of the urban environment; measures for active protection of plants from pests and diseases.

References

1. I.L. Bukharina, Izv. Samarsk. nauch. tsentra RAN, **10**, 607-612 (2008) (in Russian).
2. M.V. Larionov, N.V. Larionov, T.S. Gromova, I.S. Sirayeva, E.A. Logacheva, V.V. Soldatova, *Aktualnye problemy prirodopol'zovaniya i prirodoobustrojstva,* 179-184 (2019) (in Russian).
3. N.B. Prokhorenko, G.V. Demina, Izv. Samarsk. nauch. tsentra RAN, **18**, 177-181 (2016) (in Russian).
4. A.K. Razinkova, E.N. Perelygina, Lesotekh. Zh, **6**, 36-46 (2016) (in Russian).
5. E.A Deynega, *Dendroflora g. Dubna Moskovskoj oblasti: raznoobrazie i zhiznennoe sostoyanie zelenyh nasazhdenij*, (2016) (in Russian).
6. L.L. Kiseleva, E.A. Parahina, Zh.G. Silayeva, Izv. Samarskogo nauch. tsentra RAN, **18**, 702-706 (2016) (in Russian).
7. O.G. Bolysheva, *Ocenka sostoyaniya nasazhdenij i napravleniya optimizacii ozeleneniya malyh gorodov Lipeckoj oblasti*, (2013) (in Russian).
8. I.L. Bukharina, A.N. Zhuravleva, O.G. *Bolysheva Gorodskie nasazhdeniya: ekologicheskij aspect* (2012) (in Russian).
9. *Otsenka zelenykh nasazhdeniy goroda Lipetska.* URL: http://www.hintfox.com/article/otsenka-zelenih-nasazhdenij-goroda-lipetska.html (in Russian).
10. V.A. Alekseyev, Lesovedeniye, **4**, 51-57 (1989) (in Russian).

E3S Web of Conferences **169**, 03009 (2020) https://doi.org/10.1051/e3sconf/202016903009
APEEM 2020

Anthropogenic dynamics and recovery of disturbed landscapes at the Far East Russia permafrost area

Pavel E. Tikhmenev[1,2*], Andrey A. Smirnov[2,3], Evgeniy A. Tikhmenev[1,2], and Galina V. Stanchenko[1,2]

[1]Institute on Biological Problems of the North, Russian Academy of Science, Far Eastern Branch, Magadan, Russia
[2]Nort-East State University, Magadan, Russia,
[3]Federal Research Institute of Fishery and Oceanography, Moscow, Russia

Abstract. The results of studies of anthropogenic landscapes functioning with an assessment on the processes of natural self revegetation are summarized. The principles of accelerated restoration of the ecological and aesthetic value of disturbed landscapes at the permafrost zone are substantiated, based on the data obtained during studying the self-revegetation processes on disturbed complexes. Development of placer and ore deposits of mineral resources is leading to deep transformation of landscape and to destruction of soil-vegetable complexes. The processes caused by mining-technical activity are leading to variable mechanisms of degradation of soil and vegetables often having complex impact. Results of study of sustainability of soil-vegetable complexes to the impact of man-caused activity have showed the dependence from the structure of the soil profile and it characteristic, from the character of genetic horizons, frozen status of the landscape elements, form the structure and biomass and productivity of vegetation cover.

1 Introduction

During exploitation of deposits at the permafrost zone of Far North-East, major technogenic processes of denudation, transportation and accumulation of material occur, in a volume comparable with the naturally occurring active morpholithogenesis [1].

At the beginning of our research, there was no data on the self-revegetation potential of the landscapes disturbed during exploitation of ore deposits at the North-East, as well as the was no experience their biological reclamation. The aim of research is to create an ecological basis of restoration of disturbed sites of permafrost landscapes.

* Corresponding author: tpavel@mail.ru

2 Materials and methods

The studies were carry out in light coniferous taiga communities. Main forest – forming species of region under investigation is larch (*Larix cajanderi*), which grows on various elements of the mountainous relief. Widely distributed cider-plank communities (*Pinus pumila*), occupying the mountain slopes, so were studied. Succession processes in self-renewing vegetation of the landscape after mining operation were observed. In during field research were carried out of plant cover and soil of natural and disturbed sites after mining operation. Widely approved methods were used [2, 3]. Species given name we follow by Flora and vegetation of Magadan region [5].

3 Results and discussion

Our observations were made on the technogenic formations at the Kubaka gold ore deposit, situated in Omolon river basin, the largest tributary of the Kolyma river. It was founded that natural revegetation process of disturbed natural complexes at larch woodlands and cedar-plank communities is usually unsufficient The Kubaka mine is located at the altitudes of 650-750 m above sea level on Kolyma upland. That predefines the difficulties in the processes of natural revegetation of disturbed complexes. Our studies showed that there is practically no self-revegetation in large-fragment rock dumps and quarries. The process of plant communities natural revegetation in the areas of surface disturbances, such as geological exploration tranches and temporary roads proceeds more successful (Fig.1).

Fig. 1. Self-revegetation on the geological trench, disturbed during geological exploration (Kubaka gold ore deposit, 560-570 m above sea level).

Most common species of plants sattled on technogenic formations are perennial grasses, shrub species of willows (*Salix krylovii, S. alaxensis, S. schwerenii, S. glauca*), dog rose (*Rosa acicularis*), Kuril tea (*Potentilla fruticosa*), meadowsweet (*Spiraea beauverdiana*), lean birch and Middendorff's birch shrubs (*Betula exilis, B. middendorffii*). Chosenia (*Chosenia arbutifolia*) and poplar (*Populus suaveolens*) are also common on the disturbed lands. Larches prefer denudated areas of the landscape with a sufficient amount of fine particles. In natural conditions, self-restoration of vegetation cover is aimed to form the initial type of forest that grew here before the exploitation. This process proceeds most successfully if the time of violation of the integrity of natural complexes coincides with the larch seeding productive years, observed at intervals of once every 5-7 years in the specific conditions of permafrost landscapes. If disturbances of natural complexes occurred in not productive years, then wasteland with low biodiversity are formed, where self-healing of forest communities can last for decades. Since humification is mainly a biochemical process, this similarity of deposits of different ages confirms low rates of humus formation and inhibition of microbiological activity. If the natural zones of vegetation are preserved in close proximity, the natural distribution of seeds into technogenic zones is ensured, which accelerates the formation of sods and the stabilization of the surface of disturbed objects. The depth of violations, the graveliness of the substrate, the shape and height of the man-made dumps determine the rates of succession processes, which proceed differently in the regenerated groups of plants on objects of different genesis.

Post-technogenic landscapes at the first stages of restoration are characterized by the absence any vegetation, low microbiological activity and a slow circulation of substances (Fig. 2). The small amount of «living matter» involved in the biosynthesis and accumulation of nutrients predetermines very low rates of natural restoration of disturbed landscapes in the initial period. The data obtained during our studies allows us to determine the direction of succession processes and the duration of regeneration pause of plant groups. Self-revegetation of the disturbed vegetation cover begins 5-6 years after the removal of the technogenic press. At the same time, humus accumulation, its composition stabilization in the emerging soil and vegetation complexes of technogenic landscape elements after ten years or more, approaches its zonal type. A wide range of conditions for the settlement of pioneer plants, intrinsic of technogenic landscapes, is reflected in the structure and productivity of the vegetation.

Fig. 2. Reclaimed waste rock dumps at the Kubaka mine (light coniferous taiga).

The features of the natural revegetating soil-plant complexes of disturbed sites consist in a wide variation in the projective plant cover - from the complete absence of vegetation to clearly expressed regenerative phytocenoses. At the initial stages of natural revegetation process, its influence on the formation of the soil profile is offset by the heterogeneity of the mineral base and the variety of ecological niches of disturbed areas. Plants usually settle in places where the necessary conditions for their growth are created, mainly in micro depressions of the relief, where the accumulation of organic material occurs under the condition of sufficient moisture (Fig. 3).

Fig. 3. Reclamation of disturbed sites of the Kubaka gold ore deposit (Omolon basin river) using seeds species of native flora of region, as well adopted to hard surrounding condition of permafrost region.

Post-technogenic landscapes at the first stages of restoration are characterized by the absence of any vegetation, low microbiological activity and a slow circulation of substances. The small amount of «living matter» involved in the biosynthesis and accumulation of nutrients predetermines very low rates of restoration of disturbed landscapes in the initial period.

In self-revegetatied plant complexes there are 1-2 clearly defined dominant species, which are cereals, sedges, fireweeds, stubbleweed, aserisk, saxifrages, wormwoods, astragalus, and ostracisans, which usually do not take place prominent positions in natural plant communities. The source of seed application to disturbed lands is natural plant complexes adjacent to disturbed areas. Often there are cases of settlement of disturbed areas in the cedar-plank belt with typical types of river floodplains, such as poplar (*Populus suaveolens*) and chosenia (*Chosenia arbutifolia*). Favorable conditions for the settlement of these species in unusual habitats are provided by active intra-ground condensation of vaporous moisture [4].

Succession processes are significantly accelerated with the accumulation of fine earth material and humus in technogenic forms of relief. The depth of disturbances, the amount of coarse rock material in the substrate, the shape and height of technogenic forms determine the self-revegetation processes that occur differently in the restored plant groups. Areas of surface disturbances, overburden dumps and technological dams are actively overgrown, where usually there is sufficient fine earth, usually adjacent to natural plant communities. Considering the content of organic matter and humic acids under different types of plants, a definite connection can be noted. With an increase in the total supply of organic matter, the content of humic acids in it also increases, but it remains two times lower compared with zonal soils.

At the considerably deeply disturbed areas (Fig.2), self-revegetation proceeds extremely slowly, especially in the mountain-tundra (alpine belt). The complete absence of vegetation is characteristic of the main technogenic formations that are formed during the development of ore deposits - quarries, dumps of large waste rocks, which occupy the most areas in the mining area. They are characterized by great dryness, active aeration, lack of fine earth, slightly acidic or alkaline reaction of the substrate. The geochemical composition of waste rocks is characterized by the presence of such compounds as lead, chromium, vanadium, nickel, manganese, tin, copper, zinc, often in concentrations below the maximum permissible values. On such elements of disturbed mountainous landscape, anti-erosion land reclaimaition is required. Sites with deep lithogenic disturbances require the directed formation of the soil-vegetation complexes with the implementation of a full cycle of land reclamation work. The best results in biological stage of land reclamation of disturbed lands are achieved using seeds of local flora species that are most adapted to the harsh conditions of the permafrost zone (Fig.3).

4 Conclusions

The study of the structural and functional organization of the natural landscape environment into which the developed mineral deposits are «embedded» provides a basis for optimizing technological processes to accelerate the self-revegetating of disturbed lands. The performed studies provide a theoretical basis for the implementation of remediation work to stabilize the natural environment and accelerate the restoration of the productivity of disturbed lands. The data obtained provide an opportunity to summarize the available experience in order to accelerate the elimination of undesirable consequences of ore deposits in the zone of widespread permafrost and in other parts of the planet.

References

1. A.A. Pugachev, E.A. Tikhmenev. *Structural and functional organization and dynamics of the soil-vegetation complexes of the Far North-East of Russia* (NWSU Publishing House,Magadan, 2011) (in Russian)
2. *The program and methodology for the study of technogenic ecosystems*, (Nauka, Novosibirsk, (1978) (in Russian)
3. *Anthropogenic dynamics of the vegetation cover of the Arctic and Subarctic: principles and methods of study. Proceedings of the Botanical Institute RAS*, **15** (1995) (in Russian)
4. A.I. Zinchenko, I.M. Papernov., VNII-1, **37**, 29-36 (1977) (in Russian)
5. *Flora and vegetation of Magadan region (Checklist of vascular plants and outline of vegetation)* (IBPN FEB RAS, Magadan, 2010) (in Russian).

 https://doi.org/10.1051/e3sconf/202016904001

Manganese and cobalt levels in hair and urine of RUDN University first-year students

Anatoly A. Kirichuk[*]

Peoples' Friendship University of Russia (RUDN University), Faculty of Ecology, 6 Miklukho-Maklaya St, Moscow, 117198, Russian Federation

Abstract. The objective of the present study was to assess hair and urinary levels of cobalt (Co) and manganese (Mn) in foreign students of RUDN University in comparison to the Russian counterparts. A total of 272 first-year students of the RUDN University originating from Russia, Asia, Middle East, Africa, and Latin America were enrolled in the study. Evaluation of Mn and Co levels in urine and hair was performed using inductively-coupled plasma mass-spectrometry. The obtained data demonstrate that hair cobalt levels were found to be the highest in students originating from Russia and Africa. Similarly, urinary Co levels in subjects from Asia, Middle East, and Latin America were significantly lower than those in the Russian students. The patterns of manganese status were generally similar to those observed for cobalt. Particularly, the lowest hair Mn content was detected in subjects originating from Middle East. Manganese level in urine of students from Africa was significantly higher than that in subjects from Asia, Middle East, and Latin America, but not Russia. It is proposed that specific national dietary patterns as well as geochemistry of Mn and Co may underlie the observed differences.

1 Introduction

Essential trace elements participate in a plethora of metabolic processes thus being required for organism functioning [1]. Micronutrient deficiency is considered a significant public health concern due to both high medio-social outcome and wide prevalence exceeding 2 billion of people worldwide [2]. However, data on micronutrient deficiency are insufficient and highly contradictory for different regions. Systematic data exist on only certain micronutrients including vitamins, as well as iron, zinc, iodine, selenium [2], whereas the prevalence of inadequate intake of other essential elements is underestimated.

Manganese [3] and cobalt [4] are also considered as essential metals, the particular clinical features of their deficiency are not estimated. At the same time, both Mn [3] and Co [5] overload may exert toxic effects. Particularly, the health effects of manganese are explained by U-shaped relationship, when both deficiency and toxicity are associated with adverse health effects [6]. At the same time, data on geographic variability of cobalt and manganese status are absent.

[*] Corresponding author: Kirichuk-aa@rudn.ru

RUDN University is characterized by the highest number of foreign students from various regions including Middle East, Asia, Africa, and Latin America. Therefore, the objective of the present study was to assess hair and urinary levels of cobalt and manganese in foreign students of RUDN University in comparison to the Russian counterparts.

2 Materials and methods

A total of 272 first-year students (40% males, 60% females) of the RUDN University originating from Russia (n = 65), Asia (n = 57), Middle East (n = 84), Africa (n = 40), and Latin America (n = 28). Anthropometric examination as well as sample collection was performed during the first medical examination. The overall mean age of the examinees was 22.4±4.3 years. Due to group variability all comparisons were adjusted for age and gender values. The protocol of the current study was approved by the Institutional Ethics Committee (RUDN University, Moscow, Russia). All procedures were performed in agreement with the principles set in the Declaration of Helsinki and its later amendments (2013).

Evaluation of manganese (Mn) and cobalt (Co) levels in urine and hair after microwave digestion was performed using inductively-coupled plasma mass-spectrometry (ICP-MS) at NexION 300D (PerkinElmer Inc., USA). Laboratory quality control was performed using certified reference materials of human hair and urine.

The obtained data were processed using Statistica 10.0 (Statsoft, OK, USA) software. Due to skewed distribution as assessed by Shapiro-Wilk test, median and interquartile range (IQR) boundaries were used as descriptive statistics for hair and urinary metal levels. In order to account for variability in age and gender, assessment of the group difference was performed using analysis of covariance (ANCOVA) with Fisher's LSD post-hoc test at the level of significance of $p < 0.05$. Multiple regression analysis with adjustment for age and gender was performed in order to evaluate the association between hair Mn and Co levels (dependent variables) and urinary metal levels, as well as the region of origin (independent variables).

3 Results

The obtained data demonstrate that hair cobalt levels were found to be the highest in students originating from Russia and Africa. Particularly, hair Co content in students from Asia, Middle East, and Latin America were 53%, 79%, and 42% lower as compared to the respective values in Russian examinees. In turn, the African student values exceeded the respective ones in examinees from Asia by a factor of more than four. No significant group difference in Co levels in hair between Russian and African students was observed.

Generally, the difference in urinary cobalt levels followed similar patterns. Particularly, urinary Co levels in subjects from Asia, Middle East, and Latin America were 40%, 34%, and 26% lower than those in the Russian students. Urinary Co concentration in African students exceeded the respective values in subjects originating from Asia and Latin America by 60% and 31%. It is notable that the values of urinary Co levels were nearly similar in African and Russian students.

It is notable that the patterns of manganese status were generally similar to those observed for cobalt. Specifically, the lowest hair Mn content was detected in subjects originating from Middle East, being lower than that in students from Russia, Asia, Africa, and Latin America by 76%, 44%, 78%, and 69%, respectively. Hair Mn content in subjects originating from Africa and Latin America also exceeded the respective values found in the group of Asian students by 161% and 84%. No significant group difference was observed in hair Mn content in students from Russia, Africa, and Latin America was observed. Manganese level in urine of students from Africa was significantly higher than that in subjects from Asia, Middle East, and Latin America by 43%, 29%, and 25%, but did not differ significantly from the Russian values. At the same time, the values of urinary Co concentrations in Russian students exceeded those in subjects from Asia by 26%.

Table 1. Hair (μg/g) and urinary (ng/ml) Co and Mn levels of Russian and foreign students of RUDN University

Region	Hair Co	Urine Co	Hair Mn	Urine Mn
Russia	0.019 (0.009 - 0.049)	1.16 (0.728 - 2.057)	0.843 (0.335 - 1.179)	0.891 (0.603 - 1.309)
Asia	0.009 [1] (0.004 - 0.017)	0.705 [1] (0.53 - 1.017)	0.353 (0.238 - 1.037)	0.706 [1] (0.542 - 1.011)
East	0.004 [1] (0.003 - 0.007)	0.768 [1] (0.54 - 0.973)	0.200 [1,2] (0.155 - 0.354)	0.784 (0.564 - 1.162)
Africa	0.018 [3] (0.012 - 0.031)	1.127 [2] (0.767 - 1.717)	0.923 [2,3] (0.604 - 1.411)	1.01 [2,3] (0.723 - 1.343)
Latin America	0.011 [1] (0.005 - 0.022)	0.859 [1,4] (0.475 - 1.414)	0.651 [2,3] (0.271 - 1.176)	0.809 [4] (0.604 - 1.023)
Data are expressed as median (IQR); [1,2,3,4] – significant group difference as compared to the respective values in students from Russia (1), Asia (2), Middle East (3), and Africa (4) at $p < 0.05$, respectively (ANCOVA Fisher's LSD post-hoc)				

Multiple regression analysis demonstrated that hair Co level was inversely associated with male gender ($\beta = -0.374$; $p < 0.001$) and urinary metal excretion ($\beta = -0.131$; p 0.032). At the same time, no significant relationship between the country of origin and hair Co content was revealed. The overall model incorporating age, gender, geographic region, as well as urinary Co concentrations accounted for 15% of hair Co variability ($p < 0.001$).

In contrast, no significant relationship between hair and urinary Mn levels was observed ($\beta = 0.032$; $p = 0.545$). As in the case of cobalt, male gender was found to be a negative predictor of hair Mn content ($\beta = -0.435$; $p < 0.001$ *). It is also notable that African origin was significantly associated with higher hair Mn levels ($\beta = 0.207$; $p = 0.046$). Generally, the overall model was significantly associated with up to 24% of hair Mn variability ($p < 0.001$).

4 Conclusion

The obtained data demonstrate that the lowest levels of cobalt and manganese in RUDN University students were found in subjects originating from Asia and especially Middle East. In turn, the highest levels were observed in the Russian and African examinees. At the same time, the obtained values of hair Co content in Russian students also indicate suboptimal Co status, being lower than the reference ranges for Co estimated as 0.07-0.50 μg/g [7]. The obtained data on hair Mn levels in Russian students corresponds to the reference limit of 0.29-1.76 μg/g [7]. Lower levels of hair Co and Mn in newcomers from Middle East are in agreement with the earlier data [8].

Reduced hair Co content may be hypothetically indicative of suboptimal intake of vitamin B12, being a biologically active form of cobalt [9]. In this view, low hair Co levels correspond to high prevalence of vitamin B12 deficiency in subjects living in Middle East and Asia [10]. Reduced dietary intake of cobalt [11] and vitamin B12 [12] with low-meat diets may underlie the observed deficiencies.

At the same time, the particular causes of the observed low Mn status in students from Middle East are unclear. Cereals and leafy vegetables are considered as prior sources of manganese [13], although consumption of these foods increased significantly in Middle East populations in the last decades [14]. Hypothetically, the observed differences may occur due to geochemical aspects of manganese deposition. Particularly, manganese deposits and production are located in Africa, Asia, Russia, as well as Latin America, but not Middle East [15].

Although the particular causes of the observed differences are unclear, reduced levels of Co and Mn may significantly contribute to health and education performance. Particularly, it has been demonstrated that children with low Co status are characterized by lower adaptive reserves [16]. In turn, both high and low Mn status (U-shaped) were shown to be associated with impaired intelligence [6]. These findings indirectly corroborate the earlier observed higher rate of impaired adaptive responses in foreign students [17]. Therefore, monitoring and nutritional improvement of trace element status in foreign students may be considered as a useful tool for increasing health and performance in foreign students.

References

1. M. G. Skalnaya, A. V. Skalny, *Essential trace elements in human health: a physician's view* (Publishing House of Tomsk State University, Tomsk, 2019).
2. P. J. Magee, M. T. McCann, Proc. Nutr. Soc., **78**, 147-149 (2019)
3. P., Chen, J. Bornhorst, M. Aschner, Front. Biosci., **23**, 1655-1679 (2019)
4. K. Yamada, *Interrelations between Essential Metal Ions and Human Diseases,* 295-320(Springer, Dordrecht, 2013)
5. L. Leyssens, B. Vinck, C., Van Der Straeten, F. Wuyts, L. Maes, Toxicology, **387**, 43-56 (2017)
6. B., Milton, D., Krewski, D. R., Mattison, N. A., Karyakina, S., Ramoju, N., Shilnikova et al., Neurotoxicology, **58**, 217-225 (2017).
7. A. V., Skalny, M. G., Skalnaya, A. A., Tinkov, E. P., Serebryansky, V. A., Demidov et al. Environ. Monit. Assess., **187**, 677 (2015)
8. B. H. Kabki, M. Al-Mohalep, A. A. Kirichuk, Trace Elem. Electroly. **35**, 190 (2018)
9. G. N. Schrauzer, K. P. Shrestha, M. F. Flores-Arce. Biol. Trace Elem. Res., **34**, 161-176 (1992).
10. C. Yajnik, U. Deshmukh, P. Katre, T. Limaye, *Vitamin B_{12}*, 131-150 (CRC Press, 2017).
11. B., Hokin, M. Adams, J. Ashton, H. Louie, Asia Pac. J. Clin. Nutr., **13**, 289-291. (2004)
12. S. P., Stabler, & R. H. Allen, Annu. Rev. Nutr. **24**, 299-326 (2004)
13. EFSA. EFSA Journal, **13**, 4150 (2015)
14. R. Bahn, S. E. Labban, N. Hwalla. Sustain Sci. **14**, 1131-1146 (2019)
15. A. P. Das, L. B. Sukla, N. Pradhan, S. Nayak., Bioresour. Technol., **102**, 7381-7387 (2011)
16. V. Detkov, A. Skalny, M. Karganov, A. Cherepov, Y. Medvedeva, M. Glazov, L. Isankina, Technol. Living. Syst., **10**, 22-28 (2013)
17. A. A. Kirichuk, I. V. Radysh, A. Y. Chizhov, Human Ecol., **1**, 20-25 (2019)

Biological contamination of soils in urbanized ecosystems by Toxocara sp. eggs.

Olga Maslennikova[1], *Viktoriya* Erofeeva[2,3]

[1]Vyatka State Agricultural Academy, Kirov, 610017, Russia
[2]Peoples' Friendship University of Russia (RUDN University), Faculty of Ecology, 6 Miklukho-Maklaya St, Moscow, 117198, Russian Federation
[3]Moscow Technical University of Communications and Informatics, Moscow, 111024, Russia

Abstract. As a result of the study of three sites (two city parks and a special area for walking dogs) on the contamination of Toxocara sp. eggs it was found that 28.0±2.3% of the studied fecal samples from the territories of the Kirov Park were infected with Toxocara eggs, 21.0±3.6% of fecal samples from the territory of the Park of the 50th anniversary of the Komsomol and 7.0±0.4% of fecal samples were infected with Toxocara eggs selected at a specially equipped dog walking area in the Metrograd residential complex in Kirov. The average rate of dogs which were infected by Toxocara (EI) was 18.7±6.2%. The average rate of contamination of soil by eggs Toxocara in the study sites amounted to 34.6%. Contamination of soil samples with eggs of Toxocara sp. in parks was 45.6% and 45.8%, respectively, and in the dog walking area it was 12%. The largest number of dogs, including stray ones, were registered in the parks: 74.8±7.1 and 62.8±7.0 individuals, respectively. There were no stray individuals in the dog walking area, and there were 21.6±1.0 individuals over the entire study period.

1 Introduction

At present, biological pollution of the environment is becoming an environmental problem on the territory of large cities. Biological pollution is the contribution to the environment and reproduction of microorganisms in it, the presence of eggs of parasites that cause diseases of humans and farm animals. Biological contamination of the environment with helminth eggs from domestic animals is particularly dangerous.

The soil of cities undergoes a huge biological pollution as a result of unauthorized placement of a huge number of feces of dogs and cats, contaminated with helminth eggs. The main sources of soil contamination by feces of urbanized territories are the increasing population of decorative, service, hunting breeds of dogs, as well as cats of various breeds, and increasing populations of stray animals. Infected dogs and cats, polluting objects of the external environment by helminth eggs with feces, endanger all those around them. Carnivores, especially dogs infected with parasites, pose a direct threat of infection and disease in humans [1].

Pollution of the natural environment with helminth eggs is an acute environmental problem in many major Russian cities, including Kirov [2-6].

A person can be infected with 32 types of helminths that are common to animals and humans [2]. One of the dangerous worms to humans, especially children, are Toxocara (Toxocara canis and Toxocara mystax). Sources of infection of people are pets (cats and dogs). Invasive diseases of cats and dogs are registered everywhere, especially among stray dogs.

Toxocara eggs are very resistant to adverse environmental factors, Toxocara eggs can persist in the soil for a long time and resist the action of various chemicals. Cockroaches and flies can participate in the spread of Toxocara eggs as well as other helminths, and earthworms play the role of paratenic hosts [7-9].

The aim of our study is to assess the level of bio- contamination of environmental objects (soil) by Toxocara sp. eggs in places where dogs are walking in parks in Kirov.

2 Materials and methods

The research was conducted from 01.05.2017 to 31.05.2018 in a specialized laboratory complex.

The collection of material for research was carried out on three sites with an area of 100x100 m in the territory of Kirov: the Kirov Park; the Park of the 50th anniversary of the Komsomol; dog walking area in the residential area of Metrograd. The objects of research were 300 depersonalized canine excrement and 45 combined soil samples taken at these sites. Test sites were laid in areas with homogeneous soil and vegetation cover. Soil sampling was carried out in accordance with the methods of sanitary and parasitological research (2000).

The method of standard visual accounting was applied to account for the number of dogs on three sites [10]. The study of depersonalized dog excrement for the presence of eggs was performed by the Fulleborn method [4]. The extent of infestation (EI) was calculated. The method of N. A. Romanenko (1996) was used to study the soil for helminth eggs.

Statistical processing of the findings was carried out using the Excel computer program.

3 Results

We calculated the number of walking dogs on all three sites in Kirov from May to September 2017. The findings is shown in table 1.

Table 1. Calculated the number of walking dogs on all three sites in Kirov from May to September 2017.

Place of research	May	June	July	August	September	The average number of dogs
The Kirov Park	85.0±9.4	65.0±9,7	94.0±12.2	76.0±13.0	54.0±8.1	74.8±7.1
The Park of the 50th anniversary of the Komsomol	78.0±8.7	48.0±6.9	72.0±12.1	72.0±8.7	44.0±7.4	62.8±7.0
The dog walking area	23.0±3.0	20.0±4.3	23.0±2.7	21.0±2.6	21.0±2.9	21.6±1.0

According to these findings, the largest number of dogs were registered in the Kirov Park it is 74.8±7.1 individuals, in July this indicator reached 94.0±12.2 individuals. In the Park of the 50th anniversary of the Komsomol the average number of dogs for the accounting period was 62.8±7.0 individuals. In June and September, the smallest number of dogs were counted – 48 and 44, respectively. The number of dogs on the dog-walking site for the entire period of research was only 21.6±1.0 individuals.

Percentage of occurrence of the invasive origin of helminths (eggs) in soil samples in the Kirov Park is shown in picture 1.

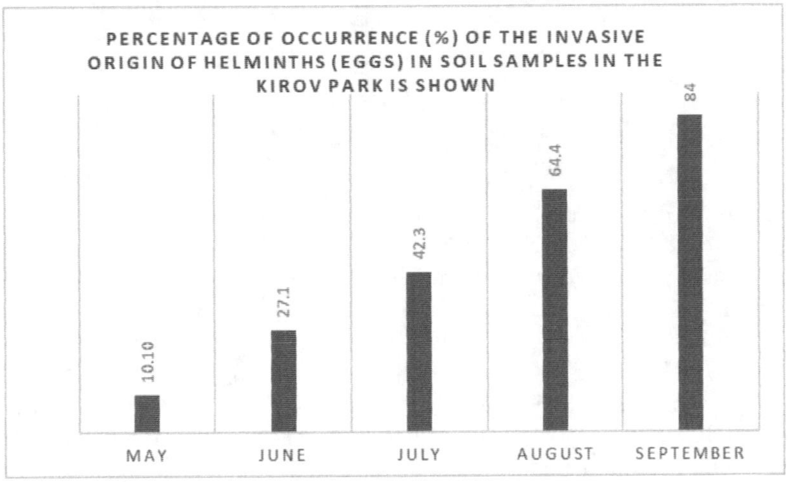

Fig.1. Percentage of occurrence of the invasive origin of *Toxocara sp.* in soil samples in the Kirov Park is shown

According to the results of the research, positive findings of toxocar eggs were recorded throughout the experiment, and there was an increase in the number of positive samples from May to September. The average rate of soil contamination in the Kirov Park is 45.6%. The highest percentage of soil contamination was registered in September 84.0 %, the lowest in may-10.1%.

The percentage occurrence of eggs of Toxocara sp. in soil samples in the Park of the 50th anniversary of the Komsomol is shown in picture 2. In this park the contamination of soil with eggs of Toxocara sp., as in the Kirov Park, increased from May to September. The average rate of soil contamination was 45.8%. The highest percentage of soil contamination was registered in August and September (70.8%; 76.1 %, respectively) and the lowest in may – 8.7%.

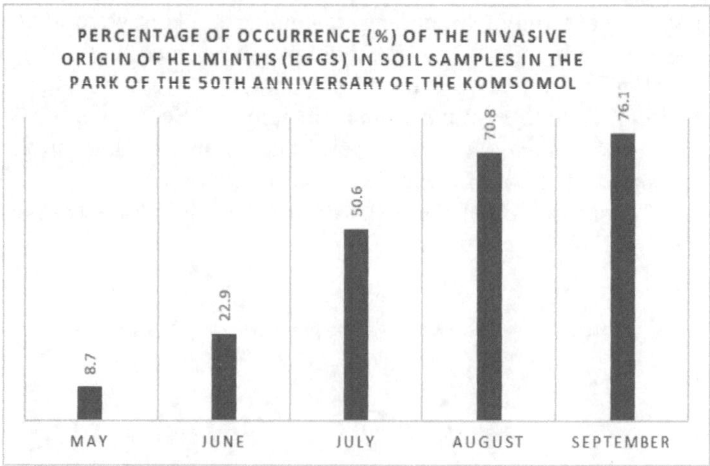

Fig.2. Percentage of occurrence of the invasive origin of *Toxocara sp.* in soil samples in the Park of the 50th anniversary of the Komsomol

The average rate of soil contamination at the dog walking area in the Metrograd residential complex is 12%. There was an increase in the number of positive samples with the registration of *Toxocara* sp eggs from May to September. The highest percentage of soil contamination was registered in August and September (17.5%; 25.7%, respectively) and the lowest in May and June (1.0%, 3.6%, respectively). In September soil contamination with *Toxocara* eggs was the highest and reached an average of 63.2% for three sites, due to the accumulation of infestation and its preservation.

The number of eggs per 1 kg of soil was calculated for three sites. The results of studies of the degree of soil contamination by *Toxocara* eggs on the territory of the study sites are presented in table 2.

Table 2. The results of studies of the degree of soil contamination by *Toxocara sp.* on the territory of the study sites

Sampling	The Kirov Park	The Park of the 50th anniversary of the Komsomol	The Metrograd residential complex in Kirov	Average value
***Toxocara* eggs, ind./kg**	11.1±2.4	6.5±2.0	1.2±1.0	6.3±2.9

According to SanPiN 2.1.7.1287-03 "Sanitary and epidemiological requirements for soil quality" [11] in the total number of helminth eggs per 1 kg of soil, the Kirov Park is classified as dangerous (more than 10 eggs per 1 kg of soil), the Park of the 50th anniversary of the Komsomol and the dog walking area in the Metrograd residential complex in Kirov are classified as moderately dangerous (less than 10 eggs per 1 kg of soil).

The study of 300 samples of depersonalized dog feces found that the average rate of infection of dogs *Toxocara* (EI) was 18.7±6.2%. The findings on contamination of dog excrement with *Toxocara* canis eggs (%) in Kirov are presented in table 3.

Table 3. The findings on contamination of dog excrement with *Toxocara sp.* eggs (%) in Kirov

Sampling	The Kirov Park	The Park of the 50th anniversary of the Komsomol	The Metrograd residential complex in Kirov
The number of positive samples of excrement, %	28.0±2.3	21.0±3.6	7.0±0.4

In the course of laboratory research, it was found that toxocar eggs infected 28±2.3% of the studied fecal samples from the territories of the Kirov Park, 21±3.6% of the studied fecal samples from the territory of the Park of the 50th anniversary of the Komsomol and 7±0.4% of the studied fecal samples were infected with *Toxocara* eggs selected at a specially equipped dog walking area in the Metrograd residential complex in Kirov.

4 Discussion

Given the high number of dogs in the parks of Kirov, according to the conducted research , it can be assumed that the city's ecosystem is experiencing a high degree of pollution from animal biological waste.

The incidence of toxocarosis in children and adults depends directly on the contamination of the soil with toxocar eggs, which are spread by dogs. The increase in the proportion of *Toxocara* eggs found indicates an increasing risk of infection of the population of Kirov with toxocarosis.

The results of the research showed that the soil of different types of objects in Kirov has characteristic differences in the degree of contamination with t *Toxocara* eggs. The average contamination of soil by *Toxocara* eggs in the tudy sites amounted to 34.6%. Soil from the territories of Kirov Park and the Park of the 50th anniversary of the Komsomol was contaminated to the maximum extent , which is due to the presence of a large number of stray and neglected dogs in these places (45.6% and 45.8% of the tested samples were positive). Low indicators of soil contamination with *Toxocara* eggs -12% were observed on a specially equipped dog walking area in the Metrograd residential complex in Kirov, which is due to the protection of the territory from the penetration of stray and homeless dogs , the novelty of the construction and compliance with sanitary and epidemiological rules of the site maintenance.

According to research by M. V. Guzeeva [3], conducted in the Zelenograd administrative district of the Moscow region, the percentage of soil contamination with toxocar eggs in the city on average is 13.5±1.1, playgrounds of residential buildings amounted to 14.8±1.7%, sandboxes of children's educational institutions – 11.0±1.5 %. The maximum indicators of the level of soil contamination with toxocar eggs were found at dog walking sites-35.1±6.3 % [3].

We found that in May, soil contamination with *Toxocara* eggs at three sites was the lowest and amounted to 6.6%, which is due to the death of a large number of eggs in the winter at low temperatures (frosts reached -38°C). In the summer, the contamination of soil by toxocar eggs increased to 34.3%, due to the high temperatures and moist soil.

Comparison of soil contamination and dog excrement with eggs of *Toxocara* sp. showed a directly proportional relationship - soil contamination with *Toxocara* eggs directly depends on soil contamination with canine excrement.

According to L. V. Shishkanova [6], the intensity of contamination of the substrate with *Toxocara* canis eggs in the Krasnodar region is 3.3 eggs/kg, in the Rostov region-3.7 eggs/kg, in the Republic of Adygeya – 5.2 eggs/kg, and their viability is 70.8 %, 62.3 % and 67.1 %.

The intensity of contamination of the external environment with helminth eggs depends largely on the sanitary improvement of settlements, on the sanitary culture and level of education of the population, the level of infection with helminths of dogs and the conditions of their maintenance.

5 Conclusions

Dog records showed a large number of stray dogs in parks. The low number of dogs and the absence of stray dogs were registered on a specially equipped dog walking area in the Metrograd residential complex.

According to the existing rules of sanitary and hygienic assessment of the soil for the total number of helminth eggs per 1 kg of soil, the Kirov Park is classified as dangerous, the Park of the 50th anniversary of the Komsomol and the dog walking area in the Metrograd residential complex in Kirov are classified as moderately dangerous. The average rate of infection of dogs with toxocariasis 18.7±6.2%.

References

1. A.N. Volichev, Tr. Vseros. Institute of helminthol., **39**, 55-64 (2003)
2. I.A. Arkhipov, O.A. Zeynalov, L.M. Kokorina, D.A. Avdanina, S.V. Likhotina, Russian Veterinary Journal, **2**, 26-30 (2005)
3. M.V. Guzeeva, *The role and place of rare helminthiases in parasitic pathology in Russia* (2009)
4. G. A. Kotelnikov, *Helminthological studies of animals and the environment*, (1984)
5. V.V. Erofeeva, O.V. Maslennikova, Modern problems of science and education, **3**, 573 (2015)
6. L.V. Shishkanova, *Toxocariasis in the south of Russia* (2011)
7. S. Okoshi, M. Usui, Jpn. J. Vet. Sci, **30**, 151-166 (1968)
8. V.V. Erofeeva, O.V. Maslennikova, Hygiene and sanitation, **8**, 897-902 (2019)
9. O.V. Maslennikova, V.V. Erofeeva, Modern problems of science and education, **5**, 683 (2015)
10. A. S. Severtsev, A.O. Vereshchagin, A.D. Poyarkov et al., *Problems of research of a domestic dog: Materials of a meeting*, 95-114 (2006)
11. *Methods of sanitary-parasitological research* (2000)

About the role of environmental factors in carcinogenesis

Sergey K. Pinaev[1*], *Olga G.* Pinaeva[1], and *Alexey Ya.* Chizhov[2]

[1] Far Eastern State Medical University, Khabarovsk, Russia
[2] Peoples' Friendship University of Russia (RUDN University), Faculty of Ecology, 6 Miklukho-Maklaya St, Moscow, 117198, Russian Federation

Abstract. There were found out the connections of the activity of the Sun and smoke of forest fires with the incidence of various forms of benign and malignant neoplasms in the population of young children 0-4 years old, in Khabarovsk Region, Russia. Thus, it was determined that so-called "sporadic" fluctuations in the incidence of neoplasms in the child population are not random. They are caused by long multi-year cycles of changes in the complex of environmental factors. The authors identified this phenomenon as "Alternative oncogenesis", meaning by it the predominant occurrence of various forms of neoplasms in a certain period of time due to a change in the parameters of the complex of environmental factors. A proposed hypothesis is: ecologically linked oxidative stress as a cause of the epigenomic modulation leading to an imbalance between semaphorins and integrins that brings to oncogenesis.

1 Introduction

Investigation of the links between environmental factors and malignant neoplasms is important both for understanding the causes of cancer and developing measures for its prevention. Our approach consists in the simultaneous study of the association of environmental factors with the incidence of benign tumors and malignant neoplasms in cohorts of young children 0-4 years, since this allows a broader out look at oncogenesis. Tumors in this period have predominantly the prenatal origin, and it makes possible to obtain information in a fairly short time. As environmental factors we have chosen solar radiation and the smoke of forest fires.

2 Materials and methods:

The incidence of benign tumors in the cohorts of young children born in Khabarovsk city in 1976-1986 years (693 causes) have been studied (per 10^5). It was studied the incidence of malignant neoplasms (per 10^6) in the cohorts of young children born in Khabarovsk Region in 1972-1988 years (572 cases of the disease).

[*] Corresponding author: pinaev@mail.ru

For the system analysis, we used the interval characteristics within the calendar year: the average of annual number of sunspots, the number of forest fires in the Khabarovsk Region, and the incidence of neoplasms. The time series were subjected to multivariate correlation and regression analysis using the IBM SPSS Statistics 23 software package in seven iterations with a 1 year lag to the year of birth of children in the cohorts (-3, -2, -1.0, + 1, + 2, + 3).

3 Results and discussion

When studying the influence of the Sun, it was established a link between the Sun activity 3 years before the birth of children and the incidence of benign tumors ($r = 0.785$; $p = 0.004$). Among the specific forms of benign neoplasms, tumors of soft tissues [1] as well as papillomas of the skin and mucous ($r = 0.720$; $p = 0.012$) showed a connection with the activity of the Sun 3 years before the birth of children. Among malignant neoplasms, a connection between solar activity 1 year before the birth of children was found for non-Hodgkin's lymphomas (Fig.1) [2] .

Fig. 1. Incidence of non-Hodgkin's lymphoma (NHL; per 10^6) in cohorts of children 0-4 years old (born 1975-1986) , and sunspots number in 1 year before their birth (Khabarovsky Region, Russia)

The smoke of forest fires 2 years before the birth of the children turned out to be associated with the incidence of lymphangiomas ($r = 0.695$; $p = 0.009$), and for teratomas a connection with smoke was established 1 year before birth ($r = 0.656$; $p = 0.014$). The study of the connection of smoke with the occurrence of malignant neoplasms revealed a link between the number of fires 2 years before the birth of children and the incidence of Hodgkin's disease [3]. The number of fires 1 year after the birth of the children turned out to be associated with incidence of tumors of the central nervous system (Fig.2) [4].

For hemangiomas and leukemia, which are the most frequent forms of benign tumors and malignant neoplasms, respectively, a significant relationship was found with both the activity of the Sun and the smoke of forest fires. Multivariate analysis showed a strong multiple correlation between the activity of the Sun 3 years before the birth of children, the number of fires 3 years after their birth, and the incidence of hemangiomas (R = 0.870; F (2.8) = 12.424; p <0.004) [1] . According to the graph (Fig. 3), the incidence of hemangiomas predominantly reflects the long-term slow change in solar activity that had an

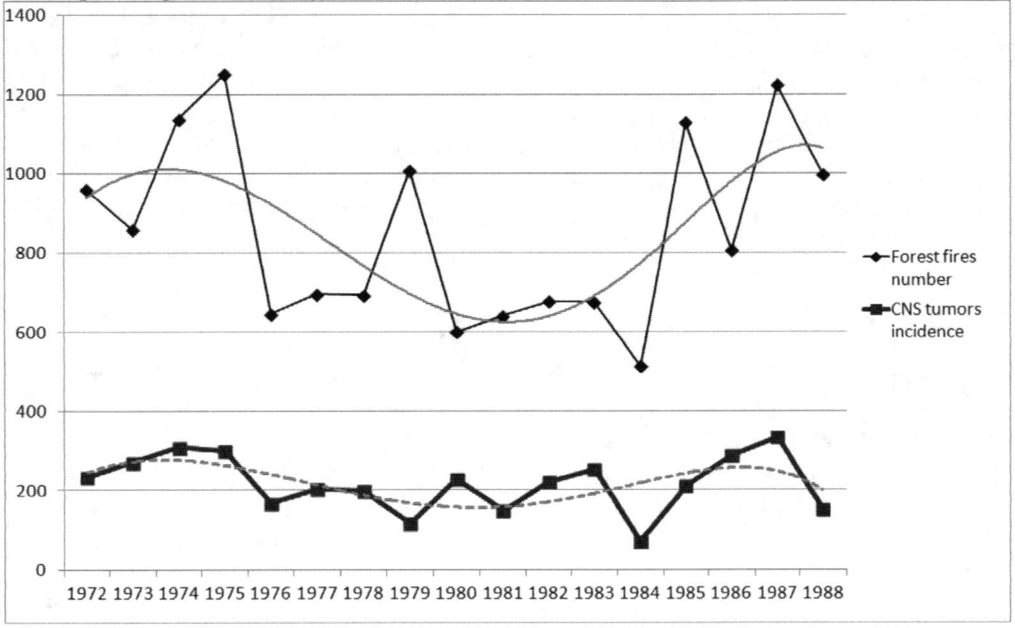

Fig. 2. Incidence of the central nervous system (CNS) tumors (per 10^6) in cohorts of children 0-4 years old (born 1972-1988) and the number of forest fires 1 year after the birth of the children (Khabarovsky Region, Russia).

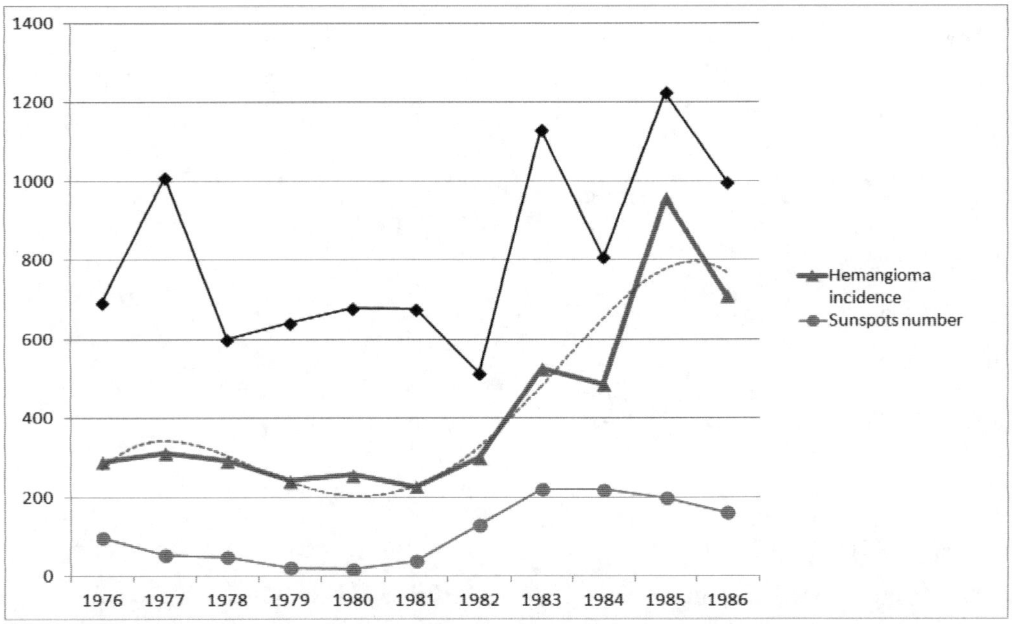

Fig. 3. Hemangioma incidence (per 10^5) in cohorts of children 0-4 years old (born 1976-1986), sunspots number 3 years before their birth, and the number of fires in a season 3 years after their birth (Khabarovsky Region, Russia).

effect on the gametes of the parents, and to a much less extent was a consequence of the postnatal effect of smoke.

For leukemia, multivariate analysis established the presence of a multiple correlation between the number of fires 2 years before the birth of children, the activity of the Sun 3 years after their birth and the incidence of this pathology [5]. As it can be seen (Fig. 4), the leukemia incidence curve reflects annual fluctuations in the number of fires, and the polynomial trend of this incidence has a high degree of similarity to long-term changes in solar activity. It can be assumed that a significant part of cases of leukemia in children occurs due to air pollution by the smoke of forest fires that had affected parental gonads 1-2 years before conception. A change in the activity of the Sun during the period of continued rapid development of the body of a child at the age of about 3 years contributes to the appearance of additional cases of the disease.

4 Conclusions

Thus, "sporadic" fluctuations in the incidence of neoplasms in the child population are not random. They are caused by long multi-year cycles of changes in the complex of environmental factors. We identified this phenomenon as "Alternative oncogenesis", meaning by it the predominant occurrence of various forms of neoplasia in a certain period of time due to a change in the parameters of the complex of environmental factors.

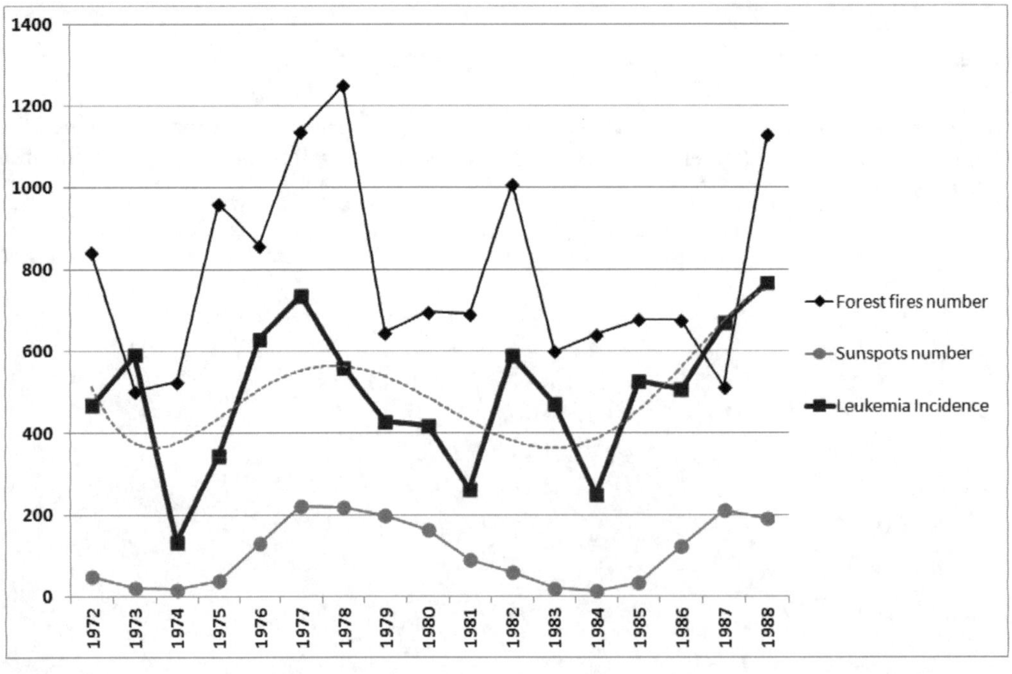

Fig.4. Leukemia incidence (per 10^6) in cohorts of children 0-4 years old (born 1972-1988), the number of fires in a season 2 years before their birth, and sunspots number 3 years after their birth (Khabarovsky Region, Russia)

The connection of neoplasms incidence with the smoke of forest fires confirms the carcinogenic effect of smoke in general. The link of solar activity with the incidence of leukemia and lymphomas may be due to changes in the Earth's magnetic field under the influence of the Sun, leading to failure of physiological rhythms. Our hypothesis is confirmed by the fact that professional exposure to low-frequency electromagnetic fields increases the risk of non-Hodgkin's lymphomas by 3 times [6], and the birth at the period of a calm Sun increases the average life expectancy by 5.2 years [7].

The oxidative stress and the epigenomic modulation caused by it, leading to an imbalance between semaphorins and integrins, with an outcome in tumor transformation can be the vector uniting environmental influences on a living organism.

In order to reduce the risk of neoplasia in children, it is recommended to the future parents, starting from the period 1-2 years before the planned conception, mothers during pregnancy and lactation, and young children to use sufficient vitamins, protect their respiratory organs with effective respirators, clean the indoor air, and make an immune rehabilitation using Transfer factor.

References

1. S.K. Pinaev, A.Ya. Chizhov, Advances in molecular oncology, **5(4)**, 18-19 (2018)
2. S.K. Pinaev, A.Ya. Chizhov, O.G. Pinaeva, *Fundamental science and technology – promising developments XVII,* **(2)**, 14-17- (Academic, North Charleston, 2018)
3. S.K. Pinaev, A.Ya. Chizhov, Advances in molecular oncology, **5(4)**, (2018).
4. A.Ya. Chizhov, S.K. Pinaev, O.G. Pinaeva, Technologies of Living Systems, **1**, 53-58. (2019)
5. A.Ya. Chizhov, S.K. Pinaev, Radiation and risk, **27(4)**, 87-94. (2018) DOI:
6. P.J. Villeneuve, D.A. Agnew, A.B. Miller, P.N. Corey, Occup. Environ. Med. Apr; **57(4)**: 249-57 (2000)
7. G.R. Skjærvø, F. Fossøy, E. Røskaft, Proc. Biol. Sci., **22**, 282(1801) (2015)

Physiological Evaluation of the Cardiology System of Nonresident Students at Different Training Periods

Yana Mayorova[1], *Viktor* Glebov[1*], *Viktoriya* Erofeeva[1,2], *Sergey* Yablochnikov[2], *Bogdan* Laver[3]

[1]Peoples ' Friendship University of Russia, Moscow, 117198, Russia
[2]Moscow Technical University of Communications and Informatics, Moscow, 111024, Russia
[3]Academy of Postgraduate Education of the Federal Medical and Biological Agency of Russia, Moscow, 125371, Russia

Abstract. The article provides a physiological assessment of the state of the cardiovascular system of students in different periods of the academic year. The study was conducted on a sample of 311 nonresident and Moscow students. An analysis of the spectral characteristics of HRV showed that during the examination session the total power of the spectrum decreased compared with the indicators during the training session. During the intercessional period, the total spectrum power in the studied group of students amounted to 4856.34 ± 645 ms^2, and during the examination session this indicator decreased by 35.8% and amounted to 3119.41 ± 522 ms^2. A comparative analysis of the HRV activity wave spectrum of the studied student sample showed significant changes ($p < 0.05$) of students' functional state indicators (SDNN, LF, HF, LF / HF; SI, ARS). According to the studied indicators, a growth trend was noted: (SDNN: from 43.89 ± 8.12 to 78.03 ± 3.86; LF: 27.78 ± 7.33 - 42.56 ± 4.51; LF / HF: 0.67 ± 0.03 - 1.72 ± 0.02; SI: 102.95 ± 11.3 - 467.77 ± 6.1; ARS: 1-2 ± 0.12 - 5-7 ± 0.01) and a decrease in HF (41.74 ± 5.3 to 24.67 ± 2.94).

1 Introduction

Students represent a special social group with specific living conditions. These conditions are characterized by a high and long-term level of psychoemotional stress and a high level of information loads [1]. Research by many authors shows that at the initial stages of training, students have a high psychoemotional level and information loads, which are associated with students not being adapted to the new educational environment [2,4,7].

The process of adaptation to the educational process among students is a complex, multi-level socio-psychological process. This process is accompanied by a significant strain of the compensatory-adaptive systems of the body. It was found that the training load causes students to adapt changes in the functional state of the main regulatory systems

* For correspondence: vg44@mail.ru

(central nervous system, cardio-respiratory system) [3].

It is important to note that the state of the cardiovascular system (CVS) of a person is a sensitive indicator of the adaptive processes of the body. Indicators of heart rate variability (HRV) show the state of human psychosomatic health [3-6].

It should also be noted that teaching students at different periods of the school year is accompanied by different intensities of the actions of psychoemotional and informational factors (anxiety, psychoemotional lability, memory loss, headaches, sleep disturbances, changes in the functioning of the cardiovascular system, information overload, etc.). These factors are complex and are not presented in our work due to the restrictive nature. They are presented in our other studies [1,3]. The main goal of this work was to study and monitor the status of the main regulatory systems of students in different periods of training (periods of the session and passing the exam). Such psychophysiological monitoring makes it possible to track the maladaptive states of students and, in the early stages of the educational process, take timely preventive and corrective actions.

2 Materials and methods

2.1 The study Sample

The studies involved 311 (156 young men and 155 girls) practically healthy nonresident and Moscow students aged 18.4 to 19.6 years of the first year of study. The sample size of students and its representativeness was calculated according to the generally accepted formula $n=Z^2 x\ p\ (1-p):t^2$ [15]. The studied sample of nonresident students was random and represented two Moscow universities: PFUR and Moscow State University named after M.V. Lomonosov. Nonresident students were represented by the Central, Volga, North Caucasian and Siberian federal districts of the Russian Federation, who had no complaints at the time of the study, and without somatic pathology, with a normal level of blood pressure.

All students based on HRV indicators were divided into three groups. The first group (n = 174), having a state of satisfactory adaptation. The second group (n = 89) was characterized by a state of increased functional stress. The third group (n = 48) was characterized by a state of high level of stress.

2.2 Research Stages

The study was conducted in two stages: Stage 1 - in the autumn intersessional period; Stage 2 - during the winter examination session (December-January).
The studies were carried out with the voluntary informed consent of the subjects, in accordance with the ethical standards of the Helsinki Declaration (2000).

2.3 Research Methods

HRV monitoring was carried out using the hardware and software complex "Psychophysiologist 1-30" (Research Center «Medicom», Taganrog). The functional state and adaptive potential were evaluated by the methods of time and spectral analysis of HRV. Registration was carried out in accordance with the "International Standard" (1996) using 5-minute records on the apparatus [6].

Assessment of CVS was carried out according to the activity of regulatory systems (ARS). ARS was calculated in points from 1 to 10 according to a special algorithm that takes into account statistical indicators, histogram indicators and data of spectral analysis of

cardio intervals. ARS allows evaluating various degrees of stress in regulatory systems. The following gradations of the degree of tension of regulatory systems exist. Normal state, or state of satisfactory adaptation to environmental conditions (ARS = 0-2), moderate stressful state of regulatory systems (ARS = 3-4), over moderate stressful state of regulatory systems or unsatisfactory adaptation state (ARS = 5-7), regulatory depletion state systems or failure of adaptation (ARS = 8-10).
The study used HRV parameters recommended by the joint Working Group of the European Cardiology Society and the North American Society of Stimulation and Electrophysiology [5,6]:

- Standard deviation of RR intervals (SDNN).
- The total power of the spectrum (TP), - powers in the frequency range (0.003 to 0.40 Hz). The indicator reflects the total activity of neurohumoral influences on the heart rhythm.
- High-frequency oscillations (HF) - fluctuations in the frequency of the heart rhythm (0.15-0.40 Hz). This indicator is mainly associated with respiration and reflects vagus control of heart rate.
-Low-frequency oscillations (LF) - oscillations in the frequency range (0.04-0.15 Hz). This indicator is influenced by changes in the tone of both the sympathetic (mainly) and parasympathetic division of the ANS.
- Stress Index - regulatory stress index (by Baevsky)

Statistical data processing was carried out using the software package "Statistica 6.0". The standard methods of variation statistics were used: calculation of average, standard errors of the mean. Reliable considered differences in performance at $p < 0.05$.

3 Results

At the first stage, an analysis of the functional state of students was carried out. The analysis showed that the first group of 174 (56%) of the studied students, in the intersessional period, was in a state of satisfactory adaptation. This group of students was characterized by the optimal mode of functioning of the CVS. The second group of students, consisting of 89 students (28.6%), was in a state of increased functional tension of adaptation mechanisms. The third group of students - 48 (15.4%) was in a state of overstrain of adaptive mechanisms, which indicated their high level of psychoemotional stress and the intense work of CVS (Table 1).

Table 1. Indicators of the functional state of students' CVS (n = 311)

Indicators	**1-st group. State of satisfactory adaptation (n = 174)**	**2-nd group. State of increased functional stress (n = 89)**	**Group 3. The state of high level of stress (n= 48)**
SDNN, ms^2	43.89± 8.12	55.04± 4.46	78.03± 3.86*
LF, %	27.78 ± 7.33	33.5 ± 5.68	42.56± 4.51*
HF, %	41.74± 5.3	35.45± 3.73*	24.67± 2.94*
LF/HF	0.67± 0.03	0.94 ± 0.02*	1.72± 0.02*
SI, c.u.**	102.95 ± 11.3	166.54± 8.23*	467.77± 6.1*
ARS*,balls**	1-2± 0.12	3-4± 0.02	5-7± 0.01*

Note: the reliability of the difference between indicators: * $p<0.05$ between 1 and 3 and 2 and 3 groups of students; c.u**. - conventional unit; *** ARS - activity of regulatory systems

We conducted a comparative analysis of HRV indicators between these groups. A comparative analysis of the HRV activity spectrum of students showed significant changes ($p < 0.05$) of students' functional state indicators (SDNN, ms; LF,%; HF,%; LF / HF; SI, cu; ARS, points) . There was an increase in indicators for: SDNN (from 43.89 ± 8.12 to 78.03 ± 3.86 ms), LF (from 27.78 ± 7.33% to 42.56 ± 4.51%), LF ratios / HF (from 0.67 ± 0.03 to 1.72 ± 0.02), SI (from 102.95 ± 11.3 to 467.77 ± 6.1 cu), ARS (from 1- 2 ± 0.12 to 5-7 ± 0.01 points). A decrease in HF values was noted (from 41.74 ± 5.3% to 24.67 ± 2.94%).

The second stage was carried out in the winter examination session. We focused on the third group of students who experienced difficulties in adapting to the educational environment and had psychoemotional difficulties (anxiety, frustration, communication difficulties, etc.). To assess the functional state of the CVS of the maladaptation group, a comparative analysis of HRV indicators was performed (Table 2).

Table 2. Heart rate variability indicators for a sample of students with adaptation problems at different periods of the educational process

Periods	**SDNN,** ms^2	**TP,** ms^2	**HF,** ms^2	**LF,** ms^2
Inter-session period (p = 48)	64.86 ± 5.3	4856.34± 645	1256.34 ± 564	2076.45 ± 433
Exam stress (p = 48)	42.72 ± 5.2*	3119.41 ± 521*	511.45 ± 343*	1741.39 ± 210*

Note: significant confidence between indicators at rest and under stress: * $p < 0.05$

4 Conclusions

Examination stress is one of the first places among the causes of psychoemotional stress in students. This has a powerful effect on the functional systems of a person, in particular on CVS [8-11].

Our method of assessing CVS using the HRV of the studied sample revealed 3 levels of students' adaptation processes: a group with satisfactory adaptation (n = 174), a group with increased stress level (n = 89), and a group with a high level of stress and maladaptation of CVS (n = 48)

The above comparative analysis of the spectrum of the HRV activity wave of the entire studied sample of students convincingly showed significant changes ($p < 0.05$) of the students' functional state indicators (SDNN, LF, HF, LF / HF; IN, ARS). According to the studied indicators, a growth trend was noted: (SDNN: from 43.89 ± 8.12 to 78.03 ± 3.86; LF: 27.78 ± 7.33 - 42.56 ± 4.51; LF / HF: 0.67 ± 0.03 - 1.72 ± 0.02; SI: 102.95 ± 11.3 - 467.77 ± 6.1; ARS: 1-2 ± 0.12 - 5-7 ± 0.01) and a decrease in HF (41.74 ± 5.3 to 24.67 ± 2.94). The presented dynamics of HRV indicators shows physiological changes in the work of students' CVS, their adaptation processes, which in some students caused a state of overstrain (in particular, students of the 3rd group) [1,2,4].

An analysis of the spectral characteristics of HRV in a maladaptive group of students showed that during the examination period, students had a decrease in the studied parameters (SDNN, ms^2 TP, ms^2 HF, ms^2 LF, ms^2). This decrease shows the powerful effect of informational overloads of the educational environment on the nervous system of students with adaptation problems. This accordingly affects the work of the CVS of such students. Our data are confirmed by the research results of N.P. Devaev (2010), A.E. Severin, V.I. Torshin, O.V. Mankaeva (2018), where it is noted that, during the examination session, students recorded pronounced violations of the autonomic regulation of the cardiovascular system. These disorders are manifested in an increase in heart rate, increased blood pressure, an increase in muscle and psychoemotional stress, which are accompanied by anxiety, fatigue, memory problems, headaches, sleep disturbance, angina pectoris [12-14].

Thus, the psychoemotional and informational loads at the university at different periods of the academic year can cause students of various levels of stress. This process may be accompanied by a violation of autonomic homeostasis, increased lability of reactions of the CVS and the growth of maladaptation processes in some students during the examination session.

The publication was financially supported by the Ministry of Education and Science of the Russian Federation (the Agreement number 02.a03.0008).

References

1. V.V. Glebov, D.S. Lyamina, V.A. Trushin, P.F. Popova, *Ecological and physiological problems of adaptation materials of the XVIII all-Russian Symposium with international participation*, 66-67 (2019)
2. D.A. Dimitrieva, E.V. Saparova, Rus. Physiol. J. **101,** 98-107 (2015)
3. V.V. Glebov, Rus. J. of Cardiology, **24**, 6-7 (2019)
4. V.A. Bodrov, *The manual for High schools* (2000)
5. R.M. Baevsky, Human Phys. J. **28,** 72-82 (2002)
6. *Heart Rate Variability*, 1043-1065 (1996)
7. N.P. Devaev, Physiology, **2**, 622-626 (2010)
8. A.E. Severin, T.E. Batotsyrenova, V.I. Torshin, *Materials of the International Scientific and Practical Conference*, 144-150 (2017)
9. A.E. Severin, V.I. Torshin, O.V. Mankayeva, N. Mansur, *Ecological and physiological problems of adaptation*, 194-195 (2017)
10. V.I. Torshin, A.E. Severin, E.A. Severina, Yu.P. Starshinov, *Scientific and practical conference*, 143-146 (2016)
11. S.I. Kudinov, S.S. Kudinov, *Psychodiagnostics of personality* (2018)
12. S.I. Kudinov, S. S.Kudinov, Psychology and Pedagogy, **15**, 38-52 (2018)
13. A.E. Severin, V.I. Torshin, O.V. Manceva, E.A. Severin, Y.M. Zhuk, O.A. Mawarid, *Materials of IV all-Russian Scientific-Practical Conference with International Participation,* 477-478 (2018)
14. A.E. Severin, L.T. Sushkova, T.E. Batotsyrenova, V.I. Torshin, Yu.P. Starshinov, V.V. Rozanov, O.M. Zubova, *The XIII International Scientific Conference with the Scientific Youth School named after* (2018)
15. N.I. Malykh, *Statistics: Theory of Statistics: a Textbook and Workshop for Secondary Vocational Education*, (2019)

https://doi.org/10.1051/e3sconf/202016904005

Effects of Almaty city ecological factors on students blood indices

Tilek Baidaulet[1,3*], *Alibek* Ydyrys[2] , *Moldir* Imanaliyeva[3], *Zifa* Askarova[3], *Gulziya* Srailova[3], *Maira* Murzakhmetova[3] and *Sultan* Tuleukhanov[3]

[1]Asfendiyarov Kazakh National Medical University, 94 Tole bi Str. 050000 Almaty, Kazakhstan
[2] Al-Farabi Kazakh National University, Biomedical research centre. 71 al-Farabi Ave. 050040 Almaty, Kazakhstan
[3] Al-Farabi Kazakh National University, 71 al-Farabi Ave. 050040 Almaty, Kazakhstan

Abstract. This research gives an insight into the main features of the cardiovascular system's statistical and dynamic, as well as young local and foreign citizens in Almaty city adaptation indices. 1500 analysis of blood indices was carried out before and after the physical workload. Hematological research were carried out based on N. M. Mykolayiv method and included the erythrocytes and leukocytes count in peripheral blood vessel and absolute number of lymphocytes . Foreign students' absolute amount of leukocytes and erythrocytes in the peripheral blood vessels and the phagocytic activity of neutrophil were different from local students' indices.

1 Introduction

Environment is the sum of all conditions and influences that affect the development and life of all organisms on earth. [1]. The environment affects our health in a variety of ways. The interactions between human health and the environment has been extensively studied and environmental risks have been proven to significantly impact human health, either directly by exposing people to harmful agents, or indirectly, by disrupting life-sustaining ecosystems [2].

The ultimate purpose for the cardiovascular system is to facilitate exchange of gases, fluid, electrolytes, large molecules and heat between cells and the outside environment [3]. Contact with unsafe drinking or bathing water can impose serious risks (both acute and delayed) to human health [4;5].

All the cardiovascular changes occurring during physical activity ensure that active muscles are supplied with increased blood flow and oxygen while maintaining normal, or even elevated, arterial pressures [3]. Sulfur dioxide irritates respiratory tissues. Chronic exposure causes a condition similar to bronchitis [6]. It is shown that the adverse ecological factors of environment have harmful effect to the human blood indices. The characters and extent of the effect was depending on the ecological condition [7].

* Corresponding author: aduinger@gmail.com

2 Materials and Methods

50 young girl students were examined and information is listed in the following table1. They were examined before and after the physical workload, time: 14:00 - 17:00, December to January. And they were estimated to make a prediction to their cardiovascular system status.

Table 1. information of objects

Group	Number of students	Age, year	Weight, kg	Height, cm	Origin
First group	20	18-22	48-65	150-170	Almaty city
Second group	15	18-21	50-66	158-169	southern Kazakhstan
Third group	15	18-22	52-70	155-170	western Kazakhstan
overall	50	18-22	48-70	150-170	-

They all were university students in Almaty, and were kept normal mode of labor and rest. The study was carried out at Keremet medical center in Almaty. Indoor temperature was 22-25C°. outdoor was -7, -10 C°.

The blood indices analysis was carried out before and after the physical workload. Therefore, we did 1500 times research among the students. The hematological research included the erythrocytes count in peripheral blood vessel, leukocytes count in peripheral blood vessel and the absolute number of lymphocytes and they were carried on based on N. M. Mykolayiv method.

The research was conducted over three years.

3 Results and Discusson

Table 2 shows the results of measuring the number of white blood cells in different groups before and after exercise.

Table 2. Leukocytes counts in the peripheral blood vessels of the students before and after the physical workload

	2017-2018		2018 - 2019		2019 - 2020	
	before	after	before	after	before	after
First group (control group)	5627.6	6200.4	5630.7	6140.3	5618.5	6112.4
Second group	5523.1	6950.7	5862.9	7163.5	5702.5	6702.4
Third group	5952.8	6968.3	5752.9	6600.7	5809.7	6849.4
($P^{*}<0,05$), ($P^{**}<0,01$), significant differences between groups						

The results of these studies showed that the volume of blood fluid students of the second group was higher than in the local group of students, the number of leukocytes in 1

MKL of blood in the second group was also lower than in the local group of students, the average level of the absolute number of leukocytes in the third group were significantly higher than other groups in the first year. It can be concluded that the climatic and geographical conditions of students in the second and third groups were more severe than those of students in the control area. The reason for this was that the students came from a region that had a lower air temperature, the body of students in this group was adapted to harsh climatic and geographical conditions than in the control groups. In subsequent years, the indicators of students in the second and third groups change, while in the first group they remain approximately at the same level. It can be concluded that the body of students has already adapted to severe and serious climatic and geographical conditions, and not to local climatic and geographical conditions.

Table 2 also shows that after exercise, the number of white blood cells in the peripheral blood vessels increased incredibly. However, this change is noticeably greater in the second and third groups. than in the control group.

The absolute number of lymphocytes in peripheral blood vessels was also analyzed in students of Almaty city, both of local origin, and from western and southern Kazakhstan, the results are shown in table 3

Table 3. Lymphocyte counts in the peripheral blood vessels of the students before and after the physical workload

	2017-2018		2018 - 2019		2019 - 2020	
	after	before	after	after	before	after
First group (control group)	23.14	25.22	23.9	25.02	24.04	25.07
Second group	25.62	28.9	23.61	27.01	25.79	27.14
Third group	25.17	29.19	25.45	28.76	27	28.36
(P^{*} <0,05), (P^{**} <0,01), significant differences between groups						

According to the research results, there were much more obvious changes in the immune system of students to adapt to the new climatic and geographical conditions of Almaty cities from foreign climatic and geographical conditions. In contrast to local students living in Almaty, students from the southern and Western parts of the Republic of Kazakhstan all indicators, such as the absolute number of white blood cells in peripheral blood vessels, the absolute number of lymphocytes in peripheral blood vessels and phagocytic activity of neutrophils, differed from the indicators of local students, and this gives us an opportunity to judge the large dynamics and activity of functional blood indicators when adapting to new extreme conditions.

4 Conclusions

In conclusion, it should be noted that the indicators of the cardiovascular system of students living in the southern and Western parts of the Republic of Kazakhstan were more sensitive than those of local students living in the city of Almaty, so their bodies faced the need to adapt. The results show that the physiological indicators of the functional state of students who arrived from other parts of the Republic of Kazakhstan, significantly change relative to the initial year of study, while the local students did not change. Students ' blood counts demonstrate the adaptation process. It should also be noted that physical activity significantly affects students from other parts of the Republic of Kazakhstan, both in the initial year and in subsequent years.

References

1. World Health Organization. Preventing disease through healthy environment: Towards an estimate of the environmental burden of disease. Available online: http://www.who.int/ quantifying_ehimpacts/ publications/preventingdisease/en/index.html. (2017)
2. K.D. Anil, *Environment and ecology*, 23-36 (New Age International (P) Ltd., New Delhi, 2009)
3. R.E. Klabunde, *Cardiovascular Physiology Concepts*, 56-62. (Lippincott Williams &Wilkins. USA, 2005)
4. *Human physiology*, 32-45. (Wikibooks contributors, 2006) Available online: http://en.wikibooks.org/wiki/Human_Physiology
5. E. Bharucha, *Textbook for environmental studies for undergraduate courses*, 35-48 (University Grants commission, New Delhi, 2004)
6. T. Baidaulet, S.T. Tuleuhanov, N.T. Ablaikhanova, A. Ydyrys, A. Baishanbo, Int. J. of Biology and Chemistry, **10(1)**, 42-49 (2017)

Environmentally-induced alternative oncogenesis: EROS arrows

Sergey K. Pinaev[1*], *Olga G.* Pinaeva[1], and *Alexey Ya.* Chizhov[2]

[1] Far Eastern State Medical University, Khabarovsk, Russia
[2] Peoples' Friendship University of Russia (RUDN University), Faculty of Ecology, 6 Miklukho-Maklaya St, Moscow, 117198, Russian Federation

Abstract. The cause of the oncological diseases in children and the specific spectrum of this pathology, with hemangiomas prevailing among benign tumors (BT), and leukemia, lymphomas, and neural tissue tumors among malignant neoplasms (MN), remains unknown. The authors studied the similarities and differences of the connections between environmental factors, benign tumors and MN, as well as the relations between BT and MN. The causality of the so-called "sporadic" fluctuations in the incidence of neoplasms and their relations with environmental factors were revealed. A hypothesis of environmentally-induced alternative oncogenesis is suggested. According to the hypothesis, fetal hemoglobin enhances environmentally-related oxidative stress (EROS), leading to epigenomic regulation disorder of SEMA 7A and MICAL family proteins in the endothelium, axon growth cone, and hematopoietic stem cells.

1 Introduction

The role of environmental factors in oncogenesis does not cause any doubts [1]. Moreover, the mechanism of development of many forms of tumors is still unclear. Thus, until now, the cause of the oncological diseases in children and the special spectrum of this pathology with the predominance of hemangiomas among benign tumors (BT), and leukemia, lymphomas, and nervous tissue tumors among malignant neoplasms (MN), remains unknown. While in adults the majority of cancer forms are caused by a prolonged exposure to specific organs and systems with a fairly well-known etiopathogenesis, oncological pathology occurs in children, as a rule, in the absence of signs of obvious oncogenic causes. Therefore, detecting the causes and ways of developing neoplasms in children is a very urgent objective aimed at working out measures for their prevention.

Objective of the study was to suggest a hypothesis of environmentally-induced oncogenesis in children.

* Corresponding author: pinaev@mail.ru

2 Materials and methods:

Our approach consists in analyzing the similarities and differences in the connections of environmental factors with benign and malignant neoplasias, as well as in the study of the relationship between BT and MN. To study the ecosystem, a system-determined dynamic balance model of the block type was worked out (Fig. 1). Balance dynamic models make it possible to evaluate the phenomena in their development as a set of processes of transfer of matter and energy, and the functional nature of the model allow extrapolating the revealed patterns and predict the future states of system objects on their base.

The incidence of BT in the population of young children (YC) 0-4 years old in the population of Khabarovsk was studied; 11 cohorts of children born in 1976-1986 were formed. The authors also studied the incidence of MN among YC in the Khabarovsk Region; 17 cohorts of the children born in 1972-1988 were formed .Using the IBM SPSS Statistics 23 software package, a regression-correlation analysis was performed between the dynamic series of the BT and MN incidence in cohorts using the time lag in relation to the year of birth in children with BT in seven iterations (0, -1, -2, -3 and + 1, +2, +3 years). The results are interpreted based on an analytical review of publications in the PubMed, eLibrary, MEDLINE databases.

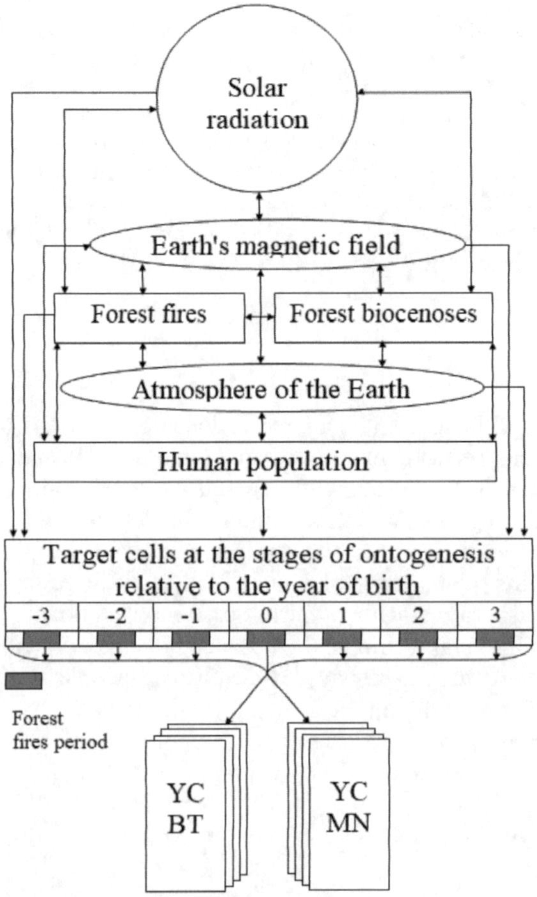

Fig. 1. The system-determined dynamic balance model of the ecosystem of the block type "Sun - smoke of forest fires - YC population"

3 Results and discussion

The relations between the incidence of hemangiomas and non-Hodgkin's lymphomas in cohorts 2 years older ($r = 0.769$, $p = 0.003$) confirmed the previously identified lag difference between the activity of the Sun, the occurrence of hemangiomas (-3 years) [2] and non-Hodgkin's lymphomas (-1 year) [3]. On the other hand, the correlation between the incidence of hemangiomas and leukemia ($r = 0.675$, $p = 0.023$) at a lag of +2 years reflects the relations of the former with postnatal exposure to smoke, while the incidence of leukemia confirms its preconceptive effect [2,4]. Generally, the negative correlation ($r = -0.705$; $p = 0.015$) between the incidence of BT and MN in cohorts a year older [2] reflects the fact that the former's respond to solar activity in earlier periods of ontogenesis than the latters (Fig. 2).

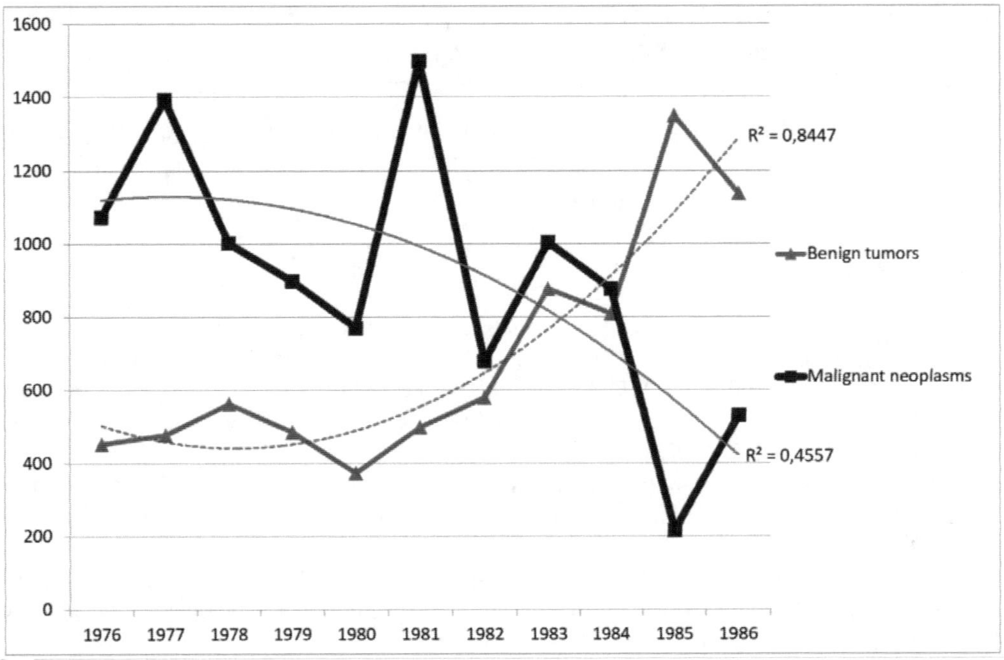

Fig. 2. BT incidence in the cohorts of YC (per 10^5) and MN (per 10^6), lag (-1) year.

4 Conclusions

Thus, the environmental factors connections with the incidence of benign tumors and MN have significant differences. For example, the preconceptive effect of solar activity is important for hemangiomas, and postnatal for leukemia. On the contrary, the smoke of forest fires affects the incidence of leukemia being exposed to ontogenesis during the preconceptive period, while for the incidence of hemangiomas, a connection with its postnatal exposure has been revealed [2,4]. On the whole, the data obtained by us allow stating the causality of the so-called "sporadic" fluctuations of neoplasms incidence and their connections with environmental factors.

We suggest a hypothesis of environmentally-induced alternative oncogenesis. According to it, the common vector of ecological factors is environmentally-related oxidative stress (EROS)[5].

https://doi.org/10.1051/e3sconf/202016904006

Its results at the preconceptive stage of ontogenesis are epigenomic regulation disorders of semaphorins, especially SEMA 7A, and the MICAL family proteins. At the stage of placental period, the sensitive targets for EROS in the embryo and fetus are the embryonic and fetal hemoglobins, which trigger, due to their instability, the amplification of it [6].

An additional increase of EROS occurs in the mitochondria of all actively proliferating cells, especially in the growth cone of axon, bone marrow, and endothelium, where semaphorins are presented. Normally, semaphorins provide axonal guidance, angiogenesis, and proliferation of immune cells. If violations occur, the transition of physiological proliferation into tumor is possible [7]. The proteins of the MICAL family represent a direct path to oncogenesis by EROS, growth factors, and oncogenes [8].

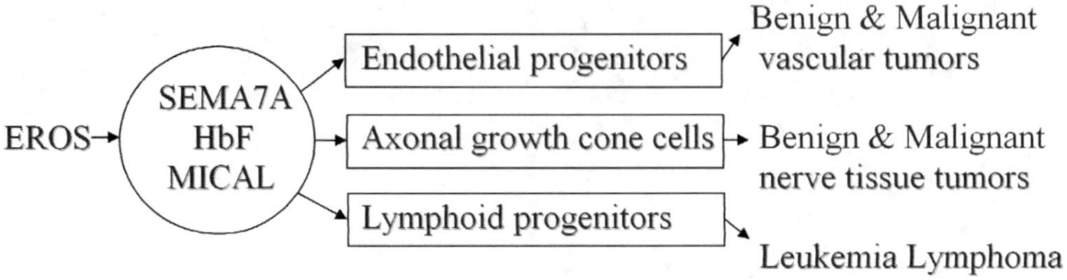

Fig. 3. "EROS arrows": Scheme of environmentally-induced alternative oncogenesis in YC.

Thus, environmentally-induced alternative oncogenesis is a consequence of amplification of EROS by fetal hemoglobin. This leads to epigenomic violations in the functions of semaphorins and MICAL in endothelial cells, bone marrow and lymphoid tissues directly contacting with red blood cells, and in the cells of the axon growth cone and its microenvironment [6].

The result of environmentally-induced alternative oncogenesis is a special spectrum of childhood neoplasms, with hemangiomas and lymphangiomas prevailing among BT, and hemoblastoses and neurogenic tumors among MN (Fig. 3). Moreover, depending on the changes in the parameters of the complex of environmental factors, certain forms of neoplasms arise predominantly during certain periods of time, forming alternative fluctuations of their incidence.

References

1. N.A. Agadzhanyan, A. Ya. Chischov, T.A. Kim, CIVILIZATION DISEASES. Human Ecology, **4**, 8-11 (2003)
2. S.K. Pinaev, A.Ya. Chizhov, Advances in molecular oncology, **5(4)**, Appendix, P.18-19 (2018)
3. S.K. Pinaev, A.Ya. Chizhov, O.G. Pinaeva, Fundamental science and technology – promising developments XVII, **2**, 14-17(Academic, North Charleston, 2018)
4. A.Ya. Chizhov, S.K. Pinaev, Radiation and risk, **27(4)**, 87-94. (2018)
5. A.Ya. Chizhov, S.K. Pinaev, S.Z. Savin, Technologies of Living Systems, **1(9)**, 47-53 (2012)
6. J. S. Fossen, Methemoglobinemia: Infants at risk. Curr. Probl. Pediatr. Adolesc. Health Care, **49(3)**, 57-67 (2019)
7. G. Neufeld, Y. Mumblat, T. Smolkin, S. Toledano, I. Nir-Zvi, K. Ziv, O. Kessler, Cell Adh Migr, **10(6)**, 652-674 (2016)
8. Yoon J., Terman J.R., Mol Cell Oncol, **5(1)**, e1384881

Impact of Urban Environment on Students' Daily Rhythms and Sleep Patterns

Elena A. Piven[1], *Nikolay P.* Piven[2], and *Nikolay N.* Bushuev[3]

[1]RUDN-University, Ecological Faculty, 117198 Miklukho-Maklaya st. 6, Moscow, Russia
[2]Pirogov Russian National Research Medical University, 117513 Moscow, Russia
[3]Bauman Moscow State Technical University, 105005 Moscow, Russia

Abstract. This article presents the results of a research of the sleep patterns of 200 students living in dormitories in Moscow. The average duration of night sleep in the surveyed group was 6h 37min. The proportion of students satisfied with the duration of night sleep was 36.5 %. Daily activity rhythms, in particular, the going to sleep and awakening time at the same time, are compiled by 47.5 % of the surveyed students. 86.5 % of students go to bed after midnight. The most late going to sleep time is observed among students of the 1st and 4th years. 37 % of students suffer from insomnia, which is most often caused by stress and work in extracurricular time. It was found that the best indicators of well-being and learning ability were observed at night sleep duration from 6 to 8 hours. When the duration of night sleep is less than 6 hours, there was a deterioration in the state of students' health, well-being and learning ability. It was revealed that 92.2 % of students who fall ill three or more times a year have a night sleep duration of less than 6 hours.

Introduction

In our century of rapid development of scientific and technological progress in an urban environment, information and mental overloads are increasing, while biological processes in the human body are changing very slowly. Urbanization with its increasing psycho-emotional loads is related to the restructuring of the organism of young students, a violation of circadian rhythm, based on genetically determined metabolism and biochemical cycles. In urban activities related to mental loads, longer working hours are required by reducing rest periods. At the same time, the shortage of time constantly continues to be felt, while the whole complex of human daily rhythms has a strict order, which ensures a stable internal synchronization of circadian rhythms [1-3].

University students are characterized by high mental loads, sharp changes of habitual way of life, formation of interpersonal relations outside the family, necessity of adaptation to new working, living and nutritive conditions, which allows classifying them to the group of significant risk of disease development [1-3]. They are developing a new, more intense daily rhythm, which includes preparation for classes, attendance of extracurricular student events, part-time work in the night shift. The formation of the daily rhythm is influenced by the experience of living in a dormitory, freedom from parental care. All of this can lead to serious sleep disturbances, and consequently to a decrease in student performance and a number of serious diseases [3-5]. Students need sufficient duration and quality of sleep due to high academic and psycho-emotional loads. It is known that without good sleep there can be no good waking [5-7].

There has been insufficient comprehensive research to assess the sleep patterns of students living in dormitories in large metropolitan cities; little research has been done on the factors that influence the duration of sleep and its impact on post-wake-up well-being. Therefore, the topic of the presented work is relevant and of great scientific and practical importance.

1 Objective

The objective of the work was to reveal the peculiarities of the sleep pattern of students of universities living in dormitories of a large metropolitan cities (the case of Moscow), associated with the impact of urban environment.

2 Research Method

A set of modern methods was used to achieve this goal: directional selection, sampling method, stratified sampling, survey (questionnaires), mathematically deductive and analytical methods. The target of the research is a random sampling of 100 female and 100 male students of universities who live in dormitories in Moscow.

3 Results and Discussion

In the sample we studied, the average age of students living in hostels is 21.3 years. The proportion of students in the 18-20 age range was 41.5 %, 31.5 % in the 21-23 years old range, 23.5 % in the 24-26 years old range, and 3.5 % in the 17 years old range.

It is established that 23.5 % of students were freshmen, 20 % were second year students, 15 % were third year students, 30 % were fourth year students, and 11.5 % were holders of Master's degree. The most part of the samples studied during the statistical analysis was for the 4th year students.

The recommended amount of night sleep for students is 7.5-8 hours per day, as the negative impact of inadequate sleep on their academic performance and health has been proven [3-5]. Among people in the survey, the average duration of night sleep was 6h 37min. At the same time, 81.5 % of students have night sleep duration ranging from 5 to 8 hours.

The duration of night sleep for students living in dormitories, which was less than 5 hours, was revealed in 8.5 % of respondents, in the range of 5-6 hours – 31.5 %, 6-7 hours – 25 %, 7-8 hours – 25 % and more than 8 hours – 10 %.

Depending on the length of the night sleep, the feeling of well-being of the young students also changed. With sleep duration less than 5 hours, 85 % of students were found to be less attentive, sleepy and irritable, and only 15 % of respondents did not notice any changes in their well-being and learning ability during the day. With sleep duration from 5 to 6 hours, the percentage of respondents who did not note any changes in their well-being and learning ability was 17.5 %. While the duration of night sleep increased to 6-7 hours, there was a sharp increase in the proportion of students with good health and learning ability which amounted to 58 %. At the same time, the best learning and well-being indicators were recorded for night sleep between seven and eight hours, and they were found in 68 % of the surveyed students. It has been established that if the duration of night sleep exceeds eight hours, the number of students with no learning and well-being problems is reduced to 40 %. Thus, taking into account learning and well-being during the day, the optimal duration of night sleep for students should be between six and eight hours.

The proportion of students satisfied with the duration of night sleep was 36.5 %, while 63.5 % of students said that sleep was not enough. The highest number of students with less than 6 hours of night sleep is observed among 1st and 4th year students.

Following the sleep pattern (falling asleep and waking up at the same time) is the most important condition for quality sleep. The optimal bed time is 10-11 p.m., and the best time to wake up in the morning is 6-7 a.m.

We studied the question of bed time. It is established that 10-11 p.m. 5 % of students go to bed, at 11 p.m. – 8.5 %, from 00 a.m. to 01 a.m. – 38.5 %, from 01 a.m. to 02 a.m. – 20 %, and after 02 a.m., 28 % of students. From this data it follows that students most often go to bed after midnight. The latest bed time is observed among 1st and 4th year students.

When studying student wake-up time, it was found that 5 % of students wake up before 6 a.m., 11.5 % wake up at 6-7 a.m., 36.5 % wake up at 7-8 a.m., 32 % wake up at 8-9 a.m., 6.5 % wake up at 9-10 a.m. and 8.5 % wake up after 10 a.m. 72 % of students felt languid and drowsy when waking up, and 28 % felt well and active.

The analysis of the morning wake-up time of the students depending on the time of leaving the dormitories to classes was carried out. The data showed that 31.5 % of students wake up more than 1 hour before leaving the dorms, 28.5 % wake up less than 30 minutes before leaving, and 40 % wake up 30-60 minutes before leaving. At the same time, 33.5 % of students have time to have breakfast, clean up and get ready, while 66.4 % do not have time.

It should be noted that 47.5 % of the surveyed students go to bed and wake up at the same time, while 52.5 % do not do that.

One of the most important sleep problems is how long does it take to fall asleep. The research found that 20 % of students fall asleep depending on their roommates, while 6.5 % of those surveyed are not able to fall asleep for an hour or more, 36.5 % fall asleep for no more than 30 minutes, and 37 % fall asleep immediately. Answering the questionnaire on whether neighbors hinder the quality of sleep, 32 % of students said yes.

Recently, a big problem has become the long stay of young people at the computer and TV before going to sleep. Thus, in our survey only 11.5 % of respondents said that they rarely watch TV and use a computer before going to bed. Less than one hour before going to bed 20 % of students spend in front of the computer and TV. More than 68.5 % of students spend more than one hour in front of the computer and watching TV before going to bed, 33.6 % of students go to bed between 00-01 a.m., and 66.4 % later than 01 a.m.

It should be noted that only a small proportion of students who rarely watch TV and do not work on the computer before going to bed or spend less than 1 hour in front of them, go to bed around 11:00 p.m.

Hard mental work should be stopped more than one hour before going to sleep. We analyzed the students' stay in front of the computer and the time when they finished their mental work before going to sleep. It was found that 8.5 % of students who rarely watch TV and do not use a computer before going to bed finished their mental work 30-60 minutes before going to sleep, and 3 % – 1 hour or more. Among students who spend less than one hour in front of a computer and watching TV, 10 % of those surveyed finish their mental work 30-60 minutes before going to sleep, and another 10 % more than 1 hour before going to sleep. Among students who spend more than one hour before going to sleep in front of a computer and TV, 39.5 % finish their mental work 30-60 minutes before going to sleep, and 29 % finish it 1 hour or more.

In general, among the group of students 58 % finish their mental work 30-60 minutes before going to sleep, and 42 % finish it 1 hour or more. At the same time, there is a clear correlation between students staying in front of the computer and TV and the end of their mental work. Thus, the more time students spend in front of the computer and TV, the less they rest from mental work before going to sleep. Mental work performed just before bed makes it difficult to fall asleep, leads to contextual dreams, languor, and feeling ill after awakening [4-6].

The study examined sleep disorder problems. It was found that only 63 % of students do not suffer from insomnia, 15 % often suffer from insomnia, and 22 % rarely suffer from it. Among students, who often suffer from insomnia, 66.7 % of those surveyed had stress as a cause, 20 % had work outside school hours, and the remaining 13.3 % had no reasons. Among students with a rare manifestation of insomnia, the most frequent causes are stress – 54.6 % of students, studies – 6.8 % of students, work outside school hours – 11.3 %, and 27.3 % – the cause is not identified.

Fresh air favours fast falling asleep, good sleep and a pleasant waking up. Only 40 % of students follow hygienic guidelines and air the room before going to bed.

In assessing food intake, 21.5 % of students had dinner more than 3 hours before bed, 16.5 % had dinner 2-3 hours, 36.5 % had dinner 1-2 hours, 17 % had dinner 30-60 minutes and 8.5 % had dinner less than 30 minutes before going to bed. Thus, students most often eat dinner 1-2 hours before going to bed at the recommended mealtime, i.e. no later than 2-3 hours before going to bed.

Morning exercises play an important role in a person's health and well-being. It is established that 3.5 % of students do morning exercises within 10 minutes, 20 % within 1-5 minutes, and 76.5 % do not do exercises. After a ten-minute morning exercise, all students felt active and good all day long. Among the students who did morning exercises for 1-5 minutes, 50 % were languid, reduced concentration and paid less attention, and 50 % were cheerful and active during the day. Among students who did not do exercises, 41 % were cheerful, while the remaining 59 % were languid, decreased concentration and paid less attention.

Violation of daily rhythms can lead to overfatigue and adversely affect health, weaken immunity and promote frequent colds, lead to somatic pathology [1-4, 6]. In a subjective assessment of their health, 85.5 % of respondents consider themselves to be healthy, while 14.5 % do not consider themselves to be, with 62 % of the last group of students surveyed having a night sleep of less than 6 hours.

Chronic diseases are noted by 8.5 % of students, 73.5 % of students do not suffer from them and 18 % do not know the objective information and it is difficult to answer the questionnaire due to lack of time for medical examination. Three or more times a year, 32 % of those surveyed had colds, 30.5 % had them twice a year, and 37.5 % of students had them once a year or less. At the same time, 92.2 % of students who fall ill three or more times a year have less than 6 hours of night sleep.

Most of the students surveyed (91.5 %) believe that good sleep supports health and is an important part of a healthy lifestyle.

4 Conclusion

A study of 200 students living in dormitories of a large city was conducted. The average sleep duration of the those surveyed was 6h 37min. 63.5 % of students consider their night sleep insufficient in duration. The best learning and well-being indicators were found when 68 % of the surveyed students had a night sleep between 7 and 8 hours. Most students (86.5 %) go to bed after midnight. The latest time to fall asleep is observed in 1st and 4th year students. At the same time, only 33.5 % of students have time to have breakfast, clean up and get ready, while 66.4 % do not have time to do that. Only 47.5 % of students go to bed and wake up at the same time, while 52.5 % do not do that. In this research, 20 % of students falling asleep time depends on their roommates, 36.5 % falling asleep time does not exceed 30 minutes, and 37 % fall asleep immediately. Roommates affect 32 % of students' sleep quality.

37 % of students suffer from insomnia. The most frequent causes of insomnia are stress, work outside school hours and study, while in some cases the cause of insomnia is not identified. Among the surveyed students, 58 % finish their mental work 30-60 minutes before going to bed, which may adversely affect the quality of sleep, and 42 % – 1 hour or more. A clear correlation between students' stay in front of the computer and the end of their mental work before going to bed has been revealed. So, the more students spend time in front of the computer, the less time they have to rest before going to bed.

Spending more than one hour before going to bed in front of the computer and watching TV is noted by 68.5 % of students, 33.6 % of whom go to bed between 00-01 a.m., and 66.4 % after 01 a.m. Only 40 % of students follow hygienic guidelines and ventilate the room before going to bed. Most often, students (36.5 %) have dinner 1-2 hours before they go to bed with an optimal last meal time not later than 2-3 hours before they go to bed.

It was found that 76.5 % of students did not do morning exercises, while students who did a ten-minute morning exercise felt active and good all day long. 91.5 % of students believe that a properly organized and adequate sleep is necessary to keep fit.

Sleep for students is a universal way to optimize human daily rhythms. It is necessary to keep fit and have good academic achievements.

References

1. A.I. Almojali, S.A. Almalki, A.S. Alothman, E.M. Masuadi, M.K. Alaqeel, J. of Epidemiology and Global Health. 2017 Sep;**7**(3):169-174.
2. G. Benham, Sleep Health. 2019 Dec; **5**(6): 587-591.
3. S.P. Becker, M.A. Jarrett, A.M. Luebbe, A.A. Garner, G.L. Burns, M.J. Kofler, Sleep Health. 2018 Apr; **4**(2):174-181.
4. L.K. Suen, K.L. Hon, W.W. Tam. Chronobiology international. 2008 Sep; **25**(5): 760-775.
5. M.E. Hartmann, J.R. Prichard. Sleep Health. 2018 Oct; **4**(5): 463-471.
6. M.C. Azad, K. Fraser, N. Rumana, A.F. Abdullah, N. Shahana, P.J. Hanly, T.C. Turin. J. of clinical sleep medicine. 2015 Jan; **11**(1): 69-74.
7. E.A. Piven, D.A. Breusov. RUDN J. of Medicine. 2017 Jan; **21**(1): 127-136.

E3S Web of Conferences **169**, 05001 (2020) https://doi.org/10.1051/e3sconf/202016905001
APEEM 2020

Comparative analysis of the application of educational technologies for sustainable development

Lyudmila I. Sokolova*, and *Dmitry S.* Ermakov

Peoples' Friendship University of Russia (RUDN University), Department of Psychology and Pedagogy, 117198, 6 Miklukho-Maklaya Street, Moscow, Russia

Abstract. It must be admitted that education is the necessary requirement, and educational technologies are the important tools of implementation for sustainable development. This study is based on a comparative and pedagogical approach. Having used the content analysis method we have considered the data of national reports (2018) from 31 countries on implementing the Strategy of education for sustainable development (ESD) of UNECE (sub-indicator 2.1.3) and defined pedagogical technologies applied at different levels of formal, non-formal and informal education. Besides, the rating of technologies according to the frequency of their usage for 0–8 levels has been made. The revealed differences we explain by age-related psychological features of students and pupils, as well as the national peculiarities of educational systems. Moreover, a brief overview about additional educational technologies and methods that are outside the framework of the current ones (non-formal and informal education in particular) has been prepared. The results of experts' survey in Russian Federation carried out in 2019 have been introduced. The general evaluation of progress in the field of ESD with the 3-point scale (0.7; in progress) and estimation of indicator 2.1 (0.9; in progress) has been defined. We also characterised the revealed educational-methodical achievements.

1 Introduction

It is a common understanding that education has been recognized as the indispensable condition and an important tool of implementation for sustainable development [1]. The quality and accessibility of education (Sustainable development goal № 4) are largely defined by the applied pedagogical technologies [2]. They include the system of training forms, methods and approaches that insure the most efficient achievement of setting goals.

To bring to the attention of the students the global values for sustainable world (diversity, respect for civic freedoms, gender equality, the urgency of combatting climate change, etc.) it is necessary for educators to master a wide range of appropriate actual teaching and learning methods that must be actively used in the educational process [3].

* Corresponding author: lsokolova777@mail.ru

2 Research Methods

We have previously analysed the national implementation reports of the United Nations Economic Commission for Europe (UNECE) Strategy for education for sustainable development (ESD) from 37 countries (phase III, 2011–2015). According to the analysis, the Pedagogical technologies used for an implementation of UNECE Strategy for ESD in the levels of formal, non-formal and informal education have been revealed and classified (traditional, project, interactive, information and communication technologies). The main pedagogical technologies in ESD have been defined and ranked according to the frequency of their application in different countries and at different levels of education [4].

The results of the similar study of the national implementation reports from 31 countries (2018) are introduced in this article [5]. Having analysed data of sub-indicator 2.1.3 "Are teaching / learning methods that support ESD addressed explicitly in the curriculum / programme of study at various levels of formal education?") we defined the appropriate teaching technologies/methods that support ESD (abbreviation is given further in the brackets): case studies (CS), conceptual and perceptual mapping (M), discussions (D), excursions and outdoor learning (OL), good practice analyses (GP), information and communicative technologies (I), learner-driven projects (P), philosophical inquiry (PI), problem-solving (PS); scenarios, modelling (SM); simulations, role playing, games (G); surveys (S), value clarification (VC), workplace experience (WE).

In order to get the complete assessment of the ESD advance, the survey of experts who are known as the specialists in the field of ESD was carried out in 2019. The same sub-indicator and 3-point scale were used in accordance with the national implementation reports.

3 Findings

The main teaching and learning methods that support ESD at various levels of formal education are presented in Table 1 (in accordance with International standard classification of education ISCED 2011) [6]: 0. Early childhood education; 1. Primary education; 2. Lower secondary education; 25. Lower secondary vocational education; 3. Upper secondary education; 35. Upper secondary vocational education; 4. Post-secondary non-tertiary education; 45. Post-secondary non-tertiary vocational education; 5. Short-cycle tertiary education; 55. Short-cycle tertiary vocational education; 6. Bachelor's or equivalent level; 7. Master's or equivalent level; 8. Doctoral or equivalent level.

Table 1. Pedagogical technologies/methods used in ESD (number of countries).

Technologies (symbol)	ISCED levels												
	0	1	2	25	3	35	4	45	5	55	6	7	8
CS	6	13	18	12	22	14	15	7	11	7	16	17	16
D	14	21	24	13	25	16	15	9	11	7	17	17	17
G	22	25	24	14	23	15	15	8	11	7	15	15	13
GP	5	9	14	10	15	11	13	6	10	6	17	17	16
I	16	22	23	14	24	17	16	9	11	7	16	16	16
M	8	16	18	10	18	12	11	7	8	7	14	15	15
OL	23	25	24	14	24	15	16	9	12	7	14	12	11
P	11	12	20	10	21	13	13	7	9	6	12	12	11
PI	6	10	11	7	19	10	14	7	7	5	16	17	17
PS	14	20	22	14	24	15	17	9	11	7	17	17	17
S	7	15	17	12	20	15	14	8	11	7	17	16	16
SM	13	16	22	10	22	13	12	6	9	5	16	16	15
VC	12	18	20	10	22	13	15	9	11	7	14	15	15
WE	5	8	10	8	17	14	16	10	11	8	15	14	13

The reporting templates have not been submitted by Canada, Cyprus, Denmark, Luxemburgh, Norway, Poland and the Ukraine yet. Italy presented the brief information from the Ministry of Environment, Land and Sea. In accordance to the analysis some countries (Austria, Finland and Switzerland) have not indicated teaching and learning methods used for ESD at the different ISCED levels that are of particular significance in their country. They have not indicated the results in the box of sub-indicator 2.1.3. The detailed information about teaching and learning methods supporting ESD at various levels of formal education is provided in reports of Belarus, Croatian, Estonia, Kyrgyz Republic, Latvia, Montenegro, Romania, Slovenia [5]. All countries except Romania have not inserted additional teaching or learning methods in the blank that are also considered to be key methods in their countries in teaching and learning for sustainable development. Romania added additional inquiry and debates [7].

Analysing the best practices of different countries, it should be noted that in Austria, for example, there are some special student programmes to enhance vision thinking and participation of students like "Future lectures" [8]. In Finland teaching and learning methods are included in the basic educational programmes of schools as well as ESD is based on doing a research [9]. In Switzerland teachers and trainers implement methods, hand-outs, and didactical recommendations on internet data-bases and platforms, or at different teacher education institutions or education publisher houses [10].

The results of ranking of relevant teaching and learning technologies for ESD from the Table 1 according to the frequency of their usage are presented in the Table 2. We composed this rank from 1 to 14 based on the mentioning of certain technologies in the national implementation reports of 2018 in which '1' corresponds to "the most frequent method" and '14' as "the least frequent method".

Table 2. Ranking of pedagogical technologies used in ESD.

Rank	ISCED levels												
	0	1	2	25	3	35	4	45	5	55	6	7	8
1	OL	PS	OL	PS	OL	VC	WE	GP	S	GP	CS	WE	I
2	G	S	PS	S	S	OL	GP	WE	GP	D	OL	OL	OL
3	I	VC	S	VC	VC	S	S	VC	WE	G	P	CS	WE
4	D	OL	VC	WE	WE	WE	VC	S	VC	M	WE	M	M
5	PS	WE	WE	OL	PS	PS	OL	OL	OL	OL	I	I	CS
6	SM	D	SM	P	M	P	PS	D	D	P	M	P	P
7	VC	G	D	M	D	M	M	PS	PS	PS	SM	VC	VC
8	P	SM	PI	G	SM	GP	D	P	P	S	VC	SM	SM
9	M	P	G	D	PI	D	P	G	M	VC	GP	PS	D
10	S	M	M	SM	P	SM	I	I	CS	WE	PS	D	G
11	PI	PI	P	PI	I	PI	PI	M	PI	CS	S	G	PS
12	CS	I	CS	CS	G	G	CS	PI	SM	PI	D	GP	GP
13	GP	CS	I	GP	GP	CS	SM	CS	G	I	G	S	S
14	WE	GP	GP	I	CS	I	G	SM	I	SM	PI	PI	PI

As we can see from the table 2, this applied technologies considerably change from one level of education to another. It must be explained by the age-related psychological features of students and pupils [11]. For example, outdoor learning is a popular technology in the kindergarten that introduce children to the nearest social natural environment as well as games. Although in higher education excursions (together with case studies) also take place, but apparently they are used for more detailed learning of environmental problems and development. However, workplace experience is the most common technology in the level of higher education. It should be noted that discussions, information and communicative technologies, learner-driven projects, problem-solving, surveys, value clarification preferably find use in all levels of education. In a small measure philosophical inquiry is in demand and in the high levels its importance is reduced.

Non-formal education related to the ESD is mentioned in the national implementation reports of Hungary, Slovenia, and Estonia 2018. Thus, the sustainability thematic weeks are realized in Hungary [12]. In Romania non-formal education includes study visits, excursions and outdoor learning, thematic projects / activities / workshops / camps, problem-solving, ICT, case studies, debates, volunteering, anti-alcohol /anti-tobacco /anti-pollution / preventing persons traffic campaigns, exchange of experience, activities dedicated to specific events like ("The European day for citizenship through education", "Earth day", "International day for environment", "The water day", "The tree day", "The world day for animals", "Let's do it!", "Meeting with the employers") [7]. Non-formal education in Slovenian is understood as a particularly strong and relevant frame for SD due to its complexity and local embeddedness. The methodological emphasis is done on the combination of experiential learning (field work and behavioural experiments) and constant reflection through the dialogue with participants based on three types of case: 1. field locations prepared according to several criteria, 2. cases selected and described by participants and 3. written learning materials [13]. In Estonia the programme "Preparation of in-service training curricula and training courses for teachers of formal education and universities and non-formal environmental education professionals" was launched in 2011. The objective of the program was to equip educators and teachers with the competences to implement active learning methods and integrate the cross-curriculum sustainable development topic in their everyday teaching. In-service programs and ESD courses providing initial teachers' education have been prepared and organised. A special in-service training curriculum for formal education teachers and non-formal education professionals was compiled [14].

Data analysis of additional educational technologies and methods presented in the national implementation reports 2018 are of great interest. They are beyond the framework of Table 1. So, problematic and research-oriented learning, creative tasks, networking are specified in the reports of Belarus [15]; in Bosnia and Herzegovina they use research and team teaching [16]; in Croatia – cooperative learning, interdisciplinary and multidisciplinary approach [17]; in Montenegro – activity, problem and process centred approaches, active teaching [18]; in Kyrgyz Republic – a result oriented teaching [19]; in Russian Federation – communicative approach (communication and interaction with adults and peers), cognitive-research approach, competence-based approach [20]. Also heuristic and research methods are used in Slovakia [21] and in Slovenia – meta-cognitive approach, cross-curricular links, learning strategies and approaches, activity days, authentic tasks, method of formative monitoring and evaluation of students' progress, holistic thinking [13].

In May 2018 after splitting the functions of state governing bodies in Russian Ministry of Education and Science between the Ministry of Education and the Ministry of Science and Higher Education the data were presented only for ISCED Levels 0–55. In this regard, in order to obtain a more complete assessment of progress in the field of ESD, we conducted a survey of experts-specialists in the field of ESD.

In the whole it must be admitted that there is a little progress in the field of ESD for Russian Federation by the state of 2019. The summary and self-assessment is estimated as 0.7 according to the scale (0 – not started; 1– in progress; 2 – developing; 3 – completed) that is higher than in 2011 (0.5), but it is lower than in 2014 [22] and on the indicator 2.1 "Sustainable development key themes are addressed in formal education" – 0.9 (0.8 и 1.1 in the years 2011 and 2014 accordingly). At the same time it is pointed out by the experts that a number of achievements related to ESD have been already developed such as the development of value and philosophical core of ESD content, involving "green" axioms and metaphors in the content, interaction of school subjects and their results in analogy with the functional system, designing the "local agenda 21" at a school level, creating internet consulting platforms and online resources for sustainable development and ESD.

4 Conclusion

The pursuit of sustainable development is difficult, and there are no unique solutions [23]. We reviewed various teaching and learning methods and technologies that demonstrate the positive result of their implementation in different levels of education and it is obvious that these forms of outcomes are desired to be applied. It is expected from the teachers to implement a wide range of methods and technologies in the educational process. In our opinion, the most promising educational technologies are the development and implementation of projects by students aimed at practical participation in solving environmental problems. Moreover, the discussions, conceptual and perceptual mapping, simulations, role playing, games, scenarios and modeling, surveys as well as ICT also have essential value. In striving for application of educational technologies for ESD we conclude that it is important to connect SD concept to all areas of human development with a focus on achieving the Sustainable Development Goals accepted in 2015 [24]. Meanwhile, ESD should be considered as a driver of innovative development of the educational system in the whole.

The reported study was funded by Russian Foundation for Basic Research according to the research project № 19-013-00722 "Education for sustainable development in action".

References

1. *Agenda 21,* Report of the United Nations Conference on environment and development, **I**, 9–479 (1993).
2. *Worldwide educating for the future index 2019: from policy to practice* (2019).
3. *UNECE Strategy for education for sustainable development.* URL: http://www.unece.org/fileadmin/DAM/env/documents/2005/cep/ac.13/cep.ac.13.2005.3.rev.1.e.pdf (2005).
4. A. Ermakov, S. Popov, N. Sabanina, A. Tokarev, D. Ermakov, Eur. Proc. Soc. & Behav. Sci. **LXXI**, 61–70 (2019).
5. *National implementation reports.* URL: http://www.unece.org/env/esd/implementation.html (2018).
6. *International standard classification of education ISCED* (2012).
7. *National implementation report. Romania.* URL: http://www.unece.org/env/esd/implementation.html (2018).
8. *National implementation report. Austria.* URL: http://www.unece.org/env/esd/implementation.html (2018).
9. *National implementation report. Finland.* URL: http://www.unece.org/env/esd/implementation.html (2018).
10. *National implementation report. Switzerland.* URL: http://www.unece.org/env/esd/implementation.html (2018).
11. R. Calfee, *Handbook of educational psychology*, 29–42 (Mahwah, Erlbaum, 2006),
12. *National implementation report. Hungary.* URL: http://www.unece.org/env/esd/implementation.html (2018).
13. *National implementation report. Slovenia.* URL: http://www.unece.org/env/esd/implementation.html (2018).
14. *National implementation report. Estonia.* URL: http://www.unece.org/env/esd/implementation.html (2018).

15. *National implementation report. Belarus.* URL: http://www.unece.org/env/esd/implementation.html (2018).
16. *National implementation report. Bosnia and Herzegovina.* URL: http://www.unece.org/env/esd/implementation.html (2018)
17. *National implementation report. Croatia.* URL: http://www.unece.org/env/esd/implementation.html (2018)
18. *National implementation report. Montenegro.* URL: http://www.unece.org/env/esd/implementation.html (2018)
19. *National implementation report. Kyrgyz Republic.* URL: http://www.unece.org/env/esd/implementation.html (2018)
20. *National implementation report. Russian Federation.* URL: http://www.unece.org/env/esd/implementation.html (2018)
21. *National implementation report. Slovakia.* URL: http://www.unece.org/env/esd/implementation.html (2018)
22. D. S. Ermakov, Pedagogy, 6, 44–48 (2013)
23. N. Giangrande, R. M. White, M. East, R. Jackson, T. Clarke, M. S. Coste, G. Penha-Lopes, Sust. Sci., **11**, 2832 (2019)
24. Transforming our world: the 2030 Agenda for sustainable development. URL: http://www.un.org/ga/search/view_doc.asp?symbol= A/RES/70/ 1&Lang=E (2015)

E3S Web of Conferences **169**, 05002 (2020) https://doi.org/10.1051/e3sconf/202016905002
APEEM 2020

Implementation trajectories of environmental education for sustainable development in formal, non-formal and informal education based on eduScrum project management methodology

Natalya Ryazanova [1], *Vladimir* Naumov [2], *Natalia* Kamennykh[2]

[1]Moscow State Institute of International Relations (MGIMO University), Department of Integrated International Ecological Problems and Wildlife Management, 119454, 76 Vernadskogo prospect, Moscow, Russia
[2]Moscow Timiryazev Agricultural Academy, Department of soil science, Geology and landscape studies, 127550, 49 Timiryazevskaya st., Moscow, Russia

Abstract. The aim of the study is to study approaches to localization of the topic of SDG in higher education in Russia and some tracks for implementation of environmental education for sustainable development (EESD). The objectives were: to formulate a goal setting for each of the approaches; to develop a methodology for the introduction of the SDG ideology into EESD; to implement the personal and collective role of research team members in working on the project; to create effective collaborations and introduce an interdisciplinary approach; to be able to present the results obtained in a visual and analytical form; to expand the professional outlook of environmentalists, to form a worldview based on the development of competencies for sustainable development. The research methodology in formal education was based on the hypothesis that modern environmental education should be based on the triad of SDG clusters, and all didactic units of university programs should be filled with this ideology. The information approach was implemented through the introduction of project laboratory technologies, case studies and youth models. Strategically, the higher education institutions participating in the experiment developed a methodology for tracking higher education for sustainable development and introduced methods for working with the subjects of SDG.

1 Introduction

The topicality of the research consists in adaptation of the subjects of Education for Sustainable Development (ESD) to different forms, methods of teaching and different didactic units in Russia [1]. The tradition of using an interdisciplinary approach in its modern understanding (integration of economic, social and environmental knowledge and information about the research subject) is only emerging. In the first years, a project approach to education is being introduced, which in higher education is introduced only in a few universities on a permanent basis and is going through the formative stage.

Most developed countries have entered the ESD process using the UNESCO recommendations to develop competencies for sustainable development. Russia supports the general trend in this direction, but there are no requirements for their introduction and quantitative indicators and a road map for their achievement. There is no localization of these requirements in state documents [2], no special national methods have been developed. The author of this work tends to see this not as an obstacle, but as a huge opportunity for creative independent work of research schools across the country, a deep exchange of experience in this direction and a positive assessment of the existing and future contribution of Russian pedagogical science to the global ESD piggy bank.

The research hypothesis. The modern volatile "VUCA" world does not allow teachers to build long chains of learning and to adapt them to the needs of the labor market for a long time. Employers have no time to wait, then adapt and give time to adapt a young employee to the rapidly changing demands of the state economy. The modern learning process in the university needs the introduction of Agile technologies and the project learning environment of eduScrum. For this purpose, it has been proposed to introduce the project approach in various types of training and extracurricular activities. Students are immersed in the themes of the SDG, conduct research on environmental issues, and determine, at the same time, what tasks the SDG correspond to, study the qualitative and quantitative parameters of the possibility of achieving the SDG by various methods. Teachers and employers should be able to assess the product (educational) result and develop mechanisms to work with it as flexibly as possible [8].

The novelty of the work is to develop a flexible functional block (consisting of blocks) model of the project approach in education for sustainable development. Students studying in the field of environmental and nature management training should get a relatively perfect and simple learning mechanism, which can be easily transformed into tasks, ways of their solution and the final expected product (educational result). Such an approach will allow teachers to input relevant baseline data, set a research objective quite simply, and only minimal participation (direction) in the learning process. The model should contribute to the maximum flexibility of the learning process, be understandable for teachers, students and potential employers, it is they who should be able to monitor the process of formation of future specialists and assess the final result in terms of the quality of skills.

The work is carried out according to Adgile methodology in the educational environment on the basis of the approach to the formation of project work skills using eduScrum project management methodology [3-5]. The scientific and pedagogical activity has been implemented by identifying the range of modern professional tasks of an ecologist and creating methodological recommendations for the implementation of the PP MFB within the framework of educational and extra-curricular activities. Practical pedagogical activity was supported by the creation of Road Maps and soft tools to support research groups to achieve them in time, and such competencies as communication, teamwork, critical thinking and creativity ("4K") were pumped up.

The paper presents various approaches to the formation of the MTF PP for the main educational process, for club and club work of students during after-hours, for major games. The main forms of presentation of results: design laboratories, youth models of international and Russian processes, business games, simulations, etc.).

The first results of the experience of implementation of the MTF PP in educational and enlightening processes are presented. Qualitative assessments of students' skills received by three categories of mentors are given: public administration sector, business sector, science sector. Representatives of these three sectors, after each event conducted using the MTF

PP, gave feedback and practical recommendations to students and their teachers to improve the Model of the project educational environment and its precise adaptation to the tasks.

2 Methods and materials

The eduScrum project management methodology was introduced in 2011 in the Netherlands and is now presented as an implemented technology in 36 countries worldwide. The eduScrum method has evolved from numerous developments in various fields of activity, which have adopted this approach from the IT sector and managed to adapt it to their own needs. [6,7]. "The agile teaching/learning methodology (ATLM) was developed back in 2004, when Agile was adapted to education methods [8,9].

In the formal approach the technology was implemented twice, within the framework of the profile discipline Basics of Nature Management in the form of setting two educational tasks. For an example of working with the method, one of the educational tasks was considered.

The educational objective: Comprehensive multi-criteria evaluation of the implementation of national projects to achieve sustainable development in the Russian Federation: analysis of the current state and potential of each federal district.

Objectives: To carry out a comprehensive analysis of sustainable development goals and national projects of the Russian Federation; to relate national projects to world development priorities (SDG); to develop a road map for achieving sustainable development goals and strategic objectives of national projects of the Russian Federation for each federal district.

Training and research tasks: Inventory, selection and justification of the selection of functional CSD clusters. Analysis and distribution by functional cluster of national projects of the Russian Federation. Analysis of the current state of the federal district and responsible government agencies, taking into account the natural resource potential. Analysis of the potential for achieving the targets of the National Projects in accordance with complementary SDGs. Creation of a functional model for achieving the SLM. Formation of criteria for assessing the achievement of national projects and SDGs within the federal district. Creation of a road map for achieving sustainable development in the federal district.

In the informal approach, the technology was implemented many times, for example, in the preparation of student and not only conferences, models, round tables, when the organizing committee of the event formulated an order to update the Agenda-2030 within the theme of the event. For example, we will demonstrate a research task, which does not have any obvious signs of training, but significantly expands the horizons of participants of project teams.

Research task: Creation of a World Water Forum Youth Model based on the objectives of the Sustainable Development Goals.

Purpose of work: 1. To create six thematic teams (as in the original event): Climate, People, Growth, Quality, Ecosystems, Guidance. Working out and filling the thematic content of each thematic section (in our case - the game team). Publication of Analytical materials on the topic of Model work.

Research tasks: 2. Distribution of authority for the content of work within your team.

3. Identification of a list of source documents and information sources for training on the subject. 4. Consultation and cooperation with the official bodies in charge of this area of work in the region. 5. Working meetings of each group (students + local government curators) to coordinate the material received and determine the quality and quantity of information that will remain in the final presentation at the event. 6. Updating of initial and

final information for the region. 7. Development of targeted recommendations for the "Customer".

In the informal approach the technology was also implemented more than 10 times. The role of the Customer was played by governors or regional governments, state executive bodies (Ministries of Education, Natural Resources and Environment, etc.), Forum Organizing Committees. The method was implemented in the form of project laboratories, youth models of international and Russian processes, business games, simulations, etc.). each time teams from different universities of the country were invited to participate in the games. Students of different faculties find application of their professional knowledge, which allows to implement the sustainable development approach when three clusters (economy, ecology, society) are actualized and the developed solutions help to achieve the SSD in an integrated manner.

Table 1. Technology of project activities implementation in all types of training

Type of training	Product Owner (Customer)	SCRUM Master	Delivery Team	Educational result
Formal education	The state, prospective employers	A teacher in cooperation with representatives of a conditional employer (from government agencies, business, science and so on)	Study groups divided into work teams (mini-groups)	Analytical materials (notes), coursework, research report, final qualification paper
Non-formal education	Student clubs, clubs, informal communities, media groups, etc.	Event Organizing Committee	Students from as many departments, faculties, institutes as possible; students of different courses, of different ages, with different scientific and life experience, with different ratings	Media analytics, publication of the event material, formation of the community interested in development and maintenance of the theme of students and teaching community
Information Education	State government bodies, executive authorities, business structures, scientific institutions, forum organizing committees and others	Customer representative - a person authorized to interact directly with the process participants, who is able to correctly orient and advise the players in the guiding documents, targets and the expected result of the simulation	Students of different higher education institutions of the country studying in different areas of training. This activity is not always taken into account in the current educational process, is poorly integrated into it and will not be appreciated by the university from which the team arrived.	Development of strategies, plans, road maps, strategic vision offers for the customer

For example, a project task is given, which is not an educational one, but an expanding horizon; it allows to look at the problem much wider than it is seen by the performers at the moment, to attract experience and knowledge of various professionals and helps to discover new sources of information and technologies of project work (Table 1).

Project task: Creation of the Youth Model of the Project Office "UN Sustainable Development Goals: the Baikal Dimension". Baikal-2030 Youth Strategy.

The objective of the IMPO (Youth Project Office Model) is to identify a focus on the complex problems related to environmental management in the regions and develop a road map to overcome them for each of the thematic working groups to achieve the UN sustainable development goals.

Project Objectives: To identify focus on complex environmental issues in the regions and develop a road map to address them for each of the thematic working groups to achieve the UN sustainable development goals:

1. Examine the peculiarities of the territorial situation and natural and geographic opportunities for economic and bio-resource development; 2. To propose additional measures for introduction of innovative approaches in the economic structure of the region to preserve and multiply the existing potential and opportunities to solve complex problems for sustainable development of the regions, while maintaining environmental quality indicators at a high level; 3. Analyze and highlight legal, economic, scientific, technical and technological measures already being implemented in the field of environmental security in the form of a decision matrix for each of the thematic working groups; 4. To propose additional measures to introduce sustainability principles into the economic structure of the region; 5. Weigh the existing short- and long-term natural and anthropogenic risks for the region; 6. Formulate criteria for public assessment and propose a basis for development of tools for long-term socio economic and ecological safety in the region.

3 Results

As a result, the Functional Block Model of the project approach was created. At realization of formal training technologies Agile and project educational environment eduScrum are introduced. Analytical notes, term papers and final qualifying papers have been received. In non-formal education, students were formed skills to create media analytics, published materials of events, formed communities interested in developing and maintaining the theme of students and teaching community. In the informal approach there were created strategies and plans of regional, federal and thematic refraction and introduction of CSD ideology, as well as road maps for their achievement; local communities and young people have formed a practical skill of integrated approach to regional problems based on the triad of clusters: economy, ecology, society.

Involvement of young people in the real social and political process in the region for active and inclusive participation of local communities in collective decision making to improve the basic indicators of people's lives.

4 Discussion

The difficulty of attracting specialists from the real sector of the economy for two reasons should be noted as a constraint in formal education: many of them do not possess methodological and teaching qualities at all and do not have free time convenient for the schedule of students in these disciplines. In informal and non-formal learning, there are other limitations: teams of players need to be so interested in the topic and the process of modeling that they will develop in a short time and will be able to work, so that none of the players will "pull the blanket over", so that players will not be considered with free time. As a rule, 1.5 months before the event, all players are deprived of the opportunity to systematically study and rest, i.e. they find themselves in the most stressful situation. But, "survived" in these project youth models claim to more serious work!

5 Conclusions

The role of expert jury in all kinds of design works is performed by representatives of potential employers, who give an independent assessment of the quality of work performed and help to constantly update the agenda, guiding documents and materials, the order for the final result. In their opinion, the Functional Block Model of the Project Approach (FBM PP) has already been created and is functioning quite successfully. 4K competences are also being successfully formed: communication, teamwork, critical thinking and creativity.

In general, it has been shown that Adgile methodologies in the educational environment on the basis of the project work skills development approach with the application of eduScrum project management methodology proved to be very effective and are very positively perceived by employers, as they obviously form practical skills of graduates, which allows to integrate their real current work process virtually seamlessly (seamless technology). For example, such technology implemented by the Laboratory of Geoecology and Sustainable Nature Management of the Moscow State Institute of International Relations of the Russian Ministry of Foreign Affairs has allowed graduates to be employed for more than 5 years in the Project Offices and consulting agencies of the Big Four (for environmental consulting projects) represented in Russia.

References

1. A.N. Zakhlebnyy, E.N. Dzyatkovskaya, *BSU*, **2**, 43-53 (2015).
2. T.A. Lanshina, V.A. Barinova, A.D. Loginova, E.P. Lavrovsky, I.V. Ponedelnik, BIO, **1**, 207-224 (2019)
3. A. Sureka, M. Gupta, D. Sarkar, V. Chaudhary, 1st IWCM for CE, pp.72-78. (2015).
4. Z. Masood, R. Hoda, K. Blincoe, J. of Syst.and Softw, **10** (**1**), 501-510 (2018).
5. I. Noguera, A-E. Guerrero-Roldan, R. Maso, Comp.and Ed., **1(1)**, 110-129 (2018).
6. M. Cubric, Int. J. of Management Education, **11(30)**, 119-131 (2013)
7. L. Tverdostupova, LP, **5**, 22-28 (2016)
8. M.A. Manokin, A.R. Ozhegova, E.A. Shenkman, University Management: Practice and Analysis, **4**, 83-96 (2018)
9. A. Chun, LNCS – Advances in Web-Based L, **8**, 11-18 (2004)

Toxic communication as a risk factor when promoting Health, Safety & Environmental culture in an organization

Elena Pavlova[1,*], *Elena* Terentyeva[2], and *Armine* Matevosyan[3]

[1]Peoples' Friendship University of Russia (RUDN University), Faculty of Ecology, 6 Miklukho-Maklaya Street, Moscow, 117198, Russian Federation
[2]Volgograd State University, 100 University Avenue, Volgograd, 400062, Russian Federation
[3]Yerevan State University, 1 Alek Manukyan Street, Yerevan, 0025, Armenia

Abstract. The study focuses on the problem of toxic communication as one of the risk factors in Health, Safety & Environmental culture development. The data obtained indicate that the effectiveness of Health, Safety & Environmental communication is reduced due to the expansion of toxic communication, which includes such phenomena as intentional confusion, topic overdiscussing, lack of dialogue on equal footing, lack of feedback, and weak corporate culture. It is also shown that in most cases, ineffective Health, Safety & Environmental leadership of managers and chiefs demonstrates the communicative behaviour of a toxic boss. The conclusion is drawn on the need for further studies of toxic communication in order to reduce its negative impact on Health, Safety & Environmental culture promotion.

1 Introduction

As a brief overview of the websites of leading Russian companies and their several subsidiaries in Armenia shows, all of them have set ambitious goals to become leaders not only in the field of production, but also in the field of industrial safety, proclaiming a firm commitment to the values of Health, Safety & Environmental (HSE) culture. As an illustration, here are some fragments of texts posted on the official websites of the companies:

LUKOIL: HSE Management System. The 2019 results of the latest in a series of recertification audits of LUKOIL and 43 LUKOIL Group organizations confirm compliance of the HSE Management System with ISO 14001 (Environmental management system) and OHSAS 18001 (Occupational health and safety management systems).

SIBUR: Holding views the health and well-being of its personnel and all stakeholders as its top priority, emphasizing it above operating achievements. Some of our production sites fall into the category of hazardous industrial facilities, therefore ensuring safety is a major focus. The Company considers the corporate health and safety management system (HSMS) an essential component of efficient operations control, and undertakes to

* Corresponding author: pavlova-eb@rudn.ru

manage production risks that may affect employees' health and safety or facilities' property and equipment.

Rosneft: HSE IMS has been successfully operating in the Company since 2006. The Company continuously confirms the system's compliance with the requirements of ISO 14001 and OHSAS 18001 standards through certification, oversight and re-certification audits conducted by the international certification bodies.

Nornickel Group: The Company's environmental priorities include: switching the environmental management system (EMS) to a risk-oriented approach to environmental management in compliance with ISO 14001:2015.

Gazprom: The Environmental Management System (EMS) is the mainstay of Gazprom's Environmental Policy (the fundamental document regulating the Company's environmental activities). Gazprom's EMS encompasses various management levels ranging from the Company's Board of Directors to subsidiary headquarters and facilities. The EMS is deployed at 34 of Gazprom's subsidiaries specialized in exploration, production, transmission, storage and processing of gas and gas condensate, as well as investment activities.

CJSC Gazprom Armenia is a wholly owned subsidiary of PJSC Gazprom. The main activity of the company is the supply and sale of natural gas in the domestic market of the Republic of Armenia

Of course, the complete minimization of any environmental impact and the reduction of injuries and industrial accidents to zero are still difficult to achieve, but nevertheless, the development of HSE culture in an organization is the main means to pursue these goals within the framework of promotion the philosophy of sustainable production [1] and sustainable organization [2]. Therefore, the concept of HSE culture continues to be relevant. It is known that there is "an innovative approach to accelerating HSE culture in an organization based on the concept that managers, particularly senior managers, through their influence on employees, can strongly affect the HSE performance and HSE culture of an organization. For this influence to be beneficial to the organization, managers need to consistently demonstrate behaviors and characteristics of strong HSE leadership" [3]. However, both in Russia and Armenia, the development of HSE culture is accompanied by a number of difficulties of a psychological [4] and organizational and communicative nature. On the one hand, employees doubt the sincerity of managers' desire of to take care of staff and the environment, they do not believe in the effectiveness of feedback from management, sometimes neglect the requirements of safety and labor protection, demonstrate resistance to these requirements and do not report the violations to the management and relevant services. On the other hand, employees face frequent manifestations of toxic communication on the part of management, which becomes an obstacle to the development of HSE culture and a breeding ground for maintaining the negative behaviors mentioned above.

Here, toxic production communication is referred to as destructive communication in workplaces which violate ethical standards, provoke a sense of dissatisfaction and anxiety in employees, and reduce motivation for work. Toxic production communication from bosses, managers, and colleagues may cause adverse health effects in staff due to high levels of stress. Research in the field of social neuroscience showed that participants who were exposed to the threat showed a consistent increase in cortisol levels, indicative of a stress response, compared to those who did not receive a threat. These findings suggest that group-based threats do indeed incur a stress-related physiological response [5].

The relevance of the work undertaken and the importance to study communication in an organization is explained by the fact that "communications are often a component of other HSE processes, but can exist as an HSE tool in its own right" [6].

The objectives of the study are to generalize the communicative aspects of the HSE culture development and promotion; to analyze the attributes of toxic communication and to identify trends which can be an obstacle to enhancing HSE culture in an organization.

2 Research materials and methods

The study was conducted in three stages. At the first stage, analytical work was carried out to identify the communicative aspects of HSE culture, and the attributes of toxic communication in Russian linguistic culture were specified. At the second stage, 50 employees of a large company's regional branch and 50 employees of a small enterprise were anonymously surveyed on awareness of HSE culture. Those employees who were aware of HSE culture were offered a questionnaire consisting of questions designed considering the principles of HSE culture and factors which impede its promotion. Several questions were aimed at establishing the quality of HSE communication and the presence of the toxic communication attributes. Those employees who were not aware of HSE culture answered the questionnaire about the type of communication in their organization and how it could be evaluated. At the third stage of the study, the questionnaires were analyzed, and conclusions were drawn.

Based on the analysis of the documents developed by the Institute for an Industrial Safety Culture (ICSI), HSE communication tools (Tool-box talks; HSE meetings; HSE alerts; HSE newsletters; Handover information) were established and the principles of Leadership in Safety (A guide to selecting appropriate tools to improve HSE culture; Safety leadership principles; Safety Culture), first of all, Principle 3 Share the Safety Vision: influence, persuade and promote the flow of information through the hierarchy and Principle 6 Be available on-site to observe, listen and communicate effectively were analysed [7]. The following attributes of toxic communication at work were identified: intentional confusion, topic overdiscussing, lack of dialogue on equal footing, lack of feedback, and weak corporate culture [8].

At the second stage of the study, all respondents were asked two main questions:

• Are you familiar with HSE culture?

• Do you know what HSE communication is?

If respondents showed awareness of these questions, they further answered the following questions:

• Do you understand the gist of HSE culture?

• Do you share the values of HSE culture?

• Does your organization give HSE the same importance as issues of production and quality?

• Has HSE culture been established in your organization?

• What prevents, in your opinion, the promotion of HSE culture in your organization?

• Do you have an opportunity to make suggestions for improving HSE culture?

• Do you report safety requirements breaches?

• Does the management respond to suggestions and complaints regarding safety in your organization?

• Does your organization analyse industrial accidents?

• Are HSE issues regularly discussed with employees?

• Do all meetings begin with HSE information (Safety Moment)?

• Do you feel annoyed when there is no clear HSE communication?

• Are HSE issues discussed with employees in an informal setting?

• Can HSE communication be considered effective in your organization?

• What prevents, in your opinion, the effectiveness of HSE communication in your organization?

If the respondents were not aware of the HSE culture and HSE communication, they were asked questions regarding the specifics of production communication and its organization.

The survey showed that the staff of a large company (78% of respondents) was more informed about HSE culture. The employees of a small organization were almost not familiar with HSE culture. The questionnaire results were processed using the Analysis ToolPak in Excel.

3 Results

Those questionnaires were especially thoroughly analysed in which respondents noted the presence of ineffective HSE communication or noted such communication features that we qualified it as toxic. The respondents' answers were summarized, and the table below shows the results:

Table 3. Types of communication based on the survey results.

Type of communication	Examples
Ineffective HSE communication	The requirements are not clear. General words are told about honesty, trust, commitment and care. HSE communication exists only on paper. The management keeps giving the same thing over and over again at meetings. No feedback. They often use words and abbreviations I don't understand. My boss only demands.
Toxic communication	Opposite and confusing orders are usually made. Long and pointless meetings. Abusive relationship with subordinates. Lack of feedback. Lack of corporate culture. Fear of the boss and fear of expressing my point of view. Surveillance and denunciations by managers and heads of services.

As can be seen from the table, ineffective HSE communication is very close to toxic communication. Where there is a formal approach and incomprehensible terminology is used, no personal interest of the leadership, no positive emotional impulse from a safety leader to promote HSE culture values and no constant open dialogue, the attributes of toxic communication are clearly manifested, which broadens the scope of its functioning. In other words, HSE communication often exists only on paper, in plans and reports, but in fact, toxic communication dominates, which negates all efforts of HSE management.

All respondents noted that ineffective communication occurred when there were "shouts and threats from the boss, conflicting orders", "no feedback", "only formal communication and no close contact and dialogue".

4 Conclusions

Thus, we have noticed a tendency of HSE communication to replace toxic communication. We have also noticed one more tendency, according to which the expansion of the toxic communication scope contributes to the transformation of a safety leader into a toxic boss. Of course, we understand that the study undertaken cannot be exhaustive and needs to be continued in order to reduce the negative impact of toxic communication on the HSE culture promotion since the results show that toxic communication remains a significant barrier for targeted changes to people's attitudes towards safety and for establishing constant contacts with them in order to ensure accurate feedback.

References

1. J. Pfeffer, *Academy of Management Perspectives*, 1–27 (2010)
2. E.B. Pavlova, E.V. Terentyeva, *IOP Conference series: MSE* (2019)
3. K. Breitsprecher, J. Hinton, W. Harris, S. Crabb, P. Jacques and B. Hoedt, SPE **2** (2012)
4. A. Guslyakova, N. Guslyakova, V. Kirsanov, M. Vetkhova, L. Poluyanova, E. Sharipova and A. Bengardt, IJCIET **10**, 2 (2019)
5. S. Sampasivam, K.A.Collins, C. Bielajew and R. Clément, *J. of Language and Social Psychology* **37**, 6 (2018)
6. A guide to selecting appropriate tools to improve HSE culture. Report No. 435 (2010)
7. ICSI https://www.icsi-eu.org
8. L. Too, M. Harvey, Journal of Corporate Real Estate **14**, 3 (2012)

"Green neology" and its role in environmental education (a media discourse approach)

Alla Guslyakova[1]*, *Nailya Valeeva*[2], *Olga Vatkova*[3]

[1] Peoples' Friendship University of Russia (RUDN University), Faculty of Ecology, 6 Miklukho-Maklaya St, Moscow, 117198, Russian Federation; Moscow Pedagogical State University, Institute of Foreign Languages, Moscow, Russia
[2] Peoples' Friendship University of Russia (RUDN University), Faculty of Ecology, 6 Miklukho-Maklaya St, Moscow, 117198, Russian Federation
[3] Secondary school "Cvetan Spasov", Pleven, Bulgaria

Abstract. The article covers the problem of ecological neologisms which mostly appear in the present-day media discourse space and actively function there by making a significant impact on people's environmental worldview. Moreover, the research also raises the problem of neologisms as an effective educational instrument for raising human ecological awareness. A comparative analysis of the most popular neologisms circulated in the Russian, British and American media discourse space is introduced in the study. Finally, it is suggested that green neology in its close correlation with media ecology provide a foundation for building and developing a new ecosystemic society with more sustainable ideology towards the future of the planet.

1 Introduction

Language of human beings has the design feature of creativity due to its duality, and the feature of duality of structure of the language system is a crucial feature of human language [1]. Modern linguistics considers the language to be a complex dynamic system, the direction and nature of the evolution of which are subject to the general laws of dialectics.

The most active development of the language is regarded in the field of vocabulary. This can be explained due to a number of its features in comparison with other language levels and, in particular, a higher degree of extra-linguistic determination [2]. The concept of extra-linguistic determination is closely related to another important notion introduced by Edward Sapir – *"the drift of a language"*, which can be characterized as "constituted by the unconscious selection on the part of its speakers of those individual variations that are cumulative in some special direction" [3]. Since the phenomenon of the present-day world are divided into products of nature, products of culture as well as the products of massive digital technologies, there is certainly a need for linguistic and learning (educational) action when newly developed objects or new concepts have to be named.

* Corresponding author: guslyakova_av@pfur.ru

At the turn of the XXth and the XXIst centuries humankind faced severely the problem of environmental pollution and a negative human impact on the natural resources of the planet because of the increasing role of scientific and technological revolution. To a varying degree, all natural components of the environment have undergone a change. Therefore, people started actively protesting against the destruction of Nature trying to stop its barbarian exploitation. As a result, a lot of new ecological words and concepts (neologisms) appeared and continue appearing in the languages, for example: eco-conscious, eco-speak, eco-tourism, eco-refugee, etc. Moreover, today, the words "ecology", "environment", "sustainability" have become one of the most widely circulated terms which are used not only by scientists, but also by economists, politicians, journalists and specialists from many branches of industrial and social spheres of human activity. All these important ecological neologisms are actively introduced in the media discourse environment and used in the written and oral speech by people in different languages and countries. Thus, the purpose of this research is to show how ecological neologisms created and popularized through the media discourse space may change people's environmental behavior and make them become more eco-friendly while keeping a more sustainable lifestyle.

2 Objects and Methods.

2.1 Neology as an important branch of modern linguistics and social sciences

In connection with the orientation of modern linguistics to the study of the processes of language development, a deep and comprehensive study of neologisms and neology are of undoubted interest. The term "neology" (from French "néologie") first appeared in France in 1758 and obtained its linguistic meaning by the end of the XIXth century. Today neology has become a special branch of linguistics that studies the most dynamic language layer – the vocabulary – and focuses on neologism which is described as "a new word or a new meaning for an established word; the use of, or the practice of creating, new words or new meanings for established words" [4]. Neologisms must be understood as words that appeared at a certain period in the language, "arising from the memory of the generation that uses them" [5]. That is, neologisms possess relative and historical properties [6].

Neologisms, arising in discourse, can turn into linguistic units or remain individual entities. They are quite often spread and popularized in the media discourse space. Newspapers, magazines, video channels, social networks have become centres of attraction for neologisms which are popular vocabulary units among different categories of media users.

Thus, neologisms are created to be used by people, which is the pragmatic aspect of neology, but the leading role in the birth of neologism remains due to the cognitive aspect and it is necessary to determine the place of a new word in the speaker's mind as well as relate it to a certain category [7]. It is essential to mention that in the recent decade a significant part of neologisms are lexical units related to the field of ecology. Some of them have become ecological terms, which get fixed in the language because of the viability of the objects of reality they denote.

2.2 Environmental neologisms and sustainable development through a media discourse perspective

At present, the media discourse space occupies a significant part of human life. Media discourse reflects the recent state of society, the latest trends and important actual problems in the world, including environmental issues. New values and social functions of media may turn out as the most fundamental question concerning the new media environment [8]. Accordingly, one of the most important questions of the media discourse speculation turns out to be about the role of digitalization of the media means for society as a whole. How is the change in the media related to the general change in society, including its sustainable development?

In the present-day media research, such comprehensive social questions are studied especially within the media ecology research tradition. Media ecology sees the media discourse not only as means of communication, but more as social environments which are undergoing significant social changes. People are constructing a new social and cultural world which should become better, greener and more sustainable. Thus, a new "ecosystemic society" model is being built [9, 10] which is actively covered in the mass media environmental discourse.

This type of discourse includes speech genres such as brief informational notes, thematic articles, comments, reporting from the scene, interviews, essays, feuilletons, appeals, speeches, posters and slogans. The expressiveness of the language of the media environmental discourse allows its creators to influence people that belong to different social groups. This media influence is represented in instilling the audience certain ideas (for example, clothes production from recycled plastic), contributing to the emergence of not only an emotional reaction in the form of an experienced attitude, but also reactions in the form of physical actions (school climate strikes across the world inspired by an Swedish activist Greta Thunberg).

Therefore, the informative content of the media discourse is presented in oral and written texts which include key environmental concepts embodied in words that quite often become "green" neologisms. The use of neologisms fill the media discourse with bright speech shades and help its creators portray events more vividly and emotionally, express their thoughts and feelings clearly as well as play an important role in people's ecological literacy and education. Everyone knows that words are singularly the most powerful force available to humanity. Thus, ecological neologisms definitely have the power and ability to help, to heal, to inspire and to teach.

2.3 "In words we trust" – the effect of green neologisms on environmental education

Since people took an active interest in the environment and sustainable development in the 20th century, environmental terminology and ecological neologisms mainly consists of concepts that have appeared in the language relatively recently. At the turn of the XXth – XXIst centuries, social and political movements that advocate for the protection of the environment become extremely popular. As a result, there appeared terms denoting these movements.

For example, in the semantic structure of the word "*green*" another lexical and semantic variant appeared: 1) related to the protection of the environment; 2) a representative of the party of environmentalists. A similar process affected the semantics of the word "*garbage*", which also means "something perceived as worthless." Such semantic changes caused the emergence of a whole class of two-component neologisms, which are based on the transfer of value from a property to a subject. An example is the metonymic transfer of a characteristic to an object in this case: a green - a person associated with the protection of the environment. There have also been the formation cases of new stable combinations formed in this way: *green consumers* - people who buy only those products that are environmentally friendly; *the green revolution* - increased crop yields worldwide; *green party* - a formally organized political party based on the principles of green politics, such as social justice, environmentalism and nonviolence; *green politics* - environmental policy. In the above phrases, the meaning of the first element clarifies the meaning of the whole phrase.

Thus, the *green* component prescribes the phrase "related to environmental protection". A group of morphological neologisms with the prefix *eco–*, which is the meaning-forming centre, is formed in a similar way: *eco-friendly* - environmentally friendly, *eco-freak* - a supporter of environmental conservation, *ecoterrorism* - environmental terrorism, *ecotourism* - ecological tourism, *ecowarriors* - activists that disrupt commercial activities that may threaten the environment.

In the process of communication, a person needs a new word that expresses his communicative intention at the time of speech. As a result, such "green" neologisms appear as *ecotage* (ecology + sabotage) - sabotage organized for environmental reasons; *environmentalism* - a direction in science that promotes environmental protection; *monkeywrencher* - a fanatical environmentalist.

To sum up, we assume that while countless ecological neologisms can spark the interests of language learners (and teachers), one of the most critical issues affecting people globally remains the environmental degradation of the planet. Themes such as global warming, deforestation, and water pollution are regularly featured in the media discourse, thereby raising a person's awareness of the troubled state of the Earth. When educational institutions - from elementary schools to universities - recognize the value of environmental education, language teachers will be in a unique position to promote environmental awareness while teaching a mother language as well as a foreign one.

Many of today's educators understand the importance of protecting our planet and raising students' awareness and perception of the sustainable future. And thus in our research we are stating an example of how ecological neologisms basically circulated in the media discourse space may promote eco-friendly philosophy in students' consciousness and teach them to "become informed about as many issues as possible that intrinsically affect their lives" [11].

2.4 Methods

Synthesis of interpretive and comparative approaches applied to the studied material allows making use of a complex of methods, with the hypothetic and deductive method, analysis, statistical data processing, analogy and generalization, being the leading ones.

3 Results

As we have already mentioned above, neologisms are those words that have appeared in the language relatively recently. At present, there are a huge number of them, but for our research we selected 50 of the most interesting English-language and 45 Russian environmental neologisms. A statistical analysis of the selected neologisms was conducted in the English-language and Russian media discourse. For that purpose, we made a request for each neologism in the Google search engine. We will consider the process of collecting statistics on the example of the English ecological neologism *garbology*. As we can see in Fig. 1 the result was 97400 documents containing the neologism under discussion.

Garbology is the study of a person or group of people by examining what they throw away. "In Garbology, everyday pieces of trash suddenly become valuable and interesting artefacts from which many inferences about their source can be drawn" [12]. Garbology involves the careful observation and study of the waste products produced by a population in order to learn about that population's activities, mainly in areas such as waste disposal and food consumption. Garbage anthropologists, known as *garbologists*, can use rubbish to draw comparisons between what is known as real behaviour (what people actually do) and ideal behaviour (what people say they do or what they'd like to think they do).

 https://doi.org/10.1051/e3sconf/202016905004

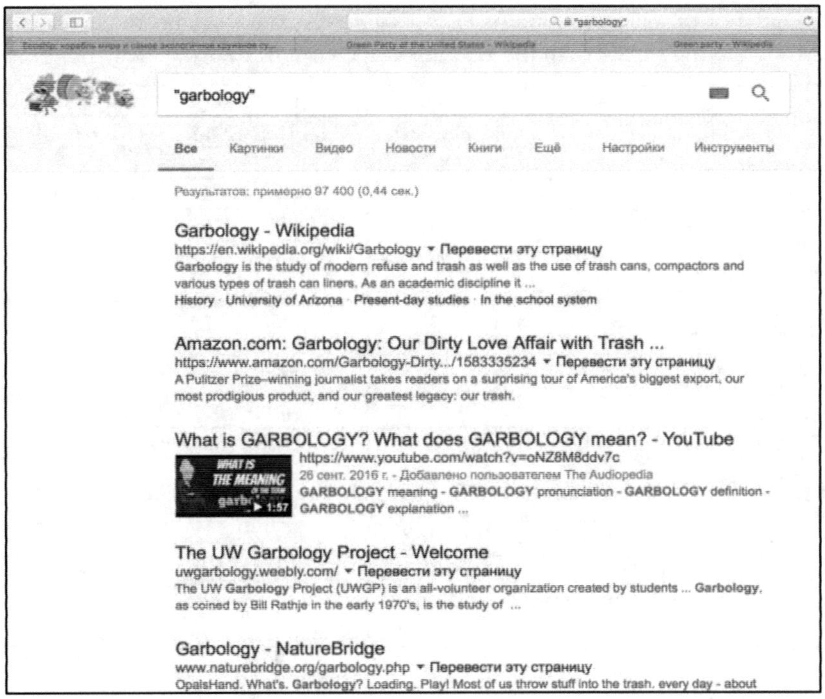

Fig. 1. The example of the request for the neologism "garbology" in Google search engine

This neologism has become very popular in the English and Russian media discourse space. Hence, we may observe how both Russian and English-speaking media are promoting garbology as a valuable method of young generation's upbringing as well as an effective way of training people to sort out their household waste.

For example,

"Where there's muck there's lessons to be learned. So says Suffolk county council, which has advertised for a garbology officer to teach pupils the value of old rubbish" [13].

Garbologists initiate the most advanced projects in the field of waste management - from Garbage 101 children's kits that teach children to be responsible for their waste, to the large-scale My Garbogy program, which polarizes the same qualities among the adult population of the planet. Garbologists daily save the Earth from the inglorious end under the mountains of human waste, and we owe it to them with such technological breakthroughs as "Energy from waste" [14].

It is important to mention that after being neologisms for a certain period of time the words "garbology" and "garbologist" have turned into ecological terms which are frequently used not only in the media discourse space but also in other types of discourse, including academic discourse at schools and higher education institutions.

In our research we also managed to analyze other popular "green" neologisms which have become the key concepts of the present-day digital epoch and are bringing an important environmental mission in the world (see tables1, 2 and 3).

Tables 1, 2 and 3 demonstrate the most popular neologisms which are circulated in the British, American and Russian media discourse space. These environmental concepts can be characterized as vital life issues and challenges that people have to tackle at the present moment.

Table 1. Statistics of search results for English-language neologisms in the most popular newspapers in the UK **(number of articles found)**

Nr.	Neologism	Title of the mass media edition			Total
		The Sun	The Daily Mail	Metro	
1	Veganism	82	4937	718	5737
2	Eco-friendly	521	4606	596	5723
3	Environmentally friendly	283	2920	639	3842
4	Pescatarian (also Pescetarian)	45	289	87	421
5	Ecotourism	45	337	30	412
6	Garbage man	5	352	16	373
7	Green revolution	7	148	48	203
8	Freecycle	44	83	36	163
9	Locavore	0	11	3	14
10	Ecofeminism	0	1	0	1

Table 2. Statistics of search results for English-language neologisms in the most newspapers in the USA (number of articles found)

Nr.	Neologism	Title of the mass media edition			Total
		The New York Times	USA Today	The Los Angeles Times	
1	Environmentally friendly	2 250	669	565	3484
2	Eco-friendly	1344	766	1044	3154
3	Veganism	290	242	2060	2592
4	Green revolution	1280	28	82	1390
5	Ecotourism	807	109	134	1050
6	Locavore	581	69	54	704
7	Garbage man	421	70	54	545
8	Pescatarian (also Pescetarian)	90	23	24	137
9	Freecycle	81	7	5	93
10	Ecofeminism	8	0	0	8

Table 3. Statistics of search results for English-language neologisms in the most popular newspapers in Russia (number of articles found)

Nr.	Neologism	Title of the mass media edition				Total
		Izvestiya	Kommersant	Vedomosti	Komsomol'skaya pravda	
1	Elektromobil'	216	33	3490	3680	7419
2	Elektrobus	37	99	181	1170	1487
3	Biotoplivo	93	7	262	138	500
4	Ekoturizm	62	0	21	192	275
5	Zelenyy ofis	10	0	11	147	168
6	Zelenaya energiya	5	1	3	23	31
7	Ekootel'	0	0	4	10	14
8	Zelenaya kompaniya	0	0	1	1	1
9	Ekostil'	0	0	0	12	12
10	Ekotovar	0	0	0	7	0

Metaphorically speaking neologisms are often referred to as "barometers" of the changes that happen in the society and the shifts in popular social tendencies [15]. On the one hand, new words reflect public consciousness but on the other hand, they influence its development [16]. The words and word combinations introduced in the tables above demonstrate people's intentions to build a more sustainable world on the planet and teach future generations to save the life on the Earth by maintaining an eco-friendly lifestyle.

4 Conclusion

Neology is a science that is directly related to almost all areas of human activity. "Green" neology is an essential language sphere which introduces a lot of important new words and collocations that describe the present-day environmental concept of human life. Since we are living in the epoch of digitalization and cutting-edge media technologies, environmental studies are closely interrelated with the media discourse resources. This symbiosis even led to the foundation of a new scientific approach that is called media ecology and it discusses a possible future, in "which the network society is deepening to become an internet-based ecosystemic society" [17]. Probably, ecological neologisms that are so frequently used in the media discourse space play important functions in the development of modern ecosystemic society. Firstly, they introduce ecological problems and estimate a current environmental situation. Secondly, they may reflect the reasons that stand behind the present-day ecological problems. Thirdly, green neologisms can give a hint on environmental problem-solving issues. Finally, they can inspire people to behave in a more eco-friendly way. Multiple media resources that contain a lot of useful information with the above mentioned neologisms have become effective mediators in people's ecological education process.

In summary, we would like to remark that the issue of green neology and ecological neologisms is highly topical and important. It is necessary to keep studying neologisms that cover the problem of sustainable development ideology through the media environmental discourse perspective so as to better understand how language, media and environmental education may change the human consciousness and lay the foundations of a more sustainable future on the planet.

References

1. C. Linhua, *An Introduction to Linguistics* (Jilin: Jilin University Press, 2008)
2. L.B. Gaztalova, *Neology as a science in a general paradigm of a modern language (based on the Russian and Ossetian languages)* (Nalchik, 2005)
3. R. Fischer, *Lexical Change in Present-day English: A Corpus-based Study of the Motivation, Institutionalization, and Productivity of Creative Neologisms* (Tübingen, Narr, 1998).
4. L. Zhou, Sino-US Engl. Teach., **13**, 4, 292-295 (2016)
5. M.A. Ryashchenko, Vest. Chelyab. gos. univer., **16 (345)**, 91, 121-123 (2014)
6. F. Saussure, *Cours de linguistique générale* (Payot, 1995)
7. V.I. Zabotkina, *Cognitive and pragmatic approach to the study of English neology* (Proceed. of a scient. confer., Moscow, MGLU, 2002)
8. I. Siles, P.J. Boczkowski, New Media Soc., **14(8)**, 1375–1394 (2012)
9. J. Ruotsalainen, S. Heinonen, Europ. Jour. of Fut. R., **3**, 9 (2015)
10. M. Lister, J. Dovey, S. Giddings, I. Grant, K .Kelly, *New media: a critical introduction* (Routledge, London, 2003).
11. H.D. Brown, The Lang. Teach., **15 (8)**, 4–5 (1991).

12. Buzzword "Garbology". URL: https://www.macmillandictionary.com/buzzword/entries/garbology.html
13. K. Scott, The Guard. (2004).URL: https://www.theguardian.com/society/localgovt/news/0,8368,1349317,00.html
14. What is Garbology and why is it so important? URL: https://w2e.ru/blog/chto-takoe-garbologiya-i-pochemu-ona-vazhna/
15. W. Croft, *Social Evolution and Language Change* (Manchester, University of Manchester, 2007)
16. N.L. Shamne, I.V. Rets, Vestn. Volgogr. gos. un-ta. Ser. 2, Yazykozn, **1 (25)**, 72-77 (2015)
17. J. Ruotsalainen, S. Heinonen, Europ. Jour. of Fut. R., **3**, 9 (2015)

E3S Web of Conferences **169**, 05005 (2020) https://doi.org/10.1051/e3sconf/202016905005
APEEM 2020

Regional case study of Sustainable Development Goals implementation: Informal education and third mission of universities

Sayana Vologzhina[1], *Natalya* Ryazanova[2], *Vasily* Eroshenko[3]

[1]Irkutsk State University, 664003, 1, Karl Marx st., Irkutsk, Russia,
[2]Moscow State Institute of International Relations (MGIMO University), Department of Integrated International Ecological Problems and Wildlife Management, 119454, 76 Vernadskogo prospect, Moscow, Russia
[3]Moscow Pedagogical State University, 119991, 1, building 1, Malaya Pirogovskaya st., Moscow, Russia

Abstract. The aim of the study is to introduce practical experience and explore the possibilities of sustainable development ideology in the region and to involve youth activists in international processes of regional reflection. Objectives: research of possibilities of youth organizations and universities in realization of non-formal education for overcoming environmental problems in the region; development of practical experience in realization of environmental education for sustainable development (ESD); development of pedagogical technology for non-formal work; creation of pedagogical model of interaction of different segments of youth organizations. Methodology: creation of the World Water Forum youth model. Work directions: adaptation of the international agenda to the regional one (Baikal region) and creation of the pedagogical model of practice-oriented work of youth communities for implementation of the third mission of HEIs. Regional ministries, youth government of Irkutsk region, colleges and universities of the region were involved in the model development. As a result of the thematic groups' work, an inventory of the region's problems was made by the activities of each working group; lists of the most promising areas of work to improve the environmental and social situation in the region were compiled; and road maps to achieve the set goals for up to 5 years were developed.

1 Introduction

The relevance of the study was to order the region to attract research work of colleges and universities, aimed at taking into account the importance of youth practices and grant work for real work to improve the environmental situation in the region. The Baikal region is a part of the UNESCO program of the World Natural and Cultural Heritage and in 2017 on the international level began to discuss the obvious lack of work and attention of the regional government on the conservation of Baikal and significant degradation of the Baikal territory. In this regard, proposals were made to remove the Baikal from the programme and strip it of its natural World Heritage status. The new government of the region was very concerned about this and appealed to local educational organizations to raise the level of university science and expertise so that grant research has practical, not only theoretical significance.

In addition, a Youth Government has been established in the region, which should become a real personnel reserve for the current generation of officials and politicians.

The essence of the program was to combine the capabilities of youth communities from different sectors: educational, public administration, political and non-profit organizations.

The main research practice was based on Agile ideology and eduScrum methodology. Scientific institutions of the region provided their expertise in the preparation of the WWF Model and scientific research conducted during the previous 5-7 years on the Baikal territory to improve the quality and maximize the approach of prepared youth solutions and proposals to the needs of the region. The scientific research was combined with social and political practices.

This work differed so much from the previous ones, that scientific organizations and the government spoke about it at the regional level as the first practice of socially responsible policy and the first practical experience of achieving SDG indicators in the region. The setting of tasks and proposed solutions for the first time "came out of the office into people".

Research hypothesis: universities and young people can act as a mediator between regional governments, academia and the public and gain valuable practical experience in working with information, creating a regional strategy and roadmaps for achieving them.

Rapidly changing environments have shaped the concept of Life Long Learning, where people are forced to learn permanently, thus creating a new culture and new approaches to education [1]. The project-based approach to learning allows us to take a completely new look at the very process of forming practice-oriented competencies of future specialists and to concern the teaching staff with the question: "Does the university fulfill its social responsibility to the needs of the region? It is this direction of HEIs activity that actualizes economic and legal aspects of HEIs support by business and state structures in the region. The same theme feeds the educational process with actual information from scientific institutions. Social responsibility makes it possible to include more actively the socially important agenda, on which NPOs in the region work, into the educational practices of students.

Such work to move the international agenda to the regional level, to create a methodology for step-by-step work and to achieve the desired result was almost the first time. The real long-term experience of youth modeling was developed within the framework of the work of the Russian UN Association in the form of the UN Youth Model. However, within the framework of this event the real international process (work of UN committees and commissions, but not the regional Russian agenda) is modeled.

2 Methods and materials

The method of work was based on the Agile approach due to its great complexity and many uncertainties [3-5]. The educational paradigm in the process approach formed congruent competences and the ability to work with texts and partners, in the project approach promoted the formation of personality and was directed at an activity-oriented paradigm.

The main technologies: personality and activity training in the method of foresight design; cognitive instruction - in the form of a directed process of step-by-step implementation of the project under the guidance of a tutor; technology of development training allowed to eliminate the fear of young people before state reports and documents and scientific literature written in complex language. Great resource potential was used in the form of scientific-methodical, reporting, reference, educational literature, international documents and media creation, which were filled in the process of the project implementation.

Table 1. Main steps of project activity implementation

Steps on the project	Pedagogical technologies	Project activity result
Step 1: Inventory of environmental problems in the region	Ecological-humanistic axiology. Ecological development education: ecosystem thinking, cognitive activity.	A relatively complete list of obvious and non-obvious, but local problems in the region has been compiled
Step 2: Literary review and retrospective analysis	Establishment of environmental thinking	A thematic catalogue of literature on the directions of environmental problems identified in Step 1 has been compiled. It is structured by the directions: state documents, scientific literature, mass media information. The general level of study of the problem and its perception by residents of the region, at the federal level by authorities, at the international level by the UNESCO Commission on World Natural and Cultural Heritage is analyzed.
Step 3: World best practices for stabilizing the environmental situation	Protection of nature, sustainable environmental management	The best international practices in other lake regions have been studied in the following areas: government administration, opportunities for large industries, business impact on the environmental situation, NGO work to improve the situation
Step 4: Visionary: the future we want.	Axiological values: love for the nature of the native land	Making the desired image of the region in the short term (2 years) and in the long term (up to 10 years)
Step 5: Roadmap for achieving the goal	Personal and action-oriented approaches	Drawing up a step-by-step road map of desirable positive changes in the following areas: finding fresh approaches and projects to eliminate accumulated environmental damage, working with local communities, transforming federal and regional legislation to achieve the desired changes in the shortest possible time
Step 6: Resource definition	Structural-functional model: purpose, content, motivation, activity, evaluation, reflexion	The necessary funding and other resources have been identified. Persons responsible for the implementation of each direction and action have been identified

The project was carried out by a temporary creative team, whose term of activity was finite and limited to the final date of presentation of the completed project - this is the main condition of the project approach.

Stages of work in the project:

(1) creation of a workable team; (2) task definition; (3) creation of a road map of work; (4) distribution of responsibilities for short periods of time; (5) coordination of work by the head of the research team; (6) assembly of ready parts of the project by the head; (7) preparation of the final report and presentation for speech; (8) presentation of the results of the project study (speech before the expert community); (9) reflection, comparison of its

results with the results of other working groups; (10) introduction of expert corrections; (11) presentation of the results of the project to the expert community.

Composition of the game teams: head - member of the youth government of the region, team members - students of local colleges and universities, studying in different areas of training, research director and tutor - teacher of the university or college and director of the department of one of the local ministries. This team structure ensures the involvement of different age groups, correct problem setting, and a significant improvement in the quality of the expected design result.

The working groups (teams of players) were named exactly as the main directions of the World Water Forum:

Climate - water security and climate change (team captain - understudy minister of ecology and nature management of the region, mentor - minister of ecology and nature management of the region). People - water supply, sanitation and health (captain of the understudy team, mentor to the Minister of Social Development). Growth - water for sustainable production (team captain, understudy of the Minister of Economic Development, mentor, Minister of Economic Development). Quality - water quality, wastewater and reuse (team captain, understudy team captain, mentor, Minister of Health). Ecosystems - water and ecosystems (team captain, understudy of the Minister of Social Development, mentor, Minister of Social Development). Management - water management (team captain - head of Youth understudy government, mentor - governor of Irkutsk region). Sharing - attraction of stakeholders at all levels to share the benefits provided by water resources (team captain - understudy minister of sports and youth, mentor - minister of social development). Capacity - education, capacity building and technology transfer (team captain - understudy minister of culture, mentor - minister of culture). Financing - implementation of decisions (team captain - understudy of the Minister of Property Relations, mentor - Minister of Property Relations).

The teams consisted of college pupils and students, master's and postgraduate students of the local three universities.

The guidelines for the model contained a presentation guide, which was approximately the same for each working group. A sample is given for the "Climate" group:

Slide 1. Title. Name of the team. Logo of the university or department. Team composition (can be from players' photos). Slide 2. Review of the world climate theme and attempts to address it at the present stage. Use the above references in the report, it is very good to know the texture of the issue for successful further discussion. Slide 3. Overview of organizations leading hydrometeorological surveillance in the Baikal region, where they publish their data. Slide 4. Relevance of climatic subjects for Baikal region. What climatic changes are really taking place and what are the observed consequences? Slide 5. Brief review of literary data on scientific journals (find and prepare independently), not from mass media, on real climate problems and forecasts of changes in natural parameters in the region. Slide 6. Overview of local and federal climate change mitigation measures, from legislation to actual stoppage or conversion or re-profiling of businesses to mitigate climate impacts. Consider literally everything: the impact of housing and utilities, transport, enterprises and other agents. Reflect the subject matter of all the paragraphs mentioned above in this paragraph. Slide 7. Analysis of measures taken and identification of their weaknesses and non-performing positions. Slide 8. Analysis of the activities of existing local regional state protected areas (Specially Protected Areas) and NGOs engaged in "climate" activities (landscaping, Bird Days, restoration of plant nurseries, conservation of natural landscapes, implementing large and long term natural projects, such as WWF and others), etc. Slide 9. Own proposals for climate change mitigation at the regional level: presentation of 1-3 specific programs (e.g. conversion of boilers to partially use alternative energy; conversion of enterprises to other products; development of volunteer tour routes with restoration of natural flora of the region, etc.) and seeking targeted funding (it is necessary to be convincing for potential sponsors, even if the state can partially sponsor). Projects should be well developed, but not disclosed in detail, this will be necessary for further polemics and discussion. Slides 10-12. Own bold and creative proposals on how to increase and qualitatively improve the Baikal region's contribution to reducing global climate risks. Proposals may be general in nature, but they look more solid when they contain step-by-step targeted recommendations to different agencies. Slide 13. Literary sources. Slide 14. Acknowledgements to managers and organizations.

In addition, for the presentation, each working group prepared an analytical review (in the form of a printed document) of its topics on the clear recommendations of the Water Forum for its real participants.

The work of the teams was carried out by the members of the expert jury, filling in the evaluation sheets on the following items: knowledge of the invoice of the main material on the topic, completeness of the disclosure of the topic, the quality and addressability of the developed recommendations, the quality of the report and presentation, updating of information for the Baikal region, the quality and informativeness of the presentation, the quality of questions asked to other teams, the work of the players during the answers to the questions. Each team was sure to receive these sheets according to the results of the Model and could carry out reflection of its work according to the records of each member of the jury.

3 Results and Discussion

The concept of the project approach [6-7] allowed consolidating society and youth community in the region around a sensitive problem - deterioration of the ecological condition of Lake Baikal and its possible removal from the World Natural and Cultural Heritage List. Critical thinking was formed, which was significantly hampered by the formal educational process and outdated approaches to learning. From the point of view of knowledge and competence approach, the technology of actual problem setting and search of rational ways of its solution were pumped up. At youth skill of modeling of a situation, bases of management of the project of the decision of problems deconsolidated on segments is formulated. The skill of collective work, communication with various audiences and people with different level of thinking and competence is formed.

The possibilities and limitations of direct and feedback relationships between all Model participants were studied; the relationships in the emerging researchers' "ecosystem" were defined; the boundaries of competence and real level of material possession by research areas (working groups) were defined.

The level of readiness of educational organizations in the region and their motivation for integration into the world educational environment (space) and their ability to constantly saturate the labor market with specialists capable of solving urgent tasks of modern economy have been determined [8].

The project was implemented on the site of the First International Baikal Water Forum in 2017, on the threshold of the WWF in Brazil.

4 Conclusions

Practical experience in implementation of environmental education for sustainable development (ESD) has been developed; pedagogical technologies for informal work have been developed; pedagogical model of interaction between different segments of youth organizations, region's management and NGOs for effective environmental work in the region has been created; innovative directions in pedagogical activity have been identified, which are relevant in connection with adaptation of the global SDG agenda to regional processes.

References

1. J. Field, *Lifelong Learning and the New Educational Order. Stoke on Trent* (2000)
2. A. Sureka, M. Gupta, D. Sarkar, V. Chaudhary, *A Case- Study on Teaching Undergraduate-Level Software Engineering Course Using Inverted-Classroom, Large-Group, Real-Client and Studio-Based Instruction Model* (1st International Workshop on Case Method for Computing Education, 2015). URL: http://ceur-ws.org/Vol-1519/paper15.pdf
3. Z. Masood, R. Hoda, K. Blincoe, J. of Systems and Software, **144**, 501-510 (2018).
4. I. Noguera, A-E. Guerrero-Roldan, R. Maso, Computers & Education, **116**, 110-129 (2018).
5. M. Cubric, The Int. J. of Management Education, **11(3)**, 119-131 (2013).
6. E.V. Protopopov, A.V. Feoktistov, M.V. Temlyblev, O.V. Gordeeva, M.B. Vasilieva, Bull. of SibGIU, **4(22)**, 63-69 (2017) (in Russian).
7. M.V. Fominykh, B.A. Uskova, V.V. Mantulenko, O.N. Kuzmina, E.N. Shuravina, Int. Electron. J. of Mathematics Education, **11(8)**, 2814-2840 (2016).
8. E.F. Zeer, A.V. Streltsov, Int. Electron. J. of Mathematics Education, **11(7)**, 2639-2650. (2016).

 https://doi.org/10.1051/e3sconf/202016905006

Legal aspect of the implementation of the UN sustainable development goals in the field of environmental management in Russia

Maria Mukhlynina[1*], and *Natalia* Vedysheva[2]

[1]Institute of State and Law of the Russian Academy of Sciences (ISL RAS), Moscow, Russia
[2]Kutafin Moscow State Law University, Moscow, Russia

Abstract. Based on the analysis of strategic documents and other regulatory legal acts, the authors of the article attempt to consider a number of issues of the state environmental policy of the Russian Federation designed to contribute to the implementation of the UN sustainable development goals in the field of environmental management. The opinion is expressed that a successful environmental policy in the implementation of the UN sustainable development goals is possible only with a systematic approach to the implementation of environmental legislation. The authors analyze the measures, undertaken by the Russian Federation in the use and protection of forests and water resources, in terms of their effectiveness and draw a number of conclusions.

1 Introduction

New global challenges of our time forced the outdated traditional model of the economy, which could no longer resist these challenges, to change. Russia, like many other developed countries of the world, is now undergoing a difficult process of reorienting to build a new model in accordance with the Sustainable Development Goals. On September 25, 2015, UN member states adopted 17 Sustainable Development Goals (SDGs) until 2030 and 169 tasks which the countries of the world must fulfill in this period. Such a reorientation of the country requires changes both at the level of the regulatory legal base, and at the organizational and managerial level in all spheres of the country's life.

In general, researchers differently define the concept of “public policy”. For example, E.V. Luneva notes that public policy is “a system of values, ideas and views that reflect the direction of state activity toward obtaining a predetermined result in any sphere of public relations” [1]. State regulation of forest and water relations during the reform period is of great importance. Especially debatable by theorists and practitioners is the issue of expanding the scope of private regulation in the field of forest relations.

Let us dwell in more detail on two natural resources that are of crucial importance for the economy and vital functions of our country - these are water and forest resources. The study of legal and organizational problems of the protection of forests and waters in the Russian Federation is reflected in the works of E.V. Luneva [1], A.I., Nikiforov, M.M. Mukhlynina, O.N. Vedysheva, E.N. Redikultseva, R.F. Martynova [5], N.G. Zhavoronkova, V.B. Agafonova, G.V. Vypkhanova [6], Y.I. Shupletsova [8] and others.

* Corresponding author: muhlyninamm975@mail.ru

2 Methods

The methodological basis of the study was general scientific methods, so analysis and synthesis were used to study the existing theoretical and methodological approaches and provisions, as well as to conduct scientific research on the protection of forests and waters.

The informational basis of the study was made up of legal acts, documents of state authorities and their officials, scientific works of Russian scientists on the problems of improving forest protection in modern conditions [2, 3, 4, 7].

3 Results and discussion

One of the goals of sustainable development is to ensure the availability and rational use of water resources. By 2030, the goal is to improve water quality by reducing pollution, eliminating waste and minimizing emissions of hazardous chemicals and materials, halving the proportion of untreated wastewater, increasing recycling and safe reuse of wastewater around the world. In addition, countries should ensure integrated water resources management at all levels, including cross-border cooperation. It is necessary to increase the efficiency of water use by reducing water consumption in economic activities. Water use efficiency is one of the indicators for assessing the economic and social use of water resources in various sectors of the economy (in agriculture, production, housing and communal services, etc.).

If we turn to the experience of legal regulation and the practice of applying the water and environmental laws of the Russian Federation, we can single out the Water Strategy of the Russian Federation for the period until 2020, the Federal Program Development of *the Water Management Complex of the Russian Federation in 2012-2020*, the State Program of the Russian Federation in 2014 *Reproduction and use of natural resources (*subprogram *Use of water resources*) [2] and others. The presented documents highlight such problems as irrational use of water resources, water shortages in some regions of the Russian Federation, unsatisfactory quality of drinking water and limited access to centralized water supply systems.

To solve these problems and implement the SDGs in the field of water use in Russia, the following tasks are set in the strategic planning documents:

- ensuring the efficient operation of water management systems and hydraulic structures;
- ensuring the effective implementation by state authorities of the Russian Federation of the powers in the field of water relations;
- restoration and environmental rehabilitation of water bodies;
- reduction of negative anthropogenic impact on water bodies; development and modernization of the system of state monitoring of water bodies.

Forests, taking into account their global environmental significance and economic potential, are also among the priority areas of Russian state policy. According to *the Strategy for the Development of the Forest Sector of the Russian Federation until 2030*, one of the goals in the field of forestry is to achieve sustainable forest management, innovative and effective development of the use, conservation, protection and reproduction of forests that ensure the growth of the forest sector of the economy, social and environmental safety of the country, and the implementation of international obligations of the Russian Federation regarding forests [3]. The 1992 Rio de Janeiro Conference, which marked the beginning of sustainable development, identified 17 Sustainable Development Goals for the period up to 2030. Goal number 15 is defined as "Protecting and restoring land ecosystems and promoting their rational use, sustainable forest management, combating desertification, stopping and reversing the process of land degradation and ending the process of biodiversity loss" [4].

According to researchers A.I. Nikiforov, M.M. Mukhlynina, O.N. Vedysheva, E.N. Redikultseva, R.F. Martynova, the main measures for the implementation of the state policy of the Russian Federation in the field of forest use, conservation, protection and reproduction include improvement of legislation regarding state forest pathological monitoring and monitoring of forest reproduction [5], state forest supervision, and compensation for damage caused to forests and objects of nature [6] and others.

In order to streamline the use of forests, reduce illegal wood circulation, control over felling and the origin of wood and its turnover there is the Federal Law of the Russian Federation No. 415 of December 28, 2013 *On Amending the Forest Code of the Russian Federation and the Code of Administrative Offenses of the Russian Federation* which introduced unified state automated information system for wood accounting and transactions with it. 2019 proved that the interagency cooperation between the Federal Forestry Agency, the Federal Customs Service, the Federal Tax Service, the Ministry of Internal Affairs of Russia, Russian Financial Monitoring and other federal bodies in terms of control over the legal circulation of wood from the cutting area to processing or export is possible. Voluntary certification is an addition to the information system regarding responsible forest management and achieving sustainable development. It supports environmentally responsible, socially oriented, cost-effective forest management.

It should be noted that there are two international forest certification systems in Russia: FSC (Forest Stewardship Council) and PEFC (Pan-European Certification Scheme, since 2003 it has been renamed the Forest Certification System Approval Program). FSC certificates (forest management, supply chain, certificate of controlled wood for forest use); PEFC forest management certificates and management chains. Certification shows that during forest management all legislatively established requirements have been met. A voluntary certification system does not replace the state forest management policy, but has an impact on its formation.

Among the effective forest management mechanisms, the Federal Agency for Forestry (Rosleskhoz) maintains a register of unscrupulous tenants of forest plots and buyers of forest stands. Forest users who avoid making forest payments, systematically failing to fulfill contractual obligations regarding forestry activities, and other requirements of forest legislation are entered in such a register. For them, the possibility of access to forest resources for entrepreneurial purposes is further excluded. According to statistics, as of March 5, 2020, the register contains 531 unscrupulous forest users. The number of terminated agreements and the facts of evading the conclusion of the agreement is 803 [7].

The main types of violations of forest legislation include:

- failure to comply with forestry regulations;
- failure to implement a forest development project regarding forest reproduction;
- failure to implement a forest development project regarding the protection of forests from fires.

From inspection of lease agreements for forest plots that were provided for logging by Rosleskhoz, 3,500 violations were identified. Inspectors issue binding orders to eliminate violations on the protection, conservation and reproduction of forests, and in case of failure to apply to the court with a request for early termination of leases of forest fund plots [6].

Thus, there is a tendency to increase the registry of unscrupulous forest users.

4 Conclusions

To sum up, it can be noted that the Federal Forestry Agency believes that one of the possible measure of preventing violations is introducing into the duties of tenants the delivery of work to protect and reproduce forests only after field inspections. In order to ensure sustainable forest management, Rosleskhoz proposes to create a specialized state-owned company, endowed with economic functions and rights for the integrated use of forests. It will help to solve the problem of sustainable forest management with the involvement of extrabudgetary sources of funding and public-private partnership opportunities.

As for the legal ways to solve problems within the framework of achieving the SDGs in the field of water use, the Russian Federation actively interacts with foreign countries that have transboundary water bodies. Russia participates in intergovernmental agreements with the following countries: Abkhazia, Azerbaijan, Belarus, Kazakhstan, China, Mongolia, Finland and Estonia.

Sustainable water use is impossible without environmental protection measures being carried out domestically. For example, 4 out of 11 federal projects in the Ecology National Project were allocated on water bodies: *Clean Water, Improvement of the Volga, Preservation of Lake Baikal, Preservation of Unique Water objects.* These projects are aimed to improve the quality of drinking water for residents of settlements that are not equipped with modern centralized water supply systems; ecological improvement of water bodies, including the Volga River, preservation of unique water systems, including Lakes Baikal and Teletskoye.

For example, the institution providing the right to use water bodies allows to optimize water use and ensure the protection of water bodies through the terms of a water use agreement. On January 1, 2019, 111,267 permits were registered in the state water registry. In addition, the Federal Water Resources Agency (Rosvodresursy) maintains a register of dishonest water users and participants of auction for the right to conclude a water use agreement, which in practice allows to prevent the use of water bodies and resources by previously identified offenders.

We agree with the opinion of Y. I. Shupletsova, who notes that the rational use of natural resources is an important environmental problem and is one of the most acute global issue requiring urgent fundamental study and solution, but, unfortunately, there are many problems with forest industry, and for their effective solution it is necessary to adopt a large number of legal acts [8].

References

1. E.V. Luneva, Lex Russica, **12**, 67 (2018)
2. *Order of the Government of the Russian Federation of August 27, 2009 No. 1235-r (as amended on April 17, 2012) On Approving the Water Strategy of the Russian Federation for the Period Until 2020*
3. *Order of the Government of the Russian Federation of September 20, 2018 No. 1989-r (as amended on February 28, 2019) On Approving the Development Strategy of the Forestry Complex of the Russian Federation until 2030*, No. 40, art . 6147 (2018)
4. The official website of the UN [Electronic resource]. URL: https: // www.un.org
5. A.I. Nikiforov, M.M. Mukhlynina, O.N. Vedysheva, E.N. Redikultceva R.F. Martynova, International Journal of Engineering and Advanced Technology, **8(5)**, 2251-2255 (2019)
6. N.O. Vedysheva, *Problems of forest legislation and ways to improve it,* 119-123 (2019)
7. The official website of Rosleskhoz [Electronic resource]. URL: https: // rosleshoz.gov.ru
8. Y.I. Shupletsova, Journal of Russian Law, **3(219)**, 72-29 (2015)

Forest conservation in Russia: problems of implementation

Dmitry Mukhlynin*[1]

[1] SIC-2 Research Institute of the Federal Penitentiary Service of Russia, 119991, 14, Zhitnyaya str., Moscow, Russia

Abstract. In the article, the author analyses the legal and organizational problems of forest conservation in the Russian Federation. It was established that certain problems related to managerial errors, insufficient financing of the forest industry, and the imperfection of regulatory legal acts regulating these public relations impeded the effective protection of forests in Russia. Based on the analysis of statistical data and reporting materials of the Ministry of Natural Resources of the Russian Federation and other departments, it is concluded that the problems identified are multi-faceted and that a whole range of measures to protect forests is necessary. The author offers options to overcome these problems.

1 Introduction

The situation with forest conservation in Russia is currently under special control of the state. The events of 2019, as litmus paper, have revealed the problems of the forest industry that have been accumulating over the years, namely: large-scale forest fires, illegal logging, the presence of systemic problems associated with imperfect management, underfunding of the industry and an unbalanced distribution of powers.

The unfavorable environmental situation in the country as a whole, the updating of environmental legislation, including forest legislation, the adoption of numerous state federal and regional programs, demonstrate the great importance and determination of the state to cope with environmental problems as quickly and efficiently as possible. Among such problems there is an urgent issue of protection forests.

The study of legal and organizational problems of forest protection in the Russian Federation is reflected in the works of M.M. Mukhlynina [3], B.A. Molchanova [8], A.R. Aberkhaev, L.I. Alexandrova, B.V. Andreeva [9], M.S. Andrianov, S.A. Bogolyubov, N.D. But, D.G. Dobretsov, K.P. Ermakova [10] A.I. Nikiforova, O.N. Vedysheva, E.N. Redikultseva, R.F. Martynova [11] and others. However, issues of a comprehensive analysis of forest conservation remain understudied.

* Corresponding author: muhlynin78@mail.ru

2 Methods

The methodological basis of the study was the following general scientific methods: analysis and synthesis were used to study existing theoretical and methodological approaches, as well as to conduct scientific research on the protection of forests. Structural and logical methods are used to systematize the factors affecting the implementation of innovations in the legal regulation of forest protection. The factor analysis method was used to determine the influence of factors on the level of satisfaction of forest legal entities on the level of law and order in this industry.

The informational basis of the study was made up of normative legal acts, documents of state authorities and their officials, scientific works of Russian scientists on the problems of improving forest protection in modern conditions [1, 2, 4, 5, 6. 7, 10].

3 Results and discussion

It should be noted that the Russian Federation is the world leader in forest area - 1184.1 million ha, or 20% of the total forest area of the world [1]. Forests in our country play a key role in various directions. They are not only "lungs of the planet" in their global environmental significance, but also export-import goods, which bring tangible budget revenues. D. Kobylkin, the head of the Ministry of Natural Resources and Ecology of the Russian Federation, notes that "the forest industry shows a statistically positive dynamics for all key indicators, but does not meet the expected results and is often criticized. In particular, from 2009 to 2019 there was more than a twofold increase in income to the country's budget for the use of forests - from 20 billion rubles up to 53 billion rubles; a significant increase in timber harvesting - from 176 million cubic meters up to 219 million cubic meters and reforestation area from 813 thousand ha to 1,127 thousand ha." [2]. Financial support from the state and the active implementation of the principle of environmental management fees contribute to the process of reforestation. According to the researcher M.M. Mukhlynina, in its content and essence, this principle of environmental management and environmental damage compensation is also connected with the solution of public legal problems of ensuring environmental well-being [3].

If we turn to the legal definition of the concept of "conservation" of forests, then the legislation, in particular, the Forest Code of the Russian Federation, contains a generalized chapter 2.4 "Conservation, protection, reforestation", which was introduced into the code structure only in 2016. The Forest Code of the Russian Federation does not contain the definition of the concept of "conservation" of the forest, revealing only the general types of protection, objects and subjects of such protection in article 50.7 of the Forest Code of the Russian Federation (hereinafter - FC RF). The concepts of "forest conservation" should also be distinguished from "forest protection", since if "conservation" is aimed at implementing protective measures against fires and pollution (radioactive and oil) (Chapter 3.2. FC RF), then "protection" involves identification of harmful organisms in forests and the prevention of their spread (Article 60.1 of the FC RF). Reforestation includes forest seed production, reproduction, and forest maintenance [4].

Thanks to the active implementation of the Ecology national project and other strategic documents by the President of the Russian Federation V.V. Putin, representatives of the authorities and the public, there is much attention paid to the problem of nature conservation and, in particular, forests in modern Russia. According to paragraph 7 of the Decree of the President of the Russian Federation dated May 7, 2018 No. 204 *On National Goals and Strategic Tasks of the Development of the Russian Federation for the Period until 2024*, it is necessary to ensure forest conservation, including their reproduction in all areas of felled and lost forest stands for the purpose of the scientific and socio-economic development of the country and improving the living standards of citizens. [5].

According to official information of the Ministry of Natural Resources and Ecology of the Russian Federation, within the framework of the federal project "Forest conservation" of the national project "Ecology", the issue of forest reproduction is now very serious. Russia is going to restore 1 million hectares of forest, 778 thousand hectares of which are on leased forest areas. The regions are assisted by forestry machinery and equipment (currently, about 150 units of machinery and more than 880 units of equipment) [6].

Violation of forest legislation is another serious problem that has to be solved, as "the volume of illegal forest management exceeds the national average in those regions of the Russian Federation where state control over the conservation of forests is weakened". Such a conclusion was made by the Council of the Federation of the Federal Assembly of the Russian Federation [7].

The legal concepts of "forest violation" among researchers differ, for example, G.N. Polyanskaya defined forest violation as "damage, destruction and theft of woody vegetation, as well as damage to land intended for natural reforestation, and all types of unauthorized use of forest land for agricultural and other purposes". Researcher B.N. Tsvetkova defines the forest violation as "an unlawful act committed by a public authority or a citizen who is not in the legal relationship of forest management and associated with an encroachment on the legal regime of forests established by a socialist state" [8].

The law and order in the field of forest protection is carried out by environmental and law enforcement agencies. The activity of environmental authorities is aimed at preserving forests, including their reproduction in all areas of deforested and lost forest plantations, as well as the implementation of many other measures in accordance with applicable law. Thanks to the active implementation of the national project Ecology and other strategic documents by the President of the Russian Federation V.V. Putin, representatives of the authorities and the public, attention to the problem of nature conservation and, in particular, forests, in modern Russia is reinforced many times.

According to official information of the Ministry of Natural Resources and Ecology of the Russian Federation, within the framework of the federal project "Forest Conservation" of the national project "Ecology", the issue of forest reproduction is now very serious. It is planned to restore 1 million hectares of forest, 778 thousand hectares of which are in leased forest areas. The problem of underfunding is currently being addressed at all levels, the regions are being helped by forestry machinery and equipment (about 150 units of machinery and more than 880 units of equipment).

The second group of bodies supporting the law and order in the field of forest protection is law enforcement agencies: their task is to carry out law enforcement activities with the aim of protecting the rights and legitimate interests of citizens, legal entities, the state and society, ensuring safety and legality. Law enforcement activity is heterogeneous, it is carried out by various bodies, which are faced with distinct, specific tasks and apply various means and methods. Let us briefly dwell on two bodies - the police and the prosecutor's office, designed to identify and stop violations of forest legislation. Police identify and investigate crimes. The powers of the prosecutor's office extend to almost all spheres of legal relations regulated by law, in particular, they exercise oversight functions, but, if necessary, take measures to restore the violated law [9]. Let us analyze the statistics of violations of forest legislation: although the volumes of illegal logging detected by state authorities of the constituent entities of the Russian Federation do not exceed 2 million cubic meters of wood, which is less than 1 percent of the volume of legal forest use, the identification of offenses is one of the serious problems. According to researchers, M.S. Andrianova, S.A Bogolyubova, N.D. But, D.G. Dobretsova, K.P. Ermakova, statistics of the use of articles of the Criminal Code of the Russian Federation, which provides for liability for environmental crimes, allows us to conclude that mainly crimes related to the seizure of natural resources are recorded and disclosed. In the same time, criminal environmental pollution practically does not become the subject of criminal prosecution for the vagueness of the wording of the elements of crimes in the Criminal Code of the Russian Federation [10].

Another problem of forest violations is weak state control. So "the volume of illegal forest management exceeds the national average in those regions of the Russian Federation where state control over the conservation of forests is weakened" - such a conclusion was made by the Federation Council of the Federal Assembly of the Russian Federation [7]. Researchers A.I. Nikiforov M.M. Mukhlynina, O.N. Vedysheva, E.N. Redikultseva and R.F. Martynova. They believe that the current form of state control, primarily focused on coercion and submission, has largely reached its limits, so the current situation requires a transition to preventive control, which acts as a service [11].

4 Conclusions

To sum up, it can be noted that at present, there are a few political and legal documents devoted to environmental protection, where the issues of rational environmental management are brought to the forefront and many issues are systematically worked out in them. Now we are reaping the benefits of civil law relations introduced in the Forest Code of the Russian Federation in the field of turnover of forest areas, which have actually relieved tenants of their duty to take care of forests, ensure fire safety. The forests that have not been leased out have completely lost real protection from negative factors. The weakening of the control function also contributed to an increase in the number of forest violations. According to the researchers, a return to the stable functioning of the management system and overcoming the legal nihilism of forest users will require years and considerable efforts on the part of the control and supervisory authorities.

The state has taken forests under its tight control. Note that more than one ministry is responsible for the law and order in this industry; many bodies are involved in the work, including law enforcement agencies. Forest protection is a very complex set of measures. The effectiveness of the mechanism of environmental and legal regulation can be promoted only by overcoming defects in legal regulation.

References

1. *Report of the Chairman of the Committee of the State Duma of the Russian Federation on Natural Resources, Property and Land Relations Nikolaev N.P. Natural Resources: Efficiency and Control* (2017) URL: http: // komitet3-1.test.km.duma.gov.ru/upload/site49/document_news/000/264/228/doklad.pdf. (accessed date 08/05/2018).
2. Website of the Ministry of Natural Resources of Russia [Electronic resource]. URL:http://www.mnr.gov.ru/press/news/glava_minprirody_rossii_predstavil_osnovy_strategii_razvitiya_lesnoy_otrasli_na_printsipakh_sinkhron/ (accessed 08.03.2020).
3. M.M. Mukhlynina, *Ensuring the implementation of the principle of environmental management paid and environmental damage in the interpretative acts of the Constitutional Court of the Russian Federation*, (Nizhny Novgorod Academy of the Ministry of Internal Affairs of the Russian Federation, Nizhny Novgorod, 2020)
4. *Forest Code of the Russian Federation dated December 4, 2006 No. 200-FZ (as amended on December 27, 2018)*

5. *Decree of the President of the Russian Federation of May 7, 2018 No. 204 On national goals and strategic objectives of the development of the Russian Federation for the period until 2024*
6. Website of the Ministry of Natural Resources of Russia [Electronic resource]. URL: http://www.mnr.gov.ru/press/news/federalnyy_proekt_sokhranenie_lesov_vosstanovleno_245_tysyach_ga_lesnykh_nasazhdeniy/ (accessed 10.03.2020).
7. *Decree of the Council of the Federation of the Federal Assembly of the Russian Federation dated January 30, 2019 No. 17-SF On Strengthening Control over the Turnover of Wood and Counteracting its Illegal Logging* [
8. B. Molchanov, *Legal protection of Russian forests of the twentieth and early twenty-first centuries: Textbook*, (Vladimir State University, Vladimir, 2005)
9. A.R. Aberkhaev, L.I. Aleksandrova, B.V. Andreev, *The prosecutor's handbook* (Moscow, 2019)
10. M.S. Andrianov, S.A. Bogolyubov, N.D. But, D.G. Dobretsov, K.P. Ermakova and others, *Legality: theory and practice: monograph* (Institute of Legislation and Comparative Law under the Government of the Russian Federation, Academy of the Prosecutor General of the Russian Federation: Law Firm Contract, Moscow, 2017) URL: https://base.garant.ru/ (access date 05/15/2019).
11. A.I. Nikiforov, M.M. Mukhlynina, O.N. Vedysheva, E.N. Redikultceva, R.F. Martynova, *Technologies and main modern economic and legal aspects of the formation of state control systems (supervision) over the use of natural resources*, International Journal of Engineering and Advanced Technology (IJEAT), **8**, 2255 (2019)

On the implementation of the state environmental policy in law-making and law-enforcement activities in Russia

Maria Mukhlynina[*]

Ph.D, Institute of State and Law of the Russian Academy of Sciences, 119019, 10, Znamenka str., Moscow, Russia

Abstract. The article considers a number of issues of the state environmental policy of the Russian Federation based on the analysis of new laws of the Russian legislation and the current socio-economic situation in the country. The author notes the state strategic documents and programs developed by the Ministry of natural resources of the Russian Federation, presents some data on environmental legislation in the country, and makes judgments that a successful environmental policy is possible only with an integrated approach to the implementation of environmental legislation.

1 Introduction

President V.V. Putin has set an ambitious goal - Russia should become one of the five largest economies in the world in the next few years with economic growth rates exceeding the world ones (according to the Decree of the President of the Russian Federation of May 7, 2018 No. 204 "On national goals and strategic objectives of the development of the Russian Federation for the period until the year 2024") Achieving this goal requires the creation of a highly productive export-oriented sector, primarily in two industries - manufacturing and agriculture, which will undoubtedly influence the country's ecology. Twelve key directions in which national projects have been developed in accordance with the Decree relate to all the main areas of life, and one of them is ecology [1]. The sector of the economy that will be involved in this process first of all is the forest industry.

The state environmental policy issues of the Russian Federation are reflected in the works of E.V. Vinogradova [2], B.A. Molchanova [3], L.V. Rashchupkina [4], G. Dieterle and A. Kushlin [6], A.P. Anisimov [8] and others. However, the issues of a comprehensive analysis of the state environmental policy of the Russian Federation remain insufficiently studied.

[*] Corresponding author: muhlyninamm975@mail.ru

2 Methods

The methodological basis of the study was the following general scientific methods: analysis, synthesis and deduction. The factor analysis method was used to determine the influence of various factors on the level of implementation of the state environmental policy of the Russian Federation and the degree of satisfaction of citizens with its results.

3 Results and discussion

Considering public policy in relation to ecology, we turn to a theoretical understanding of the concept of "public policy", which, for example, is defined by administrative law as an interconnected set of public authority decisions, drawn up into a set of legal acts implemented by all branches of government through certain tools [2]. The concept of "environmental policy", according to researchers B. Molchanov and others, was used for the first time in the 70s - 80s of the twentieth century, which was justified by the interdependence and interrelation with the socio-economic problems of society [3]. According to L. Rashchupkina, today experts in the field of agrarian, land and ecological law are mainly involved in the scientific and theoretical understanding of the legal regulation of environmental problems. Whereas, the activities of executive authorities in organizing public administration in the field of nature management and environmental protection are studied by specialists in the field of administrative law, etc. [4].

The need for a scientific analysis of state environmental policy is dictated by the fact that there are still contradictions between the process of modernization of the Russian economy and the lack of generalized theoretical models for assessing the state, dynamics of mechanisms and trends in its development.

One of the priorities of state policy in the environmental sphere is the development of forestry relations, which are important from an environmental point of view both for domestic consumption and global environmental one, as an environment-forming and protective natural object. The development and enhancement of the competitiveness of the Russian forest industry for the long term is considered by researchers as a result of the increased contribution of the forest sector to the country's socio-economic development. However, despite the huge concentration of forest resources on the territory of the Russian Federation on a global scale (25 percent of the global reserves, according to official data), Russia's share in the global production of wood and paper-based products is only 2 percent [5].

Let us dwell on one of the directions of the state environmental policy - the policy on the regulation and protection of forests. At present, the state forest policy is being implemented in accordance with a rather impressive number of adopted regulatory legal acts, such as *the Forest Code of the Russian Federation*, the Federal Law *On Industrial Policy in the Russian Federation* and a number of regulatory legal acts that determine the state's policy in relation to the forest sector: *Innovation Development Strategy of the Russian Federation for the period until 2020*, approved by order of the Government of the Russian Federation on 8 December 2011, No. 2227-r, *Forecast of scientific and technological development of the Russian Federation for the period until 2030*, approved by the Prime Minister of the Russian Federation, *Strategy for scientific and technological development of the Russian Federation*, approved by *Decree of the President of the Russian Federation* dated December 1, 2016 No. 642, *the Fundamentals of the state policy in the field of use, conservation, protection and reproduction of forests in the Russian Federation for the period up to 2030*, approved by *the Decree of the Government of the Russian Federation* of September 26, 2013 No. 1724-r., *Order of the Government of the Russian*

Federation of September 20, 2018 No. 1989-r *On the Strategy for the Development of the Forestry Complex of the Russian Federation until 2030* and etc.

The set of issues related to the state environmental policy needs scientific analysis and assessment in different planes, so we will trace such a development in forest relations.

Russia, like many countries of the world aimed at rapid development, is on the path to digitalization of the economy. Back in 2017, the national program *Digital Economy of the Russian Federation* was approved, and as a result of which 30 projects should be implemented by 2025, including innovative digitalization projects in the forest industry. One of the tasks that should be solved now is to establish an accurate accounting of timber stocks, its movement at all stages, starting from the harvesting, transportation, storage and sale of wood raw materials with a minimal influence of the human factor.

G. Diterle and A. Kushlin, relying on the views of World Bank experts, as early as 2004 highlighted several primary indicators of the Russian timber industry compared with other countries of the world: 1. The presence of a large-scale domestic market, protected from competition; 2. The world's largest stocks of coniferous wood; 3. Stable sales to Europe and China; 4. Low cost-price of wood; 5. The attractiveness of the domestic forestry industry for investment by international industrial and financial companies [6]. Unfortunately, over time, some factors began to play a negative rather than a positive role in forestry relations, and the forestry sector of the economy turned into a very criminalized industry.

Currently, the state and the environmental ministry of the Russian Federation, has begun preparing *the Forest Industry Development Strategy*. The goal of the future Strategy is to create a closed-loop economy on renewable resources, that is, a full-fledged, cost-effective forestry. The Ministry of Natural Resources and Ecology of the Russian Federation approaches the solution of a new large task in an integrated manner. Effective management of the industry requires complete reliable and constantly updated information on the forest resources available in the country, for which purpose the Unified Information System *Forests of Russia* is being created. It is connected with the databases of the Federal Register of Regions, the Federal Tax Service, the Federal Customs Service, the Ministry of Internal Affairs of Russia and other government bodies. The information coming to the *Forest of Russia* UIS will be taken from the state forest register in electronic form and updated forest inventory materials. First of all, according to Minister D. Kobylkin, forest inventory should be carried out in priority areas for use, which has already been reflected in the prepared amendments to the legislation of the Russian Federation. Another direction for the development of the forest industry will be the creation of conditions for the financial self-sufficiency of regional institutions – so called leshozes (forestry). They are supposed to be provided with additional economic functions, and their field of activity is vast - this is 70 percent of all the country's forest fund areas that are not currently leased [7].

Researchers, in particular A.P. Anisimov, estimate positively the increasing work of environmental authorities on the possibility of any interested subject in obtaining environmental and legal information in the public domain, for example, documents accepted by the Ministry of Natural Resources of Russia, statistical data, etc. Keeping an interactive map allows citizens to receive environmental information, for example, on objects of environmental control, on water, forest, biological resources, within the framework of any subject of the Russian Federation [8]. This openness and accessibility of information contributes to a positive image of public authorities for the population.

4 Conclusion

To sum up, it can be noted that the achievement of the strategic goals of the state environmental policy and environmental management in law-making and law enforcement in Russia is carried out through the formation and implementation of a long-term state policy aimed at protecting and reproducing the natural and environmental potential of our country, increasing the level of environmental education and environmental culture of citizens.

As a result of a review of the problem, we note that the measures taken by the state are multidimensional and set a high level of planned changes in the field of ecology. However, I would like to ask another question, which is more difficult to answer than to give an overview of legal and scientific literature, as well as legislative changes: do we have effective mechanisms that can implement all those tasks that are determined by the state environmental policy? There is no definite answer to this question. The specificity of environmental problems is that many efforts may not give the desired result, not because of unwillingness to act, but because of the unreality, to immediately change the situation for the better, or even over the next five to ten years. It is impossible to pollute, for example, the Volga River for decades, deplete its fish stocks, not take measures aimed at protecting the waters and the entire ecosystem of the river, and then fix it in a few years. There is an understanding that technical and industrial progress cannot be stopped by the forces of either ecologists or governments; society is not ready to abandon the usual benefits of civilization. Humanity is moving at an increasing rate along the path of increasing use of all types of natural resources, and as a natural effect of such activities - pollution of nature, the exhaustion of natural resources, and irreversible processes in different parts of the world.

The latest and most environmentally friendly technologies in all spheres of life and production, both already used and only being developed, are not able to quickly and comprehensively change the situation for the better, this requires organizational, financial resource and time. The ecological "trap" into which a man drove himself is still surmountable, but only by the joint efforts of countries around the world, since environmental problems are transboundary. We believe that it is necessary to create such a body as the environmental police, with the help of which it would be possible to implement both preventive and oversight measures related to environmental violations at the national and international levels. As for Russia, there was the experience of the functioning of the environmental police since the early 1990s until the beginning of the 2000s, and it was very successful. We note that *the Ecological Police* was a special unit existing in the Ministry of Internal Affairs system ... Employees fought against illegal actions of people over animals, landfills, waste and pollutants, etc. A strong Russian state is unthinkable without an effective national security system, and its functioning directly depends on the optimal operation of all elements, including the police system.

References

1. *Decree of the President of the Russian Federation of May 7, 2018 No. 204 On national goals and strategic objectives of the development of the Russian Federation for the period until 2024*
2. M.M Mukhlynina, E.V. Vinogradova, *State policy in the field of environmental development of the Russian Federation and analysis of the competence of entities implementing this policy?*, 38 (2017)
3. B.A. Molchanov, *Legal protection of Russian forests of the twentieth and early twenty-first centuries: Textbook allowance*, 11-12 (Publishing house Vladim. state University, Vladimir, 2005)
4. L.V. Rashchupkina, *Environmental Law and* Order (General Theoretical Analysis) thesis ... candidate of legal sciences, 3-4 (Vladim. state University, Vladimir, 2006)
5. *Decree of the Council of the Federation of the Federal Assembly of the Russian Federation dated January 30, 2019 No. 17-SF On Strengthening Control over the Turnover of Wood and Counteracting its Illegal Logging*

6. G. Diterle, A Kushlin, *Key problems of forest policy reform in Russia*, Sustainable forest management, **2 (4)**, 18 (2004)
7. Website of the Ministry of Natural Resources of Russia [Electronic resource]. URL: http://www.mnr.gov.ru/press/news/glava_minprirody_rossii_predstavil_osnovy_strategii_razvitiya_lesnoy_otrasli_na_printsipakh_sinkhron/
8. A.P. Anisimov, Siberian Legal Review, **3**, 315 (2019)

Renewable energy as a factor of sustainable development and regulation of ecological problems in Africa

Liudmila Nefedova[1,*], *Alexander* Solovyev[1], *Yulie* Rafikova[1], *Dmitriy* Solovyev[2]

[1]Lomonosov Moscow State University, 119991 Leninskie gory, 1/19, GSP-1 Moscow, Russian Federation
[2] Shirshov Institute of Oceanology, Russian Academy of Sciences, 117997, Nakhimovsky Prospect, 36, Moscow, Russian Federation

Abstract. The analysis of the main directions of renewable energy in Africa, as a factor in sustainable development and reduction of greenhouse gas emissions is performed. The ecological problems of the modern and prospective development of the energy complex of African countries are considered. For African countries the issue of ensuring reliable and environmentally friendly access to electricity for the population is extremely acute. It is shown, that the electricity demand for industry in Sub-Saharan Africa - the most problematic region - is growing on a very large scale. The construction of new large coal-fired thermal power plants in the required volumes will lead to serious environmental and climatic consequences. The study of regional data allowed us to conclude that PV solar systems are of priority importance for increasing people's access to electricity in rural Sub-Saharan Africa. Based on numerous materials from international energy structures the estimates and calculations of volumes of reduction of greenhouse gas emissions due to the use of renewable energy sources as an alternative to carbon fuel are carried out. The study has shown that of particularly great importance for reducing CO_2 emissions in Kenya is the development of geothermal energy.

1 Introduction

The carbon-free path of energy development in Africa is an important contribution to solving such problems of global warming, greenhouse gas emissions, forest destruction. The problem of access to electricity is one of the priority conditions for the economic and social development of African countries. In African countries with high potential resources of renewable energy, development of renewable energy resources is an important factor of social and economic development of these countries, increase of energy and economic independence, as well as a lever to overcome the spatial heterogeneity of the population's access to electricity, improve ecological conditions and fight against deforestation. The problem of low electricity access for people in rural Africa is one of the most acute problems

* Corresponding author: nefludmila@mail.ru

of our time. About 600 million rural people do not have access to grid electricity, of which 98% are concentrated in the Sub-Saharan region. Access to electricity is extremely low: as of 2018 - on average in Sub-Saharan Africa (SSA) – 43%. In 30 countries, the indicator is from 11% to 39%, and in 10 countries - less than 10% [1].

Without electricity, it is challenging, if not impossible, to promote economic growth, overcome poverty, expand employment, and support human development. Sustainable energy is the seventh goal of the 17 UN Sustainable Development Goals (SDGs), with a call to "ensure access to affordable, reliable, sustainable and modern energy for all." Its five targets indicate areas where policies can be designed—such as boosting the share of renewable energy (RE) in the global energy mix and doubling the global rate of improvement in energy efficiency [2]. Most of the countries in the region plan to increase 3-5 times the installed capacity of power plants and the generation of electricity by 2030 to overcome the shortage of electricity and increase public access to electricity [1]. Such a scale of building up the electricity generating capacities required for the economic development of the region when using carbon fuel will have tremendous environmental and climatic consequences.

2 Object of study and methods

The objective of the study was to assess the significances of using renewable energy sources to ensure sustainable development of the region. The authors consider renewable energy in African countries in terms of its importance for providing increased access to electricity for the population, reducing carbon dioxide emissions from electricity generation and the problem of use of wood as a fuel for cooking, which leads to a decrease the forest areas on the continent.

The present article consists of **a**n analytical review of the literature, statistical calculations, and expert estimates. As a research tool, we used analysis and synthesis methods, a statistical research method, as well as tabular methods for presenting research results. The research information base was compiled from official data of the International Renewable Energy Agency, the International Energy Agency, of the World Bank and for international ecological organizations, as well as the official websites of African countries Energy Organizations.

3 Results and Discussion

3.1. Ensure universal electricity access

The population of Africa have different possibility in access to electricity. In North African countries this index is 95-100% (tab.1). For East and Central Africa the issue of ensuring reliable and environmentally friendly access to electricity for the population is extremely acute. The use of diesel generators is not always effective, since the burning of liquid fuel has a negative impact on the environment, depends on the reliability of fuel supplies, especially taking into account factors such as imperfect transport infrastructure and the risk of fuel theft. Africa's demand for electricity and for industrial development is growing on a very large scale. The total capacity of power plants on the continent over 20 years has more than doubled [2]. The construction of new large coal-fired thermal power plants in the required volumes will lead to serious environmental and climatic consequences. In Africa the number of people gaining access to electricity doubled from 9 million a year between 2000 and 2013 to 20 million people between 2014 and 2018, outpacing population growth. As a result, the number of people without access to electricity, which peaked at 610 million in 2013, declined slowly to around 595 million in 2018. Much of this recent dynamism has been in East Africa, as Kenya, Ethiopia and Tanzania accounted for more than 50% of those gaining access.

Table 1. Population access to electricity in Africa regions [3].

	Population , mln	Percent of Population access to electricity	Average annual electricity consumption kWh /pp
North Africa	175	98%	1574
West Africa	327	47%	188
East Africa	303	23%	91
Central Africa	115	25%	167
South Africa	177	93%	2061

The access rate to electricity rose in Kenya from 25% in 2013 to 75% in 2018 due to RES development. The study of regional data allowed us to conclude that solar PV systems are of priority importance for increasing people's access to electricity in rural Africa. In rural and remote regions improved access to energy can generally be acquired in three ways:

1) grid-based centralized electrification, where the grid is extended beyond urban and peri-urban areas.

2) community-level or renewable energy-based mini- or micro-grid systems;

3) household-level use of isolated devices and stand-alone systems to generate power lighting and heat for space and water heating, cooking and productive uses.

For off-grid power supply of autonomous consumers, the most cost-effective is the creation of local mini-grids with photovoltaic installations in several settlements or agricultural facilities. New forms of financing are used, for example, Pay & Go for stand-alone installations, auctions for large facilities on renewable energy sources [4]. The majority of progress over the past decade in Africa has been made as a result of grid connections, but a rapid rise has been seen in population access to electricity via solar home systems. Kenya, Tanzania and Ethiopia accounted for around half of the 5 million people gaining access through new solar home systems according to latest estimations in 2018 (up from only 2 million in 2016) [1]. The levelized cost of electricity (LCOE) in the rural mini grids in Kenya and Tanzania analyzed in this study ranged from \$0.55 to \$0.85 per kWh, with a median cost of \$0.66, using a 22% load factor. Development partners and governments will need to crowd in private-sector investment and enabling mini grid business environments in key access-deficit countries. These combined efforts can bring down the cost of mini grid electricity to \$0.25/kWh by 2025 and \$0.20/kWh by 2030 [5].

To the 2019 the total installed capacity RE power plants in Africa countries was of 14.6 GW, including: wind farms - 5765 MW, solar power plants (PP) – 6367 MW, bioelectric PP – 1713 MW, geothermal PP - 830 MW [1]. The leader of wind energy in Sub-Saharan Africa is Ethiopia: three large wind farms - Adama I &II (205 MW), Ashegoda (120 MW). In Kenya - Lake Turkana Windfarm (310 MW). Geothermal resources are concentrated in the zone of the East-African Valley with the length of 6000 km (estimated 20000 MW) in the territories of Eritrea, Djibouti, Ethiopia, Kenya, Uganda, Rwanda, Tanzania, Zambia. The greatest success in geothermal energy was achieved by Kenya, the total installed capacity of Geothermal Power Plants (813 MW). The region has the Carbon Assets Development Fund for Africa (ACAD), a public-private partnership sponsored by UNEP. ACAD serves as a platform to stimulate investment, provides seed funding and advises green entrepreneurs on carbon credits and investing in low-carbon energy initiatives.

3.2 Biomass in energy balance and problem of deforestation

The energy balance of the countries of the region is dominated by plant and animal biomass for cooking, electricity on average is less than 10%. This leads to a process of catastrophic deforestation. Mass deforestation is one of the most important environmental problems of our time. Forest communities play a crucial role in the normal functioning of natural ecosystems. They absorb atmospheric pollution of anthropogenic origin, protect the soil from erosion, regulate the normal runoff of surface water, prevent the decrease in groundwater levels and siltation of rivers, canals and reservoirs. The decrease in forest area disrupts the cycle of oxygen and carbon in the biosphere. Despite the fact that the catastrophic consequences of deforestation are already widely known, their destruction continues. Energy from biomass accounts for more than 30% of the energy consumed in Africa and more than 80% in many sub-Saharan African countries. Indoor pollution from biomass cooking — a task usually carried out by women — will soon kill more people than malaria and HIV/AIDS combined [6]. If measures will not to be taken to change the structure of consumption of energy sources, then the population growth in the region will lead to the situation that the consumption of energy raw materials from biomass, in particular wood fuel, will increase by 40% by 2040 and will cause a significant reduction of forest areas. In Sub-Saharan Africa traditional biomass burning supplies more than 92% of national energy consumption. In the region, such biomass resource as wood, agro-industrial residues and municipal waste are available is expected to meet the countries growing energy demand and can provide huge opportunity for the adoption of the briquetting technology. In some African countries, in the Republic of the Congo, for example, biomass briquettes are sold as an alternative to charcoal to protect Virunga National Park from deforestation associated with charcoal production.

The Africa Biogas Partnership Program (ABPP) in Kenya, Tanzania and Uganda was established in 2009 to promote adoption of biodigesters by rural households in sub-Saharan region. Between 2009 and 2017 over 27,000 households installed a biodigester, half of them in Kenya. Additional objectives include improving agricultural productivity by using bio-slurry, improving health, reducing deforestation, and improving livelihoods. Households perceive higher crop yields (84%–91% of users), reduced fuel consumption (84%–94% of users), reduced eye problems and respiratory symptoms (45%–91% of users). Benefits most appreciated are “easy cooking” and “saving time and money”. Fuel consumption tests show households with biodigesters use 2.1 to 3.3 fewer tons of wood per year than similar households without biodigesters [7].

3.3. Carbon dioxide emission

In recent years, the increase in CO_2 emissions by African countries has been extremely rapid and reached by 2018 more than 1.3 bln tons per year. Since 1990 greenhouse gas emissions increased in the region by 42% (in Ethiopia - by 85%, in Kenya - by 62%) [8]. The use of diesel generators is not always effective, since the burning of liquid fuel has a negative impact on the environment. The cost of generator proliferation is an exceptionally high level of CO_2 emissions. While a coal power plant, usually considered the worst energy option for the environment, might produce ~1000 g CO_2/kWh, an individually-owned 5kW diesel generator emits twice as much. In Nigeria, which is Africa’s largest importer of diesel generators with an estimated 9 mln units in use, generators produce about 29 million metric tons of CO_2 each year [9]. Our analysis of African countries’ sources showed that the largest government programs for the use of renewable energy in Sub-Saharan Africa (except South Africa, which implements its own large-scale programs for the development of renewable energy sources), have been launched in recent years in Ethiopia and Kenya. In this study, we collected information about operating power plants and potential projects of renewable energy facilities in Kenya, a leader on the use of geothermal resources on the African continent. The construction of geothermal power plants (GeoPP) in Kenya with the injection of condensation water back into the reservoir and the practical absence of a soil layer in the drilling zones reduces their environmental impact on the surrounding territory. Based on the collected materials on the location and operation of power plants, estimates of the installed capacity utilization factor (ICUF) were performed. Using the values of this coefficient, estimates and calculations of the average annual volume of electricity generation at the scheduled power plants were carried out. The further calculations allowed the authors to estimate the volume of CO_2 emission reduction due to the rejection of the use of carbon fuel at these power plants (tab.2). As can be seen from the table 2, the Ministry of Energy and Petroleum of Kenya plans to bring the capacity of RE power plants to 4400 MW, and we have calculated that this will allow the country to avoid 13 million tCO_2 emissions per annum in the energy sector. The GeoPP will make the largest contribution in this output of more than 10 million tCO2/ year, since the ICUF is much higher than that of wind and solar power plants.

Table 2. CO_2 emission reducing due to sustainable power generation in operating RE Power Plants in 2019 and scheduled Projects in Kenya

Name of RE Power Plants	Installed capacity	Commised / Planned	Power generation (GWh/yr)	CO_2 emission reducing (Kt)
Olkaria I & II GeoPP	290MW	1981-2014	2032.3	1829.1
Olkaria III	139 MW	2008-2017	974.5	877.1
Olkaria IV	140 MW	2014	981.1	883.0
Olkaria V	172 MV	2019	1205.4	1084.9
Eburru GeoPP	2,4 MW	2017	16.8	15.2
Well head Generation	70 MW	2018	490.6	441.5
Total GeoPP	813.4 MW		5700.7	5130.9
Ngong Hills Wind Power	25.5 MW	1993-2014	60.3	157.0
Lake Turkana Windfarm	310 MW	2018	701.7	203.5
Solar Plant Garisssa	50 MW	2019	68.7	10.3
Total operating RE PP	1188.9		6531.4	5501.7
Scheduled Projects				
Olkaria VI Geo Project	140 MW	2025	980	882.0
Menengai Geo Project	400MW	2030	2800	2380.0
Baringo-Silali Geo Project	300MW	2028	2100	1890.0
Suswa Geo Project	300MW	2026	2100	1890.0
Lamu Electrawinds	90 MW	2022	220	61.6
Isiolo Wind Power	150 MW	2025	368	103.0
Kipeto Wind Project	100MW	2023	236	63.7
Meru Wind Power	400 MW	2030	876	245.3
Solar PV Plant Seven forks	40MW	2021	52,5	7.9
New Solar Projects	400 MW	2030	525	89.3
Total scheduled Projects	3220 MW		10257.5	7612.8

Source: completed by the authors based on [10], [11], [12].

Solar mini grids as an alternative to diesel generators will have also a significant positive environmental impact in carbon dioxide emission reducing. It was estimated that by 2030, mini grids of 10–15 GW of PV stations in Sub-Saharan Africa will avoid 22-35 million tCO_2/ year emissions.

4 Conclusion

Africa is one of the most problematic regions in terms of the widespread use of biomass burning for cooking and heating, leading to the reduction of forests, extremely low access of the population to electricity, as well as underdeveloped infrastructure and power lines. The countries of Africa, including those located to the south of the Sahara, have a very high potential for renewable energy sources RES. The main factor for its development is electricity deficit and extremely low access of the population to electricity, development of these resources is necessary for sustainable social and economic development of the region. The development and strengthening of the concept of sustainable development contributes to the increase of renewable energy sources role in the energy sector. A number of countries in the region introduced government measures to support the development of electricity for RES, drawn up plans for their use and carried out rural electrification programs. The process of developing renewable energy sources in the region is possible only with the support of international environmental and financial structures, which are already participating actively in this issue. Many countries in Africa have already begun the process of developing RES, receiving support from international structures in this process. It is planned to provide total access to electricity by the year 2030 at the initiative of the UN in the region, using RES at the same time. The RES development is one of the main factors to solve the problems of forest reduction and carbon dioxide emission in Africa. The government plans of Kenya plans to bring the capacity of RE power plants to 4400 MW to 2030 will allow the country to avoid 13 million t CO_2 emissions per annum in the energy sector. The GeoPP will make the largest contribution in this output of more than 10 million t CO_2/ year. Solar mini grids as an alternative to diesel generators of 10–15 GW of PV stations in Sub-Saharan Africa will to 2030 avoid 22-35 million t CO_2/ year emissions.

The study was carried out within State Assign AAAA-A16-116032810088-8 and 0149-2020-0003.

References

1. *Africa Energy Outlook: World Energy Outlook Special Report* (IEA, Paris, 2019)
2. *Africa Sustainable Development Report: Tracking Progress on Agenda 2063 and the Sustainable Development Goals. African Union* (AfDB and UNDP, Addis Ababa, 2017)
3. *Renewables 2019.Global status report UNEP* (REN21 Secretariat, Paris, 2019)
4. L.V.Nefedova, A.A.Solovyev et al., *IGI Global in Engineering*, 154–168, (Hershey, 2019)
5. *Electricity storage and renewables: costs and markets to 2030* (IRENA, 2017)
6. L. Tusting, D. Bisanzio, G. Alabaster et al., *Nature*, **568**, 391–394 (2019)
7. H. Clemens, R. Bailis, A. Nyambane et al. *Energy for Sust. Dev*, **46,** 23-31 (2018)
8. S.A. Asongu *Environ Sci Pollut Res,* **25**, 9351–9359, (2018)
9. *Mini Grids for Half a Billion People: Market Outlook and Handbook for Decision Makers. ESMAP Technical Report, World Bank* (Washington, 2019).
10. *Renewable energy statistics 2019* (IRENA, 2019)
11. Kenya Electricity Generating Company PLC (KenGen). URL: https://kengen.co.ke/index.php/business/projects/ongoing.html
12. «The Geothermal Development Company» Kenya. URL: http://www.gdc.co.ke

AUTHOR INDEX

Adarchenko I.A. 124
Agbalyan Elena 1
Akhmedinova Kamila 262
Aleinikova Anna 274
Almeida Georgina 247
Andreev Dmitriy 77
Antonenko Svetlana 236
Arias Govín Israel Jose 128
Arias Priscila 247
Askarova Zifa 353
Babkin Vasilii 192
Baeva Yulia 47
Baidaulet Tilek 232, 353
Baigushikova Gulnara 62
Barakhov Anatolii 154
Barakhov Anatoly 225, 236
Basamykina Alena 204
Bauer Tatiana 154, 225
Baymurzaev Nurlan 232
Bekmanov Bakytzhan 62
Belik Aleksandra 41, 135
Belovezhets Konstantin 286
Belyakov Timofei 24
Bica Ioan 165
Bondareva Lydia 7
Borisova Elena 298
Borodina Ekaterina S. 303
Bren Dmitry 19
Burachevskaya Marina 19, 154
Burlutskiy Valeriy A. 303
Buryukova Olga 145, 240
Bushuev Nikolay N. 361
Butova Vera 225
Buzmakov Sergey 77, 268
Byashkin Maksim 274
Chaplygin Victor 13, 145
Chelinho Sónia 176
Cherchesova Susanna K. 313
Cherednichenko Oksana 62
Chernikova Natalia 13
Chernykh Natalia 47
Chizhov Alexey Ya. 343, 357
Cueva Jhonn 247
Dabbagh Alaa 308
Deryabkina Irina 13
Diop Ameth 303
Do Tat Thinh 198
Do Thi Thu Hong 198
Drygval Anna 52
Drygval Polina 52
Dubrova Stanislav 24, 176
Dylenova Elena P. 110
Dzhansugurova Leila 62
Dzhioeva Ia E. 313
Dziuba Ekaterina 77
Egorov Pavel 24, 29
Ermakov Alexander Sergeyevich 160
Ermakov Dmitry S. 366
Ermakov Dmitry Sergeyevich 160
Erofeeva Viktoriya 337, 348
Eroshenko Vasily 391
Evdokimov Alexander 57
Fedorchenko M.D. 124
Fedorenko Aleksei 225
Fedorenko Elena 236
Fedorov Yuri 36
Frumin Grigori 29
Gabov Dmitry 90
Gaivoron Tatiana 274
Glazyrina Yuliya 268
Glebov Viktor 348
Gorbunov Roman 317
Gorovtsov Andrey 236
Gülser Coºkun 240
Guslyakova Alla 383
Ibragimova Kadrya 293
Ilchenko Yaroslav 145, 240
Imanaliyeva Moldir 353
Ivlev Vasilii 192
Kalabin Gennady 192
Kamenev Roman 240
Kamennykh Natalia 372
Kapustina Olga 98
Karl Lidiya 182
Kenzhin Zhandos 90
Kharlamova Marianna 204
Khaustov Aleksandr 69, 90
Khotyanovskaya Yuliya 77
Kirichuk Anatoly A. 333
Kiselev Sergey 182
Kiseleva Ludmila 323
Kizilkaya Ridvan 145
Klevanova Elena 213
Kokoreva Anna 41
Kolupaeva Victoria 41, 135
Konotop Nikita 298
Krasnenko Aleksander 1
Kravtsova Natalia 13
Kravtsova Natalya 36
Kucherenko Alexey 145
Lamarton Sofia F. 313
Lashchenova Tatiana 182
Latushkina Elena 141
Latushkina Elena N. 128
Laver Bogdan 348
Ledashcheva Tatiana 279
Litvinov Yuri 13, 36
Llerena Silvia 247

AUTHOR INDEX

Lobzenko Ilya....225
Lukanin Alexander....213
Lyubimov Alexander....29
Mada Sharon Yeukai....204
Mainasheva Galina....274
Maltsev Vladimir....52
Mamadzhanov Roman....219
Mandzhieva Saglara....13, 154
Marenny Albert....182
Markov Dmitriy....298
Marsheva Natalia....274
Maslennikova Olga....337
Matevosyan Armine....378
Mautenbaev Askar....232
Mayorova Yana....348
Mazarji Mahmoud....154
Mazurov Vladimir N....303
Medvedeva Anna....145, 240
Mikhaylichenko K. Yu....124
Minkina Tatiana....36
Minkina Tatyana....240
Mukhlynin Dmitry....401
Mukhlynina Maria....397, 406
Mullo Feddy....187
Mullo Freddy....103
Murzakhmetova Maira....353
Naumov Vladimir....372
Navatorov Oleg A....313
Nazarenko Olga....36
Nefedova Liudmila....411
Nesterenko Aleksey....98
Nesterenko Maksim....98
Nesterov Evgeni....24, 29
Nevidomskaya Dina....19, 36
Ngo Cao Cuong....198
Nguyen Thu Hoai....198
Nikiforov Sergey....98
Nikol'skii Alexander....286
Nuraliev Serikbay....62
Nwankwo Nnabueze Kalu....255
Nyukhina Inna....135
Oleynik Anna....141
Ospanova Aigul....128
Parakhina Elena....323
Pavlova Elena....378
Peliy Alexander F....303
Pilugina Anastassiya....62
Pinaev Sergey K....343, 357
Pinaev Vladimir....279
Pinaeva Olga G....343, 357
Piven Elena A....361
Piven Nikolay P....361
Podlipskiy Ivan....176
Pogonyshev Pavel....236
Prudnikova Elena....187
Radnaeva Larisa D....110
Rafikova Yulie....411
Rajput Vishnu....19
Rajput Vishnu D....145
Rakhimov Ilgizar....293
Redina Margarita....69, 90
Ryazanova Natalya....372, 391
Salazar Cristian....103, 247
Sannikov Pavel....268
Semeshkina Polina S....303
Sereda V.N....198
Sharakhmetov Sayat....232
Sherstnev Aleksey....13
Shinkaruk Elena....1
Shiretorova Valentina G....110
Shirokov Roy....114
Shmerko Sofia....47
Silaeva Zhanna....323
Sivukhin Alexey....298
Smirnov Andrey A....328
Sokolova Lyudmila I....366
Solovyev Alexander....411
Solovyev Dmitriy....411
Srailova Gulziya....353
Stanchenko Galina V....328
Stanis Elena....52
Stanis Elena V....128
Stepnova Varvara....47
Sushkova Svetlana....225, 236
Terentyeva Elena....378
Tikhmenev Evgeniy A....328
Tikhmenev Pavel E....328
Tsitsuashvili Victoria....19
Tsviak Aleksey....98
Tuleukhanov Sultan....353
Tusat Ekrem....225
Tynybekov Bekzat....232
Umarov Mykhadi....219
Ushakova O.V....124
Valeeva Nailya....383
Vanisova Elena....286
Vasil'ev Vasilii....192
Vatkova Olga....383
Vedysheva Natalia....397
Vinogradova Julia....298
Vologzhina Sayana....391
Yablochnikov Sergey....348
Yakovleva Evgeniya....90
Ydyrys Alibek....232, 353
Yeszhanov Birlikbay....232
Zakirova Yulia....219, 255
Zamulina Inna....154
Zelenkovskiy Pavel....176

AUTHOR INDEX

Zhidkova Elena **323**
Zhigzhitzhapova Svetlava V. **110**
Zhuzhgova Larisa **268**
Zinchenko Vladislav **236**